Introduction to
ENVIRONMENTAL
MANAGEMENT

Introduction to
ENVIRONMENTAL MANAGEMENT

Mary K. Theodore
Louis Theodore

CRC Press
Taylor & Francis Group
Boca Raton   London   New York

CRC Press is an imprint of the
Taylor & Francis Group, an **informa** business

CRC Press
Taylor & Francis Group
6000 Broken Sound Parkway NW, Suite 300
Boca Raton, FL 33487-2742

© 2010 by Taylor and Francis Group, LLC
CRC Press is an imprint of Taylor & Francis Group, an Informa business

No claim to original U.S. Government works

Printed in the United States of America on acid-free paper
10 9 8 7 6 5 4 3 2 1

International Standard Book Number: 978-1-4200-8907-3 (Hardback)

Library of Congress Cataloging-in-Publication Data

Theodore, Mary K.
   Introduction to environmental management / by Mary K. Theodore and Louis Theodore.
      p. cm.
   Includes bibliographical references and index.
   ISBN-13: 978-1-4200-8907-3
   ISBN-10: 1-4200-8907-2
   1. Environmental management. I. Theodore, Louis.

GE300.T54 2009
363.705--dc22   2009005488

Visit the Taylor & Francis Web site at
http://www.taylorandfrancis.com

and the CRC Press Web site at
http://www.crcpress.com

To
Lila
our κουκλα

Contents

Part I   Overview

Part II   Air

Part III   Water

Part IV   Solid Waste

Part V   Pollution Prevention

Part VI   Environmental Risk

Part VII   Other Areas of Interest

Preface

In the last four decades, there has been an increased awareness of a wide range of environmental issues covering all sources: air, land, and water. More and more people are becoming aware of these environmental concerns, and it is important that professional people, many of whom do not possess an understanding of environmental problems, have the proper information available when involved with environmental issues. All professionals should have a basic understanding of the technical and scientific terms related to these issues as well as the regulations involved. Hopefully, this book will serve the needs of the professional by increasing his or her awareness of (and help solve) the environmental problems that society is facing now.

This book is primarily intended for individuals who have a limited environmental technical background. It is presented in simple, understandable terms for students, practicing engineers and scientists, lawyers, news media executives, business personnel, and even the consumers who need to know the fundamentals of the many environmental issues that exist and will continue to exist in the future. The authors' objective is also to provide both background material on numerous environmental issues and information on what each individual can do to help alleviate some of these problems.

This book is divided into seven parts. Part I provides an overview that includes an introduction to environmental issues, regulations, and types of pollutants. Part II deals with issues related to air pollution. It includes material on how air pollution can be controlled, and on the quality of indoor air, which is an issue in many office buildings today. Part III discusses the problems of pollution in water and methods to control this problem. Part IV focuses on solid waste management. It examines the different types of solid waste, such as hazardous, medical, and nuclear, and treatment techniques for each. This part also includes material on the Superfund program and the result of its effort to clean up waste sites.

Part V focuses on three pollution prevention topic areas, including health, safety, and accident prevention, energy conservation, and waste reduction. Since the concern with many of the environmental issues arises because of the risks involved, Part VI examines how risks are perceived and communicated, and how individuals can be educated about these risks. Part VII provides information on other areas of interest in the environmental arena. These include many popular topics like electromagnetic fields, environmental implications of nanotechnology, and ethical issues as they relate to the environment.

Mary K. Theodore
Louis Theodore

Authors

Mary K. Theodore received her BA degree in English from Manhattan College and MA from Queens in Applied Linguistics.

Ms. Theodore is presently the president of Theodore Tutorials, a company that specializes in providing training needs to industry, government, and academia. She is the author of the chapter entitled "Domestic Solutions" in the *Handbook of Environmental Management and Technology* (Wiley-Interscience, 1993; the only nontechnical chapter in the handbook); coauthor of the unpublished, but copyrighted, work *A Citizen's Guide to Pollution Prevention*; and, coauthor of *Major Environmental Issues Facing the 21st Century* (Prentice-Hall, 1996). The latter text effectively integrated the concept of "sustainable growth without environmental degradation" at both the industrial and domestic levels. Ms. Theodore is currently an adjunct faculty member at Nassau County Community College.

Since marriage and the birth of her first child, Ms. Theodore has devoted a significant part of her life to help solve the environmental problems facing society from a domestic point of view. A proactive environmentalist with no ties to any of the environmental organizations in vogue today, she has also lectured in this area. She was recently involved with the development of a nontechnical environmental calendar that serves as a consumer and youth outreach product.

Ms. Theodore resides in East Williston, Long Island, New York, with her husband of 42 years and their three children—Georgeen, Molleen, and Patrick.

Louis Theodore is a professor of chemical engineering at Manhattan College, Riverdale, New York. Born and raised in Hell's Kitchen, he received his MChE and EngScD from New York University and his BChE from The Cooper Union. Over the past 49 years, Dr. Theodore has been a successful educator, researcher, professional innovator, and communicator in the field of engineering. He has taught courses in environmental management, hazardous waste incineration, accident and emergency management, pollution prevention, and air pollution and its control at Manhattan College.

Dr. Theodore is an internationally recognized lecturer who has provided nearly 200 courses to industry, government, and technical associations; he has served as an after-dinner or luncheon speaker on numerous occasions; and he has appeared on television as a guest commentator and a news spokesperson. He has developed training materials and has served as the principal moderator/lecturer for U.S. Environmental Protection Agency (USEPA) courses on hazardous waste incineration, pollution prevention, and air pollution control equipment. He has also served as a consultant to several industrial companies in the field of environmental management, and is presently a consultant/expert witness for the USEPA and the U.S. Department of Justice.

Dr. Theodore has written 81 text/reference books including *Pollution Prevention* (Van Nostrand Reinhold), *Engineering and Environmental Ethics* (John Wiley &

Sons), *Air Pollution Control Equipment* (Prentice-Hall), and *Introduction to Hazardous Waste Incineration* (Wiley-Interscience); he has also been a section author/editor in *Perry's Chemical Engineers' Handbook* (McGraw-Hill), *Nanotechnology: Environmental Implications and Solutions* (John Wiley & Sons), and *Nanotechnology: Basic Calculations for Engineers and Scientists* (John Wiley & Sons). He is also the cofounder of Theodore Tutorials, a company specializing in providing training needs to industry, government, and academia; included in this series of 21 texts are four tutorials concerned with the professional engineer's (PE) exam.

Dr. Theodore is the recipient of the prestigious Ripperton award by the International Air and Waste Management Association that is "presented to an outstanding educator who through example, dedication, and innovation has so inspired students to achieve excellence in their professional endeavors." He was also the recipient of the American Society of Engineering Education AT&T Foundation award for "excellence in the instruction of engineering students." Dr. Theodore was recently honored at Madison Square Garden in 2008 for his contributions to basketball and the youth of America.

Dr. Theodore is a member of Phi Lambda Upsilon, Sigma Xi, Tau Beta Pi, American Chemical Society, American Society of Engineering Education, Royal Hellenic Society, and the International Air and Waste Management Association. He is also certified to referee scholastic basketball games through his membership in the International Association of Approved Basketball Officials. He has previously served on a Presidential Crime Commission under Gerald Ford and provided testimony as a representative of the pari-mutuel wagerer (horseplayer). His column "AS I SEE IT," a monthly feature of several Long Island newspapers, addresses social, economic, political, technical, and sports issues.

Contributing Authors

Richard F. Carbonaro
Department of Civil and Environmental
  Engineering
Manhattan College
Riverdale, New York

Anna M. Daversa
Department of Chemical Engineering
Manhattan College
Bronx, New York

Lauren De Sanctis
Department of Chemical Engineering
Manhattan College
Bronx, New York

Vincenza Imperiale
Wyeth Pharmaceuticals
Pearl River, New York

Shannon O'Brien
Department of Chemical Engineering
Manhattan College
Bronx, New York

Francesco Ricci
Department of Chemical Engineering
Manhattan College
Briarcliff, New York

Part I

Overview

Part I of this book serves as an overview of the 50 major environmental issues facing the twenty-first century. Part I comprises nine chapters. A brief review of the environmental issues is presented in Chapter 1. Chapter 2—the longest and most detailed chapter in the book—focuses on environmental regulations while Chapter 3 examines international regulations. Chapter 4 provides an overview of ISO 14K. Multimedia concerns and approaches are treated in Chapter 5. Chapter 6 discusses the sources and classifications of pollutants while Chapter 7 discusses the effects of pollutions. Part I concludes with Chapters 8 and 9, which address the general subject of green chemistry and sustainability.

Also note that the acronym USEPA (United States Environmental Protection Agency) and EPA (Environmental Protection Agency) are used interchangeably throughout this as well as the remaining parts of the book. This problem arises because some of the material has been drawn directly from government publications.

1 Introduction to Environmental Issues

CONTENTS

1.1 INTRODUCTION

In the past four decades there had been an increased awareness of a wide range of environmental issues covering all sources: air, land, and water. More and more people are becoming aware of these environmental concerns, and it is important that professional people, many of whom do not possess an understanding of environmental problems or have the proper information available when involved with environmental issues. All professionals should have a basic understanding of the technical and scientific terms related to these issues as well as the regulations involved. Hopefully this book will serve the needs of the professional by increasing his or her awareness of (and help solve) the environmental problems facing society.

The past four decades have been filled with environmental tragedies as well as a heightened environmental awareness. The oil spills of the *Exxon Valdez* in 1989 and in the Gulf War of 1991 showed how delicate our oceans and their ecosystems truly are. The disclosures of Love Canal in 1978 and Times Beach in 1979 made the entire nation aware of the dangers of hazardous chemical wastes. The discovery of acquired immunodeficiency syndrome (AIDS) virus and the beach washups of 1985 brought the issue of medical waste disposal to the forefront of public consciousness. A nuclear accident placed the spotlight on Chernobyl, and to this day society is still seeing the effects of that event.

An outline of the contents of the book follows. Details on each of the chapters of the seven Parts is included in the presentation.

1.2   PART I

The "Overview" provides a general background and addresses international concerns, environmental regulations, and generators of pollutants. Degradation of the environment is not a problem that is restricted to the United States, or even to developed countries. On the contrary, underdeveloped countries are struggling with several environmental issues that have already been resolved in many developed countries. In the United States, the Environmental Protection Agency (EPA) as well as the individual states is working hard to implement regulations addressing areas of environmental concern. Generators and sources of pollutants are being identified so that solutions may be targeted to specific areas. The Part concludes with chapters that deal with the chemistry of green engineering.

1.3   PART II

"Air" management issues looks into several different areas related to air pollutants and their control. Atmospheric dispersion of pollutants can be mathematically modeled to predict where pollutants emitted from a particular source, such as a combustion facility stack, will settle to the ground and at what concentration. Pollution control equipment can be added to various sources to reduce the amount of pollutants before they are emitted into the air. Acid rain, the greenhouse effect, and global warming are all indicators of adverse effects to the air, land, and sea which result from excessive amount of pollutants being released into the air. One topic that few people are aware of is the issue of indoor air quality. Inadequate ventilation systems in homes and businesses directly affect the quality of health of the people within the buildings. For example, the episode of Legionnaires' disease, which occurred in Philadelphia in the 1970s, was related to microorganisms that grew in the cooling water of the air-conditioning system. Noise pollution, although not traditionally an air pollution topic, is included in this section. The effects of noise pollution are generally not noticed until hearing is impaired. And although impairment of hearing is a commonly known result of noise pollution, few people realize that stress is also a significant result of excessive noise exposure. The human body enacts its innate physiologic defensive mechanisms under conditions of loud noise, and the fight to control these physical instincts causes tremendous stress on the individual.

1.4   PART III

Pollutant dispersion in water systems and wastewater treatment is discussed in "Water" management issues. Pollutants entering rivers, lakes, and oceans come from a wide variety of sources, including stormwater runoff, industrial discharges, and accidental spills. It is important to understand how these substances disperse in order to determine how to control them. Municipal and industrial wastewater treatment systems are designed to reduce or eliminate problem substances before they are introduced into natural water systems, industrial use systems, drinking water supply, and other water systems. Often, wastewater from industrial plants must be pretreated before it can be discharged into a municipal treatment system.

1.5 PART IV

"Solid Waste" management issues addresses treatment and disposal methods for municipal, medical, and radioactive wastes. Programs to reduce and dispose of municipal waste include reuse, reduction, recycling, and composting, in addition to incineration and landfilling. Potentially infectious waste generated in medical facilities must be specially packaged, handed, stored, transported, treated, and disposed of to ensure the safety of both the waste handlers and the general public. Radioactive waste may have serious impacts on human health and the environment, and treatment and disposal requirements for radioactive substances must be strictly adhered to. Incineration has been a typical treatment method for hazardous waste for many years. Superfund was enacted to identify and remedy uncontrolled hazardous waste sites. It also attempts to place the burden of cleanup on the generator rather than on the federal government. Asbestos, metals, and underground storage tanks either contain or inherently are hazardous materials that require special handling and disposal. Further, it is important to realize that both small and large generators of hazardous wastes are regulated.

1.6 PART V

"Pollution Prevention" covers domestic and (primarily) industrial means of reducing pollution. This can be accomplished through (a) proper residential and commercial building design; (b) proper heating, cooling, and ventilation systems; (c) energy conservation; (d) reduction of water consumption; and (e) attempts to reuse or reduce materials before they become wastes. Domestic and industrial solutions to environmental problems arise by considering ways to make homes and workplaces more energy-efficient as well as ways to reduce the amount of wastes generated within them.

1.7 PART VI

Managers also need to be informed on how to make decisions about associated risks and how to communicate these risks and their effects on the environment to the public. "Environmental Risk" topics include short-term and long-term threats to human health and the environment. Risk assessment is the most important consideration for remediation of harmful effects stemming from the presence of a hazardous substance, and risk-based decision-making is a tool that is now routinely being used to select a cleanup alternative. This Part also provides an explanation of both how to estimate and how to avoid environmental, health, and hazard risks.

1.8 PART VII

The last part, "Other Areas of Interest," discusses nine topics that are relatively new in the area of environmental management. Included in Part VII are electromagnetic fields, noise pollution, used oil, and the environment implications of nanotechnologies. Environmental audits provide a means of assessing the environmental conduct and performance of an organization. Environmental ethics, as it relates to rules of proper environmental conduct, receives treatment in the next to the last chapter. Finally, environmental justice (last chapter) is a new term for describing the disproportionate distribution of environmental risks in minority and low-income communities. Federal attention is now focused on environmental and human health conditions in these areas, with the goal of achieving equality of environmental protection for all communities.

This book is not intended to be all-encompassing. Rather, it is to be used as a starting point. References are provided at the end of each chapter, which provide more detailed information on each topic.

2 Environmental Regulations[*]

CONTENTS

[*] Adapted from Burke et al. [1] and Stander and Theodore [2].

2.1 INTRODUCTION

Environmental regulations are not simply a collection of laws on environmental topics. They are an organized system of statutes, regulations, and guidelines that minimize, prevent, and punish the consequences of damage to the environment. This system requires each individual—whether an engineer, field chemist, attorney, or consumer—to be familiar with its concepts and case-specific interpretations. Environmental regulations deal with the problems of human activities and the environment, and the uncertainties the law associates with them.

It is now 1970, a cornerstone year for modern environmental policy. The National Environmental Policy Act (NEPA), enacted on January 1, 1970, was considered a "political anomaly" by some. NEPA was not based on specific legislation; instead it is referred in a general manner to environmental and quality of life concerns. The Council for Environmental Quality (CEQ), established by NEPA, was one of the councils mandated to implement legislation. On April 22, 1970, thousands of demonstrators gathered all around the nation to celebrate the occasion of Earth Day. NEPA and Earth Day were the beginning of a long, seemingly never-ending debate over environmental issues.

The Nixon Administration at that time became preoccupied with not only trying to pass more extensive environmental legislation, but also implementing the laws. Nixon's White House Commission on Executive Reorganization proposed in the Reorganizational Plan # 3 of 1970 that a single, independent agency be established, separate from the CEQ. The plan was sent to Congress by President Nixon on July 9, 1970, and this new U.S. Environmental Protection Agency (EPA) began operation on December 2, 1970. The EPA was officially born.

In many ways, the EPA is the most far-reaching regulatory agency in the federal government because its authority is very broad. The EPA is charged to protect the nation's land, air, and water systems. Under a mandate of national environmental laws, the EPA strives to formulate and implement actions that lead to a compatible balance between human activities and the ability of natural systems to support and nurture life [3].

The EPA works with the states and local governments to develop and implement comprehensive environmental programs. Federal laws such as the Clean Air Act (CAA), the Safe Drinking Water Act (SDWA), the Resource Conservation and Recovery Act (RCRA), and the Comprehensive Environmental Response, Compensation, and Liability Act (CERCLA), etc., all mandate involvement by state and local government in the details of implementation.

This chapter provides an overview of the eight key environmental protection laws and subsequent regulations that affect the environment in the United States.

2.2 RESOURCE CONSERVATION AND RECOVERY ACT

Defining what constitutes a "hazardous waste" requires consideration of both legal and scientific factors. The basic definitions used in this chapter are derived from the RCRA of 1976, as amended in 1978, 1980, and 1986; the Hazardous and Solid Waste Amendments (HSWA) of 1984; and the CERCLA of 1980, as amended by the Superfund Amendments and Reauthorization Act (SARA) of 1986. Within these statutory authorities, a distinction exists between a hazardous waste and a hazardous substance. The former is regulated under RCRA while the latter is regulated under the Superfund program.

Hazardous waste refers to "… a solid waste, or combination of solid wastes, which because of its quantity, concentration, or physical, chemical or infectious character-istics may [pose a] substantial present or potential hazard to human health or the environment when improperly … managed …" [RCRA, Section 1004(5)]. Under RCRA regulations, a waste is considered hazardous if it is reactive, ignitable, corro-sive, or toxic or if the waste is listed as a hazardous waste in Title 40 Parts 261.31–33 of the "Code of Federal Regulations" [4].

In addition to hazardous wastes defined under RCRA, there are "hazardous sub-stances" defined by Superfund. Superfund's definition of a hazardous substance is broad and grows out of the lists of hazardous wastes or substances regulated under the Clean Water Act (CWA), the CAA, the Toxic Substances Control Act (TSCA), and RCRA. Essentially, Superfund considers a hazardous substance to be any hazardous substance or toxic pollutant identified under the CWA and applicable regulations, any hazardous air pollutant listed under the CAA and applicable regulations, any immi-nently hazardous chemical for which a civil action has been brought under TSCA, and any hazardous waste identified or listed under RCRA and applicable regulations.

The RCRA of 1976 completely replaced the previous language of the Solid Waste Disposal Act of 1965 to address the enormous growth in the production of waste. The objectives of this Act were to promote the protection of health and the environ-ment and to conserve valuable materials and energy resources by [5,6]

1. Providing technical and financial assistance to state and local governments and interstate agencies for the development of solid waste management plans (including resource recovery and resource conservation systems) that promote improved solid waste management techniques (including more effective organizational arrangements), new and improved methods of col-lection, separation, and recovery of solid waste, and the environmentally safe disposal of nonrecoverable residues.
2. Providing training grants in occupations involving the design, operation, and maintenance of solid waste disposal systems.
3. Prohibiting future open dumping on the land and requiring the conversion of existing open dumps to facilities that do not pose any danger to the envi-ronment or to health.
4. Regulating the treatment, storage, transportation, and disposal of hazard-ous waste that have adverse effects on health and the environment.
5. Providing for the promulgation of guidelines for solid waste collection, transport, separation, recovery, and disposal practices and systems.
6. Promoting a national research and development program for improved solid waste management and resource conservation techniques; more effective organization arrangements; and, new and improved methods of collection, separation, recovery, and recycling of solid wastes and environmentally safe disposal of nonrecoverable residues.

7. Promoting the demonstration, construction, and application of solid waste management, resource recovery and resource conservation systems that preserve and enhance the quality of air, water, and land resources.
8. Establishing a cooperative effort among federal, state, and local governments and private enterprises in order to recover valuable materials and energy from solid waste.

Structurewise, the RCRA is divided into eight subtitles. These subtitles are (A) General Provisions; (B) Office of Solid Waste; Authorities of the Administrator; (C) Hazardous Waste Management; (D) State or Regional Solid Waste Plans; (E) Duties of Secretary of Commerce in Resource and Recovery; (F) Federal Responsibilities; (G) Miscellaneous Provisions; and, (H) Research, Development, Demonstration, and Information.

Subtitles C and D generate the framework for regulatory control programs for the management of hazardous and solid nonhazardous wastes, respectively. The hazardous waste program outlined under Subtitle C is the one most people associate with the RCRA [6].

A RCRA training module on hazardous waste incinerators has been developed and can be viewed at: http://www.epa.gov/epaoswer/hotline/training/incin.pdf. Other information can also be viewed at http://www.epa.gov/epaoswer/hazwaste/hazcmbst.htm#emissions.

2.3 MAJOR TOXIC CHEMICAL LAWS ADMINISTERED BY THE EPA

People have long recognized that sulfuric acid, arsenic compounds, and other chemical substances can cause fires, explosions, or poisoning. More recently, researchers have determined that many chemical substances such as benzene and a number of chlorinated hydrocarbons may cause cancer, birth defects, and other long-term health effects. Today, the hazards of new kinds of substances, including genetically engineered microorganisms are being evaluated. The EPA has a number of legislative tools to use in controlling the risks from toxic substances (Table 2.1).

The Federal Insecticide, Fungicide, and Rodenticide Act of 1972 (FIFRA) encompasses all pesticides used in the United States. When first enacted in 1947, FIFRA was administered by the U.S. Department of Agriculture and was intended to protect consumers against fraudulent pesticide products. When many pesticides were registered, their potential for causing health and environmental problems was unknown. In 1970, the EPA assumed responsibility for FIFRA, which was amended in 1972 to shift emphasis to health and environmental protection. Allowable levels of pesticides in food are specified under the authority of the Federal Food, Drug, and Cosmetic Act of 1954. FIFRA contains registration and labeling requirements for pesticide products. The EPA must approve any use of a pesticide, and manufacturers must clearly state the conditions of that use on the pesticide label. Some pesticides are listed as hazardous waste and are subject to RCRA rules when discarded.

The TSCA authorizes EPA to control the risks that may be posed by the thousands of commercial chemical substances and mixtures (chemicals) that are not regulated

TABLE 2.1
Major Toxic Chemical Laws Administered by the EPA

Statue	Provisions
Toxic Substances Control Act	Requires that the EPA be notified of any new chemical prior to its manufacture and authorizes EPA to regulate production, use, or disposal of a chemical.
Federal Insecticide, Fungicide, and Rodenticide Act	Authorizes the EPA to register all pesticides and specify the terms and conditions of their use, and remove unreasonably hazardous pesticides from the marketplace.
Federal Food, Drug, and Cosmetic Act	Authorizes the EPA in cooperation with FDA to establish tolerance levels for pesticide residues in food and food producers.
Resource Conservation and Recovery Act	Authorizes the EPA to identify hazardous wastes and regulate their generation, transportation, treatment, storage, and disposal.
Comprehensive Environmental Response, Compensation, and Liability Act	Requires the EPA to designate hazardous substances that can present substantial danger and authorizes the cleanup of sites contaminated with such substances.
Clean Air Act	Authorizes the EPA to set emission standards to limit the release of hazardous air pollutants.
Clean Water Act	Requires the EPA to establish a list of toxic water pollutants and set standards.
Safe Drinking Water Act	Requires the EPA to set drinking water standards to protect public health from hazardous substances.
Marine Protection, Research, and Sanctuaries Act	Regulates ocean dumping of toxic contaminants.
Asbestos School Hazard Act	Authorizes the EPA to provide loans and grants to schools with financial need for abatement of severe asbestos hazards.
Asbestos Hazard Emergency Response Act	Requires the EPA to establish a comprehensive regulatory framework for controlling asbestos hazards in schools.
Emergency Planning and Community Right-to-Know Act	Requires states to develop programs for responding to hazardous chemical releases and requires industries to report on the presence and release of certain hazardous substances.

as either drugs, food additives, cosmetics, or pesticides. Under TSCA, the EPA can, among other things, regulate the manufacture and use of a chemical substance and require testing for cancer and other effects. TSCA regulates the production and distribution of new chemicals and governs the manufacture, processing, distribution, and use of existing chemicals. Among the chemicals controlled by TSCA regulations are polychlorobiphenyl (PCBs), chlorofluorocarbons (CFCs), and asbestos. In specific cases, there is an interface with RCRA regulations. For example, PCB disposal is generally regulated by TSCA. However, hazardous wastes mixed with PCBs are regulated under RCRA. Under both TSCA and FIFRA, the EPA is responsible for regulating certain biotechnology products, such as genetically engineered microorganisms designed to control pests or assist in industrial processes.

Additional details can be found at:

1. http://www.epa.gov/oppt/itc/pubs/sect8a.htm
2. http://www.epa.gov/oppt/chemtest/pubs/pairform.pdf

The CAA, in Section 112, listed 189 air pollutants. The CAA also requires emission standards for many types of air emission sources, including RCRA-regulated incinerators and industrial boilers or furnaces.

The CWA lists substances to be regulated by effluent limitations in 21 primary industries. The CWA substances are incorporated into both RCRA and CERCLA. In addition, the CWA regulates discharges from publicly owned treatment works (POTWs) to surface waters, and indirect discharges to municipal wastewater treatment systems (through the pretreatment program). Some hazardous wastewaters which would generally be considered RCRA regulated wastes are covered under the CWA because of the use of treatment tanks and a National Pollutant Discharge Elimination System (NPDES) permit to dispose of the wastewaters. Sludges from these tanks, however, are subject to RCRA regulations when they are removed.

The SDWA regulates underground injection systems, including deep-well injection systems. Prior to underground injection, a permit must be obtained which imposes conditions that must be met to prevent the endangerment of underground sources of drinking water.

The Marine Protection, Research, and Sanctuaries Act of 1972 has regulated the transportation of any material for ocean disposal and prevents the disposal of any material in oceans that could affect the marine environment. Amendments enacted in 1988 were designed to end ocean disposal of sewage sludge, industrial waste, and medical wastes.

See also:

1. The U.S. EPA. 2006. Laws and Regulations. http://www.epa.gov/region5/cwa.htm
2. The U.S. EPA. 2006. Introduction to WQS. http//www.epa.gov/watertrain/cwa/cwa2.htm

2.4   WATER QUALITY LEGISLATION AND REGULATION [4]

Congress has provided the EPA and the states with three primary statutes to control and reduce water pollution: the CWA; the SDWA; and the Marine Protection, Research, and Sanctuaries Act. Each statute provides a variety of tools that can be used to meet the challenges and complexities of reducing water pollution in the nation.

2.4.1   FEDERAL WATER POLLUTION CONTROL ACT

The original FWPCA was passed in 1948. This act and its various amendments are often referred to as the CWA. It provided loans for treatment plant construction and temporary authority for federal control of interstate water pollution. The enforcement powers were so heavily dependent on the states as to make the act almost unworkable. In 1956, several amendments to the FWPCA were passed that made federal enforcement procedures less cumbersome. The provision for state consent was removed by amendments passed in 1961, which also extended federal authority to include navigable waters in the United States.

In 1965, the Water Quality Act established a new trend in water pollution control. It provided that the states set water quality standards in accordance with federal guidelines. If the states failed to do so, the federal government, subject to a review hearing, would set the standards. In 1966, the Clean Water Restoration Act transferred the Federal Water Pollution Control Administration from the Department of Health, Education and Welfare to the Department of the Interior. It also gave the Interior Department the responsibility for the Oil Pollution Act.

After the creation of EPA in 1970, the EPA was given the responsibility previously held by the Department of the Interior with respect to water pollution control. In subsequent amendments to the FWPCA in 1973, 1974, 1975, 1976, and 1977, additional Federal programs were established. The goals of these programs were to make waterways of the United States fishable and swimmable by 1983 and to achieve zero discharge of pollutants by 1985. The NPDES was established as the basic regulatory mechanism for water pollution control. Under this program, the states were given the authority to issue permits to "point-source" dischargers provided the dischargers gave assurance that the following standards would be met:

1. Source-specific effluent limitations (including new source performance standards)
2. Toxic pollutant regulations (for specific substances regardless of source)
3. Regulations applicable to oil and hazardous substance liability

In order to achieve that stated water quality goal of fishable and swimmable waters by 1983, each state was required by EPA to adopt water quality standards that met or exceeded the Federal water quality criteria. After each state submitted its own water quality standards, which were subsequently approved by EPA, the Federal criteria were removed from the Code of Federal Regulations. The state water quality standards are used as the basis for establishing both point-source-based effluent limitations and toxic pollutant limitations used in issuing NPDES permits to point-source discharges.

2.4.2　Source-Based Effluent Limitations

Under the FWPCA, EPA was responsible for establishing point-source effluent limitations for municipal dischargers, industrial dischargers, industrial users of municipal treatment works, and effluent limitations for toxic substances (applicable to all dischargers).

Standards promulgated or proposed by EPA under 40 CFR, Parts 402 through 699, prescribe effluent limitation guidelines for existing sources, standards of performance for new sources, and pretreatment standards for new and existing sources. Effluent limitations and new source performance standards apply to discharges made directly into receiving bodies of water. The new standards require best available technology (BAT) and are to be used by the states when issuing NPDES permits for all sources 18 months after EPA makes them final. Pretreatment standards apply to waste streams from industrial sources that are sent to POTW for final treatment. These regulations are meant to protect the POTW from any materials that would either harm the treatment facility or pass through untreated. They are to be enforced primarily by the local POTW. These standards are applicable to particular classes of point sources and pertain to discharges into navigable waters without regard to the quality of the receiving water. Standards are specific for numerous subcategories under each point-source category.

Limitations based upon application of the best practicable control technology (BPT) currently available apply to existing point sources and should have been achieved by July 1, 1977. Limitations based upon application of the best available technology economically achievable (BATEA) that will result in reasonable further progress toward elimination of discharges had to be achieved by July 1, 1984.

2.4.3 CLEAN WATER ACT OF 1977

The 1977 CWA directed EPA to review all BAT guidelines for conventional pollutants in those industries not already covered.

On August 23, 1978 (43 FR 37570), the EPA proposed a new approach to the control of conventional pollutants by effluent guideline limitations. The new guidelines were known as best conventional pollutants control technology (BCT). These guidelines replaced the existing BAT limitations, which were determined to be unreasonable for certain categories of pollutants.

In order to determine if BCT limitations would be necessary, the cost effectiveness of conventional pollutant reduction to BAT levels beyond BPT levels had to be determined and compared to the cost of removal of this same amount of pollutant by a POTWs of similar capacity. If it was equally cost-effective for the industry to achieve the reduction required for meeting the BAT limitations as the POTW, then the BCT limit was made equal to the BAT level. When this test was applied, the BAT limitation set for certain categories were found to be unreasonable. In these subcategories EPA proposed to remove the BAT limitations and revert to the BPT limitations until BCT control levels could be formulated.

2.4.4 CONTROL OF TOXIC POLLUTANTS

Since the early 1980s, EPA's water quality standards guidance placed increasing importance on toxic pollutant control. The Agency urged states to adopt criteria into their standards for priority toxic pollutants, particularly those for which EPA had published criteria guidance. EPA also provided guidance to help and support state adoption of toxic pollutant standards with the *Water Quality Standards Handbook* (1983) and the *Technical Support Document for Water Quality Toxics Control* (1985 and 1991).

Despite EPA's urging and guidance, state response was disappointing. A few states adopted large numbers of numeric toxic pollutant criteria, primarily for the protection of aquatic life. Most other states adopted few or no water quality criteria for priority toxic pollutants. Some relied on "free from toxicity" criteria and the so-called action levels for toxic pollutants or occasionally calculated site-specific criteria. Few states addressed the protection of human health by adopting numeric human health criteria.

State development of case-by-case effluent limits using procedures that did not rely on the statewide adoption of numeric criteria for the priority toxic pollutants frustrated Congress. Congress perceived that states were failing to aggressively address toxics and that EPA was not using its oversight role to push the states to move more quickly and comprehensively. Many in Congress believed that these delays undermined the effectiveness of the Act's framework.

2.4.5 1987 CWA Amendments

In 1987, Congress, unwilling to tolerate further delays, added Section 303 (c) (2) (B) to the CWA. The section provided that, whenever a state reviews water quality standards or revises or adopts new standards, the state had to adopt criteria for all toxic pollutants listed pursuant to Section 307 (a) (1) of the Act for which criteria have been published under Section 304 (a), discharge or presence of which in the affected waters could reasonably be expected to interfere with those designated uses adopted by the state, as necessary to support such designated uses. Such criteria had to be specific numerical criteria for such toxic pollutants. When numerical criteria are not available, wherever a state reviews water quality standards, or revises or adopts new standards, the state has to adopt criteria based on biological monitoring or assessment methods consistent with information published pursuant to Section 304 (a) (8). Nothing in this Section was to be construed to limit or delay the use of effluent limitations or other permit conditions based on or involving biological monitoring or assessment methods or previously adopted numerical criteria.

In response to this new Congressional mandate, EPA redoubled its efforts to promote and assist state adoption of numerical water quality standards for priority toxic pollutants. EPA's efforts included the development and issuance of guidance to the states on acceptable implementation procedures. EPA attempted to provide the maximum flexibility in its options that complied not only with the express statutory language but also with the ultimate congressional objective: prompt adoption of numeric toxic pollutant criteria. The Agency believed that flexibility was important so that each state could comply with Section 303 (c) (2) (B) within its resource constraints. EPA distributed final guidance on December 12, 1988. This guidance was similar to earlier drafts available for review by the states. The availability of the guidance was published in the *Federal Register* on January 5, 1989 (54 FR 346).

The structure of Section 303 (c) is to require states to review their water quality standards at least once in each 3 year period. Section 303 (c) (2) (B) instructs states to include reviews for toxics criteria whenever they initiate a triennia review. EPA initially looked at February 4, 1990, the 3 year anniversary of the 1987 CWA amendments, as a convenient point to index state compliance. The April 17, 1990 *Federal Register Notice* (55 FR 14350) used this index point for the preliminary assessment of state compliance. However, some states were very nearly completing their state administrative processes for ongoing reviews when the 1987 amendments were enacted and could not legally amend those proceedings to address additional toxics criteria. Therefore, in the interest of fairness, and to provide such states a full 13 year review period, EPA's FY 1990 Agency Operating Guidance provided that states should complete adoption of the numeric criteria to meet Section 303 (c) (2) (B) by September 30, 1990.

Section 303 (c) does not provide penalties for states that do not complete timely water quality standard reviews. In no previous case had an EPA Administrator found that state failure to complete a review within 3 years jeopardized the public health or welfare to such an extent that promulgation of Federal standards pursuant to Section 303 (c) (4) (8) was justified. However, the pre-1987 CWA never mandated state adoption of priority toxic pollutants or other specific criteria. EPA relied on its water quality standards regulation (40 CFR 131.11) and its criteria and program guidance to the states on appropriate parametric coverage in state water quality standards, including toxic pollutants. With Congressional concern exhibited in the legislative history for the 1987 Amendments regarding undue delays by states and EPA, and because states have been explicitly required to adopt numeric criteria for appropriate priority toxic pollutants since 1983, the Agency is proceeding to promulgate Federal standards pursuant to Section 303 (c) (4) (B) of the CWA and 40 CFR 131.22 (b).

States have made substantial recent progress in the adoption, and EPA approval, of toxic pollutant water quality standards. Furthermore, virtually all states have at least proposed new toxics criteria for priority toxic pollutants since Section 303 (c) (2) (B) was added to the CWA in February of 1987. Unfortunately, not all such state proposals address, in a comprehensive manner, the requirements of Section 303 (c) (2) (B). For example, some states have proposed to adopt criteria to protect aquatic life, but not human health; other states have proposed human health criteria that do not address major exposure pathways (such as the combination of both fish consumption and drinking water). In addition, final adoption of proposed state toxics criteria that would be approved by EPA in some cases has been substantially delayed due to controversial and difficult issues associated with the toxic pollutant criteria adoption process. Details of biological criteria, metal bioavailability and toxicity, cooling water cathode regulation, and water reuse are discussed in Chapter 17. The SDWA is renewed in Chapter 18.

2.4.6 TOTAL MAXIMUM DAILY LOAD (TMDL)

Under section 303 (d) of the 1972 CWA, states, territories, and authorized tribes were, required to develop list of impaired water. The impaired waters do not meet water quality standards that states, territories, and authorized tribes have set for them, even after point sources of pollution have installed the minimum required levels of pollution control technology. The law requires that these jurisdictions establish priority rankings for waters on the lists and develop TMDLs for these waters.

This part of the CWA was relatively neglected until 1996. A Federal Advisory Committee was convened and produced in 1998 a report and subsequent proposed changes for implementation of the TMDL program and associated changes in the NPDES for point sources. A number of court orders were also motivation factors in the implementation and proposed changes to the rule.

The TMDL specifies the amount of a particular pollutant that may be present in a water body, allocates allowable pollutant loads among sources, and provides the basis for attaining or maintaining water quality standards.

The TMDL regulation were issued as draft in 1999 and finally published on July 13, 2000.

See also: U.S. EPA. 2006. TMDLs. http://www.epa.gov/watertrain/cwa/cwa29.htm

2.4.7 WATER QUALITY TRADING

Within the approach to TMDL rules and subsequent management policy, the emphasis on "Pollutant Trading" or "Water Quality Trading" has evolved. Trading allows sources with responsibility for discharge reductions the flexibility to determine where reductions will occur. Within the trading approach, the economic advantages are emphasized. This is driven by the TMDL or more stringent water quality based requirement in an NPDES permit and the potential fact that discharge sources have significantly different costs to control the pollutant of concern.

In January of 2003, EPA published the *Water Quality Trading Policy*. The policy outlines the trading objectives, requirements, and elements of a trading program. While barriers exist to the implementation of an effective Water Quality Trading program, the cost of discharge quality compliance would warrant consideration of the approach.

2.4.8 BIOTERRORISM ACT OF 2003

While not directly related to water quality regulations, the security, and vulnerability of community drinking water systems was addressed in the Public Health Security and Bioterrorism Preparedness and Response Act of 2002 (Bioterrorism Act). The vulnerability assessments (VAs) were intended to examine a facility's ability to defend against adversarial actions that might substantially disrupt the ability of a water system to provide safe and reliable supply of drinking water.

For community drinking water systems serving greater than 3300 persons, it was required to conduct a VA, certify and submit a copy of the VA to the EPA administrator, prepare or revise an emergency response plan based on the results of the VA, and within 6 months certify to the EPA administrator that an emergency response plan has been completed or updated. The VA requirement was to be completed by all facilities in June of 2004.

Additional details on the CWA are provided in Chapter 17.

2.4.9 SAFE DRINKING WATER ACT

The EPA establishes standards for drinking water quality through the SDWA. These standards represent the maximum contaminant levels (MCLs), and consist of numerical criteria for specified contaminants. Local water supply systems are required to monitor their drinking water periodically for contaminants with MCLs and for a broad range of other contaminants as specified by the EPA. Additionally, to protect underground sources of drinking water, EPA requires periodic monitoring of wells used for underground injection of hazardous waste, including monitoring of the groundwater above the wells.

States have the primary responsibility for the enforcement of drinking water standards, monitoring, and reporting requirements. States also determine requirements for environmentally sound underground injection of wastes. The SDWA authorizes EPA to award grants to states for developing and implementing programs to protect drinking water at the tap and ground-water resources. These grant programs may be for supporting state public water supply, wellhead protection, and underground injection programs, including compliance and enforcement.

The CWA and the SDWA place great reliance on state and local initiatives in addressing water problems. With the enactment of the 1986 SDWA amendments and the 1987 Water Quality Act Amendments, significant additional responsibilities were assigned to the EPA and the states. Faced with many competing programs limited resources, the public sector will need to set priorities. With this in mind, the EPA is encouraging states to address their water quality problems by developing state clean water strategies. These strategies are to set forth state priorities over a multiyear period. They will help target the most valuable and/or most threatened water resources for protection.

Success in the water programs is increasingly tied to state and local leadership and decision-making and to public support. The EPA works with state and local agencies, industry, environmentalists, and the public to develop environmental agenda in the following three areas:

1. *Protection of drinking water.* Although more Americans are receiving safer drinking water than ever before, there are still serious problems with contamination of drinking water supplies and of groundwater that is or could be used for human consumption. Contaminated groundwater has caused well closings. The extent and significance of contamination by toxics has not been fully assessed for most of the nation's rivers and lakes, which are often used for drinking water supply. All of these issues are areas for continued work and improvement.

2. *Protection of critical aquatic habitats.* Contamination or destruction of previously underprotected areas such as oceans, wetlands, and near coastal waters must be addressed.

3. *Protection of surface-water resources.* The EPA and the states will need to establish a new phase of the federal–state partnership in ensuring continuing progress in addressing conventional sources of pollution [7].

Additional details can be found in Chapter 17.

See also: http://www.epa.gov/safewater/sdwa/30th/factsheets/unterstand.html

2.4.10 Marine Protection, Research, and Sanctuaries Act (Title I)

EPA designates sites and times for ocean dumping. Actual dumping at these designated sites requires a permit. The EPA and the Corps of Engineers share this permitting authority, with the Corps responsible for the permitting of dredged material (subject to an EPA review role), and the EPA responsible for permitting all other types of materials. The Coast Guard monitors the activities and the EPA is responsible for assessing penalties for violations.

2.5   THE SUPERFUND AMENDMENTS AND REAUTHORIZATION ACT OF 1986

The 1986 amendments to the CERCLA, known as the SARA, authorized $8.5 billion for both the emergency response and longer-term (or remedial) cleanup programs. The Superfund amendments focused on:

1. *Permanent remedies.* The EPA must implement permanent remedies to the maximum extent practicable. A range of treatment options will be considered whenever practicable.
2. *Complying with other regulations.* Applicable or relevant and appropriate standards from other federal, state, or tribal environmental laws must be met at Superfund sites where remedial actions are taken. In addition, state standards that are more stringent than federal standards must be met in cleaning up sites.
3. *Alternative treatment technologies.* Cost-effective treatment and recycling must be considered as an alternative to the land disposal of wastes. Under RCRA, Congress banned land disposal of some wastes. Many Superfund site wastes, therefore, are banned from disposal on the land; alternative treatments are under development and will be used where possible.
4. *Public involvement.* Citizens living near Superfund sites are involved in the site decision-making process for over 5 years. They are also able to apply for technical assistance grants that further enhance their understanding of site conditions and activities.
5. *State involvement.* States and tribes are encouraged to participate actively as partners with EPA in addressing Superfund sites. They assist in making the decisions at sites, can take responsibility in managing cleanups, and can play an important role in oversight of responsible parties.
6. *Enforcement authorities.* Settlement policies were strengthened through Congressional approval and inclusion in SARA. Different settlement tools, such as *de minimis* settlements (settlements with minor contributors) are part of the Act.
7. *Federal facility compliance.* Congress emphasized that federal facilities "are subject to, and must comply with, this Act in the same manner and to the same extent … as any non-government entity." Mandatory schedules have been established for federal facilities to assess their sites, and if listed in the National Priority List (NPL), to clean up such sites. EPA will be assisting and overseeing federal agencies with these requirements.

The amendments also expanded research and development, especially in the area of alternative technologies. They also provided for more training for state and federal personnel in emergency preparedness, disaster response, and hazard mitigation.

Additional details can be found at:

1. http://www.epa.gov/superfund/action/law/sara.htm
2. http://www.epa.gov/swerosps/bf/aai/aai_final_factsheet.htm
3. http://www.epa.gov/swerosps/bf/aai/ep_deffactsheet.htm
4. http://www.epa.gov/oilspill/ncpover.htm

2.5.1 Major Provisions of Title III of SARA (Also Known as Emergency Planning and Community Right-to-Know Act or EPCRA)

1. *Emergency planning.* EPCRA establishes a broad-based framework at the state and local levels to receive chemical information and use that information in communities for chemical emergency planning.
2. *Emergency release notification.* EPCRA requires facilities to report certain releases of extremely hazardous chemicals and hazardous substances to their state and local emergency planning and response officials.
3. *Hazardous chemical inventory reporting.* EPCRA requires facilities to maintain a material safety data sheet (MSDS) for any hazardous chemicals stored or used in the workplace and to submit those sheets to state and local authorities. It also requires them to submit an annual inventory report for those same chemicals to local emergency planning and fire protection officials, as well as state officials.
4. *Toxic release inventory reporting.* EPCRA requires facilities to report annually on routine emissions of certain toxic chemicals to the air, land, or water. Facilities must report if they are in standard industrial classification codes 20 through 39 (i.e., manufacturing facilities) with 10 or more employees and manufacture or process any of 650 listed chemical compounds in amount greater than specified threshold quantities. If the chemical compounds are considered persistent, bioaccumulative, or toxic, the thresholds are much lower. EPA is required to use these data to establish a national chemical release inventory database, making the information available to the public through computers, via telecommunications, and by other means.

2.6 THE CLEAN AIR ACT

The CAA defines the national policy for air pollution abatement and control in the United States. It establishes goals for protecting health and natural resources and delineates what is expected of Federal, State, and local governments to achieve those goals. The CAA, which was initially enacted as the Air Pollution Control Act of 1955, has undergone several revisions over the years to meet the ever-changing needs and conditions of the nation's air quality. On November 15, 1990, the president signed the most recent amendments to the CAA, referred to as the 1990 CAA Amendments. Embodied in these amendments were several progressive and creative new themes deemed appropriate for effectively achieving the air quality goals and for reforming the air quality control regulatory process. Specifically the amendments:

1. Encouraged the use of market-based principles and other innovative approaches similar to performance-based standards and emission banking and trading.
2. Promoted the use of clean low-sulfur coal and natural gas, as well as innovative technologies to clean high-sulfur coal through the acid rain program.
3. Reduced energy waste and create enough of a market for clean fuels derived from grain and natural gas to cut dependency on oil imports by one million barrels per day.
4. Promoted energy conservation through an acid rain program that gave utilities flexibility to obtain needed emission reductions through programs that encouraged customers to conserve energy.

Several of the key provisions of the act are reviewed below [8].

2.6.1 PROVISIONS FOR ATTAINMENT AND MAINTENANCE OF NATIONAL AMBIENT AIR QUALITY STANDARDS

Although the CAA brought about significant improvements in the nation's air quality, the urban air pollution problems of ozone (smog), carbon monoxide (CO), and particulate matter (PM) persist. In 1995, approximately 70 million U.S. residents were living in counties with ozone levels exceeding the EPA's current ozone standard.

The CAA, as amended in 1990, established a more balanced strategy for the nation to address the problem of urban smog. Overall, the amendments revealed the Congress's high expectations of the states and the federal government. While it gave states more time to meet the air quality standard (up to 20 years for ozone in Los Angeles), it also required states to make constant progress in reducing emissions. It required the federal government to reduce emissions from cars, trucks, and buses; from consumer products such as hair spray and window-washing compounds; and, from ships and barges during loading and unloading of petroleum products. The federal government also developed the technical guidance that states need to control stationary sources.

The CAA addresses the urban air pollution problems of ozone (smog), carbon monoxide, and PM. Specifically, it clarifies how areas are designated and redesignated "attainment." It also allows the EPA to define the boundaries of "nonattainment" areas, i.e., geographical areas whose air, quality does not meet federal ambient air quality standards designed to protect public health. The law also establishes provisions defining when and how the federal government can impose sanctions on areas of the country that have not met certain conditions.

For the pollutant ozone, the CAA established nonattainment area classifications ranked according to the severity of the area's air pollution problem. These classifications are *marginal, moderate, serious, severe, and extreme.* The EPA assigns each nonattainment areas one of these categories, thus triggering varying requirements the area must comply with in order to meet the ozone standard.

As mentioned, nonattainment areas have to implement different control measures, depending upon their classification. Marginal areas, for example, are the closest to meeting the standard. They are required to conduct an inventory of their ozone-causing emissions and institute a permit program. Nonattainment areas with more serious air quality problems must implement various control measures. The worse the air quality, the more controls these areas will have to implement.

The CAA also established similar programs for areas that do not meet the federal health standards for carbon monoxide and PM. Areas exceeding the standards for these pollutants are divided into "moderate" and "serious" classifications. Depending upon the degree to which they exceed the carbon monoxide standard, areas are then required to implement programs such as introducing oxygenated fuels and/or enhanced emission inspection programs, among other measures. Depending upon their classification, areas exceeding the PM standard have to implement reasonably available control measures (RACMs) or best available control measures (BACMs), among other requirements.

2.6.2 PROVISIONS RELATING TO MOBILE SOURCES

While motor vehicles built today emit fewer pollutants (60%–80% less, depending on the pollutant) than those built in the 1960s, cars and trucks still account for almost half the emissions of the ozone precursors (volatile organic compounds, VOCs, and nitrogen oxides, NO_x) and up to 90% of the CO emissions in urban areas. The principal reason for this problem is the rapid growth in the number of vehicles on the roadways and the total miles driven. This growth has offset a large portion of the emission reductions gained from motor vehicle controls.

In view of the continuing growth in automobile emissions in urban areas combined with the serious air pollution problems in many urban areas, Congress made significant changes to the motor vehicle provisions of the CAA and established tighter pollution standards for emissions from automobiles and trucks. These standards were set so as to reduce tailpipe emissions of hydrocarbons, carbon monoxide, and nitrogen oxides on a phased-in basis beginning in model year 1994. Automobile manufacturers also were required to reduce vehicle emissions resulting from the evaporation of gasoline during refueling.

Fuel quality was also controlled. Scheduled reductions in gasoline volatility and sulfur content of diesel fuel, for example, were required. Programs requiring cleaner (the so-called reformulated) gasoline were initiated in 1995 for the nine cities with the worst ozone problems. Higher levels (2.7%) of alcohol-based oxygenated fuels were to be produced and sold in those areas that exceed the federal standard for carbon monoxide during the winter months.

The 1990 amendments to the CAA also established a clean fuel car pilot program in California, requiring the phase-in of tighter emission limits for 150,000 vehicles in model year 1996 and 300,000 by the model year 1999. These standards were to be met with any combination of vehicle technology and cleaner fuels. The standards became even more strict in 2001. Other states were able to "opt in" to this program, through incentives, not sales or production mandates.

2.6.3   AIR TOXICS

Toxic air pollutants are those pollutants which are hazardous to human health or the environment. These pollutants are typically carcinogens, mutagens, and reproductive toxins.

The toxic air pollution problem is widespread. Information generated in 1987 from the Superfund "Right to Know" rule (SARA Section 313) discussed earlier, indicated that more than 2.7 billion pounds of toxic air pollutants were emitted annually in the United States. The EPA studies indicated that exposure to such quantities of toxic air pollutants may result in 1000–3000 cancer deaths each year.

Section 112 of the CAA includes a list of 189 substances which are identified as hazardous air pollutants. A list of categories of sources that emit these pollutants was prepared [The list of source categories included (1) major sources, or sources emitting 10 tons per year of any single hazardous air pollutants; and (2) area sources (smaller sources, such as dry cleaners and auto body refinishing)]. In turn, EPA promulgated emission standards, referred to as maximum achievable control technology (MACT) standards, for each listed source category. These standards were based on the best demonstrated control technology or practices utilized by sources that make up each source category. Within 8 years of promulgation of a MACT standard, EPA must evaluate the level of risk that remains (residual risk), due to exposure to emissions from a source category, and determine if the residual risk is acceptable. If the residual risks are determined to be unacceptable, additional standards are required.

2.6.4   ACID DEPOSITION CONTROL

Acid rain occurs when sulfur dioxide and nitrogen oxide emissions are transformed in the atmosphere and return to the earth in rain, fog, or snow. Approximately 20 million tons of sulfur dioxide is emitted annually in the United States, mostly from the burning of fossil fuels by electric utilities. Acid rain damages lakes, harms forests and buildings, contributes to reduced visibility, and is suspected of damaging health.

It was hoped that the CAA would bring about a permanent 10 million ton reduction in sulfur dioxide (SO_2) emissions from 1980 levels. To achieve this, the EPA allocated allowances in two phases, permitting utilities to emit one ton of sulfur dioxide. The first phase, which became effective January 1, 1995, required 110 power plants to reduce their emissions to a level equivalent to the product of an emissions rate of 2.5 lbs of SO_2/MM Btu × an average of their 1985–1987 fuel use. Emissions data indicate that 1995 SO_2 emissions at these units nationwide were reduced by almost 40% below the required level.

The second phase, which became effective January 1, 2000, required approximately 2000 utilities to reduce their emissions to a level equivalent to the product of an emissions rate of 1.2 lbs of SO_2/MM Btu × the average of their 1985–1987 fuel use. In both phases, affected sources were required to install systems that continuously monitor emissions in order to track progress and assure compliance.

The CAA allowed utilities to trade allowances within their systems and/or buy or sell allowances to and from other affected sources. Each source must have had sufficient allowances to cover its annual emissions. If not, the source was subject to a \$2000/ton excess emissions fee and a requirement to offset the excess emissions in the following year.

The CAA also included specific requirements for reducing emissions of nitrogen oxides.

2.6.5 OPERATING PERMITS

The Act requires the implementation of an operating permits program modeled after the National Pollution Discharge Elimination System (NPDES) of the CWA. The purpose of the operating permits program is to ensure compliance with all applicable requirements of the CAA. Air pollution sources subject to the program must obtain an operating permit; states must develop and implement an operating permit program consistent with the Act's requirements; and, EPA must issue permit program regulations, review each state's proposed program, and oversee the state's effort to implement any approved program. The EPA must also develop and implement a federal permit program when a state fails to adept and implement its own program.

In many ways this program is the most important procedural reform contained in the 1990 Amendments to the CAA. It enhanced air quality control in a variety of ways and updated the CAA, making it more consistent with other environmental statutes. The CWA, the RCRA, and the FIFRA all require permits.

2.6.6 STRATOSPHERIC OZONE PROTECTION

The CAA requires the phase out of substances that deplete the ozone layer. The law required a complete phase-out of CFCs and halons, with stringent interim reductions on a schedule similar to that specified in the Montreal Protocol, including CFCs, halons, and carbon tetrachloride by 2000 and methyl chloroform by 2002. Class II chemicals hydrochlorofluorocarbons (HCFCs) will be phased out by 2030.

The law required nonessential products releasing Class I chemicals to the banned. This ban went into effect for aerosols and noninsulating foams using Class II chemicals in 1994. Exemptions were included for flammability and safety.

The following five major rules were recently promulgated to achieve significant improvement in air quality, health, and quality of life.

1. *Clean Air Interstate Rule* (70 FR 25161, May 12, 2005)
   The Clean Air Interstate Rule provided states with a solution to the problem of power plant pollution that drifts from one state to another. The rule uses a cap and trade system to reduce the target pollutants by 70 percent.
2. *Mercury Rule* (70 FR 28605, May 18, 2005)
   EPA issued the Clean Air Mercury Rule on March 15, 2005. This rule builds on the Clean Air Interstate Rule (CAIR) to reduce mercury emissions from coal-fired power plants, the largest remaining domestic source of human-caused mercury emissions. Issuance of the Clean Air Mercury Rule marked the first time EPA regulated mercury emissions from utilities, and made the

U.S. the first nation in the world to control emissions from this major source of mercury pollution.

3. *Nonroad Diesel Rule* (69 FR 38957, May 11, 2004)

The Clean Air Nonroad Diesel Rule will change the way diesel engines function to remove emissions and the way diesel fuel is refined to remove sulfur. The Rule is one of EPA's *Clean Diesel Programs*, which were promulgated to produce significant improvements in air quality.

4. *Ozone Rules* (http://www.epa.gov/ozonedesignations/)

The Clean Air Ozone Rules (dealing with 8-hour ground-level ozone designation and implementation) designated those areas whose air did not meet the healthbased standards for ground-level ozone. The ozone rules classified the seriousness of the problem and required states to submit plans for reducing the levels of ozone in areas where the ozone standards were not being met.

5. *Fine Particle Rules* (http://www.epa.gov/pmdesignations/)

The Clean Air Fine Particles Rules designated those areas whose air does not meet the health-based standards for fine-particulate pollution. This rule required states to submit plans for reducing the levels of particulate pollution in areas where the fine-particle standards are not met.

See also 15th Anniversary (2005) of Clean Air Act Amendments of 1990 at http://www.epa.gov/air/cleanairact/

2.7   OCCUPATIONAL SAFETY AND HEALTH ACT

The Occupational Safety and Health Act (OSH Act) was enacted by Congress in 1970 and established the Occupational Safety and Health Administration (OSHA), which addressed safety in the workplace. At the same time the EPA was established. Both EPA and OSHA are mandated to reduce the exposure of hazardous substances over land, sea, and air. The OSH Act is limited to conditions that exist in the workplace, where its jurisdiction covers both safety and health. Frequently, both agencies regulate the same substances but in a different manner as they are overlapping environmental organizations.

Congress intended that OSHA be enforced through specific standards in an effort to achieve a safe and healthy working environment. A "general duty clause" was added to attempt to cover those obvious situations that were admitted by all concerned but for which no specific standard existed. The OSHA standards are an extensive compilation of regulations, some that apply to all employers (such as eye and face protection), and some that apply to workers who are engaged in a specific type of work (such as welding or crane operation). Employers are obligated to familiarize themselves with the standards and comply with them at all times.

Health issues, most importantly, contaminants in the workplace, have become OSHA's primary concern. Health hazards are complex and difficult to define. Because of this, OSHA has been slow to implement health standards. To be complete, each standard requires medical surveillance, record keeping, monitoring, and physical reviews. On the other side of the ledger, safety hazards are aspects of the work environment that are expected to cause death or serious physical harm immediately or before the imminence of such danger can be eliminated.

Probably one of the most important safety and health standards ever adopted is the OSHA hazard communication standard, more properly known as the "right to know" laws. The hazard communication standard requires employers to communicate information to the employees on hazardous chemicals that exist within the workplace. The program requires employers to craft a written hazard communication program, keep MSDSs for all hazardous chemicals at the workplace and provide employees with training on those hazardous chemicals, and assure that proper warning labels are in place.

The *Hazardous Waste Operations and Emergency Response Regulation* enacted in 1989 by OSHA addressed the safety and health of employees involved in cleanup operations at uncontrolled hazardous waste sites being cleaned up under government mandate, and in certain hazardous waste treatment, storage, and disposal operations conducted under RCRA. The standard provides for employee protection during initial site characterization and analysis, monitoring activities, training and emergency response. Four major areas are under the scope of the regulation:

1. Cleanup operations at uncontrolled hazardous waste sites that have been identified for cleanup by a government health or environmental agency.
2. Routine operations at hazardous waste transportation, storage, and disposal (TSD) facilities or those portions of any facility regulated by 40 CFR Parts 264 and 265.
3. Emergency response operations at sites where hazardous substances have or may be released.
4. Corrective action at RCRA sites.

The regulation addressed three specific populations of workers at the above operations. First, it regulates hazardous substance response operations under CERCLA, including initial investigations at CERCLA sites before the presence or absence of hazardous substance has been ascertained; corrective actions taken in cleanup operations under RCRA; and, those hazardous waste operations at sites that have been designated for cleanup by state or local government authorities. The second worker population to be covered involves those employees engaged in operations involving hazardous waste TSD facilities. The third employee population to be covered involves those employees engaged in emergency response operations for release or substantial threat of releases of hazardous substances, and post-emergency response operations to such facilities (29 CFR, 1910.120(q) [9]).

2.8   USEPA's RISK MANAGEMENT PROGRAM

Developed under the CAA's Section 112(r), the Risk Management Program (RMP) rule (40 CFR Part 68) is designed to reduce the risk of accidental releases of acutely toxic, flammable, and explosive substances. A list of the regulated substances (138 chemicals) along with their threshold quantities is provided in the Code of Federal Regulations at 40 CFR 68.130.

In the RMP rule, USEPA requires a "Risk Management Plan" that summarizes how a facility is to comply with USEPA's RMP requirements. It details methods and results of hazard assessment, accident prevention, and emergency response programs instituted at the facility. The hazard assessment shows the area surrounding the facility and the population potentially affected by accidental releases. USEPA requirements include a three-tiered approach for affected facilities. A facility is affected if a process unit manufactures, processes, uses, stores, or otherwise handles any of the listed chemicals at or above the threshold quantities. The EPA approach is summarized in Table 2.2. For example, USEPA defined Program 1 facilities as those processes that have not had an accidental release with offsite consequences in the 5 years prior to the submission date of the RMP and have no public receptors within the distance to a specified toxic or flammable endpoint associated with a worst-case release scenario. Program 1 facilities have to develop and submit a risk management plan and complete a registration that includes all processes that have a regulated substance present in more than a threshold quantity. They also have to analyze the worst-case release scenario for the process or processes; document that the nearest public receptor is beyond the distance to a toxic or flammable endpoint; complete a 5 year accident history for the process or processes; ensure that response actions are coordinated with local emergency planning and response agencies; and, certify that the source's worst-case release would not reach the nearest public receptors. Program 2 applies to facilities that are not Program 1 or Program 3 facilities. Program 2 facilities have to develop and submit the RMP as required for Program 1 facilities plus: develop and implement a management system; conduct a hazard assessment; implement certain prevention steps; develop and implement an emergency response program; and, submit data on prevention program elements for Program 2 processes. Program 3 applies to processes in standard industrial classification (SIC) codes 2611 (pulp mills), 2812 (chloralkali), 2819 (industrial inorganics), 2821 (plastics

TABLE 2.2
RMP Approach

Program	Description
1	Facilities submit RMP, complete registration of processes, analyze worst-case release scenario, complete 5 year accident history, coordinate with local emergency planning and response agencies; and, certify that the source's worst-case release would not reach the nearest public receptors.
2	Facilities submit RMP, complete registration of processes, develop and implement a management system; conduct a hazard assessment; implement certain prevention steps; develop and implement an emergency response program; and, submit data on prevention program elements.
3	Facilities submit RMP, complete registration of processes, develop and implement a management system; conduct a hazard assessment; implement prevention requirements; develop and implement an emergency response program; and, provide data on prevention program elements.

and resins), 2865 (cyclic crudes), 2869 (industrial organics), 2873 (nitrogen fertilizers), 2879 (agricultural chemicals), and 2911 (petroleum refineries). These facilities belong to industrial categories identified by USEPA as historically accounting for most industrial accidents resulting in off-site risk. Program 3 also applies to all processes subject to the OSHA Process Safety Management (PSM) standard (29 CFR 1910.119). Program 3 facilities have to develop and submit the RMP as required for Program 1 facilities plus: develop and implement a management system; conduct a hazard assessment; implement prevention requirements; develop and implement an emergency response program; and, provide data on prevention program elements for the Program 3 processes.

2.9   THE POLLUTION PREVENTION ACT OF 1990

The Pollution Prevention Act, along with the CAA Amendments passed by Congress on the same day in November 1990, represents a clear breakthrough in this nation's understanding of environmental problems. The Pollution Prevention Act calls pollution prevention a "national objective" and establishes a hierarchy of environmental protection priorities as national policy.

Under the Pollution Prevention Act, it is the national policy of the United States that pollution should be prevented or reduced at the source whenever feasible; where pollution cannot be prevented, it should be recycled in an environmentally safe manner. In the absence of feasible prevention and recycling opportunities, pollution should be treated; and, disposal should be used only as a last resort.

Among other provisions, the Act directed the EPA to facilitate the adoption of source reduction techniques by businesses and federal agencies, to establish standard methods of measurement for source reduction, to review regulations to determine their effect on source reduction, and to investigate opportunities to use federal procurement to encourage source reduction. The Act initially authorized an $8 million state grant program to promote source reduction, with a 50% state match requirement.

The EPA's pollution prevention initiatives are characterized by its use of a wide range of tools, including market incentives, public education and information, small business grants, technical assistance, research and technology applications, as well as the more traditional regulations and enforcement. In addition, there are other significant behind-the-scenes achievements: identifying and dismantling barriers to pollution prevention; laying the groundwork for a systematic prevention locus; and, creating advocates for pollution prevention that serve as catalysts in a wide variety of institutions.

2.10   FUTURE TRENDS

It is very difficult to predict future regulations. In the past, regulations have been both a moving target and confusing. What can be said (for certain?) is that there will be new regulations, and the probability is high that they will be contradictory and confusing. Past and current regulations provide a measure of what can be expected.

2.11 SUMMARY

1. Environmental regulations are an organized system of statues, regulations, and guidelines that minimize, prevent, and punish the consequences of damage to the environment. The recent popularity of environmental issues has brought about changes in legislation and subsequent advances in technology.
2. The CAA Amendments of 1990 build upon the regulatory framework of the CAA programs and expands their coverage to many more industrial and commercial facilities.
3. Hazardous waste is regulated under the RCRA for currently generated hazardous waste and under the Comprehensive Environmental, Response, Compensation, and Liability Act for past generation and subsequent remediation at hazardous waste sites.
4. Under the CWA, the permit program known as the NPDES, requires dischargers to disclose the volume and nature of their discharges as well as monitor and report to the authorizing agency the results.
5. The OSH Act was established to address safety in the workplace, which is limited to conditions that exist within the workplace, where its jurisdiction covers both safety and health.
6. The Toxic Substance Control Act of 1976 provides EPA with the authority to control the risks of thousands of chemical substances, both new and old, that are not regulated as either drugs, food additives, cosmetics, or pesticides.
7. Risk management has become a top priority by industry and federal officials. The CAA Amendments will require industries to communicate the likelihood and degree of a chemical accident to the public.

REFERENCES

1. Adapted from: Burke, G., Singh, B., and Theodore, L. *Handbook of Environmental Management and Technology*, 2nd edition, John Wiley & Sons, Hoboken, NJ, 2000.
2. Adapted from: Stander, L. and Theodore, L. *Environmental Regulatory Calculations Handbook*, John Wiley & Sons, Hoboken, NJ, 2008.
3. U.S. EPA. *EPA Journal*, 14(2), March 1988.
4. Sharp, R. Personal notes, 2007.
5. Cheremisinoff, P.N. and Ellerbusch, F. *Solid Waste Legislation, Resource Conservation & Recovery Act*, A Special Report, Washington, DC, 1979.
6. Bureau of National Affairs, Resource Conservation and Recovery Act of 1976, *International Environmental Reporter*, Bureau of National Affairs, Washington, DC, October 21, 1976.
7. U.S. EPA. *Environmental Progress and Challenges: EPA's Update*, EPA-230-07-88-033, U.S. EPA, Washington, DC, August 1988.
8. The Clean Air Act Amendments of 1990 Summary Materials, November 15, 1990.
9. Theodore, M.K. and Theodore, L. *Major Environmental Issues Facing the 21st Century*, 1st edition, Theodore Tutorials (originally published by Simon & Schuster), East Williston, NY, 1995.

3 International Regulations

CONTENTS

3.1 INTRODUCTION

Along with an increasing awareness of the importance of the environment has come a deeper understanding that the environment is a global issue. For example, the inhabitants of the Earth share the same air (although there is minimal exchange between the Northern and Southern hemispheres), and some of the air over the United States might be over India in a few months, and vice versa. The harmful effects of air pollutants—agricultural damage, global climate change, acid deposition, etc.—know no political boundaries.

The industrialized nations have led the way in regulating pollutant sources, mostly on national levels. However, progress is also being made in regional alliances. The cooperation between the United States and Canada to tackle the mutual problem of acid rain is just one example of the type of effort needed in this area. Numerous worldwide conferences have helped different countries establish priorities and set agendas for their pollution control programs. Another critical and farsighted step has been to include developing countries in the discussions of pollution problems and their possible solutions. Long-term solutions are possible only if pollution-reducing technology is freely shared and only if industrializing nations are provided with economic incentives to be nonpolluters.

Environmental pollution has transcended national boundaries and is threatening the global ecosystem. International environmental concerns such as stratospheric ozone depletion, the greenhouse effect, global warming, deforestation, acid rain, and mega-disasters such as the devastating nuclear accident at Chernobyl and the toxic methyl isocyanate gas accidentally released by a subsidiary of Union Carbide in Bhopal, India, have set the stage to address global pollution problems. The potential effects of global environmental pollution necessitate global cooperation in order to secure and maintain a livable global environment.

Pollution that crosses political boundaries, such as acid rain, has caused friction between countries for at least a decade. Now, however, people are beginning to recognize a class of pollution problems that can affect not just one region, but the entire planet [1].

Nearly 30 years have passed since 113 governments agreed in Stockholm, Sweden, to cooperate in attacking a new threat to human welfare: the degradation of the global ecosystem from environmental pollution, overpopulation, and mismanagement of the natural resource base. Since then the awareness of how human and biological systems interlock has increased significantly, resulting in a far more sophisticated grasp of what must be done.

Is the goal of an environmentally healthy world realistic? To answer that question, one has only to examine the extent and significance of what has been accomplished to date. Governments have responded to the environmental challenge at national, regional, and global levels with a broad spectrum of institutional and programmatic initiatives. Indeed, despite many false starts, setbacks, continuing constraints, and the emergence of new hazards, the spirit of international cooperation has steadily grown stronger [2].

Some pollution is found throughout the world's oceans, which cover about two-thirds of the planet's surface. Marine debris, farm runoff, industrial waste, sewage, dredge material, storm water runoff, and atmospheric deposition (acid rain) all contribute to ocean pollution. Litter and chemical contamination occur across the globe, including in such remote places as Antarctica and the Bering Sea. But the level of pollution varies a good deal from region to region and from one locality to another.

The open ocean is generally healthy, especially in comparison to the coastal waters and semienclosed seas that are most directly affected by human activities. The pressures from those activities are immense; some 50%–75% of the world's population probably will live within 50 miles of a coastline within the next 10 years.

The EPA is working through various federal laws and international agreements to reduce marine pollution. For example, the EPA is helping to carry out the London Dumping Convention, MARPOL (the International Convention for the Prevention of Pollution from Ships), the Great Lakes Agreement, the Caribbean Regional Sea Program, and other marine multilateral and bilateral agreements. The EPA is also helping to develop regional and international programs to control discharges to the oceans from the land. In a major domestic initiative, the EPA is working to clean up major estuaries and coastal areas that have suffered the most from pollution [1].

A major global problem, depletion of the ozone layer, is linked to a group of chemicals called chlorofluorocarbons (CFCs). These chemicals are used widely by industry as refrigerants and by consumers in polystyrene products. Once released into the air they rise into the stratosphere and eat away at the earth's protective ozone layer. The ozone layer shields all life on the planet from the sun's hazardous ultraviolet radiation, a leading cause of skin cancer.

The United States, which has been in the forefront of efforts to reduce CFC emission, outlawed the use of CFCs in aerosol spray products nearly two decade ago. The United States joined other nations at that time, in a treaty called the Montreal Protocol, in pledging to eliminate the use of CFCs by the year 2000. Equally important is the commitment by industry to develop products and processes that do not use CFCs, and to share these substitutes with other countries. The EPA evaluates all possible substitutes to make sure they do not present new health or environmental problems.

The Clean Air Act of 1990 contains many measures to protect the ozone layer. Most important, the law requires a gradual end to the production of chemicals that deplete the ozone layer. CFC refrigerants found in car air conditioners, household refrigerators, and dehumidifiers—also known as R-12—were no longer produced after 1995. Hydrochlorofluorocarbon (HCFC) refrigerants for windows and central air-conditioning units also known as R-22—will be produced until 2020. The production of halons ended, after 1993, while methyl bromide production was limited beginning in 1994 and was phased out in 2001. The Clean Air Act also bans the release of ozone-depleting refrigerants during the service, maintenance, and disposal of air conditioners and all other equipment that contains these refrigerants [3].

The threat to the ozone layer illustrates an important principle: It is not enough simply to outlaw an environmental problem. One also must work toward a comprehensive and economically acceptable solution.

Except for solar, nuclear, and geothermal power, the production of energy requires that something be burned. That something is usually a fossil fuel such as oil, gasoline, natural gas, or coal. But it also can be other fuels—for example, wood or municipal waste. Burning any of these substances uses up oxygen and creates carbon dioxide gas. Bodies also create and exhale carbon dioxide. Plants, algae, and plankton, on the other hand, take in carbon dioxide and produce oxygen.

Modern industrial society and its need for power create far more carbon dioxide than the planet's vegetation can consume. As this excess carbon dioxide rises into the atmosphere, it acts as a kind of one-way mirror, trapping the heat reflected from the Earth's surface. Many leading scientists expect that this "greenhouse" effect from increased levels of carbon dioxide and other heat-trapping gases has caused an increase in global temperatures. Some predict that temperatures will rise significantly within the twenty-first century, and that global climate patterns could be dramatically disrupted. (See Chapter 12 for more details.) If these experts are correct, areas in the United States that are now cropland could become desert, and ocean levels could rise by 3 ft or more. The EPA is working with other federal agencies to improve the understanding of the likely amount and possible effects of global climate change. The EPA also is looking at ways to reduce carbon dioxide and other greenhouse gas emissions. This effort, like the agreement to eliminate CFCs, will require a major commitment to international cooperation by all the countries of the world.

Tropical rainforests, by absorbing large quantities of carbon dioxide, help to retard global warming. The rapid depletion of these tropical forests as well as those in the temperate zone has become a pressing global concern in recent years.

New data suggest that tropical forests are being lost twice as fast as previously believed; at present rates of destruction, many forests will disappear within 10–15 years, In July 1990, concern for the rapid loss of the great forest systems worldwide led the United States to propose a global forest convention at an economic summit attended by most industrialized nations. The agreement addressed all forests—north temperate, temperate, and tropical—as well as mapping and monitoring research, training, and technical assistance [1].

Regarding water concerns, the province of Ontario promulgated the Safe Drinking Water Act of 2002. The Ontario government also introduced the Nutrient Management Act in 2002, and adopted the Clean Water Act in 2006. In 2004, a World Health Organization report indicated that nearly one half of the environmental risk that society faces can be attributed to drinking water, sanitation, and hygiene.

3.2 THE GREENHOUSE EFFECT [4]

The threat of global warming now forces a more detailed evaluation of the environment. It forces consideration of the sacrifices which must be made to ensure an acceptable quality of the environment for the future.

As an environmental problem, global warming must be considered on an entirely different scale from that of most other environmental issues: The effects of climate change are long-term, global in magnitude, and largely irreversible. Because of the enormity of the problem and the uncertainties involved—it may take decades to determine with absolute certainty that global warming is under way—the difficult questions faced today are how and when one should react.

Fossil-fuel burning and forestry and agricultural practices are responsible for most of the man-made contributions to the gases in the atmosphere that act like a greenhouse to raise the earth's temperature; hence the term "greenhouse effect." Most of the processes that produce greenhouse gases are common everyday activities such as driving cars, generating electricity from fossil fuels, using fertilizers, and using wood-burning stoves. Because so many of these activities are so ingrained in society, reducing emissions could be a difficult task.

The search for solutions has begun. However, there is a growing concern that the costs of reducing emissions may be too high. But to put cost concerns in proper perspective, one must ask what kind of future one wants on this planet and how much does one value the environment and the cultural heritage that depends on it [4].

A consensus has emerged in the scientific community that a global warming has occurred. Scientists are certain that the concentrations of carbon dioxide (CO_2) and other greenhouse gases in the atmosphere are increasing, and they generally agree that these gases will warm the earth. Four questions remain to be answered:

1. How will the temperature rise?
2. When will the temperature rise?
3. Can the cause be attributed to man-made acturles?
4. What are the potential greenhouses?

Recent estimates indicate that if the concentrations of these gases in the atmosphere continue to increase, the earth's average temperature could rise by as much as 1.5°C–4.5°C in the next century. While this may not sound like a tremendous increase, one must keep in mind that during the last ice age 18,000 years ago, when glaciers covered much of North America, the earth's average temperature was only 5°C cooler than today.

Certainly global cooperation is an important consideration when addressing global warming issues. No single country contributes more than a fraction of greenhouse gases, and only a concerted effort can reduce emissions. In the future, as developing nations grow and consume more energy, their share of greenhouse-gas emissions will steadily increase. It is important for other nations to offer technological assistance so that these developing nations can grow in an energy-efficient manner [4].

The sources of greenhouse gases are so numerous and diverse that no single source contributes more than a small fraction of total emissions. Similarly, no single country contributes more than a fraction of emissions.

Unlike other environmental problems that the EPA could address with the stroke of a regulation, potential climate change is a problem that needs innovative global solutions. Future trends of emissions will depend on a wide range of factors, from population and economic growth to technological development and policies to reduce emissions. Past trends show that all countries have been producing greenhouse gases at a growing rate, and many countries will continue to do so for years to come. Based on careful study of the sources and trends of greenhouse emissions around the globe, countries can begin implementing prudent measures for slowing down emission while increasing economic development.

The developed countries, currently the largest CO_2 emitters, will grow in population at approximately 1.0%–1.5% per year and are projected to emit 6.7 billion tons of carbon by the year 2025. Developed countries are likely to continue to emit more CO_2 per person than developing countries. For example, the average citizen living in the United States produced six times more CO_2 each year than the average citizen in a developing country. In developing countries, population and economic growth will lead to a substantial increase in CO_2 emissions to over 5 billion tons per year, despite anticipated improvements in efficiency of energy use.

Developing countries now contribute only a small fraction of greenhouse gases, but their share of emissions is expected to increase significantly in the next 35 years. Data show the share of CO_2 emissions from Asia (including China), Africa, Latin America, and the Middle East increasing from slightly over one-fourth of the global total in 1985 to nearly one-half the total by 2025. Technologies developed in more industrialized nations to use energy efficiently could help developing nations reduce emissions as they continue to develop, but channels to transfer this technology must be developed.

On a regional basis, energy use in Western European countries is projected to grow at a relatively slow rate because of low population growth and policies that are anticipated to be implemented over the next decade. Several countries, such as Norway, Sweden, and the Netherlands, have already adopted policies specifically designed to slow the growth rate of greenhouse-gas emissions. These measures include special taxes, energy-efficiency programs, and promotion of nuclear energy, natural gas, and renewable energy sources.

The case in Eastern Europe is quite different, largely because many of these countries are among the most energy intensive and most energy inefficient in the world.

In Eastern Europe and the countries of the former Soviet Union, energy use and CO_2 emissions are projected to grow considerably over the next 35 years, but policies aimed at restructuring the economy and improving energy efficiency in Russia could have a significant impact. If these economies and those of Eastern Europe become more energy efficient and move from heavy industrial production to production of less energy-intensive consumer goods, they may be able to increase economic growth and enjoy the added benefit of reduced greenhouse-gas emissions.

In the coming years, the technical community must reevaluate how emissions are likely to change. But given this preliminary picture of the future, it is important to take the next step of assessing the specific technologies and policy measures that can reduce emissions now at low costs. Each country will have to examine its unique situation and determine appropriate responses. However, only by acting together will the global community slow the trend toward high emissions in the next century [3]. Methods for reducing the greenhouse effect are discussed later in the chapters addressing pollution prevention approaches, waste reduction, and energy conservation.

3.3 OZONE DEPLETION IN THE STRATOSPHERE [5,6]

Increasing concentrations of the synthetic chemicals known as CFCs and halons are breaking down the ozone layer, allowing more of the sun's ultraviolet rays to penetrate to the earth's surface. Ultraviolet rays can break apart important biological molecules, including DNA. Increased ultraviolet radiation can lead to greater incidence of skin cancer, cataracts, and immune deficiencies, as well as decreased crop yields and reduced populations of certain fish larvae, phytoplankton, and zooplankton that are vital to the food chain. Increased ultraviolet radiation can also contribute to smog and reduce the useful life of outdoor paints and plastics. Stratospheric ozone protects oxygen at lower altitudes from being broken up by ultraviolet light and keeps most of these harmful rays from penetrating to the earth's surface.

CFCs are compounds that consist of chlorine, fluorine, and carbon. First introduced in the late 1920s, these gases—as noted earlier—have been used as coolants for refrigerators and air conditioners, propellants for aerosol sprays, agents for producing plastic foam, and cleaners for electrical parts. CFCs do not degrade easily in the troposphere. As a result, they rise into the stratosphere where they are broken down by ultraviolet light. The chlorine atoms react with ozone to convert it into two molecules of oxygen. In the upper atmosphere ultraviolet light breaks off a chlorine atom from a CFC molecule. The chlorine attacks an ozone molecule, breaking it apart. An ordinary oxygen molecule and a molecule of chlorine monoxide are formed. A free oxygen atom breaks up the chlorine monoxide. The chlorine is free to repeat the process. Chlorine acts as a catalyst and is unchanged in the process. Consequently, each chlorine atom can destroy as many as 10,000 ozone molecules before it is returned to the troposphere.

Halons are an industrially produced group of chemicals that contain bromine, which acts in a manner similar to chlorine by catalytically destroying ozone. Halons are used primarily in fire extinguishing foam.

Laboratory tests have shown that nitrogen oxides also remove ozone from the stratosphere. Levels of nitrous oxide (N_2O) are rising from increased combustion of fossil fuels and use of nitrogen-rich fertilizers [5].

In the early 1970s, CFCs were primarily used in aerosol propellants. After 1974, U.S. consumption of aerosols had dropped sharply as public concern intensified about stratospheric ozone depletion from CFCs. Moreover, industry anticipated future regulations and shifted to other, lower cost chemicals. In 1978, EPA and other federal agencies banned the nonessential use of CFCs as propellants. However, other uses of CFCs continued to grow, and only Canada and a few European nations followed the United States' lead in banning CFC use in aerosols.

In recognition of the global nature of the problem, 31 nations representing the majority of the CFC-producing countries signed the Montreal Protocol in 1987. The Protocol, which had to be ratified by at least 11 countries before it became official at the start of 1989, required developed nations to freeze consumption of CFCs at 1986 levels by mid-1990 and to halve usage by 1999. The Protocol came into force, on time, on January 1, 1989, when 29 countries and the EEC representing approximately 82% of world consumption had ratified it. Since then several other countries have joined. Now nearly 175 countries are parties to the Convention and the Protocol, of which well over 100 arc developing countries. The Protocol is constructively flexible and it can be tightened as the scientific evidence strengths without having to be completely renegotiated. Its control provisions were strengthened through four adjustments to the Protocol adopted in London (1990), Copenhagen (1992), Vienna (1995), and Montreal (1997). The Protocol aims to reduce and eventually eliminate the emissions of human-made ozone depleting substances.

In addition to implementing the Montreal Protocol, the EPA is working with industry, the military, and other government organizations to reduce unnecessary emissions of CFCs and halons by altering work practices and testing procedures, or by removing institutional obstacles to reductions. The EPA is working with the National Aeronautics and Space Administration, the National Oceanic and Atmospheric Administration, the Department of Energy, the National Science Foundation, and other federal agencies to better understand the effects of global warming and stratospheric ozone depletion.

In 1986 and again in 1987, research teams were sent to investigate the causes and implications of the hole in the ozone over Antarctica. In 1986, the EPA published a multivolume summary with the United Nations on the effects of global atmospheric change. In addition, in 1987 EPA published a major risk assessment of the implications of continued emission of gases that can alter the atmosphere and climate.

In December of 1987, the Agency published proposed regulations for implementing the Montreal Protocol. The provisions of the Protocol would be implemented by limiting the production of regulated chemicals and allowing the marketplace to determine their future price and specific uses.

While the Montreal Protocol represents a major step toward safeguarding the earth's ozone layer, considerable work remains to be done. The major challenge is to develop a better understanding of the effects of stratospheric ozone depletion and global warming on human health, agriculture, and natural ecosystems. Substantial scientific uncertainty still exists. More must be learned about the Antarctic ozone hole and its implications, both for that region and the rest of the earth. More must also be learned about recent evidence of global ozone losses of 2%–5% during the past years.

Efforts to develop alternatives to CFCs and halons must be expedited. The Montreal Protocol provides a clear signal for industry to shift away from these chemicals. New technologies and new chemicals that will not deplete the ozone layer and increased conservation and recovery are essential to reducing the economic effects of the Protocol both in the United States and abroad. In the time since the Protocol was signed, major advancements in alternative technologies have been announced for CFC use in food packaging and solvents. Yet these are only a beginning and more must be done.

The EPA plans to continue international efforts to protect the ozone layer and to assess the risks of future climate change. EPA will send advisory teams to several key nations to help them explore options for reducing use of CFCs, such as producing different products, substituting other chemicals, and controlling emissions.

The Clean Air Act of 1990 sets a schedule for ending the production of chemicals that destroy stratospheric ozone. Chemicals that cause the most damage will be phased out first. CFCs. Halons, HCFCs, and other ozone-destroying chemicals were listed by Congress in the 1990 Clean Air Act and must be phased out. CFCs from car air conditioners are the biggest single source of ozone-destroying chemicals. At the end of 1993, all car air conditioners systems were required to be serviced using equipment that recycles CFCs and prevents their release into the air. Only specially trained and certified repair persons are allowed to buy the small cans of CFCs used in servicing auto air conditioners. Methylchloroform, also called 1,1,1-trichloroethanc, was phased out by 1996. This had been a widely used solvent found in products such as automotive brake cleaners (often sold as aerosol sprays) and spot removers used to take greasy stains off fabrics. Replacing methyl chloroform in the workplace and consumer products has led to changes in many products and processes [7].

3.4   ACID RAIN

Acid rain is not considered a threat to the global environment. Large parts of the earth are not now, and probably never will be, at risk from the effects of man-made acidity. But concern about acid rain is definitely growing. Although acid rain comes from the burning of fossil fuels in industrial areas, its effects can be felt on rural ecosystems hundreds of miles downwind. And if the affected area is in a different country, the economic interests of different nations can come into conflict.

Such international disputes can be especially difficult to resolve because it is not yet know how to pinpoint the sources in one country that are contributing to environmental damage in another.

Concerns about acid rain tend to be raised whenever large-scale sources of acidic emissions are located unwind of international borders. Japan, for example, has not yet suffered any environmental damage due to acid rain, but the Japanese are worried about the potential downwind effects of China's rapidly increasing industrialization. A similar problem has risen on the U.S.–Mexican border, where some people were worried that Mexico's copper smelter at Nacozari could cause acid rain on the pristine peaks of the Rocky Mountains. Besides scattered instances such as these, acid rain has emerged as a serious international issue only in two places: western Europe and northeastern North America.

3.4.1 EUROPE

Diplomatic problems related to cross-boundary air pollution first surfaced in Europe in the 1950s, when the Scandinavian countries began to complain about industrial emissions traveling across the North Sea from Great Britain. Since then, acid deposition has been linked to ecological damage in Norway, Sweden, and West Germany, and low-pH rainfall has been measured in a number of other European countries.

The potential and scientific controversies over acid rain are multiplied in Europe because so many countries are involved. Some countries producing very low amounts of SO_2 are nevertheless experiencing low-pH rainfall and high rates of acid deposition. Norway, for example, produced approximately 137,000 metric tons of SO_2 in 1980, yet received depositions of about 300,000 metric tons. Clearly, Norway, like a number of other European nations, is being subjected to acid deposition that originates outside its borders. The same situation presests nearly 30 years later.

Sweden pioneered the development of extensive and consistent monitoring for acid precipitation in the late 1940s. In 1954, the Swedish monitoring program was expanded to include other European countries. The results of this monitoring revealed the high acidity of rainfall over much of western Europe.

Prompted by these findings, the U.N. Conference on the Human Environment recommended a study of the impact of acid rain, and in July 1972, the U.N. Organization for Economic Cooperation and Development (OECD) began an inquiry into "the question of acidity in atmospheric precipitation." In 1979, a U.N. Economic Commission for Europe (ECE) conference in Stockholm approved a multinational convention for addressing the problem of long-range transboundary air pollution. Both the United States and Canada joined the European signatories. Later, a number of European countries, including France. West Germany, Czechoslovakia, and all the Scandinavian countries, agreed to reduce their SO_2 emissions by at least 30% from 1980 levels.

Following the Stockholm conference. ECE members decided in 1985 to broaden their goals to include the control of nitrogen oxides, which have been gaining recognition as important acid rain precursors. Workshops helped to determine the nature and extent of NO_x, pollution in various countries, as well as possible approaches for controlling it.

3.4.2 NORTH AMERICA

The United States and Canada share the longest undefended border in the world and billions of dollars in trade every year. They also share a number of environmental problems, foremost among them the problem of acid rain. In both countries, acidic emissions are concentrated relatively close to their mutual border. Canadian emissions originate primarily in southern Ontario and Quebec, while a majority of U.S. emissions originate along the Ohio River Valley, Each country is contributing to acid rain in the other. But because of prevailing wind patterns and the greater quantities of U.S. emissions, the United States sends much more acidity to Canada than Canada sends to it. In 1980, for example, the United States produced over 23 million metric tons of SO_2 and over 20 million metric tons of NO_x; Canada produced 4.6 million metric tons of SO_2 and 1.7 million tons of NO_x.

In the early 1970s, Canadian scientists began to report on the adverse environmental effects of acidity in lake water, and to link fish kills in acidic lakes and streams in eastern Canada to U.S. emission. By the late 1970s, acid rain had become a serious diplomatic issue affecting the relationship of the two countries. In 1980, the two countries took their first joint step toward resolving the issue with a Memorandum of Intent that called for shared research and other bilateral efforts to analyze and control acid rain. One of the most spectacular projects was a high-altitude experiment called "CATEX." Trace elements of various chemicals were inserted into SO_2 plumes from coal-fired power plants in the Midwest. Their dispersion was monitored along a path extending across the northeastern United States to Canada. These and other experiments have helped scientists gain new data on the formation and distribution of acid rain.

Western Europe and North America are highly industrialized, and it is likely that acid rain will continue to be a serious concern in both areas for the foreseeable future. But the nations involved are coming to terms with their common problem. In Europe, several nations have already taken steps to reduce transboundary air pollution. In North America, the president of the United States has endorsed the proposal to invest $5 billion to demonstrate innovative technologies that can be used to reduce transboundary air pollution. And, in both Europe and North America, the diplomatic groundwork for long-term cooperative activities has been established [6].

3.5   INTERNATIONAL ACTIVITIES

Several factors are pushing environmental concerns increasingly into the international arena. More and more, pollution is transboundary and even global in scope. Pressures on shared resources, such as river basins and coastal fisheries, are mounting. Resource deterioration in many nations is so extensive that other countries are affected, e.g., when ecological refugees flee across borders. As international trade increases, commodities and merchandise become the carriers of domestic environmental policies that must be rationalized.

It is not just that there are more environmental problems like ozone depletion that must be dealt with at the international level; it is also that the line between national and international environmental problems is fast disappearing.

Nitrogen oxide emissions, for example, must be regulated locally because of ground-level ozone formation, regionally because of acid rain, and globally because ground-level ozone is an infrared-trapping greenhouse gas. Methane and indirectly, carbon monoxide also contribute to the greenhouse effect.

In these instances, domestic and global environmental concerns push in the same direction. On the other hand, a major move to methanol as a substitute for gasoline could actually increase the global warming risk. A car burning methanol made from coal would result in perhaps twice the carbon dioxide emissions per mile as one burning gasoline.

Environmental diplomacy is the logical outgrowth of the desire to protect one's own national environment, to minimize environment-related conflicts with other countries, and to realize mutual benefits, including economic progress and the protection of the common natural heritage of humankind. As such, it is not entirely new. The register of international conventions and protocols in the field of the environment has grown steadily in this century: the main multilateral treaties today number about 100, many of them having to do with the protection of the marine environment and wildlife.

What is new is the prospect that environmental issues will move from being a secondary to a primary international concern and increasingly crowd the diplomatic agendas of nations. And these diplomatic agendas in turn will increasingly affect domestic environmental policy. Efforts to give international dimensions a higher priority within the Agency should continue. The EPA has established an Office of International Affairs to address international activities. Even more important is ensuring that domestic and international activities are actually coordinated internally.

The EPA also needs a world-class capacity to follow relevant developments in other countries and in international institutions, to understand and analyze the various approaches to environmental protection being taken abroad, and to anticipate future needs and developments at the international level. Beyond EPA's internal workings, new patterns of relating to other federal agencies seem desirable. Neither global nor local atmospheric issues are likely to be solved unless energy and environmental policy are made together in the future. As environmental diplomacy increases, finding appropriate patterns of interaction will become imperative. Moreover, the future is likely to bring increasing efforts to link environmental objectives and trade policy. For example, should the United States restrict imports of products that are manufactured by processes that harm the environment, much as one restricts imports of endangered species and harmful products? Should one import copper from countries where smelters operate without serious pollution control?

Much of the EPA's international activity in the past has focused on the Organization of Economic Cooperation and Development and other trans-Atlantic matters. In the future, the North–South and East–West dimensions will rival the North–North ones in importance. It already seems clear that solutions to the most serious global environmental challenges will require a series of vital understandings between the industrial and the developing countries. For example, the developing countries will expect the industrial countries to take the first and strongest actions on global warming. They will want to see the seriousness of the threat validated, and they will conclude, quite correctly, that the industrial countries are largely responsible for the problem and have the most resources to do something about it.

A tragic stalemate will occur if above argument is carried too far. Developing countries already account for about a fourth of all greenhouse gas emissions, and their share could double by the middle of the next century. Increasingly, all countries will be pressed to adopt energy and forestry strategies that are consistent with containing the greenhouse effect within tolerable limits.

The United States and the EPA need to build a new set of relationships with developing country officials so that confidence and trust are built for the challenging times ahead. One major step in this direction would be for the United States to initiate a new program of international environmental cooperation with developing countries. Such a program would not be limited to aid-eligible countries but would extend to countries like Brazil and Mexico. It would provide technical assistance, training, access to information and expertise, and planning grants all aimed at increasing the capacity of developing countries to manage their environmental challenges [8].

The EPA's overseas activities includes negotiating international environmental treaties, maintaining liaison with other health and environment organizations, and cooperating with and encouraging the environmental initiatives of other nations, particularly Third World countries. In addition, the EPA should work to spare U.S. industry from unfair foreign competitors benefiting from pollution havens.

3.6   FUTURE TRENDS

Continued industrial development will surely generate new challenges to human health and to the well-being of the Earth's environment as a whole. All development must be accompanied by research and evaluation of the potential environmental and human health consequences of each new process, product and by-product. Addressing the associated problems requires a global perspective, and solving the problems will take a global commitment.

One area that will probably see increased activity is noise pollution. According to research by the World Health Organization, thousands of people in Britain and around the world are dying prematurely from heart disease triggered by long-term exposure to excessive noise. Coronary heart disease caused 101,000 deaths in UK in 2006, and the study suggests that 3,030 of these were caused by chronic noise exposure, including daytime traffic. The EU recently issued a directive that obligates European cities with populations greater than 250,000 to produce digitized noise maps showing where traffic noise and volume is greatest.

3.7   SUMMARY

1. Environmental pollution has transcended national boundaries and is threatening the global ecosystem. International environmental concerns such as stratospheric ozone depletion, the greenhouse effect, global warming, deforestation, and acid rain have set the stage for addressing global pollution problems. The potential effects of global environmental pollution necessitate global cooperation in order to secure and maintain a livable global environment.

2. As an environmental problem, global warming must be considered on an entirely different scale from that of most other environmental issues; the effects of climate change are long-term, global in magnitude, and largely irreversible.

3. Increasing concentrations of the synthetic chemicals known as CFCs and halons are breaking down the ozone layer, allowing more of the sun's ultraviolet rays to penetrate to the earth's surface. Ultraviolet rays can break apart important biological molecules, including DNA. Increased

ultraviolet radiation can lead to decreased crop yields and reduced populations of certain fish larvae, phytoplankton, and zooplankton that are vital to the food chain.

4. Ozone, or "smog" is just one of six major air pollutants that the EPA regulates, but it is by far the most complex, intractable, and pervasive. It is also an extremely difficult pollutant to regulate effectively.

5. Several factors are pushing environmental concerns increasingly into the international arena. More and more, pollution is transboundary and even global in scope. As international trade increases, commodities and merchandise become the vehicles of domestic environmental policies that must be rationalized across borders.

6. Acid rain is not considered a threat to the global environment. Large parts of the earth are not now, and probably never will be, at risk from the effects of man-made acidity. But concern about acid rain is definitely growing.

REFERENCES

1. U.S. EPA, Communications and Public Affairs (A-107), *Preserving Our Future Today*, 2115-1012, October 1991.
2. U.S. EPA, Office of Public Affairs (A-107), *EPA Journal*, 13(7), September 1987.
3. U.S. EPA, Office and Radiation, *Protecting the Ozone Layer*, EPA 430-F-94-007, April 1994.
4. U.S. EPA, The greenhouse effect, *EPA Journal*, 16(2), 20K–9002, March/April 1990.
5. U.S. EPA, *Environmental Progress and Challenges: EPA's Update*, EPA-230-07-88-033, August 1988.
6. U.S. EPA, The challenge of ozone pollution, *EPA Journal*, 13(8), October 1987.
7. U.S. EPA, *The Plain English Guide to the Clean Air Act, Air, and Radiation*, EPA 400-K-93-001, April 1993.
8. U.S. EPA, Protecting the Earth—Are our institutions up to it? *EPA Journal*, 15(4), July/August 1989.
9. U.S. EPA, Acid rain: Looking ahead, *EPA Journal*, 12(6), June/July 1986.

4 ISO 14000*

Contributing Author: Lauren De Sanctis

CONTENTS

4.1 INTRODUCTION

The International Organization for Standardization (ISO) is a private, nongovernmental, international standards body based in Geneva, Switzerland. Founded in 1947, ISO promotes international harmonization and development of manufacturing, product, and communications standards. It is a nongovernmental organization. However, governments are allowed to participate in the development of standards and many governments have chosen to adopt the ISO standards as their regulations. The ISO also closely interacts with the United Nations [2]. ISO has promulgated over 16,000 internationally accepted standards for everything from paper sizes to film speeds. Roughly 157 countries participate in the ISO as "Participating" members or as "Observer" members. The United States is a full-voting Participating member and is officially represented by the American National Standards Institute (ANSI).

Many people will have noticed the seeming lack of correspondence between the official title when used in full, International Organization for Standardization, and the short form, ISO. Should not the acronym be IOS? That would have been the case if it were an acronym. However, ISO is a word derived from the Greek word *isos*, meaning equal. From "equal" to "standard," the line of thinking that led to the choice of "ISO" as a name of the organization is easy to follow. In addition, the name ISO is used around the world to denote the organization, thus avoiding the plethora of acronyms resulting from the translation of "International Organization for Standardization" into

---

* See Burke et al. [1].

the different national languages of members, e.g., IOS in English or OIN in French. Whatever the country, the short form of the organization's name is always ISO [3].

The ISO continues to expand the scope of their standards to incorporate areas such as the environment, service sectors, security, and managerial and organizational practice. There are currently more than 16,000 standards applying to three areas of sustainable development: economic, environmental, and social [2]. The ISO's environmental mission is to promote the manufacturing of products in a manner that is efficient, safe, and clean [4]. The ISO hopes to achieve this goal through the dedication and participation of more countries.

The ISO 14000 is a generic environmental management standard. It can be applied to any organization and focuses on the processes and activities conducted by the company. It consists of standards and guidelines regarding environmental management systems (EMSs). The idea for it first evolved from the United Nations Conference on Environment and Development (UNCED), which took place in Rio de Janeiro in 1992. The topic of sustainable development was discussed there and the ISO made a commitment to support this subject [5].

The ISO 14000 standards were first written in 1996 and have subsequently been amended and updated. Their purpose is to assist companies and organizations to minimize their negative affects on the environment and comply with any laws, regulations, or environmental requirements that have been imposed on them. It can also help to establish an organized approach to reducing any environmental impacts the company can control. Businesses that comply with these standards are eligible for certification. This certification is awarded by third-party organizations instead of the ISO [6].

In terms of history, the United Nations' 1992 Conference on Environment and Development set in motion the basic principles of the ISO 14000. The aforementioned United Nations Rio Declaration, which was produced by this conference [7], included:

1. Sovereign rights to sustainable development
2. Sustainable development for the present and future
3. Eradicate poverty for sustainable development
4. Global partnership for sustainable development
5. Eliminate unsustainable patterns
6. Participation of all
7. Equitable environmental legislation
8. International economic cooperation
9. Liability and compensation
10. Cooperation for the environment and human health
11. Precautionary approach
12. Internal environmental impact assessment
13. Immediate notification and international response

4.2   HOW THE STANDARDS ARE DEVELOPED

The impetus toward international standards is deeply rooted in economic rewards and an expansion into a global economy. The standardization of goods and services will not only increase potential market share but also allow goods and services to be available to more consumers. Different countries producing articles using the same technologies but using different sets of standards limit the amount of trade that can be executed among countries. Industries that depend on exports realized that there is need for consistent standards in order to trade freely and extensively. International standards have been established for many technologies in different industries such as textiles, packaging, communication, energy production and utilization, distribution of goods, banking, and financial services.

With the expansion of global trade, the importance of international standards will continue to grow. For example, the advent of international standards makes it possible for one to buy a computer made in the United States and use disks made in China. International certification programs such as those developed by National Association of Corrosion Engineers (NACE) set standards acceptable all over the world. Thus, industries can be assured of qualified services in a timely manner.

The technical work of ISO is highly decentralized, carried out in a hierarchy of some 2850 technical committees (TCs) and working groups. In these committees, qualified representatives of industry, research institutes, government authorities, consumer bodies, and international organizations from all over the world come together as equal partners in the resolution of global standardization problems. Approximately 30,000 experts participate in meetings every year.

ISO standards are developed according to the following principles:

1. Consensus: The views of all interests are taken into account: manufacturers, vendors and users, consumer groups, testing laboratories, governments, and research organizations.
2. Industry-wide: Global solutions to satisfy industries and consumers worldwide.
3. Voluntary: International standardization is market-driven and therefore based on voluntary involvement of all interests in the market-place.

There are three main phases in the ISO standards development process. The first phase begins with "the need" for a standard, which is usually expressed by an industry sector. The sector then communicates this need to a national member body. The latter proposes the new work item to the ISO as a whole. Once the need for an international standard has been recognized and formally agreed upon, the first phase of development involves definition of the technical scope of the future standard. This phase is usually carried out in workgroups which are comprised of technical experts from countries interested in the subject matter.

Once agreement has been reached on which particular technical aspects are to be covered in the standard, the second phase begins, during which countries negotiate the detailed specifications within the standard. This is the consensus-building phase.

The final phase comprises the formal approval of the resulting draft international standard (the acceptance criteria stipulate approval by two-thirds of ISO members that have actively participated in the standards development process, and approval by 75% of all member that vote). Following approval the text is published as an international standard. It is now possible to publish interim documents at different stages in the standardization process.

Most standards require periodic revision. Several factors combine to render a standard out of date: technological evolution, new methods and materials, and new quality and safety requirements. To take account of these factors, ISO has established the general rule that all ISO standards should be reviewed at intervals of not more than 5 years. On occasion, it is necessary to revise a standard earlier. To date, ISO's work has resulted in some 16,000 international standards, representing more than 400,000 pages in English and French (terminology is often provided in other languages as well) [3].

4.3   DEVELOPMENT OF ENVIRONMENTAL STANDARDS

It became apparent following the Rio De Janeiro conference that there was a need for international environmental standards. Due to the success of the ISO 9000 series of standards, which address general quality management, the UNCED asked ISO to develop international environmental standards. ISO established a Strategic Advisory Group on the Environment (SAGE) to look into the global standardization of environmental management practices. This body was made up of representatives of governments, national standardization organizations, and business and environmental professionals.

The SAGE considered whether such standards would

1. Promote a common approach to environmental management similar to quality management
2. Enhance an organization's ability to attain and measure environmental performance
3. Facilitate lower trade barriers

At the conclusion of their study, SAGE recommended that an ISO TC formally consider and produce final "consensus" standards. Thus, in January, 1993, Technical Committee 207 (TC 207) was established. The number 207 was chosen because it fell in the sequence of numbers for technical committees. Canada was awarded the secretariat for TC 207 and the inaugural plenary session was held in June 1993; over 200 delegates from over 30 countries and organizations attended. The largest block of countries with voting rights is from Europe. The TC regional governmental organizations includes the European Union (EU), the General Agreement of Tariffs and Trade (GATT), and the Organization for Economic Cooperation and Development (OECD). TC 207 meets annually to review the progress of its subcommittees [8].

The ISO 14000 is made up of 23 standards, guidelines, and technical reports concerning environmental management. The ISO's TC 207 organizes the environmental management standards into working group activities in subcommittees. The standards are developed through a seven-phase system. The first phase is the selection of the work item. The second phase is the preparation of the working draft, which is performed by the working groups within the subcommittees. The third phase is the committee approval of the working draft, in which the TC votes. If it is approved, the draft becomes a committee draft. The fourth phase is ratification by all ISO members. It is sent to the members as a Draft International Standard and is subjected to a 6 month voting period. The fifth phase is final confirmation, in which it is approved by the ISO member countries and becomes a final draft international standard. The sixth phase is the publication of the ISO standard, in which the now approved standard is first published in Geneva in English, French, and Russian. The final phase is the publication in other languages, in which national committees translate the standard into their national language and the new standard is published as a national standard [9].

Since its creation in 1996, ISO 14000 certification has grown substantially in reputation, the number of countries recognizing it, and organizations receiving certification increases annually. The ISO publishes a survey periodically showing the number of ISO 14000 certificates that have been awarded, broken down by region, country, and industrial sectors. The survey is compiled by a number of assessment arrangements and while great care is taken in procuring the data, some double counting and undercounting occurs. Estimates are made in some cases and there is no distinction between accredited and nonaccredited certificates [10].

By the end of the year 2000, 5 years had passed since the start of ISO 14000 certification and about 23,000 certificates had been awarded. Europe and the Far East showed the most growth over the period holding 83% of all the awarded certificates. At that point, the ISO 14000 was implemented by 112 countries. Japan as a country had the highest growth in number of certificates and the United States had the fourth highest growth. The industries holding the greatest number of certificates were electrical and optical equipment, chemicals, chemical products, and fibers. Over the first 5 years of implementation, there was an average increase in the number of certificates by approximately 73% each year [10].

The United States has established a Technical Advisory Group (TAG) consisting of academia, industry, government, and environmental groups to participate in TC 207. The structure of U.S. TAG is shown in Table 4.1.

ISO 14000 stipulates a set of ten management principles for organizations considering an EMS as follows:

1. Recognize that environmental management is one of the highest priorities of any organization.
2. Establish and maintain communications with both internal and external interested parties.
3. Determine legislative requirements and those environmental aspects associated with the activities, products, and services.
4. Develop commitment by everyone in the organization to environmental protection and clearly assign responsibilities and accountability.
5. Promote environmental planning throughout the life cycle of the product and the process.
6. Establish a management discipline for achieving targeted performances.
7. Provide the right resources and sufficient training to achieve performance targets.
8. Evaluate performance against policy, environmental objectives and targets, and make improvements wherever possible.

TABLE 4.1

Structure of the U.S. Technical Advisory Group

ISO 14001	Environmental management systems—Specifications with guidance for use
ISO 14004	Environmental management systems—General guidelines on principles, systems, and supporting techniques
ISO 14010	Guidelines for environmental auditing—General principles on environmental management systems
ISO 14011/1	Guidelines for environmental auditing—Audit procedures—Audit of environmental management systems
ISO 14012	Guidelines for environmental auditing—Qualification criteria for environmental auditors
ISO 14015	Environmental site assessments
ISO 14020	Goals and principles of all environmental labeling
ISO 14021	Environmental labels and declarations—Self declaration environmental claims—Terms and definitions
ISO 14022	Environmental labels and declarations—Self declaration environmental claims—Symbols
ISO 14023	Environmental labels and declarations—Self declaration environmental claims—Testing and verification
ISO 14024	Environmental labels and declarations—Environmental labeling Type I – Guiding principles and procedures
ISO 14025	Environmental labels and declarations—Environmental information profiles—Type III guiding principles and procedures
ISO 14031	Evaluation of environmental performance
ISO 14040	Environmental management—Life cycle analysis—Principles and framework
ISO 14041	Environmental management—Life cycle analysis—Life cycle inventory analysis
ISO 14042	Environmental management—Life cycle analysis—Impact assessment
ISO 14043	Environmental management—Life cycle analysis—Interpretation
ISO 14050	Terms and Definitions—Guide on the Principles for ISO/TC 207/SC6 terminology work
ISO Guide 64	Guide for inclusion of environmental aspects in product standards

9. Establish a process to review, monitor, and audit the EMS to identify opportunities for improvement in performance.
10. Encourage vendors to also establish EMSs.

4.4   THE ISO 14000 STANDARDS

The ISO 14000 family of standards is comprised of 23 standards which can be broken down into seven categories: EMSs, environmental auditing, environmental labeling, environmental performance evaluation, life cycle assessment, environmental management vocabulary, and environmental aspects in product standards. A brief overview [9] of each of the standards is provided below.

ISO 14001: Environmental management systems—Specification with guidance for use
This gives the requirements for an EMS, which allows for an organization to create a policy that incorporates legal requirements and information on environmental impacts. This standard is discussed in greater detail in Section 4.5.

ISO 14004: Environmental management systems—General guidelines on principles, systems, and supporting techniques
This document provides guidance on developing and implementing EMSs and coordinating them with other managements systems. These are strictly voluntary guidelines and do not impact the certification procedure.

ISO 14061: Information to assist forestry organizations in the use of ISO 14001 and ISO 14004
This serves to help forestry organizations in the application and implementation of the EMS standards.

ISO 14010: Guidelines for environmental auditing—General principles
This document provides the general principles of environmental auditing which are universal. Anything classifiable as an environmental audit should meet the given recommendations.

ISO 14011: Guidelines for environmental auditing—Audit procedures—Auditing of environmental management systems
This helps to establish the audit procedures for planning and conducting an audit of an EMS (see Chapter 46 for details on audits). The purpose of such an audit is to ascertain if the EMS is meeting the audit criteria.

ISO 14012: Guidelines for environmental auditing—Qualification criteria for environmental auditors
This document provides guidance on the qualification criteria for environmental and lead auditors. The provisions are applicable to both internal and external auditors.

ISO 14015: Environmental management—Environmental assessment of sites and organizations
This shows how to conduct such an assessment via a systematic process which identifies the environmental aspects and issues and what consequences they might have for the business. The roles and responsibilities of each party in the assessment are discussed as well as the phases of the process. However, this is not meant to provide guidance on initial environmental reviews, audits, impact assessments, or performance evaluations.

ISO/CD.2 19011: Guidelines on quality and environmental management systems auditing
This document provides the fundamentals of auditing, how to manage auditing programs, how to conduct an environmental and quality management systems audit, and the qualifications of such auditors. This is important for all businesses that have an EMS in any stage of implementation. This standard can be used for other kinds of audits but the capability of the auditors must be determined.

ISO 14020: Environmental labels and declarations—General principles

This provides the guiding principles used in developing and applying environmental labels and declarations. Other ISO standards in this category give more specific requirements for certain types of labels and these should take precedence over the general guidelines.

ISO 14021: Environmental labels and declarations—Self-declared environmental claims

These types of labels are classified as Type II environmental labels. This standard gives specific requirements for this type of labeling, which includes claims, symbols, and products. It discusses various terms that fit in this category and the qualifications of their use it also provides a basic evaluation and verification methodology for such labeling. However, any legal requirements regarding this labeling take precedence over the standard.

ISO 14024: Environmental labels and declarations—Type I environmental labeling—Guiding principles and procedures

This gives guiding principles and practices to be applied to Type I labeling which covers multiple, criteria-based, third-party environmental labeling programs. It provides criteria procedures and guidance for the certification process and is meant to serve as a reference document intending to reduce the environmental responsibility by promoting market-driven demand for products meeting this labeling program.

ISO 14025: Environmental labels and declarations—Type III environmental declarations

This document discusses the elements and issues regarding this type of labeling. It provides guidance on technical considerations, declaration formatting and communication, and administrative considerations for the development of this labeling program.

ISO 14031: Environmental management—Environmental performance evaluation—Guidelines

This standard provides guidance on the design and execution of environmental performance evaluations within the company. However, actual performance levels are not specified.

ISO/TR 14032: Environmental management—Examples of environmental performance evaluation

This gives some examples from real companies that have conducted environmental performance evaluations to illustrate how to use the guidelines described in ISO 14031.

ISO 14040: Environmental management—Life cycle assessment—Principles and framework

This document provides the basic framework, principles, and requirements for conducting and analyzing life cycle assessments.

ISO 14041: Environmental management—Life cycle assessment—Goal and scope definition and inventory analysis

This standard describes the requirements and procedures for compiling and preparing a goal and scope for the life cycle assessment. Guidelines for performing and reporting the inventory analysis are also given.

ISO 14042: Environmental management—Life cycle assessment—Life cycle impact assessment
This provides the basic framework for the life cycle impact assessment phase and reviews the important features and drawbacks of the phase as well as the requirements for conducting one.

ISO 14043: Environmental management—Life cycle assessment—Life cycle interpretation
This document details the requirements and recommendations for conducting the life cycle interpretation phase of the study.

ISO 14048: Environmental management—Life cycle assessment—Life cycle assessment data documentation format
This standard describes the required formatting for presenting data collected from the life cycle assessment.

ISO 14049: Environmental management—Life cycle assessment—Examples of application of ISO 14041 to goal and scope definition and inventory analysis
This gives examples about practices carried out in the life cycle assessment analysis and samples of possible cases that meet the requirements of the standard.

ISO 14050: Environmental management—Vocabulary
This document reviews the definitions of the terminology related to the EMS.

ISO Guide 64: Guide for the inclusion of environmental aspects in product standards
This document discusses the environmental impacts in product standards and provides considerations relating product function and environmental impacts. It also gives an outline of how provisions in the product standards can affect the environment, techniques for identifying the impacts, and suggestions for alleviating some of the harmful impacts.

4.5 IMPLEMENTING ISO 14000

Senior management commitment is required before embarking on an ISO 14000 program. The project planning begins once senior management is committed to implementing an ISO 14000 program. This planning includes scheduling, budgeting, assigning personnel, responsibilities, and resources, and, if required, retaining specialized external assistance.

Senior management needs to provide a focus for the EMS by defining the organization's environmental policy. The policy must include, among other things, a commitment to continuous improvement, prevention of pollution, and compliance with legislation and regulations. It must be specific enough to form the basis for concrete actions. When documented by management, this environmental policy must be implemented, maintained and communicated within the organization, and made available to the public.

Next, an initial review of the organization's existing environmental program is needed. This review includes the consideration of all applicable environmental regulations, existing processes, documentation, work practices, and effects of current operations. Once the initial review is completed, a strategic or implementation plan can be developed. Implementation planning is similar to project management and the steps, scope, time-frame, costs, and responsibilities need to be defined in order to develop and implement an EMS that meets the organization's targets and objectives, and promotes continuous improvement. The strategic plan sets the framework for participation of the responsible and affected parties within the organization.

Both in the initial review and on an ongoing basis, the organization's activities, products, and services require evaluation to determine their interaction with the environment. Environmental issues such as noise, emissions, environmental impact, waste reduction, and energy must be identified. The organization then needs to identify the aspects which can interact with the environment and which ones it can control or influence. The identified impacts are then used as a basis for setting environmental objectives within the organization. Objectives also need to take into account relevant legal and regulatory requirements; financial, operational, and business requirements; and the views of interested parties. Interested parties may be people or groups, such as neighbors or interest groups, concerned with the organization's environmental performance.

Objectives of the organization need to be determined and specific targets set. An objective is an overall goal, which may be as simple as "meeting or exceeding regulations" or "reduction in energy consumption," and the targets provide quantified measurements. Objectives and targets are set by the organization, not by the ISO 14000 standard. Identifying the impacts, judging their significance, and setting reasonable objectives and targets are some of the major "environmental" challenges presented by ISO 14000.

Once the targets and objectives are set, the organizations need to implement the strategic plan. Beyond the environmental challenges, management functions will have to be adapted to meet the requirements of the EMS Standard. The level of conceptual challenge this will present to ISO 9000 firms, where the corporate culture will already be changing, will be less than for non-ISO 9000 firms, but there will be some new areas that require attention [11].

4.6 MAINTAINING AN ISO 14000 ENVIRONMENTAL MANAGEMENT SYSTEM

Once the EMS is implemented, its progress needs to be continually measured and monitored. Routine measurement and monitoring must be undertaken of the activities which have been identified as having the potential for a significant impact on the environment.

Routine auditing and review are the keys to continuous improvement. Environmental as well as management components will be required in the audit program. Audits of an organization's EMS do not replace, but rather complement, the issue-specific environmental audits that may be conducted externally by regulators and consultants or internally by environmental engineers or other qualified personnel. Where issue-specific audits address regulatory compliance, site assessment, or emissions, the EMS audits address effectiveness of the management system. Periodic EMS audits are needed to determine if the EMS conforms to the requirements of

ISO 14001, and that the program is implemented, proportional to the nonconformance, to eliminate recurrence.

To ensure the continuing effectiveness of the EMS, management needs to regularly review and evaluate information such as the results of audits, corrective action, current and proposed legislation, results of monitoring, and complaints. This review allows management to look at the system and ensure that it is, and will remain, suitable and effective. The management review may result in changes to policies or systems as the organization evolves and as technology advances. An organization's EMS is not a stagnant system but must continually evolve to meet the organization's ever changing needs.

4.7 COMPARISON BETWEEN ISO 9000 AND ISO 14000 SERIES STANDARDS

As noted earlier, the ISO 14000 series of standards is made up of one standard (ISO 14001), which organizations have to comply with, and others that provide guidance to assist organizations' compliance with ISO 14001. ISO 14001 outlines the basis for establishing an EMS. The core sections of the EMS consist primarily of the five subsections highlighted below:

1. **Environmental Policy**
2. **Planning**
   2.1 Environmental aspects
   2.2 Legal and other requirements
   2.3 Objectives and targets
   2.4 Environmental management programs
3. **Implementation and Operation**
   3.1 Structure and responsibility
   3.2 Training, awareness, and competence
   3.3 Communication
   3.4 EMS documentation
   3.5 Document control
   3.6 Operational control
   3.7 Emergency preparedness and response
4. **Checking and Corrective Action**
   4.1 Monitoring and measurement
   4.2 Nonconformance, and corrective and preventive action
   4.3 Records
   4.4 EMS audit
5. **Management Review**

Organizations that meet the ISO 14001 requirements can seek registration in a process similar to ISO 9000 registration. The ISO 14000 series is complementary to the ISO 9000 series. Whereas the ISO 9000 series deals with Quality Management Standards, the ISO 14000 series deals with Environmental Management Standards. Like the ISO 9000 series, the ISO 14000 series is voluntary and does not replace regulations, legislations, and other codes of practice that an organization has to comply with. Rather, it provides a system for monitoring, controlling, and improving performance regarding those requirements. ISO 14000 is a package that ties the mandatory requirements into a management system which is made up of objectives and targets focusing on prevention and continuous improvements. The ISO 14000 uses the same fundamental systems as ISO 9000, such as documentation control, management system auditing, operational control, control of records management policies, audits, training, statistical techniques, and corrective and preventive actions [8].

Although there are numerous similarities between ISO 9000 and ISO 14000, there are some definite differences. For example, ISO 14000 has clearer statements about communication, competence, and economics than those that are currently found in ISO 9000. ISO 14000 also incorporates the setting of objectives and quantified targets, emergency preparedness, considering the views of interested parties, and public disclosure of the organization's environmental policy.

An organization with an ISO 9000 registration will find that it is far along toward gaining ISO 14000 registration right from the outset. Even though there are differences, the management systems are generally consistent within both standards. The ISO approach to management serves as a model which needs to be adapted to meet the needs of the organization and integrated into existing management systems. The standards have been designed to be applied by any organization in any country regardless of the organization's size, process, economic situation, and regulatory requirements [11].

4.8 THE ISO 14001: 2004 EDITION

This standard is the most famous in the series and is the basis for certification. It deals exclusively with EMSs. As noted above, an EMS is a tool that allows companies to identify and control environmental impact of its products and processes, improve its environmental performance, and create a systematic approach to setting environmental goals and achieving them. The purpose of this standard is therefore to provide a framework with general requirements for implementing an EMS and a common reference for communication about EMS issues [12].

Any business or organization can follow and use these standards but they must make a solid commitment to comply with the environmental laws and regulations established by the government. Companies must also desire to implement or maintain an EMS, ensure conformance with its own environmental policy and display this conformance, and obtain certification for its EMS. The standard is divided into six requirement categories within Section 4 of the document: general requirements, environmental policy, planning, implementation and operation, checking and corrective action, and management review [9]. A brief summary [13] of each requirement and subrequirement is provided below.

Requirement 4.1: General requirements
An EMS should be established, documented, implemented, and continuously improved by the organization and the meeting of all standard requirements must be shown. The organization chooses the scope and boundaries of the EMS.

Requirement 4.2: Environmental policy
The top management of the organization has to develop a policy or commitment statement that describes the chosen scope of the EMS. This is a short statement that defines the EMS and provides a framework for the objectives and targets selected. The policy must include the decision for compliance with legal and other requirements, pollution prevention, and ongoing improvement. All employees must be made aware of the policy and it should be available to the public. The policy must be documented, implemented, maintained, and kept up-to-date.

Requirement 4.3.1: Environmental aspects
A consistent procedure must be developed for identifying environmental aspects and impacts that can be controlled by the organization. The purpose of this is to assist in organizing a methodology of identifying these aspects and impacts, and using the EMS to manage and control them. The standard does not provide guidance as to determine the relative significance of each aspect. Environmental aspects include the interaction of the company's actions, products, and services with the environment and how the environment is affected.

Requirement 4.3.2: Legal and other requirements
The organization should obtain information on the environmental laws and regulations imposed on it and make all components of the company aware of them. The purpose of this is to identify the regulations and incorporate them into the EMS.

Requirement 4.3.3: Objectives, targets, and programs
A procedure has to be developed that ensures that measurable objectives and targets established by the organization comply with the policy. The following must be taken into account: commitments to compliance, continuous improvement, pollution prevention, significant aspects, legal and other requirements, and technological, financial, and business issues. The establishment of management programs which document personnel, deadlines, and measurements are also required for the EMS.

Requirement 4.4.1: Structure and responsibility
The management responsible for establishing and monitoring the EMS must be specified. Resources such as personnel, structure, financial, and technological resources must be identified. The roles and responsibilities of involved parties must be documented and a management representative responsible for overseeing and reporting on the EMS should be established. This person reports to top management and ensures that the EMS is being carried out according to plan.

Requirement 4.4.2: Competence, training, and awareness
Anyone within the organization directly participating in an activity that has an environmental impact should have a level of competence attained from education, training, and experience. A procedure must be established that ensures significant personnel are aware of EMS requirements, legal requirements, the benefits of maintaining improved performance, and consequences of not complying with regulations and requirements.

Requirement 4.4.3: Communications

A procedure must be established that defines both internal and external communications within the organization. The company can decide on the openness of information exchanged and the decision process must be documented. External communications regarding environmental aspects must be considered and recorded. Likewise, general descriptions of the manner in which communications are received and documented must be provided. Communications among the levels of organization must also be documented.

Requirement 4.4.4: EMS documentation

The EMS must be documented and the elements of the standard and how the organization complies with each element should be discussed. The system requirements should to be verified and documentation on the following must be provided: prove that effective planning, operation, and control of processes are complying with the standard, the policy, objectives and targets, and scope of the EMS.

Requirement 4.4.5: Control of documents

The documents regarding the EMS have to be controlled and current versions of the system procedures and work instructions must be utilized. A procedure must be created to ensure that documents are reviewed and updated as needed and that the current versions are used.

Requirement 4.4.6: Operational control

Critical functions of the policy, significant aspects and requirements should be identified and procedures should be developed to ensure that activities are implemented in the appropriate way. The procedures must address communication of the EMS requirements to contractors and give enough instruction to ensure compliance with the established EMS. A procedure must be created regarding contractors that discusses the legal requirements of the products and services supplied by the contractors and communications between the organization and the contractor concerning the EMS.

Requirement 4.4.7: Emergency preparedness and response

A procedure should be created that identifies potential emergencies and how to alleviate them. Emergency situations must be dealt with and the emergency procedures should be reviewed and improved if necessary.

Requirement 4.5.1: Monitoring and measuring

Performance measurement must be taken to obtain data to determine appropriate actions. A procedure should be developed that describes how operations are monitored and measured to ascertain significant impacts, performance relating to objectives and targets, and compliance with legal requirements. Equipment used must be calibrated according to procedures and documented.

Requirement 4.5.2: Evaluation of compliance

A procedure must exist that evaluates the operations' compliance with legal requirements and records of this must be kept.

Requirement 4.5.3: Nonconformances, corrective and preventive action

A procedure must be developed that identifies how to act on nonconformances, where the system deviates from planned conditions, and includes preventive and corrective actions. Nonconformances are identified through audits, monitoring, and communications. The root of the problems should be discovered and corrected. The nonconformances must be addressed to reduce environmental impacts and actions must be taken to prevent the incident from occurring again. Corrective actions must be recorded and the effectiveness should be evaluated.

Requirement 4.5.4: Control of records

Records verify the company's compliance with the standard and their established EMS and procedures should exist that maintain records. The records should be well organized and easy to locate and analyze.

Requirement 4.5.5: Internal audit

A procedure should be established that discusses methodologies, schedules, and checklists to conduct audits. Audits are used to determine the EMS compliance with the standard and proper implementation and maintenance. The procedure should include responsibilities and requirements for planning and carrying out audits, reporting results, generating records, determining the scope of the audit, the frequency of audits, and how they will be conducted.

Requirement 4.6: Management review

Top management of the organization is required to review the EMS to ascertain its compliance with the original plan, and its overall effectiveness. The results of the audits, external communications, environmental performance, status of objectives and targets, corrective and preventive actions, management reviews, and potential improvements should be reviewed and discussed. Documentation of management review should include agendas, attendance records and detailed minutes [13].

4.9   FUTURE TRENDS

The ISO continues to expand the scope of their standards to incorporate areas such as the environment, service sectors, security, and managerial and organizational practice. There are currently more than 16,000 standards applying to three areas of sustainable development: economic, environmental, and social. Since ISO's environmental mission is to promote the manufacturing of products in a manner that is efficient, safe, and clean, more standards can be expected. The ISO hopes to achieve this goal in the future through the dedication and participation of more countries.

4.10   SUMMARY

1. Participation in ISO 14000 is becoming one of the most sought-after statuses in a move toward globalization of environmental management. In recent years, there has been heightened international interest in and commitment to improved environmental management practices by both the public and private sectors.
2. The ISO is a private, nongovernmental international standards body based in Geneva, Switzerland. ISO promotes international harmonization and the development of manufacturing, product, and communications standards.

3. The impetus toward international standards is deeply rooted in economic rewards and an expansion into a global economy. The standardization of goods and services will not only enhance potential market share but also allow goods and services be available to more consumers.
4. The technical work of ISO is highly decentralized, carried out in a hierarchy of some 2850 TCs and working groups. In these committees, qualified representatives of industry, research institutes, government authorities, consumer bodies, and international organizations from all over the world come together as equal partners in the resolution of global standardization problems.
5. Due to the success of the ISO 9000 series of standards, UNCED asked ISO to develop international environmental standards. ISO established a SAGE to look into the global standardization of environmental management practices.
6. The ISO 14000 series of standards is made up of one standard: ISO 14001, with which organizations have to comply and others which are used as guidance to assist organizations to comply with ISO 14001.

REFERENCES

1. Adapted from: Burke, G., Singh, B., and Theodore, L. *Handbook of Environmental Management and Technology*, 2nd edition, John Wiley & Sons, Hoboken, NJ, 2000.
2. ISO in Brief. August 2006. http://www.iso.org/iso/isoinbrief_2006-en.pdf
3. Introduction to ISO, www.iso.ch/infoe/intro
4. General Information on ISO. http://www.iso.org/iso/support/faqs/faqs_general_information_on iso.htm
5. Summary of ISO 14000. Lighthouse Consulting, University of Rhode Island, 2003. www.crc.uri.edu/download/12_ISO_14000_Summary_ok.pdf
6. ISO 14000. http://en.wikipedia.org/wiki/ISO_14000
7. Adapted from: Sayre, D. *Inside ISO 14000*, CRC Press, Boca Raton, FL, 1996.
8. Von Zharen, W.M. *ISO 14000—Understanding Environmental Standards*, Government Institutes, Rockville, MD, 1996.
9. Jensen, P.B. *Introduction to the ISO 14000 Family of Environmental Management Standards*, Buch Jensen Quality Management ApS, Denmark, April 2000. www.environmental-expert.com/articles/article611/articles611.htm
10. The ISO Survey of ISO 9000 and ISO 14000 Certificates—Tenth Cycle. http://www.iso.org/iso/survey10thcycle.pdf
11. Fredericks, I. and McCullum, D. *International Standards for Environmental Management Systems*. www.mgmt14k.com/ems
12. ISO 14000 Essentials. http://www.iso.org/iso/iso_catalogue/management_standards/iso_9000_iso_14000/iso_14000_essentials.htm
13. Summary of Requirements of ISO 14001:2004, February 24, 2005. www.fs.fed.us/ems/includes/sum_ems_elements.pdf

5 Multimedia Concerns

CONTENTS

5.1   INTRODUCTION

The current approach to environmental waste management requires some rethinking. A multimedia approach helps the integration of air, water, and land pollution controls and seeks solutions that do not violate the laws of nature. The obvious advantage of a multimedia pollution control approach is its ability to manage the transfer of pollutants so they will not continue to cause pollution problems. Among the possible steps in the multimedia approach are understanding the cross-media nature of pollutants, modifying pollution control methods so as not to shift pollutants from one medium to another, applying available waste reduction technologies, and training environmental professionals in a total environmental concept.

A multimedia approach in pollution control is long overdue. As described above, it integrates air, water, and land into a single concern and seeks a solution to pollution that does not endanger society or the environment. The challenges for the future environmental professional include

1. Conservation of natural resources
2. Control of air–water–land pollution
3. Regulation of toxics and disposal of hazardous wastes
4. Improvement of quality of life

It is now increasingly clear that some treatment technologies (specific technologies will be discussed in later chapters), while solving one pollution problem, have created others. Most contaminants, particularly toxics, present problems in more than one medium. Since nature does not recognize neat jurisdictional compartments, these same contaminants are often transferred across media. Air pollution control devices and industrial wastewater treatment plants prevent waste from going into the air and water, but the toxic ash and sludge that these systems produce can become hazardous waste problems themselves. For example, removing trace metals from a flue gas usually transfers the products to a liquid or solid phase. Does this exchange an air quality problem for a liquid or solid waste management problem? Waste disposed of on land or in deep wells can contaminate ground water and evaporation from ponds and lagoons can convert solid or liquid waste into air pollution problems [1]. Other examples include acid deposition, residue management, water reuse, and hazardous waste treatment and/or disposal.

Control of cross-media pollutants cycling in the environment is therefore an important step in the management of environmental quality. Pollutants that do not remain where they are released or where they are deposited move from a source to receptors by many routes, including air, water, and land. Unless information is available on how pollutants are transported, transformed, and accumulated after they enter the environment, they cannot effectively be controlled. A better understanding of the cross-media nature of pollutants and their major environmental processes—physical, chemical, and biological—is required.

5.2   HISTORICAL PERSPECTIVE [1,2]

The Environmental Protection Agency's (EPA's) own single-media offices, often created sequentially as individual environmental problems were identified and responded to in legislation, have played a part in the impeding development of cost-effective multimedia prevention strategies. In the past, innovative cross-media agreements involving or promoting pollution prevention, as well as voluntary arrangements for overall reduction in releases, have not been encouraged. However, new initiatives are characterized by their use of a wide range of tools, including market incentives, public education and information, small business grants, technical assistance, research and technology applications, as well as more traditional regulations and enforcements.

In the past the responsibility for pollution prevention and/or waste management at the industrial level was delegated to the equivalent of an environmental control department. These individuals were skilled in engineering treatment techniques but, in some instances, had almost no responsibility over what went on in the plant that generated the waste they were supposed to manage. In addition, most engineers are trained to make a product work, not to minimize or prevent pollution. There is still little emphasis (although this is changing) on pollution prevention in the educational arena for engineers. Business school students, the future business managers, also have not had the pollution prevention ethic instilled in them.

The reader should also note that the federal government, through its military arm, is responsible for some major environmental problems. It has further compounded these problems by failing to apply a multimedia or multiagency approach. The following are excerpts from a front-page article by Keith Schneider in the August 5, 1991 edition of the *New York Times*:

A new strategic goal for the military is aimed at restoring the environment and reducing pollution at thousands of military and other government military-industrial installations in the United States and abroad … the result of the environment contamination on a scale almost unimaginable. The environmental projects are spread through four federal agencies and three military services, and are directed primarily by the deputy assistant secretaries. Many of the military-industry officials interviewed for this article said that the environmental offices are not sharing information well, were suffering at times from duplicate efforts, and might not be supervising research or contractors closely enough. Environmental groups, state agencies, and the Environmental Protection Agency began to raise concerns about the rampant military-industrial contamination in the 1970s, but were largely ignored. The Pentagon, The Energy Department, the National Aeronautics and Space Administration (NASA) and the Coast Guard considered pollution on their property a confidential matter. Leaders feared not the only the embarrassment from public disclosure, but also that solving the problems would divert money from projects they considered more worthwhile. Spending on military-environmental projects is causing private companies, some of them among the largest contractors for the military industry, to establish new divisions to compete for government contracts, many of them worth $100 million to $1 billion.

This lack of communication and/or willingness to cooperate within the federal government has created a multimedia problem that has just begun to surface. The years of indifference and neglect have allowed pollutants/wastes to contaminate the environment significantly beyond what would have occurred had the responsible parties acted sooner.

5.3 ENVIRONMENTAL PROBLEMS

Environmental problems result from the release of wastes (gaseous, liquid, and solid) that are generated daily by industrial and commercial establishments as well as households. The lack of consciousness regarding conservation of materials, energy, and water has contributed to the wasteful habits of society. The rate of waste generation has been increasing in accordance with the increase in population and the improvement in living standards. With technological advances and changes in lifestyle, the composition of waste has likewise changed. Chemical compounds and products are being manufactured in new forms with different half-lives (time for half of it to react and/or disappear). It has been difficult to manage such compounds and products once they have been discarded. As a result, these wastes have caused many treatment, storage, and disposal problems. Many environmental problems are caused by products that are either misplaced in use or discarded without proper concern of their environmental impacts. Essentially all products are potential wastes, and it is desirable to develop methods to reduce the waste impacts associated with products or to produce environmentally friendly products. Environmental agencies have been lax in promoting and automating tracking mechanisms that identify sources and fate of new products.

Solving problems, however, can sometimes create problems. For example, implementation of the Clean Air Act and the Clean Water Act has generated billions of tons of sludge, wastewater, and residue that could cause soil contamination and underground water pollution problems. The increased concern over cross-media shifts of pollutants has yet to consistently translate into a systematic understanding of pollution problems and viable changes.

As indicated above, environmental protection efforts have emphasized media-specific waste treatment and disposal after the waste has already been created. Many of the pollutants which enter the environment are coming from "area or point sources" such as industrial complexes and land disposal facilities; therefore, they simply cannot be solely controlled by the end-of-pipe solutions. Furthermore, these end-of-pipe controls that tend to shift pollutants from one medium to another have often caused secondary pollution problems. Therefore, for pollution control purposes, the environment must be perceived as a single integrated system and pollution problems must be viewed holistically. Air quality can hardly be improved if water and land pollution continue to occur. Similarly, water quality cannot be improved if the air and land are polluted.

Many secondary pollution problems today can be traced in part to education, i.e., the lack of knowledge and understanding of cross-media principles for the identification and control of pollutants. Neither the Clean Air Act nor the Clean Water Act enacted in the early 1970s adequately addresses the cross-media nature of environmental pollutants. More environmental professionals now realize that pollution legislation is too fragmented and compartmentalized. Only proper education and training will address this situation and hopefully lead to more comprehensive legislation of the total environmental approach.

5.4   MULTIMEDIA APPROACH

The environment is the most important component of life support systems. It is comprised of air, water, soil, and biota through which elements and pollutants cycle. This cycle involves the physical, chemical, or biological processing of pollutants in the environment. It may be short, turning hazardous into nonhazardous substances soon after they are released, or it may continue indefinitely with pollutants posing potential health risks over a long period of time. Physical processes associated with pollutant cycling include leaching from the soil into the ground water, volatilization from water or land to air, and deposition from air to land or water. Chemical processes include decomposition and reaction of pollutants to products with properties that are possibly quite different from those of the original pollutants. Biological processes involve microorganisms that can break down pollutants and convert hazardous pollutants into less toxic forms. However, these microorganisms can also increase the toxicity of a pollutant, e.g., by changing mercury into methyl-mercury in soil [3].

Although pollutants sometimes remain in one medium for a long time, they are most likely mobile. For example, settled pollutants in river sediments can be dislodged by microorganisms, flooding, or dredging. Displacement such as this earlier constituted the PCB problem in New York's Hudson River. Pollutants placed in landfills have been transferred to air and water through volatilization and leaching. About 200 hazardous chemicals were found in the air, water, and soil at the Love Canal land disposal site in New York State. The advantages of applying multimedia approaches lie in their ability:

1. To manage the transfer of pollutants
2. To avoid duplicating efforts or conflicting activities
3. To save resources by consolidation of environmental regulations, monitoring, database management, risk assessment, permit issuance, and field inspection

In recent years, the concept and goals of multimedia pollution prevention have been adopted by many regulatory and other governmental agencies, industries, and the public in the United States and abroad. Multimedia efforts in the United States have been focused on the EPA's Pollution Prevention Office, which helps coordinate pollution prevention activities across all EPA headquarter offices and regional offices. The current EPA philosophy recognizes that multimedia pollution prevention is best achieved through education and technology transfer rather than through regulatory imposition of mandatory approaches. But the progress of implementing multimedia pollution prevention has been slow (see Chapters 30 through 34 for more details on waste reduction/pollution prevention).

Recognition of the need for multimedia pollution prevention approaches has been extended from the government, industry, and the public to professional societies. The Air Pollution Control Association (APCA) was renamed as the Air and Waste Management Association (AWMA) to incorporate waste management. The American Society of Civil Engineers (ASCE) has established a multimedia management committee under the Environmental Engineering Division. The American Institute of Chemical Engineers (AIChE) has reorganized its Environmental Division to include a section devoted to pollution prevention. The Water Pollution Control Federation (WPCF) has also adopted a set principle addressing pollution prevention.

5.5   MULTIMEDIA APPLICATION [1,2]

Perhaps a meaningful understanding of the multimedia approach can be obtained by examining the production and ultimate disposal of a product or service. A flow diagram representing this situation is depicted in Figure 5.l. Note that each of the ten steps in the overall process has potential inputs of mass and energy, and may produce an environmental pollutant and/or a substance or form of energy that may be used in a subsequent or later step. Traditional approaches to environmental management can provide some environmental relief, but a total systems approach is required if optimum improvements—in terms of pollution/waste reduction—are to be achieved.

One should note that a product and/or service is usually conceived to meet a specific market need with little thought given to the manufacturing parameters. At this stage of consideration, it may be possible to avoid some significant waste generation problems in future operations by answering a few simple questions:

1. What raw materials are used to manufacture the product?
2. Are any toxic or hazardous chemicals likely to be generated during manufacturing?
3. What performance regulatory specifications must the new product(s) and/or service(s) meet? Is extreme purity required?

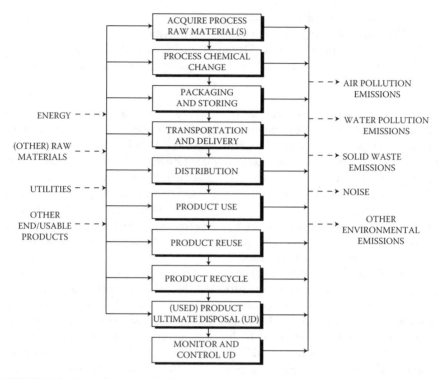

FIGURE 5.1   Overall multimedia flow diagram.

4. How reliable will the delivery manufacturing/distribution process be? Are all steps commercially proven? Does the company have experience with the operations required?
5. What types of waste are likely to be generated? What is their physical and chemical form? Are they hazardous? Does the company currently manage these wastes on-site or off-site?

5.6   EDUCATION AND TRAINING

The role of environmental professionals in waste management and pollution control has been changing significantly in recent years. Many talented, dedicated environmental professionals in academia, government, industry, research institutions, and private practice need to cope with this change, and extend their knowledge and experience from media-specific, "end-of-pipe," treatment-and-disposal strategies to multimedia pollution prevention management. The importance of this extension and reorientation in education, however, is such that the effort cannot be further delayed. Many air pollution, water pollution, and solid waste supervisors in government agencies spend their entire careers in just one function because environmental quality supervisors usually work in only one of the media functions. Some may be reluctant to accept such activities. This is understandable given the fact that such a reorientation requires time and energy to learn new concepts and that time is a premium for them. Nevertheless they must support such education and training in order to have well-trained young professionals.

Successful implementation of multimedia pollution prevention programs will require well-trained environmental professionals who are fully prepared in the principles and practices of such programs. These programs need to develop a deep appreciation of the necessity for multimedia pollution prevention in all levels of society which will require a high priority for educational and training efforts. New instructional materials and tools are needed for incorporating new concepts in the existing curricula of elementary and secondary education, colleges and universities, and training institutions. The use of computerized automation offers much hope.

Government agencies need to conduct a variety of activities to achieve three main educational objectives:

1. Ensure an adequate number of high quality environmental professionals.
2. Encourage groups to undertake careers in environmental fields and to stimulate all institutions to participate more fully in developing environmental professionals.
3. Generate databases that can improve environmental literacy of the general public and especially the media.

These objectives are related to, and reinforce, one another. For example, improving general environmental literacy should help to expand the pool of environmental professionals by increasing awareness of the nature of technical careers. Conversely, steps taken to increase the number of environmental professionals should also help improve the activities of general groups and institutions. Developing an adequate human resource base should be the first priority in education. The training of environmental professionals receive should be top quality.

There is significant need to provide graduate students with training and experience in more than one discipline. The most important and interesting environmental scientific/technological questions increasingly require interdisciplinary and/or multidisciplinary approaches. Environmental graduate programs must address this aspect. Most practicing environmental professionals face various types of environmental problems that they have not been taught in the universities. Therefore, continuing education opportunities and cross-disciplinary training must be available for them to understand the importance of multimedia pollution prevention principles and strategies, as well as to carry out such principles and strategies.

The education and training plan of multimedia pollution prevention may be divided into technical and nontechnical areas. Technical areas include

1. Products—Lifecycle analysis methods, trends-in-use patterns, new products, product lifespan data, product substitution, and product applicability. (A product's lifecycle includes its design, manufacture, use, maintenance and repair, and final disposal.)
2. Processes—Feedstock substitution, waste minimization, assessment procedures, basic unit process data, unit process waste generation assessment methods, materials handling, cleaning, maintenance, and repair.

3. Recycling and reuse—Market availability, infrastructure capabilities, new processes and product technologies, automated equipment and processes, distribution and marketing, management strategies, automation, waste stream segregation, on-site and off-site reuse opportunities, close-loop methods, waste recapture, and reuse.

Nontechnical areas include

1. Educational programs and dissemination of information
2. Incentive and disincentives
3. Economic cost and benefits
4. Sociological human behavioral trends
5. Management strategies including coordination with various concerned organizations

5.7 FUTURE TRENDS

Environmental quality and natural resources are under extreme stress in many industrialized nations and in virtually every developing nation as well. Environmental pollution is closely related to population density, energy, transportation demand, and land use patterns, as well as industrial and urban development. The main reason for environmental pollution is the increasing rate of waste generation in terms of quantity and toxicity that has exceeded society's ability to properly manage it. Another reason is that the management approach has focused on the media-specific and the end-of-pipe strategies. There is increasing reported evidence of socioeconomic and environmental benefits realized from multimedia pollution prevention [5,6]. The prevention of environmental pollution in the twenty-first century is going to require not only enforcement of government regulations and controls, but also changes in manufacturing processes and products as well as in lifestyles and behavior throughout society. Education is a key in achieving the vital goal of multimedia pollution prevention.

5.8 SUMMARY

1. The current approach to environmental waste management requires some rethinking. A multimedia approach facilitates the integration of air, water, and land pollution controls and seeks solutions that do not violate the laws of nature.
2. The EPA's own single-media offices, often created sequentially as individual problems were identified and responded to in legislation, have played a role in impeding development of cost-effective multimedia prevention strategies.
3. Environmental problems result from the release of wastes (gaseous, liquid, and solid) that are generated daily by industrial and commercial establishments as well as households. The lack of consciousness regarding conservation of materials, energy, and water has contributed to the wasteful habits of society.

4. The environment is the most important component of life support systems. It is comprised of air, water, soil, and biota through which elements and pollutants cycle.

5. Traditional partitioned approaches to environmental management can provide some environmental relief, but a total systems approach is required if optimum improvements—in terms of pollution/waste reduction—are to be achieved.

6. The role of environmental professional in waste management and pollution control has been changing in recent years. Many talented, dedicated environmental professionals in academia, government, industry, research institutions, and private practice need to cope with the change, and extend their knowledge and experience from media specific, "end-of-pipe," treatment-and-disposal strategies to multimedia pollution prevention management.

7. The prevention of environmental pollution in the twenty-first century is going to require not only enforcement of government regulations and controls but also changes in manufacturing processes and products as well as in lifestyles and behavior throughout society. Education is the key in achieving the vital goal of multimedia pollution prevention.

REFERENCES

1. Burke, G., Singh, B., and Theodore, L. *Handbook of Environmental Management and Technology*, 2nd edition, John Wiley & Sons, Hoboken, NJ, 2000.
2. L. Theodore, Personal notes, 1990.
3. Shen, T. The role of environmental engineers in waste minimization, *Proceedings of the First International Conference on Waste Minimization and Clean Technology*, Geneva, Switzerland, 1989.
4. Theodore, M.K. and Theodore L. *Major Environmental Issues Facing the 21st Century*, 1st edition, Theodore Tutorials (Originally published by Simon and Shuster), East Williston, NY, 1996.
5. *Chemecology,* 17 (1), and 19 (2), 1990.
6. Schecter, N. and Hunt, G. *Case Summaries of Waste Reduction by Industries in the Southeast*, Waste Reduction Resource Center, Raleigh, NC, 1989.

6 Classification and Sources of Pollutants

CONTENTS

6.1 INTRODUCTION

Not long ago, the nation's natural resources were exploited indiscriminately. Waterways served as industrial pollution sinks; skies dispersed smoke from factories and power plants; and the land proved to be a cheap and convenient place to dump industrial and urban wastes. However, society is now more aware of the environment and the need to protect it. The American people have been involved in a great social movement known broadly as "environmentalism." Society has been concerned with the quality of the air one breathes, the water one drinks, and the land on which one lives and works. While economic growth and prosperity are still important goals, opinion polls show overwhelming public support for pollution controls and pronounced willingness to pay for them. This chapter presents the reader with information on pollutants and categorizes their sources by the media they threaten.

6.2   AIR POLLUTANTS

Since the Clean Air Act was passed in 1970, the United States has made impressive strides in improving and protecting air quality. As directed by this Act, the Environmental Protection Agency (EPA) set National Ambient Air Quality Standards (NAAQS) for those pollutants commonly found throughout the country that posed the greatest overall threats to air quality. These pollutants, termed "criteria pollutants" under the Act, include ozone, carbon monoxide, airborne particulates, sulfur dioxide, lead, and nitrogen oxide. Although the EPA has made considerable progress in controlling air pollution, all of the six criteria except lead and nitrogen oxide are currently a major concern in a number of areas in the country. The following subsections focus on a number of most significant air quality challenges: ozone and carbon monoxide, airborne particulates, airborne toxics, sulfur dioxide, acid deposition, and indoor air pollutants.

6.2.1   Ozone and Carbon Monoxide

Ozone is one of the most intractable and widespread environmental problems. Chemically, ozone is a form of oxygen with three oxygen atoms instead of the two found in regular oxygen. This makes it very reactive, so that it combines with practically every material with which it comes in contact. In the upper atmosphere, where ozone is needed to protect people from ultraviolet radiation, the ozone is being destroyed by man-made chemicals, but at ground level, ozone can be a harmful pollutant.

Ozone is produced in the atmosphere when sunlight triggers chemical reactions between naturally occurring atmospheric gases and pollutants such as volatile organic compounds (VOCs) and nitrogen oxides. The main source of VOCs and nitrogen oxides is combustion sources such as motor vehicle traffic.

Carbon monoxide is an invisible, odorless product of incomplete fuel combustion. As with ozone, motor vehicles are the main contributor to carbon monoxide formation. Other sources include wood-burning stoves, incinerators, and industrial processes. Since auto travel and the number of small sources of VOCs are expected to increase, even strenuous efforts may not sufficiently reduce emissions of ozone and carbon monoxide [1].

6.2.2 AIRBORNE PARTICULATES

Particulates in the air include dust, smoke, metals, and aerosols. Major sources include steel mills, power plants, cotton gins, cement plants, smelters, and diesel engines. Other sources are grain storage elevators, industrial haul roads, construction work, and demolition. Wood-burning stoves and fireplaces can also be significant sources of particulates. Urban areas are likely to have windblown dust from roads, parking lots, and construction work [1].

6.2.3 AIRBORNE TOXICS

Toxic pollutants are one of today's most serious emerging problems that are found in all media. Many sources emit toxic chemicals into the atmosphere: industrial and manufacturing processes, solvent use, sewage treatment plants, hazardous waste handling and disposal sites, municipal waste sites, incinerators, and motor vehicles. Smelters, metal refiners, manufacturing processes, and stationary fuel combustion sources emit such toxic metals as cadmium, lead, arsenic, chromium, mercury, and beryllium. Toxic organics, such as vinyl chloride and benzene, are released by a variety of sources, such as plastics and chemical manufacturing plants, and gas stations. Chlorinated dioxins are emitted by some chemical processes and the high-temperature burning of plastics in incinerators [1].

6.2.4 SULFUR DIOXIDE

Sulfur dioxide can be transported long distances in the atmosphere due to its ability to bond to particulates. After traveling, sulfur dioxide usually combines with water vapor to form acid rain (see Chapter 21). Sulfur dioxide is released into the air primarily through the burning of coal and fuel oils. Today, two-thirds of all national sulfur dioxide emissions come from electric power plants. Other sources of sulfur dioxide include refiners, pulp and paper mills, smelters, steel and chemical plants, and energy facilities related to oil shale, syn (synthetic) fuels, and oil and gas production. Home furnaces and coal-burning stoves are sources that directly affect residential neighborhoods [1].

6.2.5 ACID DEPOSITION

Acid deposition is a serious environmental concern in many parts of the country. The process of acid deposition begins with the emissions of sulfur dioxide (primarily from coal-burning power plants) and nitrogen oxides (primarily from motor vehicles and coal-burning power plants). As described in Section 6.2.4, these pollutants interact with sunlight and water vapor in the upper atmosphere to form acidic compounds. During a storm, these compounds fall to the earth as acid rain or snow; the compounds may also join dust or other dry airborne particles and fall as "dry deposition" [2].

6.3 INDOOR AIR POLLUTANTS

Indoor air pollution is rapidly becoming a major health issue in the United States. Indoor pollutant levels are quite often higher than outdoors, particularly where buildings are tightly constructed to save energy. Since most people spend 90% of their time indoors, exposure to unhealthy concentrations of indoor air pollutants is often inevitable. The degree of risk associated with exposure to indoor pollutants depends on how well buildings are ventilated and the type, mixture, and amounts of pollutants in the building. Indoor air pollutants of special concern are described below (more detailed information on indoor air quality can be found in Chapter 14).

6.3.1 RADON

Radon is a unique environmental problem because it occurs naturally. Radon results from the radioactive decay of radium-226, found in many types of rocks and soils. Most indoor radon comes from the rock and soil around a building and enters structures through cracks or openings in the foundation or basement. Secondary sources of indoor radon are wellwater and building materials [2].

6.3.2 ENVIRONMENTAL TOBACCO SMOKE

Environmental tobacco smoke is smoke that nonsmokers are exposed to from smokers. This smoke has been judged by the Surgeon General, the National Research Council, and the International Agency for Research on Cancer to pose a risk of lung cancer to nonsmokers. Tobacco smoke contains a number of pollutants, including inorganic gases, heavy metals, particulates, VOCs, and products of incomplete burning, such as polynuclear aromatic hydrocarbons [2].

6.3.3 ASBESTOS

Asbestos has been used in the past in a variety of building materials, including many types of insulation, fireproofing, wallboard, ceiling tiles, and floor tiles. The remodeling or demolition of buildings with asbestos-containing materials frees tiny asbestos fibers in clumps or clouds of dust. Even with normal aging, materials may deteriorate and release asbestos fibers. Once released, these asbestos fibers can be inhaled into the lungs and can accumulate [2]. The reader is referred to Chapter 28 for a more expanded discussion of asbestos.

6.3.4 FORMALDEHYDE AND OTHER VOLATILE ORGANIC COMPOUNDS

The EPA has found formaldehyde to be a probable human carcinogen. The use of formaldehyde in furniture, foam insulation, and pressed wood products, such as some plywood, particle board, and fiberboard, makes formaldehyde a major indoor air pollutant.

VOCs commonly found indoors include benzene from tobacco smoke and perchloroethylene emitted by dry-cleaned clothes. Paints and stored chemicals, including certain cleaning compounds, are also major sources of VOCs. VOCs can also be emitted from drinking water; 20% of water supply systems have detectable amounts of VOCs [2] (see Section 6.4.1).

6.3.5 PESTICIDES

Indoor and outdoor use of pesticides, including termiticides and wood preservatives, are another cause of concern. Even when used as directed, pesticides may release VOCs. In addition, there are about 1200 inert ingredients added to pesticide products for a variety of purposes. While not "active" in attacking the particular pest, some inert ingredients are chemically or biologically active and may cause health problems. EPA researchers are presently investigating whether indoor use of insecticides and subsurface soil injection of termiticides can lead to hazardous exposure [2].

6.4 WATER POLLUTANTS

The EPA, in partnership with state and local governments, is responsible for improving and maintaining water quality. These efforts are organized around three themes. The first is maintaining the quality of drinking water. This is addressed by monitoring and treating drinking water prior to consumption and by minimizing the contamination of the surface water and protecting against contamination of ground water needed for human consumption. The second is preventing the degradation and destruction of critical aquatic habitats, including wetlands, nearshore coastal waters, oceans, and lakes. The third is reducing the pollution of free-flowing surface waters and protecting their uses. The following is a discussion of various pollutants categorized by these themes.

6.4.1 DRINKING WATER POLLUTANTS

The most severe and acute public health effects from contaminated drinking water, such as cholera and typhoid, have been eliminated in America. However, some less acute and immediate hazards remain in the nation's tap water. These hazards are associated with a number of specific contaminants in drinking water. Contaminants of special concern to the EPA are lead, radionuclides, microbiological contaminants, and disinfection byproducts.

The primary source of lead in drinking water is corrosion of plumbing material, such as lead service lines and lead solders, in water distribution systems and in houses and larger buildings. Virtually all public water systems serve households with lead solders of varying ages, and most faucets are made of materials that can contribute some lead to drinking water.

Radionuclides are radioactive isotopes that emit radiation as they decay. The most significant radionuclides in drinking water are radium, uranium, and radon, all of which occur naturally in nature. While radium and uranium enter the body by ingestion, radon is usually inhaled after being released into the air during showers, baths, and other activities, such as washing clothes or dishes. Radionuclides in drinking water occur primarily in those systems that use ground water. Naturally occurring radionuclides seldom are found in surface waters (such as rivers, lakes, and streams).

Water contains many microbes—bacteria, viruses, and protozoa. Although some organisms are harmless, others can cause disease. The Centers for Disease Control reported 112 waterborne disease outbreaks from 1981 to 1983. Microbiological contamination continues to be a national concern because contaminated drinking water systems can rapidly spread disease.

Disinfection byproducts are produced during water treatment by the chemical reactions of disinfectants with naturally occurring or synthetic organic materials present in untreated water. Since these disinfectants are essential to safe drinking water, the EPA is presently looking at ways to minimize the risks from byproducts [2].

6.4.2　Critical Aquatic Habitat Pollutants

Critical aquatic habitats that need special management attention include the nation's wetlands, near coastal waters, oceans, and lakes. In recent years, the EPA has been focusing on addressing the special problems of these areas. The following is a discussion of pollutants categorized by the habitats they affect.

Wetlands in urban areas frequently represent the last large tracts of open space and are often a final haven for wildlife. Not surprisingly, as suitable upland development sites become exhausted, urban wetlands are under increasing pressure for residential housing, industry, and commercial facilities.

Increasing evidence exists that our nation's wetlands, in addition to being destroyed by physical threats, also are being degraded by chemical contamination. The problem of wetland contamination received national attention in 1985 due to reports of waterfowl deaths and deformities caused by selenium contamination. Selenium is a trace element that occurs naturally in soil and is needed in small amounts to sustain life. However, for years it was being leached out of the soil and carried in agricultural drainwater used to flood the refuge's wetlands, where it accumulated in dangerously high levels.

Coastal water environments are particularly susceptible to contamination because they act as sinks for the large quantities of pollution discharged from municipal sewage treatment plants, industrial facilities, and hazardous waste disposal sites. In many coastal areas, nonpoint source runoff from agricultural lands, suburban developments, city streets, and combined sewer and stormwater overflows poses an even more significant problem than point sources. This is due to the difficulty of identifying and then controlling the source of the pollution.

Physical and hydrological modifications from such activities as dredging channels, draining and filling wetlands, constructing dams, and building shorefront houses may further degrade near coastal environments. In addition, growing population pressures will continue to subject these sensitive coastal ecosystems to further stress.

The Great Lakes provide an inevitable resource to the 45 million people living in the surrounding basin. A 1970 study by the International Joint Commission identified nutrients and toxic problems in the lakes. They suffered from eutrophication problems caused by excessive nutrient inputs. Since then, the United States and Canada have made joint efforts to reduce nutrient loadings, particularly phosphorus. However, contamination of the water and fish by toxics from pesticide runoff, landfill leachates, and in-place sediments remains a major problem.

Ocean dumping of dredged material, sewage sludge, and industrial wastes is a major source of ocean pollution. Sediments dredged from industrialized urban harbors are often highly contaminated with heavy metals and toxic synthetic organic chemicals like polychlorinated biphenyls (PCBs) and petroleum hydrocarbons. Although ocean dumping of dredged material, sludge, and industrial wastes is now less of a threat, persistent disposal of plastics from land and ships at sea have become a serious problem. Debris on beaches from sewer and storm drain overflows, or mismanagement of trash poses public safety and aesthetic concerns [2].

6.4.3 SURFACE WATER POLLUTANTS

Pollutants in waterways come from industries or treatment plants discharging wastewater into streams or from waters running across urban and agricultural areas, carrying the surface pollution with them (nonpoint sources). The following is a discussion of surface water pollutants categorized by their main sources.

Raw or insufficiently treated wastewater from municipal and industrial treatment plants still threatens water resources in many parts of the country. In addition to harmful nutrients, poorly treated wastewater may contain bacteria and chemicals.

Sludge, the residue left from wastewater treatment plants, is a growing problem. Although some sludges are relatively "clean," or free from toxic substances, other sludges may contain organic, inorganic, or toxic pollutants and pathogens.

An important source of toxic pollution is industrial wastewater discharged directly into waterways or indirectly through municipal wastewater treatment plants. Industrial wastes discharged indirectly are treated to remove toxic pollutants. It is important that those wastes be treated because toxics may end up in sludge, making them harder to dispose of safely.

Nonpoint sources present continuing problems for achieving national water quality in many parts of the country. Sediment and nutrients are the two largest contributors to nonpoint source problems. Nonpoint sources are also a major source of toxics, among them are pesticide runoff from agricultural areas, metals from active or abandoned mines, gasoline, and asbestos from urban areas. In addition, the atmosphere is a source of toxics since many toxics can attach themselves to dust, later to be deposited in surface waters hundreds of miles away through precipitation [2].

6.5 LAND POLLUTANTS

Historically, land has been used as the dumping ground for wastes, including those removed from the air and water. Early environmental protection efforts focused on cleaning up air and water pollution. It was not until the 1970s that there was much public concern about pollution of the land. It is now recognized that contamination of the land threatens not only future use of the land itself, but also the quality of the surrounding air, surface water, and ground water. There are five different forms of land pollutants. These include

1. Industrial hazardous wastes
2. Municipal wastes
3. Mining wastes
4. Radioactive wastes
5. Underground storage tanks

A short description of each is provided below. More detailed descriptions can be found in separate chapters later in the book.

6.5.1 INDUSTRIAL HAZARDOUS WASTES

The chemical, petroleum, and transportation industries are major producers of hazardous industrial waste. Ninety-nine percent of the hazardous waste is produced by facilities that generate large quantities (more than 2200 lb) of hazardous waste each month.

A much smaller amount of hazardous waste, about one million tons per year, comes from small quantity generators (between 220 and 2200 lb of waste each month). These include automotive repair shops, construction firms, laundromats, dry cleaners, printing operations, and equipment repair shops. Over 60% of these wastes are derived from lead batteries. The remainder includes acids, solvents, photographic wastes, and dry cleaning residue [2].

6.5.2 MUNICIPAL WASTES

Municipal wastes include household and commercial wastes, demolition materials, and sewage sludge. Solvents and other harmful household and commercial wastes are generally so intermingled with other materials that specific control of each is virtually impossible.

Sewage sludge is the solid, semisolid, or liquid residue produced from treating municipal wastewater. Some sewage sludges contain high levels of disease-carrying microorganisms, toxic metals, or toxic organic chemicals. Because of the large quantities generated, sewage sludge is a major waste management problem in a number of municipalities [2].

6.5.3 MINING WASTES

A large volume of all waste generated in the United States is from mining coal, phosphates, copper, iron, uranium, other minerals, and from ore processing and milling. These wastes consist primarily of overburden, the soil and rock cleared away before mining, and tailings, the material discarded during ore processing. Runoff from these wastes increases the acidity of streams and pollutes them with toxic metals [2].

6.5.4 Radioactive Wastes

Radioactive materials are used in a wide variety of applications, from generating electricity to medical research. The United States has produced large quantities of radioactive wastes that can pose environmental and health problems for many generations [2].

6.5.5 Pollutants from Underground Storage Tanks

Leaking underground storage tanks are another source of land contamination that can contribute to ground water contamination. The majority of these tanks do not store waste, but instead store petroleum products and some hazardous substances. Most of the tanks are bare steel and subject to corrosion. Many are old and near the end of their useful lives. Hundreds of thousands of these tanks are presently thought to be leaking, with more tanks expected to develop leaks in the next few years [2].

6.6 HAZARDOUS POLLUTANTS

Before the early 1970s, the nation paid little attention to industrial production and the disposal of the waste it generated, particularly hazardous waste. As a result, billions of dollars must now be spent to clean up disposal sites neglected through years of mismanagement. The EPA often identifies a waste as hazardous if it poses a fire hazard (ignitable); dissolves materials or is acidic (corrosive); is explosive (reactive); or otherwise poses danger to human health or the environment (toxic). Most hazardous waste results from the production of widely used goods such as polyester and other synthetic fibers, kitchen appliances, and plastic milk jugs. A small percentage of hazardous waste (less than 1%) is comprised of the used commercial products themselves, including household cleaning fluids or battery acid.

Definitions of hazardous substances are not as straightforward as they appear. For purposes of regulation, Congress and the EPA have defined terms to describe wastes and other substances that fall under regulation. The definitions below show the complexity of the EPA's regulatory task.

1. Hazardous Substances (Comprehensive Environmental Response, Compensation and Liability Act [CERCLA], or "Superfund")—Any substance that, when released into the environment, may cause substantial danger to public health, welfare, or the environment. Designation as a hazardous substance grows out of the statutory definitions in several environmental laws: the CERCLA, the Resource Conservation and Recovery Act (RCRA), the Clean Water Act (CWA), the Clean Air Act (CAA), and the Toxic Substances Control Act (TSCA). Currently there are 717 CERCLA hazardous substances.

2. Extremely Hazardous Substance (CERCLA as amended)—Substances that could cause serious, irreversible health effects from a single exposure. For purposes of chemical emergency planning, EPA has designated 366 substances extremely hazardous. If not already so designated, these also will be listed as hazardous substances.

3. Solid Waste (RCRA)—Any garbage, refuse, sludge, or other discarded material. All solid waste is not solid; it can be liquid, semisolid, or contained gaseous material. Solid waste results from industrial, commercial, mining, and agricultural operations from community activities. Solid waste can be either hazardous or nonhazardous. However, it does not include solid or dissolved material in domestic sewage, certain nuclear material, or certain agricultural wastes.

4. Hazardous Waste (RCRA)—Solid waste, or combinations of solid waste, that because of its quantity, concentration, or physical, chemical or infectious characteristics, may pose a hazard to human health or the environment.

5. Nonhazardous Waste (RCRA)—Solid waste, including municipal wastes, household hazardous waste, municipal sludge, and industrial and commercial wastes that are not hazardous [2].

6.7   TOXIC POLLUTANTS

Today's high standard of living would not be possible without the thousands of different chemicals produced. Most of these chemicals are not harmful if used properly. Others can be extremely harmful if people are exposed to them even in minute amounts. The following is a discussion of four toxic chemicals under control of the Toxic Substance Control Act of 1976.

PCBs provide an example of the problems that toxic substances can present. PCBs were used in many commercial activities, especially in heat transfer fluids in electrical transformers and capacitors. They also were used in hydraulic fluids, lubricants, dye carriers in carbonless copy paper, and in paints, inks, and dyes. Over time, PCBs accumulated in the environment, either from leaking electrical equipment or from other materials such as inks.

Like PCBs, asbestos was widely used for many purposes, such as fireproofing and pipe and boiler insulation in schools and other buildings. Asbestos was often mixed with a cement-like material and sprayed or plastered on ceilings and other surfaces. Now these materials are deteriorating, releasing the asbestos.

Dioxins refer to a family of chemicals with similar structure, although it is common to refer to the most toxic of these—2,3,7,8-tetrachlorodinitro-p-dioxin or TCDD—as dioxin. Dioxin is an inadvertent contaminant of the chlorinated herbicides 2,4,5-T and silvex, which were used until recently in agriculture, forest management, and lawn care. It is also a contaminant of certain wood preservatives and the defoliant Agent Orange used in Vietnam. Dioxins and the related chemicals known as furans also are formed during the combustion of PCBs.

Several other sources of dioxin contamination have been identified in recent years. These include pulp and paper production, and the burning of municipal wastes containing certain plastics or wood preserved by certain chlorinated chemicals.

6.8   SUMMARY

1. All of the criteria pollutants (ozone, carbon monoxide, airborne particulates, sulfur dioxide, lead, nitrogen oxide) except lead and nitrogen oxide are currently a major concern in a number of areas in the country.

2. Indoor air pollutants of special concern include radon, environmental tobacco smoke, asbestos, formaldehyde and other VOCs, and pesticides.
3. The EPA focuses its water pollution control efforts on three themes: maintaining drinking water quality, preventing further degradation and destruction of critical aquatic habitats (wetlands, nearshore coastal waters, oceans, and lakes), and reducing pollution of free-flowing surface waters and protecting their uses.
4. Land pollutants discussed include industrial hazardous wastes, municipal wastes, mining wastes, radioactive wastes, and leaking underground storage tank pollutants.
5. Hazardous pollutants are generally identified as such if they are ignitable, corrosive, reactive, or toxic.
6. Toxic pollutants include PCBs, asbestos, dioxin, and CFCs.

REFERENCES

1. Burke, G., Singh, B., and Theodore, L. *Handbook of Environmental Management and Technology*, 2nd edition, John Wiley & Sons, Hoboken, NJ, 2000.
2. U.S. EPA, Environmental progress and challenges, *EPA's Update*, August 1988.

7 Effects of Pollutants

CONTENTS

7.1 INTRODUCTION

Pollutants are various noxious chemicals and refuse materials that impair the purity of the water, soil, and the atmosphere. The area most affected by pollutants is the atmosphere or air. Air pollution occurs when wastes pollute the air. Artificially or synthetically created wastes are the main sources of air pollution. They can be in the form of gases or particulates which result from the burning of fuel to power motor vehicles and to heat buildings. More air pollution can be found in densely populated areas. The air over largely populated cities often becomes so filled with pollutants that it not only harms the health of humans, but also plants, animals, and materials of construction.

Water pollution occurs when wastes are dumped into the water. This polluted water can spread typhoid fever and other diseases. In the United States, water supplies are disinfected to kill disease-causing germs. The disinfection, in some instances, does not remove all the chemicals and metals that may cause health problems in the distant future.

Wastes that are dumped into the soil are a form of land pollution, which damages the thin layer of fertile soil that is essential for agriculture. In nature, cycles work to keep soil fertile. Wastes, including dead plants and wastes from animals, form a substance in the soil called humus. Bacteria then decays the humus and breaks it down into nitrates, phosphates, and other nutrients that feed growing plants.

This chapter will review the effects of air pollutants, water pollutants, and land (solid waste) pollutants on

1. Humans
2. Plants
3. Animals
4. Materials of construction

For reasons hopefully obvious to the reader, the material will key in on the effects on humans. It will also primarily focus on air pollutants since this has emerged as the leading environmental issue with the passage of the Clean Air Act Amendments of 1990.

7.2 AIR POLLUTION

7.2.1 HUMANS

Humans are in constant contact with pollutants, whether they are indoors or outdoors. The pollutants, primarily air pollutants, may have negative effects on human health. In some instances humans adapt and do not realize that they are being affected. For example, people living in smog-covered cities know that smog is bad for their health, but just consider it "normal." There are still some who do not think that there is anything that can be done about it.

A definite correlation seems to exist between some of the most important indoor activities and the resulting pollutants that are generated. Some examples of these are smoking, the use of personal products, cleaning, cooking, heating, maintenance of hair and facial care, hobbies, and electrical appliances such as washing machines and dryers [1,2]. Fumes from these activities can get trapped in the home or workplace, and the buildup of these over time will cause health problems in the short- and long-term future.

When people go outside, they usually say they are going to "get some fresh air." This "fresh air" to them usually means breathing in the air from a different location. Although the common term for the air outside is "fresh air," the air may not necessarily be very "fresh." The outside air can be full of air pollutants that can cause negative effects on the health of humans.

The influence of air pollution on human productivity has not been firmly established. In addition, a number of authorities suspect (and some are convinced)

TABLE 7.1

Health Effects of the Regulated Air Pollutants

Criteria Pollutants	Health Concerns
Ozone	Respiratory tract problems such as difficult breathing and reduced lung function. Asthma, eye irritation, nasal congestion, reduced resistance to infection, and possibly premature aging of lung tissue
Particulate matter	Eye and throat irritation, bronchitis, lung damage, and impaired visibility
Carbon monoxide	Ability of blood to carry oxygen impaired, cardiovascular, nervous and pulmonary systems affected
Sulfur dioxide	Respiratory tract problems, permanent harm to lung tissue
Lead	Retardation and brain damage, especially in children
Nitrogen dioxide	Respiratory illness and lung damage
Hazardous Air Pollutants	
Asbestos	A variety of lung diseases, particularly lung cancer
Beryllium	Primary lung disease, although it also affects liver, spleen, kidneys, and lymph glands
Mercury	Several areas of the brain as well as the kidneys and bowels affected
Vinyl chloride	Lung and liver cancer
Arsenic	Causes cancer
Radionuclides	Cause cancer
Benzene	Leukemia

that air pollution is associated with an increasing incidence of lung and respiratory ailments and heart disease [2]. Table 7.1 shows some of the health effects of the regulated air pollutants.

"Air toxics" is the term generally used to describe cancer-causing chemicals, radioactive materials, and other toxic chemicals not covered by the National Ambient Air Quality Standards (see Chapter 2) for conventional pollutants. Air toxics result from many activities of modern society, including driving a car, burning fossil fuel, and producing and using industrial chemicals or radioactive materials. The latter is one of the highest health risk problems the Environmental Protection Agency (EPA) is wrestling with [3].

Some major contributors to pollution that affect human health are sulfur dioxide, carbon monoxide, nitrogen oxides, ozone, carcinogens, fluorides, aeroallergens, radon, smoking, asbestos, and noise. These are treated in separate paragraphs below.

"Sulfur dioxide" (SO_2) is a source of serious discomfort, and in excessive amounts is a health hazard, especially to people with respiratory ailments. In the United States alone, the estimated amount of SO_2 emitted into the atmosphere is 23 million tons per year. SO_2 causes irritation of the respiratory tract; it damages lung tissue and promotes respiratory diseases; the taste threshold limit is 0.3 parts per million (ppm); and, SO_2 produces an unpleasant smell at 0.5 ppm concentration. In fact, sulfur dioxides in general have been considered as prime candidates for an air pollution index.

Such an index would be a measure reflecting the presence and action of harmful environmental conditions. This would aid in rendering meaningful analyses of the effect of air pollutants on human health, especially since health effects are most probably due to the complementing action of pollutants and meteorological variables.

SO_2 is more harmful in a dusty atmosphere. This effect may be explained as follows: The respiratory tract is lined with hair-like cilia, which by means of regular sweeping action, force out foreign substances entering the respiratory tract through the mouth. SO_2 and H_2SO_4 (sulfuric acid) molecules paralyze the cilia, rendering it ineffective in rejecting these particulates, causing them to penetrate deeper into the lungs. Alone, these molecules are too small to remain in the lungs; but, some SO_2 molecules are absorbed on larger particles, which penetrate to the lungs and settle there, bringing concentrated amounts of the irritant SO_2 into prolonged contact with the fine lung tissues. SO_2 and the other sulfur dioxide-particulate combinations are serious irritants of the respiratory tract. In high-pollution intervals they can cause death. Their action of severely irritating the respiratory tract may cause heart failure due to the excessive laboring of the heart in its pumping action to circulate oxygen through the body.

"Carbon monoxide" (CO) levels have declined in most parts of the United States since 1970, but the standards are still exceeded in many cities throughout the country [4]. Carbon monoxide pollution is the basic concern in most large cities of the world where traffic is usually congested and heavy. CO cannot be detected by smell or sight, and this adds to its danger. It forms a complex with hemoglobin called carboxy-hemoglobin (COHb). The formation of this complex reduces the capability of the bloodstream to carry oxygen by interfering with the release of the oxygen carried by remaining hemoglobin. Also, since the affinity of human hemoglobin is 210 times higher for CO than it is for oxygen, a small concentration of CO markedly reduces the capacity of the blood to act as an oxygen carrier. The threshold limit value (TLV), or maximum allowable concentration (MAC), of CO for industrial exposure is 50 ppm; concentrations of CO as low as 10 ppm produce effects on the nervous system and give an equilibrium level of COHb larger than 2%. A concentration of 30 ppm produces a level greater than 5% COHb, which affects the nervous system and causes impairment of visual acuity, brightness discrimination, and other psychomotor functions. Carbon monoxide concentrations of 50–100 ppm are commonly encountered in the atmosphere of crowded cities, especially at heavy traffic rush hours. Such high concentrations adversely affect driving ability and cause accidents. In addition, an estimated average concentration of CO inhaled into the lungs from cigarette smoking is 400 ppm [5].

Two major pollutants among "nitrogen oxides" (NO_x) are nitric oxide (NO), and nitrogen dioxide (NO_2). Emissions from stationary sources are estimated to be 16 million tons of NO_x per year. Mobile sources of NO_x pollution are automobiles emitting an estimated average of 10.7 million tons per year. NO is colorless, but it is photochemically converted to nitrogen dioxide, which is one of the components of smog. Nitrogen dioxide also contributes to the formation of aldehydes and ketones through the photochemical reaction with hydrocarbons of the atmosphere. Nitrogen dioxide is an irritant; it damages lung tissues, especially through the formation of nitric acid. Breathing nitrogen dioxide at 25 ppm for 8 h could cause spoilage of lung tissues, while breathing it for 1.5 h at 100–150 ppm could produce serious pulmonary edema, or swelling of lung tissues. A few breaths at 200–700 ppm may cause fatal pulmonary edema.

"Ozone" (O_3) is produced from the activation of sunlight on nitrogen dioxide, pollutants such as volatile organic compounds (VOCs), and atmospheric gases such as oxygen. It is an irritant to the eyes and lungs, penetrating deeper into the lungs than sulfur dioxide. It forms complex organic compounds in air; dominant among these are aldehydes and peroxyacetyl nitrate (PAN), which also causes eye and lung irritation. Rural areas have concentrations of 2–5 parts per hundred million (pphm) of ozone, which is distinguished by an odor of electrical shorting. A few good smells and the individual's sensitivity for this odor disappears. At 5–10 pphm, the odor is unpleasant and pungent. Exposure to ozone for 30 min at 10–15 pphm, which is normally encountered in large cities, causes serious irritation of the mucous membranes and reduces their ability to fight infection. At 20–30 pphm it affects vision, and exposure to concentrations of 30 pphm for a few minutes brings a marked respiratory distress with severe fatigue, coughing, and choking. When volunteers were exposed intermittently for 2 weeks to a 30 pphm ozone atmosphere, they experienced severe headaches, fatigue, wheezing, chest pains, and difficulty in breathing. It reduces the activity of individuals, especially those with previous heart conditions. Even young athletes tire on smoggy days.

"Carcinogens," which are often polycyclic hydrocarbons inducing cancer in susceptible individuals, are present in the exhaust emissions of the internal combustion engine, be it diesel or gasoline. Two major carcinogens are benzopyrene, which is a strong cancer-inducing agent, and benzanthracene, which is a weak one. They are essentially nonvolatile organic compounds associated with solids or polymeric substances in the air. These compounds are not very stable, and they are destroyed at varying rates by other air pollutants and by sunlight. However, as a result of industrialization and urbanization, these substances are discharged into the atmosphere in significant quantities, reportedly causing a steady increase in the frequency of human lung cancer in the world [5].

"Aeroallergens" are airborne substances causing allergies. These are predominantly of natural origin, but some are industrial. Allergic reactions in sensitive persons are caused by allergens such as pollens, spores, and rusts. A large percentage of the population is affected by hay fever and asthma each year; ragweed pollen may be the worst offender—it is about 20 μm in diameter, and under normal conditions, nearly all of it will be deposited near the source. Organic allergens come from plants, yeasts, molds, and animal hair, fur, or feathers. Fine industrial materials in the air cause allergies; for example, the powdered material given off in the extraction of oil from castor beans causes bronchial asthma in people living near the factory [5].

"Radon" is a radioactive, colorless, odorless, naturally occurring gas that is found everywhere at very low levels. It seeps through the soil and collects in homes. Radon problems hive been identified in every state, and millions of homes throughout the country have elevated radon levels. Radon in high concentrations has been determined to cause lung cancer in humans.

"Smoking" can be categorized as voluntary pollution. The smokers not only create a health hazard for themselves, but also for the nonsmokers in their company. Cigarette smoke causes lung cancer, and in pregnant women it may cause premature birth and low birthweight in newborns.

"Asbestos" is a mineral fiber that has been used commonly in a variety of building construction materials for insulation as a fire-retardant. The EPA and other organizations have banned several asbestos products. Manufacturers have also voluntarily limited the use of asbestos. Today asbestos is most commonly found in older homes in pipe and furnace insulation materials, asbestos shingles, millboard, textured paints, and floor tiles. The most dangerous asbestos fibers are too small to see. After the fibers are inhaled, they can remain and accumulate in the lungs. Asbestos can cause lung cancer, cancer of the chest and abdominal linings, and asbestosis (irreversible lung scarring that can be fatal). Symptoms of these diseases do not show up until many years after exposure. Most people with asbestos-related disease were exposed to elevated concentrations on the job, and some developed disease from clothing and equipment brought home from job sites [4]. The reader is referred to Chapter 28 for additional details on asbestos.

"Noise" pollution is not usually placed among the top environmental problems facing the nation; however, it is one of the more frequently encountered sources of pollution in everyday life. Recent scientific evidence shows that relatively continuous exposures to sound exceeding 70 dB can be harmful to hearing. Noise can also cause stress reactions which include (1) increases in heart rate, blood pressure, and blood cholesterol levels, and (2) negative effects on the digestive and respiratory systems. With persistent, unrelenting noise exposure, it is possible that these reactions will become chronic stress diseases such as high blood pressure or ulcers [4]. Additional details on noise pollution can be found in Chapter 44.

7.2.2 PLANTS

Pollutants, especially in the air, cover a wide spectrum of particulate and gaseous matter, damaging and effecting the growth of many types of vegetation [1]. Whether the particulate matter is harmful to vegetation depends upon the type of particulate matter predominating, upon the concentration of particulate matter versus time, the type of vegetation under consideration, climatic conditions, the duration of exposure, and similar factors [2].

Different types of plants are affected differently by pollutants. The three major types of plants are trees, vegetative plants (crops), and flowers. Only a few kinds of trees can live in the polluted air of a big city. Sycamores and Norway maples seem to resist air pollution best. That is why those trees are planted among most city streets. However, air pollution can kill even sycamores and Norway maples. The danger to the trees is greatest at street corners. That is where cars and buses may have to stop and wait for traffic lights to change. While they are waiting, exhaust pours out of their tailpipes, resulting in tree-kills. Pine trees do not resist air pollution as well as sycamores and Norway maples. Air pollutants, even in small amounts, are very harmful to pine trees. For example, the San Bernardino forest was a beautiful forest about 60 miles east of Los Angeles. Most of the trees in the forest were pines. Winds usually blow from west to east. The winds carried polluted air from the streets of Los Angeles to the San Bernardino forest and harmed the pine trees [6]. Crops and flowers cannot be planted within many miles of industry because they will not grow due to the pollution emitted from the factories.

The major contributors to plant pollution are sulfur dioxide, ethylene, acid deposition, smog, ozone, and fluoride.

Metallurgical smelting processes emit substantial quantities of "sulfur dioxide," and they are in general associated with a good degree of defoliation. Serious damage from sulfur dioxide is usually characterized by loss of chlorophyll and suppression of growth. Leaf and needle tissues are damaged, and die as the time of exposure increases. The attack starts at the edges, moving progressively toward the main body of the leaf or needle. A concentration as low as 2 pphm could suppress growth. Cereal crops, especially barley, are readily damaged at concentrations less than 50 pphm. The presence of soot particles in the air can increase the damage because sulfur dioxide and sulfuric acid mist are enriched at the surface of particles. It has been determined that pine trees cannot survive the damage when the mean annual concentrations of sulfur dioxide exceed 0.07–0.08 ppm [5].

"Ethylene" in the air causes injury to many flowers, whether they are orchids, lilacs, tulips, or roses. The first symptom of ethylene damage is the drying of the sepals, which are leaf-like formations located at the bottom of the flower bloom. This attack destroys the beauty of the flowers and contributes to extensive economic losses to growers. In addition, accidental escape of ethylene from a polyethylene plant caused 100% damage to cotton fields a mile away [5].

The process of "acid deposition" begins with emissions of SO_2 and NO_x. These pollutants interact with sunlight and water vapor in the upper atmosphere to form acidic compounds. When it rains (or snows) these compounds fall to the earth. Forests and agriculture may be vulnerable because acid deposition can leach nutrients from the ground, killing nitrogen-fixing microorganisms that nourish plants and release toxic metals.

"Smog" damage to vegetation is serious, especially in locations such as Los Angeles; the lower leaf surfaces of petunias and spinach become silvery or bronze in color. The most toxic substance of the Los Angeles air has been identified as PAN, formed by photochemical reactions of hydrocarbons and nitrogen oxides emanating mostly from automobile exhausts [5].

"Ozone" is a major component of the Los Angeles smog; it is phytotoxic at concentrations of 0.2 ppm, even when exposure time is only a few hours. Its effect on spinach is strong and destructive, causing whitening or bleaching of the leaves. Certain tobaccos are damaged by concentrations as small as 5–6 ppm. Ozone hinders plant growth even if bleaching or other distinctive marks are not found.

"Fluorides" are given off by factories that make aluminum, iron, and fertilizer. Due to these factories, growers have complained about the damage to fruits and leaves of peach, plum, apple, fig, and apricot trees. Fluorides also damage grapes, cherries, and citrus. It has been observed that the average yield of fruit per tree decreases 27% for every increase of 50 ppm of fluoride in the leaves [5].

7.2.3 ANIMALS

Animals are also affected by pollutants in the air. There are many similarities between the effects on humans and the effects on animals. For example, animals in zoos suffer the same effects of air pollution as humans. They also are beset with lung disease, cancer, and heart disease. Their babies have more birth defects than those of wild animals. Details of each pollutant will not be included in this section, except for the ones that differ from humans: fluorides and insecticides.

Air that is polluted with "fluorides" can be deadly for sheep, cows, and some other animals. However, inhaling the polluted air is not what causes the damage. Some plants that are eaten by the animals store up the fluoride that they have taken from the air, and after a while, contain a dangerous amount of fluoride. Animals become ill and even die after eating these plants [5].

The bald eagle is the U.S. national bird, but it is being killed mainly with "insecticides." The eagles are not killed by breathing the polluted air, but are dying because they cannot reproduce. When an insecticide is sprayed on plants, some of it misses the plants and gets into the air as a pollutant. Certain kinds of insecticides do not change into harmless substances; they are referred to as "persistent" insecticides because they remain harmful for years. Rain washes these insecticides out of the air and into the water. Small animals bioaccumulate the insecticides, and these animals are often eaten by larger animals, who also have absorbed the insecticide, thereby doubling their insecticide intake. Bald eagles eat large animals (fish), and they may store enough insecticide to kill them. Even if they do not die, the insecticide prevents the bird from reproducing. The eggs that the females lay either have very thin shells or no shells at all, causing the inability of baby eagles to hatch [6].

7.2.4 MATERIALS OF CONSTRUCTION

Air pollution has long been a significant source of economic loss in urban areas. Damage to nonliving materials may be exhibited in many ways, such as corrosion of metal, rubber cracking, soiling and eroding of building surfaces, deterioration of works of art, and fading of dyed materials and paints.

An example of the deterioration of works of art is Cleopatra's Needle standing in New York City's Central Park. It has deteriorated more in 80 years in the park than in 3000 years in Egypt. Another example is the Statue of Liberty located on Liberty Island in New York Harbor. When the statue arrived from France in 1884 it was copper, and 100 years later it has turned a greenish color. It was so deteriorated that the internal and external structures had to be renovated. More recently, the steady deterioration of the Acropolis in Athens, Greece is yet another example.

7.3 WATER POLLUTION

Pollution in waterways impairs or destroys aquatic life, threatens human health, and simply fouls the water such that recreational and aesthetic potential are lost. There are several different types of water pollution and there are several different ways in which water can be polluted. This section will focus on:

1. Drinking water and its sources (ground water and tap water)
2. Critical aquatic habitats (wetlands, near coastal waters, the Great Lakes, and oceans)
3. Surface water (municipal wastes, industrial discharge, and nonpoint sources)

7.3.1 Drinking Water

Half of all Americans and 95% of rural Americans use ground water for drinking water. Pollutants were found in the drinking water through testing water in different areas and at different times. Several public water supplies using ground water exceeded EPA's drinking water standards for inorganic substances (fluorides and nitrates). Major problems were reported from toxic organics in some wells in almost all states east of the Mississippi River. Trichloroethylene, a suspected carcinogen, was the most frequent contaminant found. The EPA's Ground Water Supply Survey showed that 20% of all public water supply wells and 29% in urban areas had detectable levels of at least one VOC. At least 13 organic chemicals that are confirmed animal or human carcinogens have been detected in drinking water wells.

The most severe and acute public health effects from contaminated drinking water from the tap, such as cholera and typhoid, have been eliminated in America. However, some less acute and immediate hazards still remain in the nation's tap water. Contaminants of special concern to the EPA are lead, radionuclides, microbiological contaminants, and disinfection byproducts. Each of these is discussed below.

"Lead" in drinking water is primarily due to the corrosion of plumbing materials. The health effects related to the ingestion of too much lead are very serious and can lead to impaired blood formation, brain damage, increased blood pressure, premature birth, low birth weight, and nervous system disorders. Young children are especially at high risk (see Chapter 29 for additional details).

"Radionuclides" are radioactive isotopes that emit radiation as they decay. The most significant radionuclides in drinking water are radium, uranium, and radon, all of which occur in nature. Ingestion of uranium and radium in drinking water can cause cancer of the bone and kidney. Radon can be ingested and inhaled. The main health risk due to inhalation is lung cancer.

"Microbiological contaminants" such as bacteria, viruses, and protozoa may be found in water. Although some organisms are harmless, others may cause disease. Microbiological contamination continues to be a national concern because contaminated drinking water systems can rapidly spread disease.

"Disinfection byproducts" are produced during water treatment by chemical reactions of disinfectants with naturally occurring or synthetic materials. These byproducts may pose health risks and these risks are related to long-term exposure to low levels of contaminants.

7.3.2 Critical Aquatic Habitats

"Wetlands" are the most productive of all ecosystems, but the United States is slowly losing them. There are many positive effects of wetlands: converting sunlight into plant material or biomass that serve as food for aquatic animals that form the base of the food chain, habitats for fish and wildlife, and spawning grounds; maintains and improves water quality in adjacent water bodies; removes nutrients to prevent eutrophication; filters harmful chemicals; traps suspended sediments; controls foods; prevents shoreline erosion with vegetation; and, contributes $20–$40 billion annually to the economy.

"Coastal waters" are home to many ecologically and commercially valuable species of fish, birds, and other wildlife. Coastal waters are susceptible to contamination because they act as sinks for the large quantities of pollution discharged from industry. The effects include toxic contamination, eutrophication, pathogen contamination, habitat loss and alteration, and changes in living resources. Coastal fisheries, wildlife, and bird populations have been declining, with fewer species being represented.

"The Great Lakes" are all being affected by toxics that are contaminating fish and the water. Lake Ontario and Lake Erie are also being affected by eutrophication.

"Oceans" are being polluted with sediments dredged from industrialized urban harbors that are often highly contaminated with heavy metals and toxic synthetic organic chemicals. The contaminants can be taken up by marine organisms. In addition, persistent disposal of plastics from land and sea has become a serious problem. The most severe effect of the debris floating in the ocean is injury and death of fish, marine animals, and birds. Debris on beaches can affect the public safety, the beauty of the beach, and the economy.

7.3.3 SURFACE WATERS

"Municipal wastewater and industrial discharges" produce nutrients in sewage that foster excessive growth of algae and other aquatic plants. Plants then die and decay, depleting the dissolved oxygen needed by fish. Wastewater that is poorly treated may contain chemicals harmful to human and aquatic life.

"Nonpoint source pollution" consists of sediment, nutrients, pesticides, and herbicides. Sediment causes decreased light transmission through water resulting in decreased plant reproduction, interference with feeding and mating patterns, decreased viability of aquatic life, decreased recreational and commercial values, and increased drinking water costs. Nutrients promote the premature aging of lakes and estuaries. Pesticides and herbicides hinder photosynthesis in aquatic plants, affect aquatic reproduction, increase organism susceptibility to environmental stress, accumulate in fish tissues, and present a human health hazard through fish and water consumption.

7.3.4 HUMANS

Humans are not affected similarly by the presence of water pollution as they may be by the presence of polluted air. Humans are affected by water pollution through consuming contaminated water or animals (fish). Due to contaminated drinking water, lakes, and oceans, humans are inflicted with diseases, impaired blood formation, brain damage, increased blood pressure, premature birth, low birth weight, nervous system disorders, and cancer (bone, kidney, and lung).

7.3.5 PLANTS

Plants are affected by wastewater, sewage, sediments, pesticides, and herbicides found mainly in surface water. Effects on plants in these areas are:

1. Decreased plant reproduction
2. Hindrance of photosynthesis in aquatic plants
3. Excessive growth of algae and other aquatic plants
4. Ultimate death of plants

7.3.6 ANIMALS

Animals, especially those that live in or near the water, are directly affected by water pollution. Chemical and solid waste disposal in the water can affect animals in many ways, varying from waste/pollutant accumulation to death. Animals such as fish, marine mammals, and birds can be injured or killed due to floating debris in the ocean. Contaminants can be taken up by marine organisms and accumulate there. The accumulations increase as the larger fish consume contaminated smaller fish. This cycle interferes with animals feeding and mating patterns, affects aquatic reproduction, and decreases the viability of aquatic life.

7.3.7 INTERNATIONAL EFFECTS

In Japan, contamination of seawater with organic mercury became concentrated in fish, and produced a severe human neurologic disorder called Minamata disease. The epidemic occurred in the mid-1950s. Almost 10 years passed before it was realized that there was an accompanying epidemic of congenital cerebral palsy due to a transplacental effect, e.g., pregnant women who ate contaminated fish gave birth to infants who were severely impaired neurologically [7].

7.4 LAND POLLUTION

Land has been used as dumping grounds for wastes. Improper handling, storage, and disposal of chemicals can cause serious problems. Several types of wastes that are placed in the land are:

1. Industrial hazardous wastes
2. Municipal wastes
3. Mining wastes
4. Radioactive wastes
5. Leakage from underground storage tanks

7.4.1 HUMANS

Potential health effects in humans range from headaches, nausea, and rashes to acid burns, serious impairment of kidney and liver functions, cancer, and genetic damage. Underground storage tank leaks may contaminate local drinking water systems, or may lead to explosions and fires causing harm and injury to the people in the vicinity (see Chapter 26 for more details).

7.4.2 Plants

Trees are usually not planted around landfills, and if they were, they would have difficulty growing due to the contaminated soil in the vicinity of the landfill. Vegetative plants also have difficulty growing around landfills. This is due to the fact that the hazardous wastes from industry are usually dumped in the landfills (see Section 7.2). Flowers also do not normally grow near landfills for similar reasons.

7.4.3 Animals

Animals are essentially affected in the same ways as humans. They may experience the effects of drinking contaminated water, and suffer from acid burns, kidney, liver and genetic damage, and cancer. Interestingly, at the close of his administration, President Bush chose to exempt factory farms from regulation. Additional details are available at: http://us.oneworld.net/article/359152-us-exempts-factory-farms-from-regulation.

7.5 FUTURE TRENDS

Pollution prevention has recently become a major environmental concern to everyone, everywhere, and appears to be an issue that will take priority in the future. Since the enactment of the Pollution Prevention Act of 1990, there has been a clear breakthrough in the nation's understanding of environmental problems. The EPA's Pollution Prevention Strategy (see Chapter 30) establishes the EPA's future direction in pollution prevention. The strategy indicates how pollution prevention concepts will be incorporated into the EPA's ongoing environmental protection efforts. The EPA is calling pollution prevention a "national objective" [4]. As more people become aware of the dangerous effects of pollutants on themselves, plants, animals, and materials of construction, they will be more conscious of the ways that they may contribute to air, water, and land pollution.

7.6 SUMMARY

1. Pollutants are various noxious chemicals and refuse materials that impair the purity of the air, water, and land. Humans and animals are affected in the same way by air pollution. Some sources of pollution affecting them are sulfur dioxide, carbon monoxide, nitrogen oxides, ozone, carcinogens, fluorides, aeroallergens, radon, cigarette smoke, asbestos, and noise. Those affecting only animals are insecticides. Plants are also affected by air pollutants such as sulfur dioxide, ethylene, acid deposition, smog, ozone, and fluoride.
2. Pollution in waterways impairs or destroys aquatic life, threatens human health, and fouls the water such that recreational and aesthetic potential are lost.
3. Land has been used as dumping grounds for wastes. Improper handling, storage, and disposal of chemicals can cause serious problems.

4. Pollution prevention has recently become a major environmental management issue to everyone and everywhere, and appears to be an issue that will take priority in the future to help reduce the effects of pollutants.

REFERENCES

1. Stern, A. *Air Pollution: The Effects of Air Pollution Vol. II*, Academic Press, New York, 1977.
2. Parker, H. *Air Pollution*. National Research Council, Washington, DC, 1977.
3. U.S. EPA. *Meeting the Environmental Challenge*. EPA's Review of Progress and New Directions in Environmental Protection, Washington, DC, December 1991.
4. Burke, G., Singh, B., and Theodore, L. *Handbook of Environmental Management and Technology*, John Wiley & Sons, Hoboken, NJ, 2000.
5. Shaheen, E. *Environmental Pollution: Awareness and Control*, Washington, DC, 1974.
6. Blaustein, E., Blaustein, R., and Greenleaf, J. *Your Environment and You: Understanding the Pollution Problem*, Dobbs Ferry, New York, 1974.
7. Zoeteman, B. *Aquatic Pollutants and Biological Effects: With Emphasis on Neoplasia*, New York Academy of Sciences, New York, 1977.

8 Green Chemistry and Green Engineering

Contributing Author: Vincenza Imperiale

CONTENTS

8.1 INTRODUCTION

Activities in the field of green engineering and green chemistry are increasing at a near exponential rate. For example, prior to the preparation of this chapter, EPA Region 2 hosted a conference on September 27, 2007, in New York City entitled, *Seize the Moment: Opportunities for Green Chemistry and Green Engineering in the Pharmaceutical Industry.* The purpose of this workshop was to discuss opportunities to encourage environmental stewardship by way of "greening" the fields of chemistry and engineering, particularly in the pharmaceutical industry. Various sessions, held throughout the morning and into the afternoon, discussed different arenas of "greening" in the industry. One particular session examined the environmental footprint of manufacturing processes. Although related to the pharmaceutical industry, it provided valuable information and challenges in this emerging area. As noted, the overall theme of the conference was to present various means to help achieve sustainability in the industry [1].

This chapter aims to familiarize the reader with both green chemistry and green engineering by defining and giving principles to each; future trends are also discussed. Before beginning this chapter, it is important that the term "green" should not be considered a new method or type of chemistry or engineering. Rather, it should be incorporated into the way scientists and engineers design for categories that include the environment, manufacturability, disassembly, recycle, serviceability, and compliance.

8.2 GREEN CHEMISTRY

Green chemistry, also called "clean chemistry," refers to that field of chemistry dealing with the synthesis, processing, and use of chemicals that reduce risks to humans
and the environment [2]. It is defined as the invention, design, and application of
chemical products and processes to reduce or to eliminate the use and generation of
hazardous substances [3]. Anastas offers these comments [4]:

1. Looking at the definition of green chemistry, one sees the concept of "invention" and "design." By requiring that the impacts of chemical products and
   chemical processes are included as design criteria, performance criteria
   are inextricably linked to hazard considerations in the definition of green
   chemistry.
2. Another part of the definition of green chemistry is found in the phrase
   "use and generation." Green chemistry includes all substances that are part
   of the process, rather than focusing only on those undesirable substances
   that might be inadvertently produced in a process. Therefore, green chemistry is a tool for minimizing the negative impact of those procedures
   aimed at optimizing efficiency, although clearly both impact minimization and process optimization are legitimate and balancing objectives of
   the subject.
3. Green chemistry, however, is also the recognition of significant consequences
   to the use of hazardous substances that span from regulatory, handling, and
   transport, to liability issues, to mention a few. Limiting the definition to deal
   with waste only would be addressing part of the problem.
4. The second to last term in the definition of green chemistry is the term "hazardous." Anastas notes that green chemistry is a way of dealing with risk
   reduction and pollution prevention by addressing the intrinsic hazards of the
   substances rather than those circumstances and conditions of their use that
   might increase their risk. Why is it important for green chemistry to adopt
   a hazard-based approach? To understand this, one must visit the concept of
   risk (see Chapter 37 for more details). Risk, in its most fundamental terms,
   is the product of hazard and exposure, as shown below.

$$\text{Risk} = (\text{Hazard})(\text{Exposure}) \qquad (8.1)$$

Virtually all general approaches to risk reduction center on reducing exposure to
hazardous substances.

Whether it is due to regulatory decree or to a desire to decrease environmental
management costs or to be perceived by the public as being more environmentally conscious, many industries are exploring the uses of green chemistry. Bishop
offers the following, "Green chemistry involves a detailed study of the by-products
from the synthesis and the effects these by-products have. Green chemistry concepts can also be used to evaluate the inputs to a synthesis pathway and determine
whether it is possible to reduce the use of endangered resources by switching to
more plentiful or renewable ones." [5] Thus, industrial chemists can no longer

concern themselves only with the chemical they are producing. They must also be mindful of [5]:

1. Hazardous wastes that will be generated during product synthesis.
2. Toxic substances that will need to be handled by the workers making the product.
3. Regulatory compliance issues to be followed in making the product.
4. Liability concerns arising from the manufacture of this product.
5. Waste treatment costs that will be incurred.
6. Alternative product synthesis pathways or processes that may be available.

The last point above will be focused on when outlining the Principles of Green Chemistry (see next paragraph). These principles provide a framework for scientists to use when designing new materials, products, processes, and systems. Why are the principles so important? Firstly, the principles focus one's thinking in terms of sustainable design criteria, and secondly, they have proven time and again to be the source of innovative solutions to a wide range of problems. Systematic integration of these principles is crucial to achieving genuine sustainability for the simultaneous benefit of the environment, economy, and society [6].

A baker's dozen Principles of Green Chemistry are provided below [3].

1. *Prevention*—It is better to prevent waste than to treat or clean up waste after it has been generated.
2. *Atom economy*—Synthetic methods should be designed to maximize the incorporation of all materials used in the process through to the final product.
3. *Less hazardous chemical syntheses*—Whenever practicable, synthetic methods should be designed to use and generate substances that possess little or no toxicity to human health and the environment.
4. *Designing safer chemicals*—Chemical methods should be designed to preserve efficacy of function while minimizing toxicity.
5. *Safer solvents and auxiliaries*—The use of auxiliary substances (e.g., solvents, separation agents, etc.) should be made unnecessary whenever possible and, innocuous when being used.
6. *Design for energy efficiency*—Energy requirements should be recognized for their environmental and economic impacts should be minimized. Synthetic methods should be conducted at ambient temperature and pressure whenever possible.
7. *Use of renewable feedstocks*—A raw material or feedstock should be renewable rather than depleting, wherever and whenever technically and economically practicable.
8. *Reduce derivatives*—*Unnecessary derivatization* (blocking group, temporary modification of physical/chemical processes) should be avoided whenever possible because such steps require additional reagents and can generate waste.

9. *Catalysis*—Catalytic reagents (that should be as selective, or discriminating, as possible) are superior to stoichiometric reagents.

10. *Biocatalysis*—Enzymes and antibodies that are used to mediate reactions.

11. *Design for degradation*—Chemical products should be designed in a way that at the end of their function they break down into innocuous degradation products and do not persist in the environment.

12. *Real-time analysis for pollution prevention*—Analytical methods need to be further developed to allow for real-time, in process monitoring and control prior to the formation of hazardous substances.

13. *Inherently safer chemistry for accident prevention*—Substances and the form of a substance used in a chemical process should be chosen so as to minimize the potential for chemical accidents, including releases, explosions, and fires.

8.2.1 GREEN CHEMISTRY RESEARCH NEEDS

While much has been accomplished in recent years to design products and chemical processes that are more environmentally sound, enough has not been done. The Council for Chemical Research has put together a list of the most needy research areas [7]. The list includes the following:

1. Replace chromium in corrosion protection, which will require development of new redox chemistry.

2. Recycle rubber more effectively, which will require new ways to reverse cross-linking and vulcanization.

3. Replace traditional acid and base catalysts in bulk processes by (perhaps) using new zeolites.

4. Develop new water-based synthesis and processing methods to minimize use of volatile organic solvents.

5. Develop new catalytic processes, based on light or catalytic antibodies, to replace traditional heavy metal catalysts.

6. Devise better chelates to separate and recycle heavy metal catalysts.

Computer assistance will become a requirement due to the complexity of the chemistry involved in developing more benign alternative synthetic pathways. Computer programs are now being made available which have the potential for proposing alternative reaction pathways that may subsequently be evaluated for their relative risk and economic viability [5].

8.3 GREEN ENGINEERING

Green engineering is similar to green chemistry in many respects, as witnessed by the underlying urgency of attention to the environment seen in both sets of the principles. According to the U.S. Environmental Protection Agency (EPA) [8]:

Green engineering is the design, commercialization, and use of processes and products which are feasible and economical while minimizing the

1. Generation of pollution at the source, and
2. Risk to human health and the environment.

Green engineering embraces the concept that decisions to protect human health and the environment can have the greatest impact and cost effectiveness when applied in the very beginning or early in the design and development phase of a process or product.

Therefore, green engineering also supports incremental improvements in materials, machine efficiencies, and energy use which can often be implemented more quickly than novel design approaches [9].

A baker's dozen Principles of Green Engineering are provided below [10].

1. *Benign rather than hazardous*—Designers need to strive to ensure that all material and energy inputs and outputs are as inherently nonhazardous as possible.
2. *Prevention instead of treatment by recycle/reuse*—It is better to prevent waste than to treat or clean up waste after it is generated.
3. *Design for separation*—Separation and purification operations should be a component of the design framework.
4. *Maximize efficiency*—System components should be designed to maximize mass, energy, and temporal efficiency.
5. *Output-pulled versus input-pushed*—Components, processes, and systems should be output-pulled rather than input-pushed through the use of energy and materials.
6. *Conserve complexity*—Energy conservation must also consider entropy (see Chapter 34 for more details). Embedded entropy and complexity must be viewed as an investment when making design choices on recycle, reuse, or beneficial disposition.
7. *Durability rather than immortality*—Targeted durability, not immortality, should be a design goal.
8. *Meet need, minimize excess*—Design for unnecessary capacity or capability should be considered a design flaw; this includes engineering "one size fits all" solutions.
9. *Minimize material diversity*—Multicomponent products should strive for material unification to promote disassembly and value retention (minimize material diversity).
10. *Integrate material and energy flows*—Design of processes and systems must include integration of interconnectivity with available energy and materials flows.
11. *Design for "afterlife"*—Performance metrics include designing for performance in (commercial) afterlife.

12. *Renewable rather than depleting*—Design should be based on renewable and readily available inputs throughout the life cycle.
13. *Engaging communities*—Actively engage communities and stakeholders in development of engineering solutions.

8.4   GREEN CHEMISTRY VERSUS GREEN ENGINEERING

What is the difference between green engineering and green chemistry?

From the definitions given previously one would conclude that green engineering is concerned with the design, commercialization, and use of all types of processes and products, whereas green chemistry covers just a very small subset of this— the design of chemical processes and products. Therefore, green chemistry may be viewed as a subset of green engineering. It is, in fact, a very broad field, encompassing everything from improving energy efficiency in manufacturing processes to developing plastics from renewable resources.

One important aspect in this area is the development of mathematically based tools that aid in decision making when faced with alternatives. Another is the discovery and development of new technology that makes the design, commercialization, and use of processes and products that reduce or eliminate pollution possible. In particular, one major focus of both green chemistry and green engineering is developing alternatives to the volatile organic solvents used so pervasively in chemical and manufacturing processes which was also addressed at the aforementioned 2007 EPA Conference. Solvents comprised 66% of all industrial emissions in 1997 in the United States [11]. The EPA Office of Pollution Prevention and Toxics reported, however, that there has been some progress from 1998 to 2002, including 91% decrease in stack air releases, 88% decrease in fugitive air releases, and 79% decrease in water releases. They also reported that 50% of greenhouse gases are from solvents [1].

Efforts to address this pressing need of developing alternative solvents for synthesis, separation, and processing are being studied. For example, supercritical carbon dioxide (CO_2) can be used to replace the copious amounts of organic and aqueous solvents used in the microelectronic industries. One new supercritical fluid technology utilizes CO_2 (bought from waste) as the best solvent for chromatography. Also, CO_2 presents unique technical advantages in device fabrication. Other studies demonstrate how strong mineral acids can be eliminated by choosing a solvent (either hot water or a CO_2-expanded liquid) where the acid catalyst can be produced reversibly in situ. In all these cases, the new solvent system presents some real technological advantage over conventional systems, instead of just solvent substitution. Hence, green chemistry and green engineering represent slightly different shades of a seamless continuum that ranges from discovery through design and decision making all the way to commercialization and use of products and processes that prevent pollution [12].

Even a handful of society's cinema heroes and heroines have committed themselves to green activities. Whether it's ranging from sporting hybrid automobiles, through public campaigning and announcing, or adjusting their lifestyles to be more eco-friendly by installing home solar panels, these entertainment personalities are setting a broader public awareness of green chemistry and engineering benefit the environment. Table 8.1 is a listing of some of the most noted green actors and actresses.

TABLE 8.1
Green Actors and Actresses

1. Leonardo DiCaprio	4. Cate Blanchett	7. Daryl Hannah
2. Cameron Diaz	5. George Clooney	8. Amitabh Bachchan
3. Robert Reford	6. Edward Norton	9. Julia Louis-Dreyfus

8.5 SUSTAINABILITY AT THE DOMESTIC LEVEL (ADAPTED FROM [13])

Many in society have grown accustomed to "reusing" and "recycling" glass, plastic, paper, etc. Both reuse and recycling have come to mean different things to different people. For purposes of this chapter, reuse, loosely defined, is the recovery and distribution of discarded, yet perfectly usable materials that provides an excellent environmental and economical alternative to exportation and landfilling. Recycling utilizes additional time, money, energy, resources, and an extensive organizational effort to extract, sort, and redistribute a discarded item's raw materials. Reuse preserves these resources, including the value of the materials, labor, technology, and energy incorporated into the manufacturing process. There are numerous "green" options available to the individual, many of them imposed by business. Six of the options in the reuse/recycle category are provided below.

1. Repair and overhaul—American businesses are employing reuse at the domestic level in several ways: most extensively through remanufacturing. Remanufacturing involves the collection of valuable parts which are refurbished in a factory and set to meet the same specifications as new products. Examples of this include the collection of "one-use" cameras or toner cartridges, which the company then reloads, repackages, and resells.
2. Deposit refund—Another method of reuse is a deposit refund scheme in which a company offers the consumer a financial incentive to return packaging for reuse, e.g., glass bottle and aluminum can collection are the most common applications of this method.
3. Cradle to cradle—By this reuse concept, the entire life cycle of a product is considered and becomes an intrinsic part of the product's design process, and is thus an area of intense interest among forward-thinking manufacturers. According to this sector, the mindset of the Industrial Revolution, with its reliance on a seemingly never-ending abundance of resources, must be replaced. In its stead, cradle to cradle applications encourage product and packaging makers to manufacture designs and employ processes which mimic the natural processes of growth, use, and decay. These associated "closed loop" schemes are not typically visible to the average consumer, but are increasingly utilized in American businesses. *Note*: A "closed loop system" is one in which the manufacturer or retailer provides packaging that is returnable and/or reusable, but does not address the waste product generated (if any). Two examples of this resource recovery system with respect to the packaging industry are returnable plastic grocery containers and a dry cleaner's wire hangers.

4. Other methods—Refillable packaging and an environmental tax are two other practices employed by businesses aiming to reduce resource consumption. Refilling an empty package at a discounted price from a store's discounted bulk supply encourages consumers to purchase one reusable item instead of several disposable items, thereby allowing a company to reduce some of the transportation and packaging costs associated with that item. Conversely, an environmental tax or surcharge imposed by a regulatory agency on a manufacturer (and ultimately, the consumer) offsets the negative impact of the product and/or encourages manufacturers to reduce associated pollution. Also referred to as "sin tax," it is applied most often to alcohol and cigarettes.

5. Regiving—There is the more familiar "regiving" option. Regiving runs the gamut from the simple exchange of outgrown or unused items among friends or family, to patronizing antique or secondhand stores, to donating to charity or posting items on Web sites (such as eBay, Freecycle, or craigslist). According to charity industry sources, the average American throws away 67.9 lb of used clothing and rags every year, which translates into an annual total of 20 billion lb of used clothing and textiles that are tossed into landfills.

6. Waste exchange—"One man's trash is another man's treasure" is among the many adages being revived with fresh significance. While still fairly limited in practice, waste exchange uses waste product from one process as the raw material for another. This practice allows businesses to avoid the environmental costs of waste disposal while obtaining new raw material, thus keeping the waste out of the landfill and environmental treatment facilities. Waste exchange has come into existence approximately 30 years ago.

8.6   ADDITIONAL RESOURCES

Internet Sources
U.S. EPA Office of Pollution and Prevention and Toxics
Green Chemistry Program Web site
http://www.epa.gov.oppt/greenchemistry

U.S. EPA Office of Pollution and Prevention and Toxics
Green Engineering Program Web site
http://www.epa.gov.oppt/greenengineering

U.S. EPA Office of Pollution and Prevention and Toxics
Exposure Assessment Tools and Models Web site
http://www.epa.gov.oppt/exposure

U.S. EPA Office of Pollution and Prevention and Toxics
Design for the Environment (DfE)
http://www.epa.gov./dfe

U.S. EPA Terminology Reference System (TRS)
http://www.epa.gov/trs/index/htm
American Institute of Chemical Engineers (AIChE)
http://www.aiche.org

National Institute of Occupational Safety & Health (NIOSH) Pocket Guide to Chemical Hazards
http://www.cdc.gov/niosh

8.7 FUTURE TRENDS

Chemists and engineers have the unique ability to affect the design of molecules, materials, products, processes, and systems at the earliest possible stages of their development. With much of the research occurring now in these two fields, the reality is that chemists and engineers must ask themselves the following questions [14]:

1. What will be the human health and the environmental impacts of the chemicals put into the marketplace?
2. How efficiently will the systems be which manufacture products?
3. What will tomorrow's innovations look like, and from what materials will they be created?

Three problem areas stand out [15]:

1. Inventing technology to support the expanded availability and use of renewable energy
2. Developing renewable feedstocks and products based on them
3. Creating technology that does not produce pollution

Some very pivotal steps that must be taken in the near future must include implementing greatly improved technologies for harnessing the fossil and nuclear fuels in order to ensure that their use, if continued, creates much lower environmental and social impact; developing and deploying the renewable energy sources on a much wider scale; and, making major improvements in the efficiency of energy conversion, distribution, and use [16].

Green chemistry and green engineering are emerging issues which come under the larger multifaceted spectrum of sustainable development. Sustainable development represents a change in consumption patterns toward environmentally more benign products, and a change in investment patterns toward augmenting environmental capital [17]. In this respect, sustainable development is feasible. It requires a shift in the balance of the way economic progress is pursued. Environmental concerns must also be properly integrated into economic policy from the highest (macroeconomic) level to the most detailed (microeconomic) level. The environment must be seen as a valuable, frequently essential input to human well-being. The field of green chemistry and engineering is rising to solve problems that are of great significance to the future of humanity.

8.8 SUMMARY

1. Green chemistry is the invention, design, and application of chemical products and processes to reduce or to eliminate the use and generation of hazardous substances.

2. Green engineering is the design, commercialization, and use of processes and products which are feasible and economical while minimizing the generation of pollution at the source, and risk to human health and the environment.
3. The Baker's Dozen Principles of green chemistry and green engineering help people achieve what is laid out by their definitions in (1) and (2).
4. There are profound benefits associated with such green fields that include inherently safer processes and production steps to people and the environment, cost effectiveness, reduced liability, and enhanced public image.
5. Green chemistry may be viewed as a subset to green engineering in the sense that green chemistry is primarily focused on the design of chemical processes and products, whereas green engineering covers a much broader field.

REFERENCES

1. EPA Region 2 Conference. *Seize the Moment: Opportunities for Green Chemistry and Green Engineering in the Pharmaceutical Industry*. New York, September 27, 2007.
2. Anastas, P.T. and Williamson, T.C. Green chemistry: An overview. In *Green Chemistry: Designing Chemistry for the Environment*, eds. P.T. Anastas and T.C. Williamson, ACS Symposium Series 626, American Chemical Society, Washington, DC, 1996, pp. 1–17.
3. Anastas, P.T. and Warner, J.C. *Green Chemistry: Theory and Practice*, Oxford University Press, New York, 1998.
4. Anastas, P.T., Black, D.StC., Breen, J., Collins, T., Memoli, S., Miyamoto, J., Polyakoff, M., Tumas, W., and Tundo, P. Synthetic pathways and processes in green chemistry. Introductory overview. *Pure and Applied Chemistry*, 72 (7), 1207–1228, 2000.
5. Bishop, P.L. *Pollution Prevention*, Waveland Press, Inc., Prospect Heights, IL, 2000, 357 p.
6. www.greenchemistryinstitute.org, 2007
7. Hancock, D.G. and Cavanaugh, M.A. Environmentally benin chemical synthesis and processing for the economy and the environment. In *Benign by Design*, eds. P.T. Anastas and C.A. Farris, ACS Symposium Series 577, American Chemical Society, Washington, DC, 1994, pp. 23–30.
8. http://www.epa.gov/oppt/greenengineering/pubs/whats_ge.html
9. http://www.eng.vt.edu/ green/Program.php
10. Anastas, P.T. and Zimmerman, J. Design through the twelve principles of green engineering, *Environmental Science and Technology*, 37, 94A–101A, 2003.
11. Allen, D.T. and Shonnard, D.R. *Green Engineering: Environmentally Conscious Design of Chemical Processes*, Prentice-Hall, Englewood Cliffs, NJ, 2002.
12. Brennecke, J.F. Department of Chemical and Biomolecular Engineering, University of Notre Dame, Notre Dame, IN. http://www.rsc.org/delivery/ArticleLinking/Display HTMLArticleforfree.cfm?JournalCode=GC&Year=2004&ManuscriptID=b411954c& Iss=8
13. http://greenlivingideas.com/reuse/the-green-basics-of-reuse-philosophy.html
14. http://portal.acs.org/portal/acs/corg/content?_nfpb=true&_pageLabel=PP_SUPER ARTICLE&node_id=1415&use_sec=false&sec_url_var=region1
15. http://www.chem.cmu.edu/groups/Collins/ethics/ethics06.html
16. Boyle, G., Everett, B., and Ramage, J. *Energy Systems and Sustainability*, Oxford University Press, Oxford, U.K., 2003.
17. Pearce, D.W., Markandya, A., and Barbier, E.B. *Blueprint for a Green Economy*, Earthscan, London, U.K., 1989.

9 Sustainability

CONTENTS

9.1 INTRODUCTION

The term "sustainability" has many different meanings to different people. To sustain is defined as to "support without collapse." Discussion of how sustainability should be defined was initiated by the Bruntland Commission. This group was assigned a mission to create a "global agenda for change" by the General Assembly of the United Nations in 1984. They defined sustainable very broadly [1]: Humanity has the ability to make development sustainable—to ensure that it meets the needs of the present without compromising the ability of future generations to meet their own needs [2].

"Sustainability" involves simultaneous progress in four major areas: human, economic, technological, and environmental. The United Nations [2] defined sustainable development as

> Development that meets the need of the present without compromising the ability of future generations to meet their own needs.

Sustainability requires conservation of resources, minimizing depletion of non-renewable resources, and using sustainable practices for managing renewable resources. There can be no product development or economic activity of any kind without available resources. Except for solar energy, the supply of resources is finite. Efficient designs conserve resources while also reducing impacts caused by material extraction and related activities. Depletion of nonrenewable resources and overuse of otherwise renewable resources limits their availability to future generations.

Another principal element of sustainability is the maintenance of the ecosystem structure and function. Because the health of human populations is connected to the health of the natural world, the issue of ecosystem health is a fundamental concern to sustainable development. Thus, sustainability requires that the health of all diverse species as well as their interrelated ecological functions be maintained. As only one species in a complex web of ecological interactions, humans cannot separate their survivability from that of the total system.

9.2 HISTORICAL PERSPECTIVE

To develop an understanding of why sustainability is a topic of urgency today, one should understand the history behind it. As agriculture developed, social structure supporting agriculture grew as well. Social stratification became increasingly widespread as humanity proceeded from agriculture to industry. Eventually, a new class-based society led to differences in standards of living between the rich and the poor. As population grew, technical development spiraled up as well. The increase in demand for goods and more powerful machines led to increased extraction of natural resources at the expense of the environment. Environmental effects build up slowly, gaining momentum as the problem worsened. Due to the populations' uncertainty and limited understanding when a problem is identified, it is often so bad that even an immediate response may not be able to solve it. Examples of such lag and momentum have been exhibited by damage to the ozone layer and global warming [1]. This is discussed in more detail in Chapter 12.

As noted earlier, activity in the sustainability area was born with the World Commission on Environment and Development (WCED). It was formally known as the Brundtland (named after its Chair Gro Harlem Brundtland), and was convened by the United Nations in 1983. The commission was created to address growing concern "about the accelerating deterioration of the human environment and natural resources and the consequences of that deterioration for economic and social development." In establishing the commission, the UN General Assembly recognized that environmental problems were global in nature. It was determined that it was in the common interest of all nations to establish policies for sustainable development [2].

Later, the United Nations Conference on Environment and Development, also known as the Earth Summit, was held in Rio de Janeiro in June 1992. A total of 178 governments participated, with 118 sending their heads of state or government [3]. Some 2,400 representatives of nongovernmental organizations (NGOs) attended, with 17,000 people at the parallel NGO Forum, who had the so-called Consultative Status were also present. One of the issues addressed, which deals with carbon dioxide related global warming, was alternative sources of energy to replace the use of fossil fuels which are linked to global climate change. An important achievement was an agreement on the Climate Change Convention, which in turn led to the Kyoto Protocol. The Earth Summit resulted in the following documents: Rio Declaration on Environment and Development; Agenda 21; Convention on Biological Diversity; Forest Principles; and, Framework Convention on Climate Change [3]. The trends in Sustainable Development Report, published by the U.N. Department of Economic and Social Affairs, highlighted key developments and recent trends in the areas of energy for sustainable development, industrial development, atmosphere/air pollution, and a host of other related topics.

9.3   RESOURCE LIMITATIONS

Most have defined the Earth as consisting of four parts:

1. Atmosphere
2. Lithosphere
3. Hydrosphere
4. Barysphere

The atmosphere is the gaseous envelope that surrounds the solid body of the planet. The lithosphere is the solid rocky crust of the earth, extending to a depth of perhaps 40 km (25 miles). The hydrosphere is the layer of water, in the form of the oceans, covers approximately 70% of the surface of the earth. The barysphere, sometimes called the centrosphere, is below the lithosphere. It is the heavy interior of the earth constituting more then 99.6% of the Earth's mass.

From a sustainability prospective, the two major resources available to humans are (2), and to a lesser degree (3). The two resources are finite and for all intents and purpose are nonrenewable. Both are briefly discussed below.

The rocks of the lithosphere primarily consist of 11 elements, which together account for about 99.5% of its mass. The most abundant is oxygen (about 46.60% of the total), followed by silicon (about 27.72%), aluminum (8.13%), iron (5.0%), calcium (3.63%), sodium (2.83%), potassium (2.59%), magnesium (2.09%), and titanium, hydrogen, and phosphorus (totaling less than 1%). In addition, 11 other elements are present in trace amounts of 0.1%–0.02%. These elements, in order of abundance, are carbon, manganese, sulfur, barium, chlorine, chromium, fluorine, zirconium, nickel, strontium, and vanadium. The elements are present in the lithosphere almost entirely in the form of compounds rather than in their free state. The most common compounds of the earth's crust are silicates and aluminosilcates of the various metals. In addition, the surface of the earth is largely covered with sedimentary rocks and soil.

The hydrosphere consists chiefly of the oceans, but technically includes all water surfaces in the world, including inland seas, lakes, rivers, and underground waters.

Traditionally, humans have viewed Earth's resources as a source of economic wealth—minerals, food, forests, and land on which to place buildings and other structures. These were looked upon as assets to be exploited, not necessarily as precious attributes to be used sustainably and preserved insofar as possible. The loss of these resources would be catastrophic. For example, the loss of Earth's food productivity would certainly adversely affect sustainability and, in the worst case, could lead to massive starvation of human populations. Although a number of human activities have adversely affected food productivity, these effects have been largely masked by remarkable advances in agriculture, including increased use of fertilizer, development of highly productive hybrid crops, and widespread irrigation.

In addition to food, humans obtain shelter, health, security, mobility, and other necessities through activities involving resources that are carried out by individuals, businesses, and government entities. By their very nature, these utilize resources (renewable and nonrenewable) and all tend to produce wastes. A number of minerals and metals are important resources. There are so many of these that a discussion is beyond the scope of this text.

The "energy resource" is a topic within itself. Consider fossil fuels. One of the greatest challenges facing humanity during the twenty-first century will surely be that of providing everyone on the planet access to safe, clean, and sustainable energy supplies. The use of energy has been central to the functioning and development of human societies throughout history. However, in recent years fossil fuel energy usage has run amuck. World petroleum resources are presently strained as prices for petroleum reached painfully high levels. (The price of crude oil had exceeded $100 a barrel at the time of the preparation of this manuscript.) Natural gas and crude oil supplies have been extended. Furthermore, the International Energy Agency projected that more than 80% of the world energy demand will continue to be met by fossil fuels in 2030. Therefore, there is an immediate need to increase the present efficiency of fossil fuel usage. This can include

1. Increasing the mileage efficiency of transportation sources
2. Improving the energy efficiency of new power plants
3. Developing "green buildings" and sustainable communities

As noted earlier, natural resources were initially abundant relative to needs. In the earlier years of the industrial revolution, production was limited by technology and labor. However, population is in surplus and technology has reduced the need for human labor. Increasingly, production is becoming limited by the Earth's natural environment that includes the availability of natural resources. The demand for most resources has increased at a near exponential rate. The emergence of newly developing economies, particularly those in the highly populated countries of China and India, has further increased the demand for resources. Humans need to realize that reduced material demand, particularly those from nonrenewable sources, is essential to sustainability. There are some elaborate changes in place to reduce material demand and the potential exists for much greater reductions. Naturally, wherever possible, materials should come from renewable sources and materials should be recyclable insofar as possible (see Part V for more details).

9.4   SUSTAINABLE DEVELOPMENT CONSIDERATIONS

Sustainable development demands change. Consumption of energy, natural resources, and products must eliminate waste. The manufacturing industry can develop green products that can meet the sustainability requirements. Life cycles analysis (see Part VI), design for environment and toxic use reduction are elements that help sustainability. Sustainable manufacturing, for example, extends the responsibility of industry into material selection, facility and process design, marketing, cost accounting, and waste disposal. Extending the life of a manufactured product is likely to minimize waste generation. Design engineers must consider many aspects of the product including its durability, reliability, remanufacturability, and adaptability.

Designing a product that can withstand wear, stress, and degradation extends its useful life. This, in many cases, reduces the cost and impact on the environment. Reliability is the ability of a product or system to perform its function for the length of an expected period under the intended environment. Reducing the number of components in a system and simplifying the design can enhance the reliability. Screening out potentially unreliable parts and replacing with more reliable parts helps to increase the system reliability.

Adaptable designs rely on interchangeable parts. For example, consumers can upgrade components as needed to maintain state-of-the-art performance. In remanufacturing, used worn products are restored to "like-new" condition. Thus, remanufacturing minimizes the generation of waste. Products that are expensive, but not subject to rapid change, are the best candidates for remanufacturing. Design continuity between models in the same product line increases interchangeable parts. The parts must be designed for easy disassembly to encourage remanufacturing.

Design of products that emphasizes efficient use of energy and materials reuse and recycling reduces waste and supports sustainability. By effective recycling, material life can be extended. Materials can be recycled through open-loop or closed-loop pathways. For example, postconsumer material is recycled in an open loop one or more times before disposal. However, in a closed-loop pathway, such as with solvents, materials within a process are recovered and used as substitutes for virgin material. Minimizing the use of virgin materials supports sustainability. Thus, resource conservation can reduce waste and directly lower environmental impact. Manufacturing a less material-intensive product not only saves materials and energy but will also be lighter, thus reducing energy and costs related to product transportation. Process modifications and alterations specifically focused on replacing toxic materials with more benign ones minimize the health risk and the environmental impact and safety of employees. Process redesign may also yield "zero discharge" by completely eliminating waste discharges. Thus, sustainability can be accomplished through several different approaches. Evaluating these options up-front will aid in developing truly sustainable processes and products, and is much more desirable than implementing control measures after unacceptable waste releases occur.

Finally, responsible businesses can begin moving toward sustainability by taking six steps:

1. Foster a company culture of sustainability.
2. Initiate voluntary performance improvements.
3. Apply eco-efficiency (material and energy conservation, toxic use reduction, recycling, etc.) concepts.
4. Grasp opportunities for sustainable business growth.
5. Invest in creativity, innovation, and technology for the future.
6. Reward employee commitment and action.

9.5 SUSTAINABLE DESIGN CONSIDERATIONS

Current design practices for suitability projects usually fall into the category of state of the art and pure empiricism. Past experience with similar applications is commonly used as the sole basis for the design procedure. In designing a new process, files are consulted for similar applications and old designs are heavily relied on.

By contrast, the engineering profession in general, and the chemical engineering profession in particular, has developed well-defined procedures for the design, construction, and operation of chemical plants. These techniques, tested and refined for better than a half-century, are routinely used by today's engineers. These same procedures should be used in the design of sustainable "facilities."

Regarding sustainability projects, a process engineer is usually involved in one of two activities: building/designing the plant/project or deciding whether to do so. The skills required in both cases are quite similar, but the money, time, and detail involved are not as great in the latter situation. It has been estimated that only 1 out of 15 proposed new processes ever achieves the implemented stage. Thus, project knowledge at the preliminary stage is vital to prevent financial loss on one hand and provide opportunity for success on the other. In well-managed process organizations, the engineer's evaluation is a critical activity that usually involves considerable preliminary research on the proposed process. Successful process development consists of a series of actions and decisions, the most significant of which takes place well before projected implementation.

It is important to determine whether a sustainability project has promise as early in thee development stage as possible. In the chemical process industry, there may be an extended period of preparatory work required if the proposed process is a unique or first-time application. This can be involved in bench-scale work by chemists to develop and better understand the process chemistry and the impact of implementing suitability principles. This is often followed by pilot experimentation by process and/or development engineers to obtain scale-up and equipment performance information. However, these two steps are usually not required in the design of an established system. This many not be the situation with most sustainability projects so some bench-scale or pilot work may be necessary and deemed appropriate by management.

Without the tools to completely document sustainability benefits, these opportunities have often been difficult to support when competing against traditional projects or life cycle analysis (LCA). LCAs has developed over the past 20 years to provide decision makers with analytical tools that attempt to accurately and comprehensively account for the environmental consequences and benefits of competing projects, including those in the sustainability arena. LCA is a procedure to identify and evaluate "cradle-to-grave" natural resource requirements and environmental releases associated with processes, products, packaging, and services. LCA concepts can be particularly useful in ensuring that identified sustainable opportunities are not causing unwanted secondary impacts by shifting burdens to other places within the life cycle of a product or process. LCA is an evolving tool undergoing continued development. Nevertheless, LCA concepts can be useful in gaining a broader understanding of the true environmental effects of current practices and of any proposed project [4].

LCA is a tool to evaluate all environmental effects of a product of process throughout its entire life cycle. This includes identifying and quantifying energy and materials used and wastes released to the environment, assessing their environmental impact, and evaluating opportunities for improvement [4]. Addition details are provided in Part V.

9.6 ECONOMIC FACTORS

Corporations are recognizing the benefits of sustainability activities. Sustainability openly allows companies to reduce the cost of doing business, create consistency, improve public image, and to be recognized on a national level as environmental leaders. However, before the cost of a project can be evaluated, the factors contributing to the cost must be recognized.

There are two major contributing factors: capital costs and operating costs; these are discussed in Chapter 47. Once the total cost of the project/process has been estimated, the engineer must determine whether or not it will be profitable. This involves converting all cost contributions to an annualized basis, a method that is also discussed in Chapter 47; if more than one project proposal is under study, this method provides a basis for comparing alternate proposals and for choosing the best proposal. In addition, a brief description of a perturbation analysis for project optimization is presented; other consideration, including regulatory compliance, reduction in liability, enhanced public image, etc. (as noted above), should also be included in the analysis [5].

Finally, corporate sustainability strategies can be grouped into four different approaches, each with different levels of financial risk and potential rewards. Table 9.1 describes each of these four approaches.

TABLE 9.1
Corporate Sustainability Strategies and Financial Impacts

	Sustainable Development Strategies			
Financial Impacts	**Franchise Protection**	**Process Changes**	**Product Changes**	**New Market Development**
Business value	Right to operate	Cost and liability reduction reputation	Customer loyalty reputation	New markets
Focus	Compliance	Efficiency	Value chain	Innovation
Main financial impact	Reduces earnings and risks and can open new markets	Increases margins and reduces risks, and often increases capital efficiency	Increases competitive advantage	Increases revenues, competitive advantage, and diversification

Sources: Data from Metzger, B. and Salmond, D., Managing for sustainability, *EM*, Air & Waste Management Association, Pittsburgh, PA, June 2004; and Reed, D., *Stalking the Elusive Business Case for Corporate Sustainability*, World Resources Institute, Washington, DC, 2001.

In general design practice, there are usually five levels of sophistication for financial evaluating and estimating. Each is discussed in the following list.

1. The first level requires little more than identification of products, raw materials, and utilities. This is what is known as an "order of magnitude estimate" and is often made by extrapolating or interpolating from data on similar existing processes. The evaluation can be done quickly and at minimum cost but with a probable error exceeding ±50%.

2. The next level of sophistication is called a "study estimate" and requires a preliminary process flow sheet (to be discussed in the next section) and a first attempt at identification of equipment, utilities, materials of construction, and other processing units. Estimation accuracy improves to within ±30% probable error, but more time is required and the cost of the evaluation can escalate to over $30,000 for a $5 million plant. Evaluation at this level usually precedes expenditures for site selection, market evaluation, pilot plant work, and detailed equipment design. If a positive evaluation results, pilot plan and other activities may also begin.

3. A "scope or budget authorization," the next level of economic evaluation, requires a more defined process definition, detailed process flow sheets, and prefinal equipment design. The information required is usually obtained from pilot plant, marketing, and other studies. The scope authorization estimate could cost upward of $80,000 for a $5 million project/process with a probable error exceeding ±20%.

4. If the evaluation is positive at this stage, a "project control estimate" is then prepared. Final flow sheets, site analyses, equipment specifications, and architectural and engineering sketches are employed to prepare this estimate. The accuracy of this estimate is about ±10% probable error. A project control estimate can serve as the basis for a corporate appropriation, for judging contractor bids, and for determining construction expenses. Due to increased intricacy and precision, the cost for preparing such an estimate for the process can approach $150,000.

5. The final economic analysis is called a "firm or contractor's estimate." It is based on detailed specifications and actual equipment bids. It is employed by the contractor to establish a project cost and has the highest level of accuracy, ±5% probable error. A cost of preparation results from engineering, drafting, support, and management/labor expenses. Because of unforeseen contingencies, inflation, and changing political and economic trends, it is impossible to assure actual costs for even the most precise estimates.

For sustainability projects, data on similar existing systems are normally available and economic estimates or process feasibility are determined from these data. It should be pointed out again that most processes in real practice are designed by duplicating or "mimicking" similar existing systems. Simple algebraic correlations that are based on past experience are the rule rather than the exception.

9.7 BENCHMARKING SUSTAINABILITY [8]

Recently a variety of sustainability indices have been published that mostly measure a companies' corporate responsibility and environmental performance. Starting in 2001, the American Institute of Chemical Engineers (AIChE) decided to strike out on a new strategic direction and a number of new initiatives were begun. These new areas included biotechnology, materials technology, and sustainable development; the AIChE ultimately formed the Institute for Sustainability (IfS) in 2004 to promote the societal, economic, and environmental benefits of sustainable and green engineering. IfS serves the needs—and influences the efforts of—professionals in industry, academia, and government. Scientists and engineers working with IfS have defined sustainability as the "path of continuous improvement, wherein the products and services required by society and delivered with progressively less negative impacts upon the Earth."

IfS established an industry group, Center for Sustainable Technology Practices (CSTP), to address practical issues of sustainability implementation with the member companies, including BASF, Dow, Cytec, Honeywell, DuPont, Air Products, FMC, and Shell. One area of focus for CSTP is the development of a Sustainability Roadmap, which is designed to improve decision making relative to sustainability.

The AIChE SI is composed of seven critical elements:

1. Strategic commitment to sustainability
2. Safety performance
3. Environmental performance
4. Social responsibility
5. Product stewardship
6. Innovation
7. Value-chain management

Details are provided in Table 9.2.

9.8 RESOURCES FOR SUSTAINABILITY

Ten key resources for sustainability are:

1. National Institute of Standards and Technology's (NIST) Building for Environmental and Economic Sustainability (BEES) Lifecycle Tool
   www.bfrl.nist.gov/oae/software/bees.html
2. U.S. Environmental Protection Agency's (EPA) Tool for the Reduction and Assessment of Chemical and Other Environmental Impacts (TRACI) Tool
   www.epa.gov/ORD/NRMRL/std/sab/traci
3. The Ecology of Commerce: A Declaration of Sustainability
   Paul Hawken
   HarperBusiness, 1994

TABLE 9.2

Examples of Indicator Areas

Strategic Commitment to Sustainability
Stated commitment
Presence and extent of sustainability goals

Safety Performance
Process safety
Employee safety

Environmental Performance
Resource use
Waste and emissions (including greenhouse gases)
Compliance history

Social Responsibility
Community investment
Stakeholder partnership and engagement

Product Stewardship
Product safety and environmental assurance process
System in place for compliance to emerging regulations (e.g., research)

Innovation
R&D in place to address societal needs (e.g., Millennium Development Goals)
Integration of sustainability concepts and tools in R&D
New products related to sustainability

Value Chain Management
Environmental management systems
Supplier standards and management process

4. Industrial Ecology: An Introduction
   University of Michigan's National Pollution Prevention Center for Higher
   Education
   www.umich.edu/~nppcpub/resources/compendia/ind.ecol.html
5. "Industrial Ecology and 'Getting the Prices Right'"
   Resources for the Future
   www.rff.org/resources_archive/1998.htm
6. *Journal of Industrial Ecology*
   The MIT Press
   mitpress.mit.edu/JIE
7. Mid-Course Correction: Toward a Sustainable Enterprise: The Interface
   Model
   Ray Anderson
   Chelsea Green Publishing Company

8. Natural Capitalism: Creating the Next Industrial Revolution
   Paul Hawken, Amory Lovins, and L. Hunter Lovins
   Rocky Mountain Institute, 1999
   www.naturalcapitalism.org
9. The Next Bottom Line: Making Sustainable Development Tangible
   World Resources Institute
   www.igc.org/wri/meb/sei/nbl.html
10. "The NEXT Industrial Revolution"
   The Atlantic Monthly
   October 1998
   www.theatlantic.com/issues/98oct/industry.htm

9.9   FUTURE TRENDS

Over the next 50 years, projections suggest that the world's population could increase by 50%. Global economic activity is expected to increase by 500%. Concurrently, global energy consumption and manufacturing activity are likely to rise to three times current levels. These trends could have serious social, economic, and environmental consequences unless a way can be found to use fewer resources in a more efficient way. The task ahead is to help shape a sustainable future in a cost-effective manner, recognizing that economic and environmental considerations, supported by innovative science and technology, can work together and promote societal benefits. However, unless humans embrace sustainability, they will ultimately deplete Earth's resources and damage its environment to an extent that conditions for human existence on the planet will be seriously compromised or even become impossible.

As stated above, sustainable development is feasible. Sustainable development means a change in consumption patterns toward environmentally more benign products, and a change in investment patterns. It will require a shift in the balance of the way economic progress is pursued. Environmental concerns must be properly integrated into rearrangement policies and the environment must be viewed as an integral part of human well-being.

Finally, some very pivotal steps that must be taken in the near future must include implementing greatly improved technologies for harnessing the fossil and nuclear fuels in order to ensure that their use, if continued, creates much lower environmental and social impact; developing and deploying the renewable energy sources on a much wider scale; and, making major improvements in the efficiency of energy conversion, distribution and use.

9.10   SUMMARY

1. The term "sustainability" has many different meanings to different people. To sustain is defined as to "support without collapse."
2. Humanity has the ability to make development sustainable—to ensure that it meets the needs of the present without compromising the ability of future generations to meet their own needs.

3. Activity in the sustainability area was born with the WCED.

4. Traditionally, humans have viewed Earth's resources as a source of economic wealth—minerals, food, forests, and land on which to place buildings and other anthrospheric structures.

5. In recent years fossil fuel energy use has run amuck. World petroleum resources are presently strained as prices for petroleum have reached painfully high levels.

6. Humans need to realize that reduced material demand, particularly those from nonrenewable sources, is essential to sustainability.

7. LCA concepts can be particularly useful in ensuring that identified sustainable opportunities are not causing unwanted secondary impacts by shifting burdens to other places within the life cycle of a product or process.

8. Benchmarking sustainability indices have been published that provide a measure of a company's responsibility and environmental performance.

9. Sustainable development is feasible. It will require a shift in the balance of the way economic progress is pursued.

REFERENCES

1. Bishop, P.L. *Pollution Prevention*, Waveland Press, Inc., Prospect Heights, IL, 2000.

2. United Nations. Report of the World Commission on Environment and Development. General Assembly Resolution 42/187, December 11, 1987. Retrieved October 31, 2007.

3. Schneider, K. White House Snubs U.S. Envoy's Plea to sign Rio Treaty, *New York Times*, June 5, 1992; Brooke, J. U.N. Chief Closes Summit with an Appeal for Action, *New York Times*, June 15, 1992.

4. Adopted from: Dupont, R., Ganesan, K., and Theodore, L. *Pollution Prevention*, CRC/Lewis Publishers, Boca Ration, FL, 2000.

5. Santoleri, J., Reynolds, J., and Theodore, L. *Introduction to Hazard Waste Incineration*, 2nd edition, John Wiley & Sons, Hoboken, NJ, 2000.

6. Metzger, B. and Salmoned, D. *Managing for Sustainability*, EM, Air & Waste Management Association, Pittsburgh, PA, 2004.

7. Reed, D. *Stalking the Elusive Business Case for Corporate Sustainability*, World Resources Institute, Washington, DC, 2001.

8. Adapted from: Cobbetal, C. *Benchmarking Sustainability*, CEP, New York, June 2007.

Part II

Air

Part II of this book serves as an introduction to air pollution. Air pollution control equipment is described for both gaseous and particulate pollutants in Chapter 10. Chapter 11 is concerned with atmospheric dispersion modeling, i.e., how pollutants are dispersed in the atmosphere.* A comprehensive examination of indoor air quality is provided in Chapter 14. Part II concludes with Chapter 15, with one of the new "hot" topics—vapor intrusion.

"Clean" air, which is found in few (if any) places on earth, is composed of nitrogen (78.1%), oxygen (20.9%), argon (0.9%), and other components (0.1%). The other components include carbon dioxide (330 parts per million by volume, or ppmv), neon (18 ppmv), helium (5 ppmv), methane (1.5 ppmv), and very small amounts (less than 1.0 ppmv) of other gases. Air often also carries water droplets, ice crystals, and dust, but they are not considered part of the composition of the air. Air exhibits the properties of a fluid, flowing to fill corners, holes, nooks, and crannies. On earth, air is essentially everywhere except in places where it has been intentionally pumped out to create a partial vacuum. Because air is invisible, it is easy to forget that it occupies space.

Air also has mass. The aforementioned can also include tiny solid particles and water droplets. Each of these tiny molecules, particles, and droplets has weight. The combined weight of all of them is quite significant; the earth's atmosphere has been estimated to weigh over 5,000,000,000,000,000 tons (4.5×10^{18} kg).

---

* Greenhouse effect and global warming receives extensive headline in Chapter 12 while Chapter 13 treats air toxics.

10 Air Pollution Control Equipment[*]

CONTENTS

10.1 INTRODUCTION

In solving an air pollution control equipment problem an engineer must first carefully evaluate the system or process in order to select the most appropriate type(s) of collector(s). After making preliminary equipment selection, suitable vendors can be contacted for help in arriving at a final answer. An early and complete definition of the problem can help to reduce a poor decision that can lead to wasted pilot trials or costly inadequate installations.

---

[*] This chapter is a condensed, revised, and updated version of material first appearing in the 1981 USEPA Training Manuals titled *Air Pollution Control Equipment of Particulates* and *Control Equipment for Gaseous Pollutants*, and the 1993 Theodore tutorial titled *Air Pollution Control Equipment* by L. Theodore and R. Allen.

Selecting an air pollution control device for cleaning a process gas stream can be a challenge. Some engineers, after trying to find shortcuts, employ quick estimates for both gas flow and collection efficiency that may be the entire extent of the collector specification. The end result can be an ineffective installation that has to be replaced. Treating a gas stream, especially to control pollution, is usually not a moneymaker, but costs—both capital and operating (see Chapter 47)—can be minimized, not by buying the cheapest collector but by thoroughly engineering the whole system as is normally done in process design areas.

Controlling the emission of pollutants from industrial and domestic sources is important in protecting the quality of air. Air pollutants can exist in the form of particulate matter or gases. Air-cleaning devices have been reducing pollutant emissions from various sources for many years. Originally, air-cleaning equipment was used only if the contaminant was highly toxic or had some recovery value. Now with recent legislation, control technologies have been upgraded and more sources are regulated in order to meet the National Ambient Air Quality Standards (NAAQS). In addition, state and local air pollution agencies have adopted regulations that are in some cases more stringent than the federal emission standards.

Equipment used to control particulate emissions are gravity settlers (often referred to as settling chambers), mechanical collectors (cyclones), electrostatic precipitators (ESPs), scrubbers (venturi scrubbers), and fabric filters (baghouses). Techniques used to control gaseous emissions are absorption, adsorption, combustion, and condensation. The applicability of a given technique depends on the physical and chemical properties of the pollutant and the exhaust stream. More than one technique may be capable of controlling emissions from a given source. For example, vapors generated from loading gasoline into tank trucks at large bulk terminals are controlled by using any of the above four gaseous control techniques. Most often, however, one control technique is used more frequently that others for a given source–pollutant combination. For example, absorption is commonly used to remove sulfur dioxide (SO_2) from boiler flue gas.

The material presented in this chapter regarding air pollution control equipment contains, at best, an overview of each control device. Equipment diagrams and figures, operation and maintenance procedures, and so on, have not been included in this development. More details, including predictive and design calculational procedures [1] are available in the literature.

10.2   AIR POLLUTION CONTROL EQUIPMENT FOR PARTICULATES

As described above, the five major types of particulate air pollution control equipment are:

1. Gravity settlers
2. Cyclones
3. Electrostatic precipitators
4. Venturi scrubbers
5. Baghouses

Each of these devices is briefly described below [2].

10.2.1 GRAVITY SETTLERS

Gravity settlers, or gravity settling chambers, have long been utilized industrially for the removal of solid and liquid waste materials from gaseous streams. Advantages accounting for their use are simple construction, low initial cost and maintenance, low pressure losses, and simple disposal of waste materials. Gravity settlers are usually constructed in the form of a long, horizontal parallelepiped with suitable inlet and outlet ports. In its simplest form, the settler is an enlargement (large box) in the duct carrying the particle-laden gases: the contaminated gas stream enters at one end, while the cleaned gas exits from the other end. The particles settle toward the collection surface at the bottom of the unit with a velocity at or near their settling velocity. One advantage of this device is that the external force leading to separation is provided free by nature. Its use in industry is generally limited to the removal of large particles, i.e., those larger than 40 microns (or micrometers).

10.2.2 CYCLONES

Centrifugal separators, commonly referred to as cyclones, are widely used in industry for the removal of solid and liquid particles (or particulates) from gas streams. Typical applications are found in mining and metallurgical operations, the cement and plastics industries, pulp and paper mill operations, chemical and pharmaceutical processes, petroleum production (cat-cracking cyclones), and combustion operations (fly ash collection).

Particulates suspended in a moving gas stream possess inertia and momentum and are acted upon by gravity. Should the gas stream be forced to change direction, these properties can be utilized to promote centrifugal forces to act on the particles. In a conventional unit, the entire mass of the gas stream with the entrained particles enter the unit tangentially and is forced into a constrained vortex in the cylindrical portion of the cyclone. Upon entering the unit, a particle develops an angular velocity. Because of its greater inertia, it tends to move across the gas streamlines in a tangential rather than rotary direction; thus, it attains a net outward radial velocity. By virtue of its rotation with the carrier gas around the axis of the tube (main vortex) and its high density with respect to the gas, the entrained particles are forced toward the wall of the unit. Eventually the particle may reach the outer wall, where they are carried by gravity and assisted by the downward movement of the outer vortex and/or secondary eddies toward the dust collector at the bottom of the unit. The flow vortex is reversed in the lower (conical) portion of the unit, leaving most of the entrained particles behind. The cleaned gas then passes up through the center of the unit (inner vortex) and out of the collector.

Multiple-cyclone collectors (multicones) are high efficiency devices that consist of a number of small-diameter cyclones operating in parallel with a common gas inlet and outlet. The flow pattern differs from a conventional cyclone in that instead of bringing the gas in at the side to initiate the swirling action, the gas is brought in at the top of the collecting tube and the swirling action is then imparted by a stationary vane positioned in the path of the incoming gas. The diameters of the collecting tubes usually range from 6 to 24 inches. Properly designed units can be constructed and operated with a collection efficiency as high as 90% for particulates in the 5–10 micron range. The most serious problems encountered with these systems involve plugging and flow equalization.

10.2.3 ELECTROSTATIC PRECIPITATORS

ESPs are satisfactory devices for removing small particles from moving gas streams at high collection efficiencies. They have been used almost universally in power plants for removing fly ash from the gases prior to discharge.

Two major types of high-voltage ESP configuration currently used are tubular and plate. Tubular precipitators consist of cylindrical collection tubes with discharge electrodes located along the axis of the cylinder. However, the vast majority of ESPs installed are the plate type. Particles are collected on a flat parallel collection surface spaced 8–12 inches apart, with a series of discharge electrodes located along the centerline of the adjacent plates. The gas to be cleaned passes horizontally between the plates (horizontal flow type) or vertically up through the plates (vertical flow type). Collected particles are usually removed by rapping.

Depending on the operating conditions and the required collection efficiency, the gas velocity in an industrial ESP is usually between 2.5 and 8.0 ft/s. A uniform gas distribution is of prime importance for precipitators, and it should be achieved with a minimum expenditure of pressure drop. This is not always easy, since gas velocities in the duct ahead of the precipitator may be 30–100 ft/s. It should be clear that the best operating condition for a precipitator will occur when the velocity distribution is uniform. When significant maldistribution occurs, the higher velocity in one collecting plate area will decrease efficiency more than a lower velocity at another plate area will increase the efficiency of that area.

The maximum voltage at which a given field can be maintained depends on the properties of the gas and the dust being collected. These parameters may vary from one point to another within the precipitator, as well as with time. In order to keep each section working at high efficiency, a high degree of sectionalization is recommended. This means that many separate power supplies and controls will produce better performance in a precipitator of a given size than if there were only one or two independently controlled sections. This is particularly true if high efficiencies are required.

10.2.4 VENTURI SCRUBBERS

Wet scrubbers have found widespread use in cleaning contaminated gas streams because of their ability to effectively remove both particulate and gaseous pollutants. Specifically, wet scrubbing involves a technique of bringing a contaminated gas stream into intimate contact with a liquid. Wet scrubbers include all the various types of gas absorption equipment (to be discussed later). The term "scrubber" will be restricted to those systems that utilize a liquid, usually water, to achieve or assist in the removal of particulate matter from a gas stream. The use of wet scrubbers to remove gaseous pollutants from contaminated streams is considered in the next section.

Another important design consideration for the venturi scrubber (as well as absorbers) is concerned with suppressing the steam plume. Water-scrubber systems removing pollutants from high-temperature processes (i.e., combustion) can generate a supersaturated water vapor that becomes a visible white plume as it leaves the stack.

Although not strictly an air pollution problem, such a plume may be objectionable for aesthetic reasons. Regardless, there are several ways to avoid or eliminate the steam plume. The most obvious way is to specify control equipment that does not use water in contact with the high-temperature gas stream, (i.e., ESP, cyclones or fabric filters). Should this not be possible or practical, a number of suppression methods are available:

1. Mixing with heated and relatively dry air
2. Condensation of moisture by direct contact with water, then mixing with heated ambient air
3. Condensation of moisture by direct contact with water, then reheating the scrubber exhaust gas

10.2.5 BAGHOUSES

The basic filtration process may be conducted in many different types of fabric filters in which the physical arrangement of hardware and the method of removing collected material from the filter media will vary. The essential differences may be related, in general, to

1. Mode of operation
2. Cleaning mechanism
3. Type of fabric
4. Equipment

Gases to be cleaned can be either pushed or pulled through the baghouse. In the pressure system (push through), the gases may enter through the cleanout hopper in the bottom or through the top of the bags. In the suction type (pull through), the dirty gases are forced through the inside of the bag and exit through the outside.

Baghouse collectors are available for either intermittent or continuous operation. Intermittent operation is employed where the operational schedule of the dust-generating source permits halting the gas cleaning function at periodic intervals (regularly defined by time or by pressure differential) for removal of collected material from the filter media (cleaning). Collectors of this type are primarily utilized for the control of small-volume operations such as grinding and polishing, and for aerosols of a very coarse nature. For most air pollution control installations and major particulate control problems, however, it is desirable to use collectors that allow for continuous operation. This is accomplished by arranging several filter areas in a parallel flow system and cleaning one area at a time according to some preset mode of operation.

Baghouses may also be characterized and identified according to the method used to remove collected material from the bags. Particle removal can be accomplished in a variety of ways, including shaking the bags, blowing a jet of air on the bags, or rapidly expanding the bags by a pulse of compressed air. In general, the various types of bag cleaning methods can be divided into those involving fabric flexing and those involving a reverse flow of clean air. In pressure-jet or pulse-jet cleaning, a momentary burst of compressed air is introduced through a tube or nozzle attached at the top of the bag. A bubble of air flows down the bag, causing the bag walls to collapse behind it.

A wide variety of woven and felted fabrics are used in fabric filters. Clean felted fabrics are more efficient dust collectors than are woven fabrics, but woven materials are capable of giving equal filtration efficiency after a dust layer accumulates on the surface. When a new woven fabric is placed in service, visible penetration of dust within the fabric may occur. This normally takes from a few hours to a few days for industrial applications, depending on the dust loadings and the nature of the particles.

Baghouses are constructed as single units or compartmental units. The single unit is generally used on small processes that are not in continuous operation, such as grinding and paint-spraying processes. Compartmental units consist of more than one baghouse compartment and are used in continuous operating processes with large exhaust volumes such as electric melt steel furnaces and industrial boilers. In both cases, the bags are housed in a shell made of rigid metal material.

10.3   AIR POLLUTION CONTROL EQUIPMENT FOR GASEOUS POLLUTANTS

As described in Section 10.1, the four generic types of gaseous control equipment include:

1. Absorbers
2. Adsorbers
3. Combustion units
4. Condensers

Each of these devices is briefly described below [2].

10.3.1 ABSORBERS

Absorption is a mass transfer operation in which a gas is dissolved in a liquid. A contaminant (pollutant exhaust stream) contacts a liquid and the contaminant diffuses (is transported) from the gas phase into the liquid phase. The absorption rate is enhanced by (1) high diffusion rates, (2) high solubility of the contaminant, (3) large liquid–gas contact area, and (4) good mixing between liquid and gas phases (turbulence).

The liquid most often used for absorption is water because it is inexpensive, is readily available, and can dissolve a number of contaminants. Reagents can be added to the absorbing water to increase the removal efficiency of the system. Certain reagents merely increase the solubility of the contaminant in the water. Other reagents chemically react with the contaminant after it is absorbed. In reactive scrubbing, the absorption rate is much higher, so in some cases a smaller, economical system can be used. However, the reactions can form precipitates that could cause plugging problems in the absorber or in associated equipment.

If a gaseous contaminant is very soluble, almost any of the wet scrubbers will adequately remove this contaminant. However, if the contaminant is of low solubility, the packed tower or the plate tower [1] is more effective. Both of these devices provide long contact time between phases and have relatively low pressure drops. The packed tower, the most common gas absorption device, consists of an empty shell filled with packing. The liquid flows down over the packing, exposing a large film area to the gas flowing up the packing. Plate towers consist of horizontal plates placed inside the tower. Gas passes up through the orifices in these plates while the liquid flows down across the plate, thereby providing desired contact.

10.3.2 ADSORBERS

Adsorption is a mass transfer process that involves removing a gaseous contaminant by adhering to the surface of a solid. Adsorption can be classified as physical or chemical. In physical adsorption, a gas molecule adheres to the surface of the solid due to an imbalance of natural forces (electron distribution). In chemisorption, once the gas molecule adheres to the surface, it reacts chemically with it. The major distinction is that physical adsorption is readily reversible whereas chemisorption is not.

All solids physically adsorb gases to some extent. Certain solids, called adsorbents, have a high attraction for specific gases; they also have a large surface area that provides a high capacity for gas capture. By far the most important adsorbent for air pollution control is activated carbon. Because of its unique surface properties, activated carbon will preferentially adsorb hydrocarbon vapors and odorous organic compounds from an airstream. Most other adsorbents (molecular sieves, silica gel, and activated aluminas) will preferentially adsorb water vapor, which may render them useless to remove other contaminants.

For activated carbon, the amount of hydrocarbon vapors that can be adsorbed depends on the physical and chemical characteristics of the vapors, their concentration in the gas stream, system temperature, system pressure, humidity of the gas stream, and the molecular weight of the vapor. Physical adsorption is a reversible process; the adsorbed vapors can be released (desorbed) by increasing the temperature, decreasing the pressure or using a combination of both. Vapors are normally desorbed by heating the adsorber with steam.

Adsorption can be a very useful removal technique, since it is capable of removing very small quantities (a few parts per million) of vapor from an airstream. The vapors are not destroyed; instead, they are stored on the adsorbent surface until they can be removed by desorption. The desorbed vapor stream is normally highly concentrated. It can be condensed and recycled, or burned in an ultimate disposal technique.

The most common adsorption system is the fixed bed adsorber. These systems consist of two or more adsorber beds operating on a timed adsorbing/desorbing cycle. One or more beds are adsorbing vapors, while the other bed(s) is being regenerated. If particulate matter or liquid droplets are present in the vapor-laden airstream, this stream is sent to pretreatment to remove them. If the temperature of the inlet vapor stream is high (much above 120°F), cooling may also be required. Since all adsorption processes are exothermic, cooling coils in the carbon bed itself may also be needed to prevent excessive heat buildup. Carbon bed depth is usually limited to a maximum of 4 ft, and the vapor velocity through the adsorber is held below 100 ft/ min to prevent an excessive pressure drop.

10.3.3 Combustion Units

Combustion is defined as a rapid, high-temperature gas-phase oxidation. Simply, the contaminant (a carbon–hydrogen substance) is burned with air and converted to carbon dioxide and water vapor. The operation of any combustion source is governed by the three T's of combustion: temperature, turbulence, and time. For complete combustion to occur, each contaminant molecule must come in contact (turbulence) with oxygen at a sufficient temperature, while being maintained at this temperature for an adequate time. These three variables are dependent on each other. For example, if a higher temperature is used, less mixing of the contaminant and combustion air or shorter residence time may be required. If adequate turbulence cannot be provided, a higher temperature or longer residence time may be employed for complete combustion.

Combustion devices can be categorized as flares, thermal incinerators, or catalytic incinerators. Flares are direct combustion devices used to dispose of small quantities or emergency releases of combustible gases. Flares are normally elevated (from 100 to 400 ft) to protect the surroundings from the heat and flames. Flares are often designed for steam injection at the flare tip. The steam provides sufficient turbulence to ensure complete combustion; this prevents smoking. Flares are also very noisy, which can cause problems for adjacent neighborhoods.

Thermal incinerators are also called afterburners, direct flame incinerators, or thermal oxidizers. These are devices in which the contaminant airstream passes around or through a burner and into a refractory-line residence chamber where oxidation occurs. To ensure complete combustion of the contaminant, thermal incinerators are designed to operate at a temperature of 700°C–800°C (1300°F–1500°F) and a residence time of 0.3–0.5 s. Ideally, as much fuel value as possible is supplied by the waste contaminant stream; this reduces the amount of auxiliary fuel needed to maintain the proper temperature.

In catalytic incineration the contaminant-laden stream is heated and passed through a catalyst bed that promotes the oxidation reaction at a lower temperature. Catalytic incinerators normally operate at 370°C–480°C (700°F–900°F). This reduced temperature represents a continuous fuel savings. However, this may be offset by the cost of the catalyst. The catalyst, which is usually platinum, is coated on a cheaper metal or ceramic support base. The support can be arranged to expose a high surface area which provides sufficient active sites on which the reaction(s) occur. Catalysts are subject to both physical and chemical deterioration. Halogens and sulfur-containing compounds act as catalyst suppressants and decrease the catalyst usefulness. Certain heavy metals such as mercury, arsenic, phosphorous, lead, and zinc are particularly poisonous.

10.3.4 CONDENSERS

Condensation is a process in which the volatile gases are removed from the contaminant stream and changed into a liquid. Condensation is usually achieved by reducing the temperature of a vapor mixture until the partial pressure of the condensable component equals its vapor pressure. Condensation requires low temperatures to liquify most pure contaminant vapors. Condensation is affected by the composition of the contaminant gas stream. The presence of additional gases that do not condense at the same conditions—such as air—hinders condensation.

Condensers are normally used in combination with primary control devices. Condensers can be located upstream of (before) an incinerator, adsorber, or absorber. These condensers reduce the volume of vapors that the more expensive equipment must handle. Therefore, the size and the cost of the primary control device can be reduced. Similarly, condensers can be used to remove water vapors from a process stream with a high moisture content upstream of a control system. A prime example is the use of condensers in rendering plants to remove moisture from the cooker exhaust gas. When used alone, refrigeration is required to achieve the low temperatures required for condensation. Refrigeration units are used to successfully control gasoline vapors at large gasoline dispensing terminals.

Condensers are classified as being either contact condensers or surface condensers. Contact condensers cool the vapor by spraying liquid directly on the vapor stream. These devices resemble a simple spray scrubber. Surface condensers are normally shell-and-tube heat exchangers. Coolant flows through the tubes, while vapor is passed over and condenses on the outside of the tubes. In general, contact condensers are more flexible, simpler, and less expensive than surface condensers. However, surface condensers require much less water and produce nearly 20 times less wastewater that must be treated than do contact condensers. Surface condensers also have an advantage in that they can directly recover valuable contaminant vapors.

10.4 HYBRID SYSTEMS

Hybrid systems are defined as those types of control devices that involve combinations of control mechanisms—for example, fabric filtration combined with electrostatic precipitation. Unfortunately, the term hybrid system has come to mean different things to different people. The two most prevalent definitions employed today for hybrid systems are

1. Two or more different air pollution control equipment connected in series, e.g., a baghouse followed by an absorber.
2. An air pollution control system that utilizes two or more collection mechanisms simultaneously to enhance pollution capture, e.g., an ionizing wet scrubber (IWS), that will be discussed shortly.

The two major hybrid systems found in practice today include IWSs and dry scrubbers. These are briefly described below.

10.4.1   Ionizing Wet Scrubbers

The IWS is a relatively new development in the technology of the removal of particulate matter from a gas stream. These devices have been incorporated in commercial incineration facilities [3,4]. In the IWS, high-voltage ionization in the charge section places a static electric charge on the particles in the gas stream, which then passes through a crossflow packed-bed scrubber. The packing is normally polypropylene: in the form of circular-wound spirals and gear-like wheel configurations, providing a large surface area. Particles with sizes of 3 microns or larger are trapped by inertial impaction within the bed. Smaller charged particles pass close to the surface of either the packing material or a scrubbing water droplet. An opposite charge on that surface is induced by the charged particle, which is then attracted to an ion attached to the surface. All collected particles are eventually washed out of the scrubber. The scrubbing water also can function to absorb gaseous pollutants.

According to Celicote (the IWS vendor), the collection efficiency of the two-stage IWS is greater than that of a baghouse or a conventional ESP for particles in the 0.2–0.6 micron range. For 0.8 micron and above, the ESP is as effective as the IWS [2]. Scrubbing water can include caustic soda or soda ash when needed for efficient adsorption of acid gases. Corrosion resistance of the IWS is achieved by fabricating its shell and most internal parts with fiberglass-reinforced plastic (FRP) and thermoplastic materials. Pressure drop through a single-stage IWS is approximately 5 in H_2O (primarily through the wet scrubber section). All internal areas of the ionizer section are periodically deluge-flushed with recycled liquid from the scrubber system.

10.4.2   Dry Scrubbers

The success of fabric filters in removing fine particles from flue gas streams has encouraged the use of combined dry-scrubbing/fabric filter systems for the dual purpose of removing both particulates and acid gases simultaneously. Dry scrubbers offer potential advantages over their wet counterparts, especially in the areas of energy savings and capital costs. Furthermore, the dry-scrubbing process design is relatively simple, and the product is a dry waste rather than a wet sludge.

There are two major types of so-called dry scrubber systems: spray drying and dry injection. The first process is often referred to as a wet–dry system. When compared to the conventional wet scrubber, it uses significantly less liquid. The second process has been referred to as a dry–dry system because no liquid scrubbing is involved. The spray-drying system is predominately used in utility and industrial applications.

The method of operation of the spray dryer is relatively simple, requiring only two major items: a spray dryer similar to those used in the chemical food-processing and mineral-preparation industries, and a baghouse or ESP to collect the fly ash and entrained solids. In the spray dryer, the sorbent solution, or slurry, is atomized into the incoming flue gas stream to increase the liquid–gas interface and to promote the mass transfer of the SO_2 (or other acid gases) from the gas to the slurry droplets where it is absorbed. Simultaneously, the thermal energy of the gas evaporates the water in the droplets to produce a dry powdered mixture of sulfite–sulfate and some unreacted alkali. Because the flue gas is not saturated and contains no liquid carryover, potentially troublesome mist eliminators are not required. After leaving the spray dryer, the solid-bearings gas passes through a fabric filter (or ESP), where the dry product is collected and where a percentage of unreacted alkali reacts with the SO_2 for further removal. The cleaned gas is then discharged through the fabric-filter plenum to an induced draft (ID) fan and to the stack.

Among the inherent advantages that the spray dryer enjoys over the wet scrubbers are:

1. Lower capital cost
2. Lower draft losses
3. Reduced auxiliary power
4. Reduced water consumption
5. Continuous, two-stage operation, from liquid feed to dry product

The sorbent of choice for most spray-dryer systems is a lime slurry.

Dry-injection processes generally involve pneumatic introduction of a dry, powdery alkaline material, usually a sodium-base sorbent, into the flue gas stream with subsequent fabric filter collection. The injection point in such processes can vary from the boiler-furnace area all the way to the flue gas entrance to the baghouse, depending on operating conditions and design criteria.

10.5 FACTORS IN CONTROL EQUIPMENT SELECTION

There are a number of factors to be considered prior to selecting a particular piece of air pollution control hardware [1]. In general, they can be grouped into three categories: environmental, engineering, and economic. These are detailed below.

Environmental

1. Equipment location
2. Available space
3. Ambient conditions
4. Availability of adequate utilities (i.e., power, water, etc.) and ancillary system facilities (i.e., waste treatment and disposal, etc.)
5. Maximum allowable emissions (air pollution regulations)
6. Aesthetic considerations (i.e., visible steam or water vapor plume, impact or scenic vistas, etc.)
7. Contribution of air pollution control system to wastewater and solid waste
8. Contribution of air pollution control system to plant noise levels

Engineering

1. Contaminant characteristics (i.e., physical and chemical properties, concentration, particulate shape and size distribution, etc.; in the case of particulates, chemical reactivity, corrosivity, abrasiveness, toxicity, etc.)
2. Gas stream characteristics (i.e., volume flow rate, temperature, pressure, humidity, composition, viscosity, density, reactivity, combustibility, corrosivity, toxicity, etc.)

3. Design and performance characteristics of the particular control system (i.e., size and weight, fractional efficiency curves, mass transfer and/or contaminant destruction capability, pressure drop, reliability and dependability, turndown capability, power requirements, utility requirements, temperature limitations, maintenance requirements, flexibility toward complying with more stringent air pollution regulations, etc.)

Economic

1. Capital cost (equipment, installation, engineering, etc.)
2. Operating cost (utilities, maintenance, etc.)
3. Expected equipment lifetime and salvage value

Proper selection of a particular system for a specific application can be extremely difficult and complicated. In view of the multitude of complex and often ambiguous pollution regulations, it is in the best interest of the prospective user to work closely with regulatory officials as early in the process as possible. Finally, previous experience on a similar application cannot be overemphasized.

10.6   COMPARING CONTROL EQUIPMENT ALTERNATIVES

The final choice in equipment selection is usually dictated by that equipment's capability of achieving compliance with the regulatory codes at the lowest uniform annual cost (amortized capital investment plus operation and maintenance costs). The reader is once again referred to Chapter 47 for details on the general subjects of economics. In order to compare specific control equipment alternatives, knowledge of the particular application and site is essential. A preliminary screening, however, may be performed by reviewing the advantages and disadvantages of each type of air pollution control equipment. For example, if water or a waste stream treatment is not available at the site, this may preclude use of a wet scrubber system and one would instead focus on particulate removal by dry systems, such as cyclones or baghouses and/or ESP. If auxiliary fuel is unavailable on a continuous basis, it may not be possible to combust organic pollutant vapors in an incineration system. If the particle-size distribution in the gas stream is relatively fine, cyclone collectors would probably not be considered. If the pollutant vapors can be reused in the process, control efforts may be directed to adsorption systems. There are many more situations where the knowledge of the capabilities of the various control options, combined with common sense will simplify the selection procedure.

General advantages and disadvantages of the most popular types of air pollution control equipment for gases and particulates are too detailed to present here but are available in literature [1,5].

10.7   FUTURE TRENDS

The basic design of air pollution control equipment has remained relatively unchanged since first used in the early part of the twentieth century. Some modest equipment changes and new types of devices have appeared in the last 20 years, but all have essentially employed the same capture mechanisms used in the past. One area that has recently received some attention is hybrid systems (see earlier section)—equipment that can in some cases operate at higher efficiency more economically than conventional devices. Tighter regulations and a greater concern for environmental control by society have placed increased emphasis on the development and application of these systems. The future will unquestionably see more activity in this area.

Recent advances in this field have been primarily involved in the treatment and control of metals. A dry scrubber followed by a wet scrubber has been employed in the United States to improve the collection of fine particulate metals in hazardous-waste incinerators; the dry scrubber captures metals that condense at the operating temperature of the unit and the wet scrubber captures residue metals (particularly mercury) and dioxin/furan compounds. Another recent application in Europe involves the injection of powdered activated carbon into a flue gas stream from a hazardous waste incinerator at a location between the spray dryer (the dry scrubber) and the baghouse (or ESP). The carbon, mixing with the lime particulates from the dry-scrubbing system and the gas stream itself, adsorb the mercury vapors and residual dioxin/furan compounds and are separated from the gas stream by a particulate control device. More widespread use of these types of systems is anticipated in the future.

10.8 SUMMARY

1. Controlling the emission of pollutants from industrial and domestic sources is important in protecting the quality of air. Air pollutants can exist in the form of particulate matter or as gases.
2. Equipment used to control particulate emissions are gravity settlers (often referred to as settling chambers), mechanical collectors (cyclones), ESPs, scrubbers (venturi scrubbers), and fabric filters (baghouses).
3. Techniques used to control gaseous emissions are absorption, adsorption, combustion, and condensation.
4. Hybrid systems are defined as those types of control devices that involve combinations of control mechanisms—for example, fabric filtration combined with electrostatic precipitation. Two of the major hybrid systems found in practice today include IWSs and dry scrubbers.
5. There are a number of factors to be considered prior to selecting a particular piece of air pollution control hardware. In general, they can be grouped into three categories: environmental, engineering, and economic.
6. The final choice in the equipment selection is usually dictated by that equipment capable of achieving compliance with regulatory codes at the lowest uniform annual cost (amortized capital investment plus operation and maintenance costs).
7. One area that has recently received some attention is hybrid systems, equipment that can in some cases operate at higher efficiency more economically than conventional devices.

REFERENCES

1. Reynolds, J.P., Jeris, J.S., and Theodore, L. *Handbook of Chemical and Environmental Engineering Calculations*, John Wiley & Sons, Hoboken, NJ, 2002.
2. Burke, G., Singh, B., and Theodore, L. *Handbook of Environmental Management and Technology*, 2nd Edition, John Wiley & Sons, Hoboken, NJ, 2000.
3. U.S. EPA. *Engineering Handbook for Hazardous Waste Incineration*, Monsanto Research Corporation, Dayton, OH, EPA Contract No. 68-03-3025, September 1982.
4. U.S. EPA. *Revised U.S. EPA Engineering Handbook for Hazardous Waste Incineration*, first draft.
5. Theodore, L. and Buonicore, A.J. *Industrial Air Pollution Control Equipment for Particulates*, CRC Press, Boca Raton, FL, 1992.

11 Atmospheric Dispersion Modeling[*]

CONTENTS

11.1 INTRODUCTION

Stacks discharging to the atmosphere have long been one of the methods available to industry for disposing waste gases. The concentration to which humans, plants, animals, and structures are exposed at ground level can be reduced significantly by emitting the waste gases from a process at great heights. This permits the gaseous pollutants to be dispersed over a much larger area and will be referred to as control by dilution. Although tall stacks may be effective in lowering the ground-level concentration of pollutants, they still do not in themselves reduce the amount of pollutants released into the atmosphere. However, in certain situations, tall stacks can be the most practical and economical way of dealing with an air pollution problem.

Atmospheric contamination arises primarily from the exhausts generated by industrial plants, power plants, refuse disposal plants, domestic activities, commercial heating, and transportation. These pollutants—which are in the form of particulates, smog, odors, and others—arise mostly from combustion processes and also contain varying amounts of undesirable gases such as oxides of sulfur, oxides of

---

[*] This chapter is a condensed, revised, and updated version of the chapter "Design of stacks" appearing in the 1975 CRC Press text titled *Industrial Control Equipment for Gaseous Pollutants, Vol. II,* by L. Theodore and A.J. Buoniocore, and the chapter "Atmospheric dispersion" from the 1994 Lewis Publishers text titled, *Handbook of Air Pollution Control Technology,* by J. Mycock, J. McKenna, and L. Theodore (chapter contributing authors: R. Lucas and A. Tseng).

nitrogen, hydrocarbons, and carbon monoxide. The expanding needs of society for more energy and advanced transportation technology, coupled with the rapid growth of urban areas, have led to ever-increasing amounts and concentrations of pollutants in the atmosphere.

Just as a river or stream is able to absorb a certain amount of pollution without the production of undesirable conditions, the atmosphere can also absorb a certain amount of contamination without "bad" effects. The self-purification of a discharge stream is primarily the result of biological action and dilution. Dilution of air contaminants in the atmosphere is also of prime importance in the prevention of undesirable levels of pollution. In addition to dilution, several self-purification mechanisms are at work in the atmosphere, such as sedimentation of particulate matter, washing action of precipitation, photochemical reactions, and absorption by vegetation and soil.

This chapter focuses on some of the practical considerations of the dispersion of pollutants in the atmosphere. Both continuous and instantaneous discharges are of concern to individuals involved with environmental management. However, the bulk of the material here has been presented for continuous emissions from point sources—for example, a stack. This has traditionally been an area of significant concern in the air pollution field because stacks have long been one of the more common industrial methods of disposing waste gases.

11.2 NATURE OF DISPERSION

The release of pollutants into the atmosphere is a traditional technique for disposing of them. Although gaseous emissions may be controlled by various sorption processes (or by combustion) and particulates (either solid or aerosol) by mechanical collection, filtration, electrostatic precipitators, or wet scrubbers, the effluent from the control device must still be dispersed into the atmosphere. Fortunately, one of the important properties of the atmosphere is its ability to disperse such streams of pollutants. Of course, the atmosphere's ability to disperse such streams is not infinite and varies from quite good to quite poor, depending on the local meteorological and geographical conditions. Therefore, the ability to model atmospheric dispersion and to predict pollutant concentrations from a source are important parts of air pollution engineering.

A continuous stream of pollutants released into a steady wind in an open atmosphere will first rise (usually), then bend over and travel with the mean wind, which will dilute the pollutants and carry them away from the source. This plume of pollutants will also spread out or disperse both in the horizontal and vertical directions from its centerline [1]. In doing so, the concentration of the gaseous pollutant is now contained within a larger volume. This natural process of high concentration spreading out to lower concentration is the process of dispersion. Atmospheric dispersion is primarily accomplished by the wind movement of pollutants, but the character of the source of pollution requires that this action of the wind be taken into account in different ways.

The dilution of air contaminants is also a direct result of atmospheric turbulence and molecular diffusion. However, the rate of turbulent mixing is so many thousand times greater than the rate of molecular diffusion that the latter effect can be neglected in the atmospheric dispersion analysis. Atmospheric turbulence and, hence, atmospheric diffusion vary widely with the weather conditions and topography. Taking all these factors into account, a four-step procedure is recommended for performing dispersion health effect studies:

1. Estimate the rate, duration and location of the release into the environment.
2. Select the best available model to perform the calculations.
3. Perform the calculations and generate downwind concentrations resulting from the source emission(s).
4. Determine what effect, if any, the resulting discharge has on the environment, including humans, animals, vegetation, and materials of construction.

11.3 METEOROLOGICAL CONCERNS

The atmosphere has been labeled the dumping ground for air pollution. Industrial society can be thankful that the atmosphere cleanses itself (up to a point) by natural phenomena. Atmospheric dilution occurs when the wind moves because of wind circulation or atmospheric turbulence caused by local sun intensity. As described earlier, pollutants are removed from the atmosphere by precipitation and by other reactions (both physical and chemical) as well as by gravitational fallout.

The atmosphere is the medium in which air pollution is carried away from its source and diffuses. Meteorological factors have a considerable influence over the frequency, length of time, and concentrations of effluents to which the general public may be exposed. The variables that affect the severity of an air pollution problem at a given time and location are wind speed and direction, insolation (amount of sunlight), lapse rate (temperature variation with height), mixing depth, and precipitation. Unceasing change is the predominant characteristic of the atmosphere; for example, temperatures and winds vary widely with latitude, season, and surrounding topography [2,3].

Atmospheric dispersion depends primarily on horizontal and vertical transport. Horizontal transport depends on the turbulent structure of the wind field. As the wind velocity increases, the degree of dispersion increases with a corresponding decrease in the ground-level concentration of the contaminant at the receptor site. This is a result of the emissions being mixed into a larger volume of air. The dilute effluents may, depending on the wind direction, be carried out into essentially unoccupied terrain away from any receptors. Under different atmospheric conditions, the wind may funnel the diluted effluent down a river valley or between mountain ranges. If an inversion (temperature increases with height) is present aloft that would prevent vertical transport, the pollutant concentration may build up continually.

One can define atmospheric turbulence as those vertical and horizontal convection currents or eddies that mix process effluents with the surrounding air. Several generalizations can be made regarding the effect of atmospheric turbulence on the effluent dispersion. Turbulence increases with increasing wind speed and causes a corresponding increase in horizontal dispersion. Mechanical turbulence is caused by changes in wind speed and wind shear at different altitudes. Either of these conditions can lead to significant changes in the concentration of the effluent at different elevations.

Topography can also have a considerable influence on the horizontal transport and thus pollutant dispersion. The degree of horizontal mixing can be influenced by sea and land breezes. It can also be influenced by man-made and natural terrain features such as mountains, valleys, or even a small ridge or a row of hills. Low spots in the terrain or natural bowls can act as sites where pollutants tend to settle and accumulate because of the lack of horizontal transport in the land depression(s). Other topographical features that can affect horizontal transport are city canyons and isolated buildings. City canyons occur when the buildings on both sides of a street are fairly close together and are relatively tall. Such situations can cause funneling of emissions from one location to another. Isolated buildings or the presence of a high-rise building in a relatively low area can cause redirection of dispersion patterns and route emissions into an area in which many receptors live.

Dispersion of air contaminants is strongly dependent on the local meteorology of the atmosphere into which the pollutants are emitted. The mathematical formulation for the design of pollutant dispersal is associated with the open-ground terrain free of obstructions. Either natural or man-made obstructions alter the atmospheric circulation and with it the dispersion of pollutants. In addition are the effects of mountain valley terrain, hills, lakes, shorelines, and buildings.

11.4 PLUME RISE

A plume of hot gases emitted vertically has both a momentum and a buoyancy. As the plume moves away from the stack, it quickly loses its vertical momentum (owing to drag by and entrainment of the surrounding air). As the vertical momentum declines, the plume bends over in the direction of the mean wind. However, quite often the effect of buoyancy is still significant, and the plume continues to rise for a long time after bending over. The buoyancy term is due to the less-than-atmospheric density of the stack gases and may be temperature or composition induced. In either case, as the plume spreads out in the air (all the time mixing with the surrounding air), it becomes diluted by the air.

Modeling the rise of the plume of gases emitted from a stack into a horizontal wind is a complex mathematical problem. Plume rise depends not only on such stack gas parameters as temperature, molecular weight, and exit velocity, but also on such atmospheric parameters as wind speed, ambient temperature, and stability conditions [1].

The behavior of plumes emitted from any stack depends on localized air stability. Effluents from tall stacks are often injected at an effective height of several hundred feet to several thousand feet above the ground because of the added effects of buoyancy and velocity on the plume rise. Other factors affecting the plume behavior are the diurnal variations in the atmospheric stability and the long-term variations that occur with changing seasons [4].

11.5 EFFECTIVE STACK HEIGHT

Reliance on atmospheric dispersion as a means of reducing ground-level concentrations is not foolproof. Inversions can occur with a rapid increase in ground-level pollutant concentrations. One solution to such situations is the tall stack concept. The goal is quite simple: Inject the effluent above any normally expected inversion layer. This approach is used for exceptionally difficult or expensive treatment situations because tall stacks are quite expensive. To be effective, they must reach above the inversion layer so as to avoid local plume fallout. The stack itself does not have to penetrate the inversion layer if the emissions have adequate buoyancy and velocity. In such cases, the effective stack height will be considerably greater than the actual stack height [3]. The effective stack height (equivalent to the effective height of emission) is usually considered the sum of the actual stack height, the plume rise due to velocity (momentum) of the issuing gases and the buoyancy rise which is a function of the temperature of the gases being emitted and the atmospheric conditions.

The effective stack height depends on a number of factors. The emission factors include the gas flow rate, the temperature of the effluent at the top of the stack, and the diameter of the stack opening. The meteorological factors influencing plume rise are wind speed, air temperature, shear of the wind speed with height, and the atmospheric stability. No theory on plume rise presently takes into account all these variables, and it appears that the number of equations for calculating plume rise varies inversely with one's understanding of the process involved. Even if such a theory was available, measurements of all of the parameters would seldom be available. Most of the equations that have been formulated for computing the effective height of an emission stack are semiempirical in nature. When considering any of these plume rise equations, it is important to evaluate each in terms of assumptions made and the circumstances existing at the time the particular correlation was formulated. Depending on the circumstances, some equations may definitely be more applicable than others.

The effective height of an emission rarely corresponds to the physical height of the source or the stack. If the plume is caught in the turbulent wake of the stack or of buildings in the vicinity of the source or stack, the effluent will be mixed rapidly downward toward the ground. If the plume is emitted free of these turbulent zones, a number of emission factors and meteorological factors will influence the rise of the plume. The influence of mechanical turbulence around a building or stack can significantly alter the effective stack height. This is especially true with high winds when the beneficial effect of the high stack gas velocity is at a minimum and the plume is emitted nearly horizontally.

Details regarding a host of plume rise models and calculation procedures are available in literature [5,6].

11.6 ATMOSPHERIC DISPERSION MODELS

The initial use of dispersion modeling occurred in military applications during World War I. Both sides of the conflict made extensive use of poison gases as a weapon of war. The British organized the Chemical Defense Research Establishment at Porton Downs during the war. Research at this institute dominated the field of dispersion modeling for more than 30 years through the end of World War II.

With the advent of the potential use of nuclear energy to generate electrical power, the United States Atomic Energy Commission invested heavily in understanding the nature of atmospheric transport and diffusion processes. Since about 1950 the United States has dominated researching the field. The U.S. Army and Air Force have also studied atmospheric processes to understand the potential effects of chemical and biological weapons.

The Pasquill–Gifford model has been the basis of many models developed and accepted today [7–9]. This model has served as an atmospheric dispersion formula from which the path downwind of emissions can be estimated after obtaining the effective stack height. There are many other dispersion equations (models) presently available, most of them semiempirical in nature. Calculation details regarding the use of this equation are available in the literature [5,6].

The problem of having several models is that various different predictions can be obtained. In order to establish some reference, a standard was sought by the government. The *Guideline on Air Quality Models* [10,11] is used by the Environment Protection Agency (EPA), by the states, and by private industry in reviewing and preparing prevention of significant deterioration (PSD) permits and in state implementation plans (SIP) revisions. The guideline serves as a means by which consistency is maintained in air quality analyses. On September 9, 1986 (51 FR 32180), EPA proposed to include four different changes to this guideline: (1) addition of specific version of the rough terrain diffusion model (RTDM) as a screening model, (2) modification of the downwash algorithm in the industrial source complex (ISC) model, (3) addition of the offshore and coastal dispersion (OCD) model to EPA's list of preferred models, and, (4) addition of the AVACTA II model as an alternative model in the guideline. In industry today, the ISC models are the preferred models for permitting and therefore are used in many applications involving normal or "after the fact" releases, depending on which regulatory agency must be answered to.

The ISC model is available as part of UNAMAP (Version 6). The computer code is available on magnetic tape from the National Technical Information Service (NTIS) or via modem through their Bulletin Board Services (BBS). It can account for the following: settling and dry deposition of particulates; downwash; area, line, and volume sources; plume rise as a function of downwind distance; separation of point sources; and, limited terrain adjustment.

In order to prepare for and prevent the worst, screen models are applied in order to simulate the worse-case scenario. One difference between the screening models and the refined models mentioned earlier is that certain variables are set that are estimated to be values to give the worst conditions. In order to use these screen models, the parameters of the model must be fully grasped.

In short, these models are necessary to somewhat predict the behavior of the atmospheric dispersions. These predictions may not necessarily be correct; in fact, they are rarely completely accurate. In order to choose the most effective model for the behavior of an emission, the source and the models have to be well understood.

11.7 STACK DESIGN

As experience in designing stack has accumulated over the years, several guidelines have evolved:

1. Stack heights should be at least 2.5 times the height of any surrounding buildings or obstacles so that significant turbulence is not introduced by these factors.
2. The stack gas exit velocity should be greater than 60 ft/s so that stack gases will escape the turbulent wake of the stack. In many cases, it is good practice to have the gas exit velocity on the order of 90 or 100 ft/s.
3. A stack located on a building should be set in a position that will assure that the exhaust escapes the wakes of nearby structures.
4. Gases from the stacks with diameters <5 ft and heights <200 ft will hit the ground part of the time, and the ground concentration may be excessive. In this case, the plume becomes unpredictable.
5. The maximum ground concentration of stack gases subjected to atmospheric dispersion occurs about 5–10 effective stack heights downwind from the point of emission.
6. When stack gases are subjected to atmospheric diffusion and building turbulence is not a factor, ground-level concentrations on the order of 0.001% to 1% of the stack concentration are possible for a properly designed stack.
7. Ground concentrations can be reduced by the use of higher stacks. The ground concentration varies (approximately) inversely as the square of the effective stack height.
8. Average concentrations of a contaminant downwind from a stack are directly proportional to the discharge rate. An increase in discharge rate by a given factor increases ground-level concentrations at all points by the same factor.
9. In general, increasing the dilution of stack gases by the addition of excess air in the stack does not effect ground-level concentrations appreciably. Practical stack dilutions are usually insignificant in comparison to the later atmospheric dilution by plume diffusion. Addition of diluting gas will increase the effective stack height, however, by increasing the stack exit velocity. This effect may be important at low wind speeds. On the other hand, if the stack temperature is decreased appreciably by the dilution, the effective stack height may be reduced. Stack dilution will have an appreciable effect on the concentration in the plume close to the stack.

These nine guidelines represent the basic design elements of a pollution control system. An engineering approach suggests that each element be evaluated independently and as part of the whole control system. However, the engineering design and evaluation must be an integrated part of the complete pollution control program.

11.8 FUTURE TRENDS

The future promises more sophisticated models to describe:

1. Plume rise
2. Effective stock heaglil
3. Atmospheric dispersion

Stacks design procedure showed remain exactly the same; only a few minor changes have occurred in the fast 50 years.

11.9   SUMMARY

1. Stacks discharging to the atmosphere have long been one of the methods available to industry for disposing waste gases. The concentration to which humans, plants, animals, and structures are exposed at ground level can be reduced significantly by emitting the waste gases from a process at great heights.

2. A four-step procedure is recommended for performing dispersion health effect studies:

   a. Estimate the rate, duration, and location of the release into the environment.

   b. Select the best available model to perform the calculations.

   c. Perform the calculations and generate downstream concentrations resulting from the source emission(s).

   d. Determine what effect, if any, the resulting discharge has on the environment, including humans, animals, vegetation, and materials of construction.

3. Major meteorological concerns include horizontal transport, vertical transport, topography, wind speed and direction, and temperature and humidity.

4. The effective stack height is usually considered the sum of the actual stack height, the plume rise due to velocity of the issuing gases, and the buoyancy rise which is a function of the temperature of the gases being emitted and the atmospheric conditions.

5. The Pasquill–Gifford model has been the basis of most models developed and accepted today. This model has served as an atmospheric dispersion formula from which the path downwind of emissions can be estimated after obtaining the effective stack height.

6. There are numerous design suggestions for stacks. One of the key recommendations is that stack heights should be at least 2.5 times the height of any surrounding building or obstacles so that significant turbulence is not introduced by these factors.

REFERENCES

1. Cooper, C.D. and Alley, F.C. *Air Pollution Control: A Design Approach*, Waveland Press, Prospect Heights, IL, 1986, pp. 493–515, 519–552.
2. Hesketh, H.E. *Understanding and Controlling Air Pollution*, Ann Arbor Science Publishers, Ann Arbor, MI, 1972, pp. 33–70.
3. Gilpin, A. *Control of Air Pollution.* Butterworth, New York, 1963, pp. 326–333.
4. Bethea, R.M. *Air Pollution Control Technology*, Van Nostrand Reinhold, New York, 1978, pp. 39–59.
5. Theodore, L. and Allen, R. *Air Pollution Control Equipment.* Theodore Tutorials E. Williston, New York, 1994.
6. Reyolds, J., Jeris, J., and Theodore, L. *Handbook of Chemical and Environmental Engineering Calculations*, John Wiley & Sons, Hoboken, NJ, 2004.
7. Pasquill, F. *Meterology Magazine*, 90 (33), 1063, 1961.

8. Gifford, F.A. *Nuclear Safety*, *2* (4), 47, 1961.

9. Cota, H. *Journal of the Air Pollution Control Association*, *31* (8), 253, 1984.

10. U.S. EPA. *Guideline on Air Quality Models*, Publication No. EPA-450/2-78-027, Research Triangle Park, NC, August 1978 (OAQPS No. 1.2-08).

11. U.S. EPA. *Industrial Source Complex (ISC) Dispersion Model User's Guide*, 2nd edition, Vols. 1 and 2, Publication Nos. EPA-450/4-86-005a and EPA-450/4-86-005b. Research Triangle Park, NC: Author unknown, 1986 (NRTIS PB 86 234259 and 23467).

12 Greenhouse Effect and Global Warming

Contributing Author: Shannon O'Brien

CONTENTS

12.1 INTRODUCTION

The "greenhouse effect" is a phrase properly used to describe the increased warming of the Earth due to increased levels of carbon dioxide and other atmospheric gases, called greenhouse gases (GHGs). Just as the glass in a botanical greenhouse traps heat for growing plants, GHGs trap heat and warm the planet. The greenhouse effect, a natural phenomenon, has been an essential part of Earth's history for billions of years. The greenhouse effect is the result of a delicate and non-fixed balance between life and the environment. Yet, the greenhouse effect may be leading the planet to the brink of disaster. Since the Industrial Revolution, the presence of additional quantities of GHGs threatens to affect global climate and the predicted effects of this increase are still debated among scientists.

The greenhouse effect works as follows. The energy radiated from the sun to Earth is absorbed by the atmosphere, and is balanced by a comparable amount of long-wave energy emitted back to space from the Earth's surface. Carbon dioxide molecules (and GHGs) absorb some of the long-wave energy radiating from the planet. Because of the greenhouse heat trapping effect, the atmosphere itself radiates a large amount of long-wave energy downward to the surface of the Earth and makes the Earth warmer than if warmed by solar radiation alone. The GHGs trap heat because of their chemical makeup and, in particular, their triatomic nature. They are relatively transparent to visible sunlight, but they absorb long wavelength, infrared radiation emitted by the Earth [1].

12.2 GLOBAL CARBON CYCLE

Carbon dioxide comprises only a very small portion of the Earth's atmosphere, but has risen from about 280 to 380 ppmv (parts per million by volume) recently (approximately 0.03% by volume). However, it is undoubtedly one of the most essential molecules for life on Earth [2]. Without this colorless, odorless gas, animals and plants would not be able to survive. Nevertheless, too much carbon dioxide in the atmosphere arising due to the burning of fossil fuels might prove to be harmful. The exchange of carbon throughout the world, termed the "global carbon cycle," is of particular importance to studying the greenhouse effect.

The transfer of carbon occurs naturally through both the terrestrial and oceanic cycles. The terrestrial cycle starts with photosynthetic plants, which use sunlight, water, carbon dioxide, and a pigment called chlorophyll to form glucose and oxygen. It has been estimated that this consumes approximately 500 billion tons of carbon dioxide each year, converting it into organic compounds and oxygen. Carbon dioxide is naturally returned into the atmosphere by respiration and decay. The oceans, which cover about 70% of the globe, consist of all of the gases contained in the atmosphere. There is an exchange of carbon dioxide between the oceans and surrounding air until equilibrium is reached. Carbon dioxide is then transferred to deeper water by convective transport cycles, which lower the concentration at the surface, thereby allowing more of the gas from the atmosphere to diffuse in the ocean.

Surprisingly, about 25% of the increase of atmospheric carbon dioxide over the past century and a half has come from changes in land use, such as deforestation and soil cultivation. During the last 150 years, about 155 billion metric tons of carbon was released to the atmosphere from such actions. The amount released each year generally increased over the period, and by the 1990s the rate of release averaged about 2 billion metric tons of carbon per year [3]. Deforestation increases carbon dioxide levels in two basic ways. First, when trees decay or are burned, they release carbon dioxide. Second, without the forest, carbon dioxide that would have been absorbed by photosynthesis remains in the atmosphere. For example, a rainforest can hold 1 or 2 kg of carbon per square meter per year, as compared to a field of crops, which can absorb less than 0.5 kg per square meter every year [4].

In addition to deforestation and other such actions, the burning of hydrocarbons, has caused a severe man-made influx of carbon dioxide into the global carbon cycle. The International Panel on Climate Change (IPCC) used knowledge of this natural process plus data on fuel consumption and land use in order to predict the concentration of carbon dioxide in the atmosphere. According to their model, approximately 3.3 gigatons more carbon dioxide is entering than leaving. This abundance of carbon dioxide has caused a disruption of the global carbon cycle. Without a "quick fix," some fear that this increase will result in catastrophic events, detailed in a later section.

Recently, the EPA appropriately replaced global warming with the term "climate change" to account for all of the environmental changes that are occurring in the world, not just the increase in temperature. Along with the federal government, the EPA has created a number of initiatives and incentives to encourage major polluters, particularly large corporations, go "green" and reduce their environmental impacts. The United States government has also enacted many climate change programs to help curb greenhouse gas emissions as well as prevent the release of other pollutants.

The most recent program (2009) is the Climate Change Technology Program (CCTP) which is discussed in more detail in Section 12.8—Future Trends.

12.3   CARBON DIOXIDE REMOVAL TECHNOLOGIES

Currently, scientists are working hard to develop new technologies to safely reduce and dispose of carbon dioxide. There are a number of proposed advancements, many of which are still in the research and development stage. Absorption, adsorption, and reaction chemistry are the processes upon which these technologies are based. Some of these are discussed below.

One existing technology used to harvest carbon dioxide is called a scrubber. Amine solutions extract carbon dioxide from flue gases through the processes of absorption. The spent amine solution is then sent to a regeneration process, where the CO_2 is collected, compressed, and liquefied [5].

Another process uses semipermeable membranes or catalyst-coated monoliths to collect carbon dioxide through adsorption. Membranes can either function according to molecular size of the entrained molecules or ease of diffusion through the membrane. The latter poses an effective solution for CO_2 capturing. The membrane selectively allows certain gases that diffuse easily to pass through, while certain gases such as carbon dioxide are captured. In this case, the ease of diffusion of the flue gas through the membrane is dependent of the material of construction, which is polymer based. Polyimide has some potential for this application [6].

A recent patent uses reaction chemistry for CO_2 removal. The process, called accelerated weathering of limestone, typically involves reacting CO_2 in a flue gas stream with water and calcium carbonate (limestone). Instead of releasing CO_2, a wastewater rich stream of bicarbonate ions is produced, which can be directly released into the ocean. The coinventor of this project claims that carbon waste in the form of dissolved bicarbonate will have minimal adverse effects on the ocean and may even be beneficial for coral reefs. The Department of Energy has patented this technology, and several power plants along coastlines will serve as the testing grounds for this CO_2 sequestering method [7].

In general, it is not very difficult to isolate carbon dioxide. However, safe disposal of the GHGs has proven to be a daunting task. Many possibilities exist, but economic consequences are a major consideration when determining the feasibility of these technologies. Because industrial and commercial uses of CO_2 are minimal compared to amounts produced (approximately 1%), there needs to be safe, cost-effective disposal methods available. Such disposal methods that are likely to provide the first large-scale opportunity are direct injection to oil and gas reservoirs, deep, unmineable coal seams, or saline aquifers. These environments for injection have the ability to sequester large amounts of carbon dioxide as well as store and separate CO_2 from areas where injection would be harmful. Also, it could contribute to oil and methane recovery by the displacement method. Furthermore, ocean sequestering can be effective by the method of direct injection. This must be performed at 1000–2000 m deep, where the temperature gradient is great enough where it could not mix with surface waters. It could also be injected over 3000 m deep, where CO_2 becomes negatively buoyant, which in turn would form a CO_2 lake. In this case, piping could be a major cost. For all of these methods, many technical, safety, liability, economic, and environmental issues remain unresolved [8].

Researchers in Australia are about to conduct the biggest test of carbon disposal to date. They are planning to inject this material underground into a 1.3 mi hole in the ground, located at the Otway Basin. The base of this hole is a natural gas reservoir covered by several layers of impermeable rock. In 2008, 100,000 tons of carbon dioxide will be introduced into the reservoir, and then the hole will be sealed. Scientists will monitor and test the surrounding areas to test for any leaks. A small leak could deem this project useless, while a large leak could potentially lead to disaster [9].

12.4 OTHER GREENHOUSE GASES

Carbon dioxide is not the only GHG heating up the planet. Methane is another odorless, colorless gas found in the atmosphere, present in traces of less than 2 ppm [2]. Like carbon dioxide, methane is a natural product and is over 20 times more effective at trapping heat than carbon dioxide. It is produced by anaerobic bacteria microorganisms that live without oxygen in wetlands, rice fields, cattle, termites, and ocean sediments. Methane's annual growth rate in the atmosphere is approximately 2.0% per year [4].

Nitrous oxide is produced both naturally and artificially. The atmosphere is 79% nitrogen and although plants need nitrogen for food, they cannot use it directly from the air. It must first be converted by soil bacteria into ammonia and then into nitrates before plants can absorb it. In the process, bacteria release nitrous oxide gas. In addition, farmers add chemical fertilizers containing nitrogen to the soil.

Another way of forming nitrous oxide is by combustion. When anything burns, whether it is a tree in the rainforest, natural gas in a stove, coal in a power plant, or gasoline in a car, nitrogen combines with oxygen to form nitrous oxide. This harmful molecule also destroys the ozone layer, which filters out dangerous ultraviolet radiation from the sun and protects life on Earth. The concentration of nitrous oxide is increasing at an approximate rate of 0.3% each year [4].

Chlorofluorocarbons (CFCs) are used as coolants: (CFC-12) for refrigerators and air conditioners, as blowing agents (CFC-11) in packing materials and other plastic foams, and as solvents. CFCs trap heat 20,000 times more effectively than carbon dioxide and are virtually indestructible. They are not destroyed or dissolved by any of the natural processes that normally cleanse the air. In addition, they may reside in the lower atmosphere. These high-power GHGs (CFCs) also attack the ozone layer; each CFC molecule can destroy 10,000 or more molecules of ozone, a gas consisting of oxygen. Due to an earlier rapid increase in the concentration of CFCs, a worldwide phase-out of CFCs production is in effect.

12.5 KYOTO PROTOCOL

Political restrictions of greenhouse emissions have proven to be vital for the fight against global warming. Therefore, it is no surprise that this environmental issue has become a major concern for the world's governments.

The Kyoto Protocol was a major international and political treaty that came about through a series of legal processes beginning in 1988. During this year, the IPCC started to express concerns about the world's changing environment. In addition, the UN General Assembly's first debate regarding global warming occurred, resulting in the adoption of the "Protection of the global climate for present and future generations of mankind." Four years later, a major step in the fight against global warming took place when the UN Framework Convention on Climate Change (UNFCCC) opened for signing in Rio de Janeiro at the Earth Summit. Countries all over the world were encouraged to help reduce greenhouse emissions. On March 21, 1994, the Convention came into force with a total of 186 governments. The following year, The Berlin Conference of the Parties (COP) began to propose the environmental requirements for industrialized countries.

The Kyoto Protocol was officially adopted on December 11, 1997 with signatures from a total of 87 countries, all of whom agreed to reduce GHGs by 5.2% within 10 years. The guidelines for this treaty were discussed all over the world until a rulebook, officially named the Marrakech Accords, was finalized in 2001. The Kyoto Protocol came into effect on February 16, 2005, 90 days after it was ratified by Russia. This worldwide agreement required a 55% global reduction of carbon dioxide, based on 1990 levels. For the first time in history, countries were committed to reducing GHGs, instead of only being requested to do so.

Meeting the new demands of the Kyoto Protocol has proven to be a very daunting task for industrialized countries. As the world's top producer of GHGs, the United States accounts for 36% of all carbon dioxide emissions. Since the country has pulled out of the agreement, emissions have increased 15% above 1990 levels—21% above the initial goal. Therefore, ratification of the treaty may mean enormous changes across the country—changes that then President Bush felt the nation was not quite ready to make. James L. Connaughton, chairman of the White House Council of Environmental Quality, told senators "The Kyoto Protocol would have cost our economy up to $400 billion and caused the loss of up to 4.9 million jobs, risking the welfare of the American people and American workers." Meanwhile, the Protocol would put no regulations on GHGs emissions in developing countries.

Several recent events may foreshadow a change in U.S. position on global warming. Environmental leaders in some states are already promoting legislation that supports many of the goals set out by the Kyoto Protocol. For example, the California Air Resources Board (CARB) has been directed to strictly regulating GHG emission. The Chicago Climate Exchange is a voluntary organization of North American municipalities, companies, and organizations that have committed to decrease GHG emissions in the coming years. In addition, 10 Northeast states are creating a mandatory cap and trade emission reduction applicable to most power plants beginning the 2009. The recent 2006 elections have placed many in office who are concerned about environmental issues such as global warming and may lead to revisions in the U.S. position on Kyoto. The 2008 presidential election sealed the deal.

On the other side of the Atlantic, the European Union (EU) became a strong proponent of the treaty and has insisted that every aspect be enforced through an Emission Trading Scheme (EUETS). Many European countries were offended at the United States' refusal to sign the Protocol and some believe that this may have fueled their drive. The EU refuses to forgive countries that have failed to meet the terms of the agreement and was even hesitant to give credit for maintaining forests which store carbon, called "carbon sinks." Despite all of these strong opinions and a seemingly resilient commitment, the EU has only reduced greenhouse emissions to 2.9% lower than 1990 levels.

In 2004, China proposed to generate 10% of its power from renewable energy sources by 2010, only 2 years after ratifying the Kyoto Protocol. Despite this commitment, there are still many lingering concerns about China's dedication due to the country's status as a developing country, even though it has the world's largest population and a rapidly expanding economy. This status exempts China from mandatory emission reduction objectives even though it is the world's largest coal producer and its oil consumption has doubled in 20 years. China does not show signs of complying with any requests for reductions in GHG emissions.

One of the biggest contributing factors to the acceptance of the Kyoto Protocol was Russia's support of the 55% reduction in GHG emissions. However, recent events have raised questions about Russia's real reasons for ratifying the treaty. Many believe that this political action was used as way to gain membership into the World Trade Organization (WTO). Since 1990, the country's industry has experienced a significant decline. With acceptance into the WTO, the country would be able to acquire billions of dollars through emissions trading. This enables Russia to sell its unused emissions to other more industrialized countries that do not meet the standards of the protocol This has brought much criticism to the Kyoto Protocol.

As a foremost member of the Kyoto Protocol, Japan was anticipated to be one of the first nations to ratify the treaty. However, the United States' refusal caused the Japanese to think twice before signing. Its eventual ratification in June 2002 was important because Japan accounts for 8% of global GHG emissions and it agreed to reduce emissions by 6% of the published 1990 levels. Despite this high goal, Japans emission actually increased to 11% over the 1990 levels by 2002. Nevertheless, this slip up has not stopped Japan from supporting clean air technology or from manufacturing hybrid cars.

In 2002, India ratified the Kyoto Protocol when its delegates realized the major impact that their country has on global warming. Even with a population of 1 billion people, India was deemed a developing nation, like China. Therefore, they avoided the strict GHG emission standards and regulations facing developed nations. India's prime minister insists that per-capita emission rates of developing countries are insignificant compared to industrialized countries.

The Kyoto Protocol places heavy burdens on industrialized countries, such as those described above because they are deemed responsible for the problem. These strict regulations are crucial in order to achieve global regulation of GHG emissions. The Kyoto Protocol, which expires in 2012, has clearly set the groundwork for future environmental agreements.

In December 2007, a 2 week U.N. climate conference took place in Bali, Indonesia. Countries like Japan and the United States petitioned against immediate guidelines, arguing that the restrictions should begin at the conclusion of the two year talks. In the end, the U.S. compromised and approved a plan to combat global warming by 2009. Now, representative from almost 190 countries must agree upon emission restrictions for industrialized nations and initiate a plan to help developing nations, while still cutting back on GHG production and saving their valuable forest land [10].

12.6 GREENHOUSE DEBATE

In early 2000, supporters of the Kyoto Protocol drew support from the National Research Council reported that Earth's surface temperature has been on the rise since 1980. Unfortunately for these advocates, the 11 members also concluded that the atmospheric temperature has not changed over this same time period. Contrary to the climate model of the Kyoto Treaty, satellites and weather balloons have shown no distinct rise in atmospheric temperature. However, the 85-page report was very vague in its reasoning, stating that "major advances" would be necessary in order to settle the controversies over the realities of global warming [11].

There are three main questions surrounding global warming. First and foremost is the Earth really heating up? If so, is this natural or man-made? And finally, what are the effects of this occurrence? Scientists have debated over these three funda-mental issues since the 1980s and have not reached a clear consensus to answer to the last two questions. However, they have concluded that global warming is a real and current potential threat. As for the latter two questions, the latest IPCC report—issued on 2007 and signed by hundreds of scientists—stated that there is a 90% chance that man-made GHGs contribute to climate changes.

Although most scientists are in agreement that higher levels of trace GHGs in the atmosphere are, at least to some degree, causing global warming, some argue against this idea. Some insist that current temperature variations are natural and harmless while others believe global warming is caused by natural phenomena. Still others believe that the planet is actually entering another ice age.

In 1988, James Hansen of the National Aeronautics and Space Administration (NASA) Goddard Institute for Space Studies testified before the Senate Committee on Energy and Natural Resources [12]. He told the committee he was 99% certain that a 1°F rise in world temperatures since the 1850s has been caused by an increas-ing greenhouse effect. "It is time to stop waffling so much and say that the evidence is pretty strong that the greenhouse effect is here," he said. But others argue that the Earth has a natural control mechanism that keeps the Earth's climate in balance and that the increasing greenhouse effect may naturally trigger events that will cool the Earth (e.g., increased clouds) and hold the climate in balance. However, many believe that the greenhouse effect is so strong as to overcome these dangerous effects.

A large majority of the scientists in the field feel that the climate models have been reliable enough to conclude that the greenhouse effect is causing global warm-ing. Dr. Hansen, the leading spokesperson on the greenhouse effect, says that "it is just inconceivable that the increase of GHGs in the atmosphere is not affecting our climate" [12].

A minority of researchers believe that the warming of the Earth over the last 100–150 years is part of a long-term, natural cycle that has little to do with the production of GHGs. They remain unconvinced that the accumulation in the atmo-sphere of GHGs is concrete evidence of any rise in the average Earth temperature. On this lack of evidence, three scientists from the G.C. Marshall Institute in Washington DC reported that any warming of the Earth in the last 100 years is better explained by the variation in natural climate and solar activity [13]. According to this theory, the most probable source of global warming appears to be variations in solar activity.

The amount of solar rays reaching the Earth is controlled by three elements that vary cyclically over time. The first element is the tilt of the Earth's axis, which varies 22°–24.5° and back again every 41,000 years. The second element is the month of the year in which the Earth is closest to the sun, which varies over cycles of 19,000 and 24,000 years. Finally, the third element is the shape of the Earth's orbit, which, over a period of 100,000 years, changes from being more elliptical to being almost fully circular.

Scientists have also stated that changes in the Earth's temperature have followed changes in solar activity over the last 100 years. When solar activity increased from 1880s to 1940s, global temperatures increased. The observed global temperature rise of 1°F was during this period, before 67% of global GHG emissions had even occurred. When it declined from the 1940s to the 1960s, temperatures also declined. During this time period, some environmentalists spoke of doomsday tales as a result of "global cooling," blaming this event on the use of hydrocarbon fuels. When temperatures began to climb again with an increase of solar activity and sunspot numbers in the 1970s and 1980s, environmentalists began singing a different tune. Instead of the devastating effects of worldwide temperature drops, media campaigns began stressing the importance of regulating GHG emissions.

Obviously, the debate among these so-called experts continues to rage. What may be needed is to bring together a group of qualified experts (e.g., a Delphi Panel)—with no interests in the results—to impartially examine this problem analytically [14].

12.7   EFFECTS OF GLOBAL WARMING

A rise in average global temperature is expected by many to have profound effects. The need to study future situations had led the EPA and other environmental scientists to investigate what would happen to the planet after a 3°F–8°F warming. These climate changes will have a significant effect on weather patterns. There will be changes in precipitation, storms, wind direction, etc. Rising temperatures are expected to increase tropical storm activity. The hurricane season in the Atlantic and Caribbean is expected to start earlier and last longer. Storms will be more severe. Some researchers believe that the planet will be a wetter place. Global circulation models (GCM) predicted that a doubling of carbon dioxide could increase humidity 30%–40% [15]. However, such increases will not occur uniformly around the world. Perhaps humid tropical areas will become wetter while some semiarid regions will become drier. It is possible that some farmers will experience a longer growing season, while others would suffer from more frequent droughts.

As the Earth gets warmer, the Intergovernmental Panel on Climate Change (IPCC) predicts that there will be a one half to three foot rise in the average water level of oceans. As the ocean water is heated, it expands, or increases in volume. In addition, the polar ice caps will continue to diminish in size and contribute to the rising sea levels, even after taking into account the accumulation due to an increase in precipitation. Eventually, low-lying coastal areas are expected flood. Beach erosions will be an increasing problem. The EPA has estimated that if the sea levels rise 3 ft (0.9 m), the nation will lose an area the size of Massachusetts, even if it spends more than $100 billion to protect critical shorelines. In addition to lost beaches, houses and other buildings that sit close to the water's edge will be engulfed and destroyed. This could bring havoc to the real estate and insurance industries, as well as the lives many people.

If polar and temperate zones become warmer, there will be a poleward shift of ecological zones. Animal and plant species that now live in a particular area may no longer be able to survive there. The EPA predicts an increase in extinction rates as well as changes in migration patterns. As the ecological zones shift polewards, there may be a decrease in the amount of area suitable for forests, with a corresponding increase in grasslands and deserts. This means an overall loss in productive land, both for agriculture and for habitats of a broad range of plants and animals. For example, the United States' major crops grown—corn, wheat, and soybeans—are strongly affected by precipitation and temperatures. The warming trend may also cause changes in water quantity and quality in some areas. This will affect drinking supplies as well as the water needs of industry and agriculture. Rising sea levels may also contaminate water supplies as seawater migrates up rivers. For example, a 2 ft (0.6 m) rise in sea levels would inundate Philadelphia's water intakes along the Delaware River, making the water too salty to drink.

The IPCC stated that "climate change is likely to have wide-ranging and mostly adverse impacts on human health, with significant loss of life." Heat waves extract a physical toll on people. The heat puts a strain on the heart, as the body tries to cool itself. Studies have shown that the number of cases of heart and lung disease increases when temperature rises. Contagious diseases, such as malaria, influenza and pneumonia, and allergic diseases, such as asthma, are also affected by the weather and become more prolific, and are expected to spread among the public. The lifecycles of mosquitoes and other disease-carrying insects are also extended in warmer weather. Warming is likely to allow tropical diseases, such as malaria, to spread northward in some areas of the world.

Climate changes, an increase in sea levels, and other direct effects of global warming can already be seen in many parts of the world. These changes are expected by many to have major social, economic, and political consequences for the future.

12.8 FUTURE TRENDS

Regarding future trends, there appears to be three approaches to the response to this problem. The first approach is the "wait-and-see" approach, whose countermeasures may be inappropriate. The second course of action is the "adaptation to incurable changes" attitude, which is based on the assumption that there will be plenty of time in which to decide and act on climactic change. The third line of attack is the "act now" approach, which is the only one that demands an immediate legislative (and industrial) response. But the problem lies in the legal and economic systems which normally respond only to immediate and certain threats [16]. Which course of action the world will adopt regarding the greenhouse problem is difficult to predict at this time. At this point in time, the answer appears to be the third approach. Whether it will be wise or successful is another story.

The EPA recently (2009) introduced the Climate Change Technology Program (CCTP). The purpose of this initiative is to attain the sustainable technology necessary to provide clean, safe, and reasonably priced energy sources on a global scale. The CCTP has set forth a Strategic Plan that outlines six major goals to be accomplished using the $3 billion in federal spending that has been set aside for this program. These objectives are summarized below [17]:

1. Reduce emissions from energy end-use and infrastructure;
2. Reduce emissions from energy supply, particularly by development and commercialization of no- or low-emission technologies;
3. Capture, store, and sequester carbon dioxide;
4. Reduce emissions of greenhouse gases other than carbon dioxide;
5. Enhance the measurement and monitoring of greenhouse gas emissions;
6. Strengthen the contributions of basic science to climate change technology.

12.9 SUMMARY

1. The greenhouse effect describes the increase of GHG in the atmosphere and the absorption and reemission of long-wave radiation by these gases.
2. There is universal consensus that there is strong evidence to show that unprecedented global warming has already begun; and others feel the planet is actually entering another ice age.
3. A large majority of the scientific world believes that the main cause of global warming is the greenhouse effect. However, these are some contrary studies. The need to study future situations has led the EPA to consider what would happen to the planet after a 2°C–5°C warming.
4. At present, both national and international politicians are noticing the important implications of the greenhouse effect, and treaties such as the Kyoto Protocol are being drawn up in order to restrict the emission of GHG.
5. The world is on a course to attempt to curb and roll back GHG emissions. Whether the current and planned rules and policies will be successful is unknown at this time.

REFERENCES

1. Thompson, S. *The Greenhouse Effect*, Lucent Books, New York, 1993.
2. Raynaud, D., Jouzel, J., Barnola, J.M., Chappellaz, J., Delmas, R.J., and Lorius, C. The ice record of greenhouse gases, *Science*, 259(5097), 926–934, 1993.
3. Global carbon cycle. The Woods Hole Research Center. http://www.whrc.org/carbon/index.htm
4. Lyman, F. *World Resources Institute: The Greenhouse Trap*, Beacon Press, West Palm Beach, FL, 1991.
5. Shao, J. Amine purification system (AmiPur)—Continuous heat stable salts removal from amine solutions. Oil & Gas, Eco-Tec. Inc., Canada. http://www.eco-tec.com/techpapers/TP%20166%20AmiPur.pdf
6. Dortmundt, D. and Kishore, D. Recent developments in CO_2 removal membrane technology. UOPLLC, 1999. http://www.uop.com/objects/84CO2RemvbyMembrn.pdf
7. Patented technology captures dioxide from power plants. *PhysOrg.com*. June 1, 2005. http://www.physorg.com/news4353.html

8. Narula, R., Wen, H., Himes, K., and Power, B. Technical and economic comparison of CO_2 reducing technologies for power plants. *CEPSI*. November 5, 2002. http://www.bechtel.com/pdf/bip/23093.pdf

9. Tesler, P. Air repair, *Current Science*, 93, 6–7, 2008.

10. At last minute, U.S. OK's climate plan at Bali talks. *The Palm Beach Post*. December 16, 2007, Sec. A: 3.

11. Robinson, A. and Robinson, N. Global warming is 300-year-old news, *Wall Street Journal*, 2000.

12. Kerr, R.A. No way to cool the ultimate greenhouse, *Science*, 262(5134), 648, 1993.

13. Tesar, J. *Our Fragile Planet: Global Warming*, Blacksmith Graphic Books, Prospect Heights, IL, 1990.

14. Theodore, L. Personal notes, 2001.

15. Boyle, R. *Dead Heat*, Basic Books, New York, 1990.

16. Schneider, S. *Global Warming*, Sierra Club Books, New York, 1993.

17. U.S. EPA. Climate change technology program. http://www.epa.gov/climatechange/policy/cctp.html. Accessed April 28, 2009.

13 Air Toxics

CONTENTS

13.1 INTRODUCTION

Although air is generally considered as approximately 20% oxygen and 80% nitrogen (by mole or volume), other substances get into the air, and some of these are referred to as pollutants. Some of the pollutants that have the potential to adversely affect human health at certain concentrations are known as toxic air pollutants (TAPs), or air toxics. The dimensions of toxic exposure are staggering. Not until well after World War II was public attention drawn to toxic exposure. Not until the 1970s did the United States begin to address toxic contamination resulting from the common use of synthetic chemicals that also affected water and food. Incidents, such as Love Canal, in which a major toxic waste dump was discovered beneath a residential community, and Bhopal, India, in which methyl isocyanate was accidentally released into the atmosphere, have sounded a warning. Contaminated communities, or residential areas that are located within the boundaries of a known exposure to some form of pollution are causing a gradual deterioration of the relationship between humans and the ecosystem. An increased awareness of the implications of toxic pollution has led society to confront a new type of threat, that of toxic exposure. The need for information on the toxicity of environmental pollutants is based on the need to protect human health. Toxic exposure may now be considered to be "the plague of our time" [1]. For example, the most comprehensive national study to date of toxic air pollutants shows that the Puget Sound region's urban counties have exceptionally high concentrations of airborne toxins that environmental regulators say put residents at greater risk of cancer. See also:

http://www.seattlepi.com/local/67321_air20.shtml

13.2 CLASSIFICATION OF AIR TOXICS

A host of complications arise in the attempt to identify toxic airborne chemicals and their effects. A few substances are clearly hazardous in even the smallest amounts. Soon after the U.S. Environmental Protection Agency (USEPA) was formed in 1970, emission standards were set for some of the most hazardous air pollutants (HAPs): arsenic, asbestos, benzene, beryllium, mercury, vinyl chloride, and radionuclides. The dangers posed by the great majority of harmful substances, though, are not as clear-cut.

One complicating issue is that air pollutants themselves are affected by numerous other factors, both during their production and after their release into the atmosphere. They can react with each other to create new substances that can be either more or less toxic than the original substances. Also, some chemicals that are harmless individually can be dangerous in combination. This effect is called synergism. The vast possibilities with chemical interactions make it difficult at best to pinpoint specific causes of some of the harmful effects of air pollution.

Adding to the complexity is the fact that many potentially hazardous substances appear to produce few or no ill effects at lower concentrations. Researchers suspect, however, that even for many of these seemingly safe chemicals, extended exposure to low levels can contribute to health and environmental problems. What's more, the risks to different individuals are rarely the same—the general population exhibits a wide range of sensitivity to any specific chemical. Identifying the critical, harmful levels of air pollutants and the duration of exposure that presents danger usually involves years and years of epidemiological studies. Good, hard evidence of cause-and-effect relationships is hard to produce. Nevertheless, government regulators and industry officials have taken steps to reduce or eliminate exposure to airborne substances that are suspected of being hazardous to ones health. Some manufacturing plants place work restrictions on their employees based on the total amount of chemical exposure that is considered safe. Special ventilation systems, hazard detection and warning systems, and protective clothing and equipment are just a few of the measures being taken to help ensure the safety of workers in areas with possible air hazards [2].

There are three major criteria for a compound to be included under the heading of "TAP":

1. It is measurable in the air.
2. It is for the most part produced by the activities of man.
3. It is not a primary air quality pollutant as currently defined by the EPA.

There are literally over a thousand candidate TAP chemicals that fit the above categories that are used commercially in the United States and emitted into the atmosphere. An estimated 70,000 chemicals are in regular use in the United States and another thousand are added every year. This includes one billion pounds of pesticides, herbicides, and fungicides that are used everyday in the United States. Beyond the toxic exposure due to the manufacture, transportation, storage, and use of these materials, this country generates between 255 million and 275 million metric tons of hazardous waste annually, of which as much as 90% may be improperly disposed of. Some facts about the causes of residential toxic exposure are provided below[1].

1. There are some 600,000 contaminated sites in the country.
2. There are some 400,000 municipal landfills.
3. There are more than 100,000 liquid waste impoundments.
4. There are millions of septic tanks.
5. There are hundreds of thousands of deep-well injection sites.
6. Some 300,000 leaking underground storage tanks threaten groundwater.

TAPs need to be prioritized based on risk analysis, so that those posing the greatest threats to health can be regulated. A risk analysis is the scientific activity of evaluating the toxic properties of a chemical and the conditions of human exposure to it in order to determine the extent to which exposed humans will be adversely affected, and to characterize the nature of the effects that they may experience. The risk analysis may contain some or all of the following four steps:

1. Hazard identification—the determination of whether a particular chemical is or is not casually linked to particular health effects.
2. Dose-response assessment—the determination of the relation between the magnitude of exposure and the profitability of occurrence of the health effects in question.
3. Exposure assessment—the determination of the extent of human exposure.
4. Risk characterization—the description of the nature and often the magnitude of human risk.

Once completed, the risk analysis should be a significant aid in determining the potential risks associated with TAPs. A more detailed presentation on risk analysis can be found in Chapter 35.

13.3 CAUSES OF TOXIC AIR POLLUTION

Industrial success commonly results in relatively high population density and has produced the problems of air, water, and soil pollution. Petrochemical facilities, motor vehicles, metal processing industries, and home space heaters are just a few of the many pollution sources that have led to contamination in the environment. Toxic organic compound emission sources can be categorized into seven major source groupings:

1. Process sources (chemical production)
2. Fugitive sources—all on-site emissions resulting from leaks in pumps, valves, flanges, and similar connections
3. Storage tanks
4. Transport—usually by railcars or trucks
5. Surface coating—paints and coatings
6. Other solvent use—degreasing, dry cleaning, and printing
7. Nonindustrial sources—motor vehicles

There are three types of TAP emissions: continuous, intermittent, and accidental. Both routine emissions associated with a batch process or a continuous process that is operated only occasionally can be intermittent sources. An example of an accidental emission was the release of methyl isocyanate in Bhopal, India.

Many of the modern industrial and commercial processes utilized by society involve the application of organic solvents. Through the transport, storage, transfer, and use of these materials, releases can occur into the atmosphere. In addition, the use of liquid fuels by motor vehicles can also result in evaporative and tailpipe losses of organic substances to the air environment. A significant quantity of the organic materials emitted into the atmosphere as solvents or through the use of liquid fuels can be classified as volatile organic compounds (VOCs). VOCs are widely used in industrial and commercial operations, and they play an important role in the formation of ozone and smog aerosols. Ozone (O_3) is the most powerful oxidizing agent among common pollutant gases and is known to be highly toxic.

The atmosphere is the medium by which air pollutants are transported away from their sources of emission. The most important parameter in the movement of pollutants by the atmosphere is the wind. Meteorological conditions will affect the levels of pollutants accumulated in the atmosphere. The greater the wind speed, the greater the turbulence and the more rapid and complete is the dispersion of pollutants in the atmosphere. Atmospheric dispersion, however, does not remove air pollution, but merely dilutes it through an increasing volume [3] (see also Chapter 11).

Polluted air environments, community or industrial, usually contain complex and ever-changing mixtures of contaminants, not all of which can be monitored adequately. The elimination of, or large reductions in, air pollution can only be accomplished by controlling the sources of emission. When a potential emission is suspected of being extremely toxic or containing a cancer-suspect material, exceptional measures are needed in the control strategy. Hazardous and toxic chemicals may require removal down to levels of a few parts per million by volume (ppmv). Two basic approaches are available for removing hazardous and volatile organics from vent streams. The pollutant may be recovered in concentrated form for use in the process or used for process heat. The other approach is to destroy the toxic material before it reaches the atmosphere.

Five control methods make up the most common methods for controlling hazardous pollutants: absorption, adsorption, condensation, chemical reaction, and incineration. It is possible to combine two or more of these methods to achieve a desired goal. The selection of the best method depends on effluent quantity, pollutant concentration, required efficiency, desired ultimate disposal, economic factors, and chemical and physical characteristics of the stream.

13.4 IMPACTS OF TOXIC AIR POLLUTION

Basic air pollutant toxicology, or the science that treats the origins of toxics, must be considered in terms of entering the body through inhalation. This makes the respiratory tract the first site of attack. Among the primary air pollutants, only lead and carbon monoxide exert their major effects beyond the lung. The more reactive a compound, the less likely it is to penetrate the lung. However, many individuals breathe a mixture of air contaminants, and many of the TAP compounds are known to cause cancer. The total nationwide cancer incidence due to outdoor concentrations of air toxics in the United States was estimated to range from approximately

1700–2700 excess cancer cases per year. This is roughly equivalent to between 7 and 11 annual cancer cases per million population (data obtained from a 1986 population of 240 million). The EPA initiated a broad "scoping" study with a goal of gaining a better understanding of the size and causes of the health problems caused by outdoor exposure to air toxics. This broad scoping study was referred to as the 6 month study. The objective was to assess the magnitude and nature of the air toxics problem by developing quantitative estimates of the cancer risks posed by selected air pollutants and their sources from a national and regional perspective. The main conclusion of the 6 month study was that the air toxic problem is widely thought to be related to the elevated cancer mortality. Table 13.1 provides a summary of the estimated annual cancer cases by pollutant [4].

The EPA classifications used in this report are A = proven human carcinogen; B = probable human carcinogen (B1 indicates limited evidence from human studies and sufficient evidence from animal studies; B2 indicates sufficient evidence from animal studies, but inadequate evidence from human studies); C = possible human carcinogen.

Toxic exposures during a disaster, such as Love Canal, or Bhopal, India, can occur in three stages:

1. Predisaster stages: origin and incubation—during the incubation stage, the community is unaware that the disaster is developing. Therefore, there are no preparations.
2. Disaster stages: discovery, acceptance, community action—the community, defined by the pollution boundaries, becomes isolated from its surroundings.
3. Postdisaster stages: mitigation and lasting impacts—toxic exposure may be chronic and indefinite. A site may be contaminated so that it will remain unsafe for generations due to the persistence of the toxic hazard. Recovery is difficult.

13.5 RESPONSE TO TOXIC EXPOSURE

With the discovery and announcement of contamination, toxic victims suddenly find themselves in a complicated institutional complex made up of the various local, state, and federal agencies having control over their contamination incident. Their lives are, in a sense, captured by agencies upon which they become dependent for clarification and assistance.

As a result of the accident at Bhopal, the U.S. Congress created Title III, a freestanding statute included in the Superfund Amendments and Reauthorization Act (SARA) of 1986. Title III provides a mechanism by which the public can be informed of the existence, quantities, and releases of toxic substances, and requires the states to develop plans to respond to accidental releases of these substances. Further, it requires anyone releasing specific toxic chemicals above a certain threshold amount to annually submit a toxic chemical release inventory (TRI) form to the EPA. At present, there are 308 specific chemicals subject to Title III regulation.

In the 1970 Clean Air Act (CAA) Amendments, the U.S. Congress established a program that was to regulate a category of pollutants that it considered to be more

TABLE 13.1

Summary of Estimated Annual Cancer Cases by Pollutant

Pollutant	EPA Classification[a]	Estimated Annual Cancer Cases
1. Acrylonitrile	B	113
2. Arsenic	A	68
3. Asbestos	A	88
4. Benzene	A	181
5. 1,3-Butadiene	B	2266
6. Cadmium	B	110
7. Carbon tetrachloride	B	241
8. Chloroform	B	2115
9. Chromium (hexavalent)	A	147–265
10. Coke oven emissions	A	7
11. Dioxin	B	22–125
12. Ethylene dibromide	B	268
13. Ethyl dichloride	B	245
14. Ethlene oxide	B	1–26
15. Formaldehyde	B	1124
16. Gasoline vapors	B	219–276
17. Hexachlorobutadiene	C	9
18. Hydrazine	B	26
19. Methylene chloride	B	25
20. Perchloroethylene	B	26
21. PIC[b]		433–1120
22. Radionuclides	A	3
23. Radon[c]	A	2
24. Trichloroethylene	B	27
25. Vinyl chloride	A	25
26. Vinylidene chloride	C	10
27. Miscellaneous[d]		15
Totals		1726–2706

[a] For a discussion of how EPA evaluates suspect carcinogens and more information on these classifications, refer to "Guidelines for Carcinogen Risk Assessment" (51 Federal Register 33992).

[b] EPA has not developed a classification for the group of pollutants that compose products of incomplete combustion (PIC), although EPA has developed a classification for some components, such as benzo(a)pyrene (BaP), which is a B2 pollutant.

[c] From sources emitting significant amounts of radionuclides (and radon) to outdoor air. Does not include exposure to indoor concentrations of radon due to radon in soil gases entering homes through foundations and cellars.

[d] Includes approximately 68 other individual pollutants, primarily from the TSDF study and the Sewage Sludge Incinerator study.

hazardous or more toxic than those regulated by the application of air quality standards. The HAP concept recognized a need to regulate pollutants that were unique because of the nature of their toxic or hazardous properties and the localized contamination problems they posed. The 1970 CAA required that the EPA provide an ample margin of safety to protect against HAPs by establishing national emissions standards for hazardous air pollutants (NESHAP), However, in actual practice, the designation and subsequent regulation of hazardous pollutants has been very slow. Initially three pollutants (asbestos, mercury, and beryllium) were designated and regulated in the 1970s. After a considerable pause, the EPA assigned regulations for vinyl chloride, benzene, radioactive isotopes, and arsenic. From 1970 to 1990, over 50 chemicals were considered for designation as HAPs, but the EPA's review process was completed for only 28 chemicals. In a period of 20 years, the EPA was only able to designate and regulate NESHAPs for a total of eight substances: beryllium, mercury, vinyl chloride, asbestos, benzene, radionuclides, inorganic arsenic, and coke oven emissions.

Because the EPA was so slow in setting standards for HAPs, many states had gone their own ways in regulating air toxics. States developed and implemented their own TAP control programs. Such programs, as well as the pollutants they regulate, differ widely from state to state. Up until 1990, state agencies had established some type of emission standard for over 800 toxic chemicals. The slow federal pace in regulating air toxics was in part due to the fact that the EPA, under NESHAP provisions, was required, in setting emission standards, to provide an ample margin of safety. Because many air toxics are carcinogenic, the EPA has at various times interpreted the statutory language of the 1970 CAA amendments as requiring an emission standard of zero for carcinogens. It was therefore reluctant to regulate emissions of economically important substances that are potentially carcinogenic in humans because such regulation could have required a total ban on their production [5].

The 1990 CAA amendments deal with the problem of HAPs or air toxics in a substantial way. Congress lists approximately 190 toxic pollutants for which the EPA is to designate emission standards by enforcing maximum achievable control technologies (MACT). The amendments mandated that the EPA issue MACT standards for all sources of the 190 substances in phased stages by the year 2000. These are pollutants that are known to be, or reasonably anticipated to be, carcinogenic, mutagenic, teratogenic, neurotoxic, cause reproductive dysfunctions, or are acutely or chronically toxic. In addition, the EPA must determine the risk remaining after MACT is in place and develop health-based standards that would limit the cancer risk to one case in one million exposures. Emission standards are intended to achieve maximum reduction taking into account the cost of control measures. The benchmark for gaseous air toxics is a 90% average reduction.

13.6 FUTURE TRENDS

The use of solvents that are integral to many chemical process industry operations are changing. Industry and regulatory players are rethinking solvent processes, compounds, and equipment. Regulations are targeting industries such as food processing, wastewater treatment, electronics manufacturing, and forest products processing. New adsorbents, catalysts, and recovery systems have been added to the arsenal of control technologies for these hazardous solvents. New and redesigned systems are promising even more less expensive choices in the years ahead. Volatiles can be burned in order to provide extra energy or they can be recycled for resale. In the case of halogenated organics, the use of catalytic systems is allowing for safer incineration of these compounds.

Over the last decade and in the decades to come, regulations and economics will continue to drive the increased use of technology to control toxic pollutants.

In addition to improved control technologies, the future is certain to find widespread use of pollution prevention principles (see Part V) for managing air toxics. These include:

1. Retrofit, don't change a process—many units can be retrofitted to reduce emissions and solvent use.
2. Reuse waste solvents—high-quality solvents are used once for precision cleaning and are then disposed of. They can often be reused, untreated, for applications that require lower standards of purity, such as general purpose cleaning.
3. Use replacement solvents.
4. Make a process solventless.

13.7 SUMMARY

1. Pollutants that have some potential to adversely affect human health at certain concentrations are known as TAPs, or air toxics.
2. There are three major criteria for a compound to be included under the heading of TAP and over a thousand candidate TAP chemicals that fit these criteria.
3. Industrial success has mainly produced the problems of air, water, and soil pollution. Toxic organic compound emission sources can be categorized into seven major source groupings with three types of TAP emissions.
4. Many of the TAP compounds are known to cause cancer. The air toxics problem is widely thought to be related to the elevated cancer mortality.
5. Toxic victims suddenly find themselves in a complex institutional context made up of the various local, state, and federal agencies having control over their contamination incident.
6. The use of solvents that are integral to many chemical process industries are changing. New and redesigned systems are promising less expensive choices in the years ahead.

REFERENCES

1. Edelstein, M. *Contaminated Communities*, Westview Press, London, 1988.
2. Adapted from: AWMA. *Air Quality Resources Guide*, AWMA, Pittsburgh, PA, 2008.
3. Reynolds, J., Jeres, J., and Theodore, L. *Handbook of Chemical and Environment Engineering Calculations*, John Wiley & Sons, Hoboken, NJ, 2004.
4. U.S. EPA. *Cancer Risk from Outdoor Exposure to Air Toxics*, EPA, Washington, DC: Author unknown, 1990.
5. Godish, T. *Air Quality*, 2nd edition, Lewis Publishers, Boca Raton, FL, 1991.

14 Indoor Air Quality

CONTENTS

14.1 INTRODUCTION

Indoor air pollution is rapidly becoming a major health issue worldwide. Although research efforts are still under way to better define the nature and extent of the health implications for the general population, recent studies have shown significant amounts of harmful pollutants in the indoor environment. The serious concern over pollutants in indoor air is due largely to the fact that indoor pollutants are not easily dispersed or diluted as are pollutants outdoors. Thus, indoor pollutant levels are frequently higher than outdoors, particularly where buildings are tightly constructed to save energy. In some cases, these indoor levels exceed the Environmental Protection Agency (EPA) standards already established for outdoors. Research by the EPA in this area, called the Total Exposure Assessment Methodology (TEAM) studies, has documented the fact that levels indoors for some pollutants may exceed outdoor levels by 200%–500% [1].

Since most people spend 90% of their time indoors, many may be exposed to unhealthy concentrations of pollutants. People most susceptible to the risks of pollution—the aged, the ill, and the very young—spend nearly all of their time indoors. These indoor environments include such places as homes, offices, hotels, stores, restaurants, warehouses, factories, government buildings, and even vehicles. In these environments, people are exposed to pollutants emanating from a wide array of sources.

Some common indoor air contaminants are

1. Radon
2. Formaldehyde

3. Volatile organic compounds (VOCs)
4. Combustion gases
5. Particulates
6. Biological contaminants

In addition to air contaminants, other factors need to be observed in indoor air quality (IAQ) monitoring programs to fully understand the significance of contaminant measurements. Important factors to be considered in IAQ studies include:

1. Air exchange rates
2. Building design and ventilation characteristics
3. Indoor contaminant sources and sinks
4. Air movement and mixing
5. Temperature
6. Relative humidity
7. Outdoor contaminant concentrations and meteorological conditions

Designers, builders, and homeowners must make crucial decisions about the kinds and potential levels of existing indoor air pollutants at proposed house sites. Building structure design, construction, operation, and household furnishings, all rely on specific design parameters being set down to handle the reduction of these pollutants at their sources.

The health effects associated with IAQ can be either short or long term. Immediate effects experienced after a single exposure or repeated exposures include irritation of the eyes, nose, and throat; headaches; dizziness; and, fatigue. These short-term effects are usually treatable by some means, oftentimes by eliminating the person's exposure to the source of pollution.

The likelihood of an individual developing immediate reactions to indoor air pollutants depends on several factors, including age and preexisting medical conditions. Also, individual sensitivity to a reactant varies tremendously. Some people can become sensitized to biological pollutants after repeated exposures, and it appears that some people can become sensitized to chemical pollutants as well. Other health effects may show up either years after exposure has occurred, or only after long or repeated periods of exposure. These effects range from impairment of the nervous system to cancer; emphysema and other respiratory diseases; and, heart disease which can be severely debilitating or fatal. Certain symptoms are similar to those of other viral diseases and difficult to determine if it is a result of IA pollution. Therefore, special attention should be paid to the time and place symptoms occur.

Further research is needed to better understand which health effects can arise after exposure to the average pollutant concentrations found in homes. These can arise from the higher concentrations that occur for short periods of time. Yet, both the amount of pollutant, called the dose, and the length of time of exposure are important in assessing health effects. The effects of simultaneous exposure to several pollutants are even more uncertain. IAQ can be severely debilitating or even fatal. Indoor air pollutants of special concern are described below in separate sections.

It is not possible to provide estimates of typical mixtures of pollutants found in residences. This is because the levels of pollutants found in homes vary significantly depending on location, use of combustion devices, existing building materials, and use of certain household products. Also, emissions of pollutants into the indoor air may be sporadic, as in the case of aerosols or organic vapors that are released during specific household activities or when woodstoves or fireplaces are in use. Another important consideration regarding indoor pollutant concentrations is the interaction among pollutants. Pollutants often tend to attach themselves to airborne particles that get caught more easily in the lungs. In addition, certain organic compounds released indoors could react with each other to form highly toxic substances.

The data provided in this chapter consists of approximate ranges of indoor pollutants based on studies conducted around the United States. These provide an overview of several major pollutants that have been measured in residences at levels that may cause health problems ranging from minor irritations or allergies to potentially debilitating diseases.

14.2 RADON

Radon is a unique environmental problem because it occurs naturally. Radon results from the radioactive decay sequence of uranium-238, a long-lived precursor to radon. The isotope of most concern, radon-222, has a half-life (time for half to disappear) of 3.8 days. Radon itself decays and produces a series of short-lived decay products called radon progeny or daughters. Polonium-218 and polonium-214 are the most harmful because they emit charged alpha particles more dangerous than x-rays or gamma rays [2]. They also tend to adhere to other particles (attachment) or surfaces (plate out). These larger particles are more susceptible to becoming lodged in the lungs when inhaled and cause irreparable damage to surrounding lung tissue (which may lead to lung cancer).

Radon is a colorless, odorless gas that is found everywhere at very low levels. Radon becomes a cause for concern when it is trapped in buildings and concentrations build up. In contrast, indoor air has approximately 2–10 times higher concentrations of radon than outdoor air. Primary sources of radon are from soil, well water supplies, and building materials.

Most indoor radon comes from the rock and soil around a building and enters structures through cracks or openings in the foundation or basement. High concentrations of radon are also found in wells, where storage, or holdup time, is too short to allow time for radon decay. Building materials, such as phosphate slag (a component of concrete used in an estimated 74,000 U.S. homes) has been found to be high in radium content [3]. Studies have shown concrete to have the highest radon content when compared to all other building materials, with wood having the least.

It is becoming increasingly apparent that local geological factors play a dominant role in determining the distribution of indoor radon concentrations in a given area. To date, no indoor radon standard has been promulgated for all residential housing in the United States. However, various organizations have proposed ranges of guidelines and standards.

Data taken from various states suggest an average indoor radon-222 concentration of 1.5 pCi/L (picocuries per liter, a concentration radiation term), and approximately 1 million homes with concentrations exceeding 8 pCi/L [3]. One curie is equal to a quantity of a material with 37 billion radioactive decays per second. One trillionth of a curie is a pCi. Assuming residents in these homes spend close to 80% of their time indoors, their radon exposure would come close to the level for recommended remedial action set by the U.S. National Council on Radiation Protection and Measurements. The EPA believes that up to 8 million homes may have radon levels exceeding 4 pCi/L air, the level at which the EPA recommends corrective action. In comparison, the maximum level of radon set for miners by the U.S. Mine Safety and Health Administration is as high as 16 pCi/L.

Radon may be the leading cause of lung cancer among nonsmokers. Several radiation protection groups have approximated the number of annual lung cancer deaths attributable to indoor radon. The EPA estimates that radon may be responsible for 5,000–20,000 lung cancer deaths among nonsmokers. Also, scientific evidence indicates that smoking, coupled with the effects of exposure to radon, increases the risk of cancer by 10 times that of nonsmokers [1].

A variety of measures can be employed to help control indoor concentrations if radon and/or radon progeny. Mitigation methods for existing homes include placing barriers between the source material and living space itself using several techniques, such as:

1. Covering exposed soil inside a structure with cement
2. Eliminating and sealing any cracks in the floors or walls
3. Adding traps to underfloor drains
4. Filling concrete block walls

Soil ventilation prevents radon from entering the home by drawing the gas away before it can enter the home. Pipes are inserted into the stone aggregate under basement floors or onto the hollow portion of concrete walls to ventilate radon gas accumulating in these locations. Pipes can also be attached to underground drain tile systems drawing the radon gas away from the house. Fans are often attached to the system to improve ventilation. Crawl space ventilation is also generally regarded as an effective and cheap method of source reduction. This allows for exchange of outdoor air by placing a number of openings in the crawl space walls.

Home ventilation involves increasing a home's air exchange rate—the rate at which incoming outdoor air completely replaces indoor air—either naturally (by opening windows or vents) or mechanically (through the use of fans). This method works best when applied to houses with low initial exchange rates. However, when indoor air pressure is reduced, pressure-driven radon entry is induced, increasing levels in the home instead of decreasing them. The benefit of increased ventilation can be achieved without raising radon exposure by opening windows evenly on all sides of the home.

Mechanical devices can also be used to help rid indoor air of radon progeny. Air-cleaning systems use high-efficiency filters or electronic devices to collect dust and other airborne particles, some with radon products attached to them. These devices decrease the concentration of airborne particles, but do not decrease the concentration of smaller unattached radon decay products which can result in a higher radiation dose when inhaled.

14.3 FORMALDEHYDE

Formaldehyde is a colorless, water-soluble gas that has a pungent, irritating odor noticeable at less than 1 ppm. It is an inexpensive chemical with excellent bonding characteristics that is produced in high volume throughout the world. A major use is in the fabrication of urea-formaldehyde (UF) resins used primarily as adhesives when making plywood, particleboard, and fiberboard. Formaldehyde is also a component of UF foam insulation, injected into sidewalls primarily during the 1970s. Many common household cleaning agents contain formaldehyde. Other minor sources in the residential environment include cigarette smoke and other combustion sources such as gas stoves, woodstoves, and unvented gas space heaters. Formaldehyde can also be found in paper products such as facial tissues, paper towels, and grocery bags, as well as stiffeners and wrinkle resisters [2].

Although information regarding emission rates is limited, in general, the rate of formaldehyde release has been shown to increase with temperature, wood moisture content, humidity, and with decreased formaldehyde concentration in the air.

UF foam was used as a thermal insulation in the sidewalls of many buildings. It was injected directly into wall cavities through small holes that were then sealed. When improperly installed, UF foam emits significant amounts of formaldehyde. The Consumer Product Safety Commission (CPSC) measured values as high as 4 ppm and imposed a nationwide ban on UF foam, but it was later overturned.

The superior bonding properties and low cost of formaldehyde polymers make them the resins of choice for the production of building materials. Plywood is composed of several thin sheets of wood glued together with UF resin. Particleboard (compressed wood shavings mixed with UF resin at high temperatures), can emit formaldehyde continuously from several months to several years. Medium density fiberboard was found to be the highest emitter of formaldehyde.

Indoor monitoring data on formaldehyde concentrations are variable because of the wide range of products that may be present in the home. However, elevated levels are more likely to be found in mobile homes and new homes with pressed-wood construction materials. Indoor concentrations also vary with home age since emissions decrease as products containing formaldehyde age and cure. In general, indoor formaldehyde concentration exceed levels found outdoors.

Although individual sensitivity to formaldehyde varies, about 10%–20% of the population appears to be highly sensitive to even low concentrations. Its principal effect is irritation of the eyes, nose, and throat, as well as asthma-like symptoms. Allergic dermatitis may possibly occur from skin contact. Exposure to higher concentrations may cause nausea, headache, coughing, constriction of the chest, and rapid heartbeat [1].

One of the most promising techniques for reducing indoor formaldehyde concentrations is to modify the source materials to reduce emission rates. This can be accomplished by measures performed during manufacture or after installation. A variety of production changes, i.e., changes in raw materials, processing times, and temperatures, are promising methods for reducing emission rates. Applying vinyl wallpaper or nonpermeable paint to interior walls, venting exterior walls, and increased ventilation are other methods employed after installation.

14.4 VOLATILE ORGANIC COMPOUNDS

In addition to formaldehyde, many other organic compounds may be present in the indoor environment. More than 800 different compounds can be attributed to volatile vapors alone. Common sources in the home are building materials, furnishings, pesticides, gas or wood burning devices, and consumer products (cleaners, aerosols, and deodorizers). In addition, occupant activities such as smoking, cooking, or arts and crafts activities can contribute to indoor pollutant levels.

Organic contaminants in the home are usually present as complex mixtures of many compounds at low concentrations. Thus, it is very difficult to provide estimates of typical indoor concentrations or associated health risks. It is likely, however, that organic compounds may be responsible for health-related complaints registered by residents where formaldehyde and other indoor pollutants are found to be low or undetectable. The sources of three major types of organic contaminants include solvents, polymer components, and pesticides.

Volatile organic solvents commonly pollute air. Exposure occurs when occupants use spot removers, paint removers, cleaning products, paint adhesives, aerosols, fuels, lacquers and varnishes, glues, cosmetics, and numerous other household products. Halogenated hydrocarbons such as methyl chloroform and methylene chloride are widely used in a variety of home products. Aromatic hydrocarbons such as toluene have been found to be present in more than 50% of samples taken on indoor air [3]. Alcohols, ketones, ethers, and esters are also present in organic solvents. Some of them, especially esters, emit pleasant odors and are used in flavors and perfumes, yet are still potentially harmful.

Polymer components are found in clothes, furniture, packages, and cookware. Many are used for medical purposes—for example, in blood transfusion bags and disposable syringes. Fortunately, most polymers are relatively nontoxic. However, polymers contain unreacted monomers, plasticizers, stabilizers, fillers, colorants, and antistatic agents, some of which are toxic. These chemicals diffuse from the polymers into air. Certain monomers (acrylic acid esters, toluene-diisocynate, and epichlorohydrin) used to produce plastics, polyurethane, and epoxy resins in tile floors, are all toxic.

Most American households use pesticides in the home, garden, or lawn, and many people become ill after using these chemicals. According to an EPA survey, 9 out of 10 U.S. households use pesticides and another study suggests that 80%–90% of most people's exposure to pesticides has been found in the air inside homes. Pesticides used in and around the home include products to control insects, termites, rodents, and fungi. Chlordane, one of the most harmful active ingredients in pesticides, has been found in structures up to 20 years after its application. In addition to the active ingredient, pesticides are also made up of inerts that are used to carry the active agent. These inerts may not be toxic to the targeted post, but are capable of causing health problems. Methylene chloride, discussed earlier as an organic pollutant, is used as an inert [1].

Human beings can also be significant sources of organic emissions. Human breath contains trace amounts of acetone and ethanol at 20°C and 1 atmosphere. Measurements taken in schoolrooms while people were present averaged almost twice the amount of acetone and ethanol present in unoccupied rooms. At least part of this increase for ethanol was presumed to be due to perfume and deodorant, in addition to breath emissions [2].

As mentioned earlier, large number of organic compounds have been identified in residences. Studies have shown that of the 40 most common organics, nearly all were found at much higher concentrations indoors than outdoors. Another EPA study identified 11 chemicals present in more than half of all samples taken nationwide. Although individual compounds are usually present in low concentrations which are well below outdoor air quality standards, the average total hydrocarbon concentration can exceed both outdoor concentrations and ambient air quality standards [3].

Little is known of the short- and long-term health effects of many organic compounds at the low levels of exposure occurring in nonindustrial environments. Yet cumulative effects of various compounds found indoors have been associated with a number of symptoms, such as headache, drowsiness, irritation of the eyes and mucous membranes, irritation of the respiratory system, and general malaise. In general, VOCs are lipid soluble and easily absorbed through the lungs. Their ability to cross the blood–brain barrier may induce depression of the central nervous system and cardiac functions. Some known and suspected human and animal carcinogens found indoors are benzene, trichloroethane, tetrachloroethylene, vinyl chloride, and dioxane.

One of the best methods to reduce health risks from exposure to organic compounds is for residents or consumers to increase their awareness of the types of toxic chemicals present in household products. Attention to warnings and instructions for storage and use are important, especially regarding ventilation conditions. In some instances, substitution of less hazardous products is possible, as in use of a liquid or dry form of a product rather than an aerosol spray. Consumers should also be wary of the simultaneous use of various products containing organic compounds, since chemical reactions may occur if products are mixed, and adverse health effects may result from the synergism between/among components.

14.5 COMBUSTION GASES

Combustion gases, such as carbon monoxide, nitrogen oxides, and sulfur dioxide, can be introduced into the indoor environment by a variety of sources. These sources frequently depend on occupant activities or lifestyles and include the use of gas stoves, kerosene and unvented gas space heaters, woodstoves, and fireplaces. In addition, tobacco smoke is a combustion product that contributes to the contamination of indoor air. More than 2000 gaseous compounds have been identified in cigarette smoke, and carbon monoxide and nitrogen oxide are among them [1].

This section focuses on nitrogen oxides (primarily nitrogen dioxide) and carbon monoxide because they are frequently occurring products of combustion often found at higher indoor concentrations than outdoors. Other combustion products such as sulfur dioxide, hydrocarbons, formaldehyde, and carbon dioxide are produced by combustion sources to a lesser degree or only under unusual or infrequent circumstances.

Unvented kerosene and gas space heaters can provide an additional source of heat for homes in cold climates or can serve as a primary heating source when needed for homes in warm climates. There are several basic types of unvented kerosene and gas space heaters which can be classified by the type of burner and type of fuel. Unvented gas space heaters can be convective or infrared and can be fueled by natural gas or propane. Kerosene heaters can be convective, radiant, two-stage, and wickless. A recent study found that emission rates from the various types of heaters fall into three distinct groups. The two-stage kerosene heaters emitted the least CO and the least NO_2. The radiant/infrared heater group emitted the most CO under well-tuned conditions; and the convective group emitted the most NO_2. Many studies have also noted that some heaters have significantly higher emission rates than heaters of other brands or models of the same type. Older or improperly used heaters will also increase emission rates.

The kitchen stove is one of the few modern gas appliances that emit combustion products directly into the home. It is estimated that natural gas is used in over 45% of all U.S. homes, and studies show that most of these homes do not vent the combustion-produced emissions to the outside. Combustion gas emissions vary considerably and are dependent upon factors such as the fuel consumption rate, combustion efficiency, age of burner, and burner design, as well as the usage pattern of the appliance. An improperly adjusted gas stove is likely to have a yellow-tipped flame rather than a blue-tipped flame, which can result in increased pollutant emissions (mostly NO_2 and CO).

Increasing energy costs, consumer concerns about fuel availability, and desire for self-reliance, are some of the factors that have brought about an upswing in the use of solid fuels for residential heating. These devices include woodburning stoves, furnaces, and fireplaces. Although woodstoves and fireplaces are vented to the outdoors, a number of circumstances can cause combustion products to be emitted to the indoor air: improper installation (such as insufficient stack height), cracks or leaks in stovepipes, negative air pressure indoors, downdrafts, refueling, and accidents (as when a log rolls out of a fireplace). The type and amount of wood burned also influences pollutant emissions, which vary from home to home. Although elevated levels of CO and NO_2 have been reported, the major impact of woodburning appears to be on indoor respirable suspended particles (RSP).

The term nitrogen oxides (NO_x) refers to a number of compounds, all of which have the potential to affect humans. NO_2 and NO have been studied extensively as outdoor pollutants, yet cannot be ignored in the indoor environment. There is evidence that suggests these oxides may be harmful at levels of exposure that can occur indoors. Both NO and NO_2 combine with hemoglobin in the blood, forming methemoglobin, which reduces the oxygen-carrying capacity of the blood. It is about four times more effective than CO in reducing the oxygen-carrying capacity of the blood. NO_2 produces respiratory illnesses that range from slight burning and pain in the throat and chest to shortness of breath and violent coughing. It places stress on the cardiovascular system and causes short- and long-term damage to the lungs. Concentrations typically found in kitchens with gas stoves do not appear to cause chronic respiratory diseases, but may affect sensory perception and produce eye irritation.

Carbon monoxide (CO) is a poisonous gas that causes tissue hypoxia (oxygen starvation) by binding with blood hemoglobin and blocking its ability to transport oxygen. CO has in excess of 200 times more binding affinity for hemoglobin than oxygen does.

The product, carboxyhemoglobin, is an indicator of reduction in oxygen-carrying capacity. A small amount of CO is even produced naturally in the body, producing a concentration in unexposed persons of about 0.5% CO-bound hemoglobin. Under chronic exposure (for example, cigarette smoking), the body compensates somewhat by increasing the concentration of red blood cells and the total amount of hemoglobin available for oxygen transport. The central nervous system, cardiovascular system, and liver are most sensitive to CO-induced hypoxia. Hypoxia of the central nervous system causes a wide range of effects in the exposure range of 5%–15% carboxyhemoglobin. These include loss of alertness and impaired perception, loss of normal dexterity, reduced learning ability, sleep disruption, drowsiness, confusion, and at very high concentrations, coma and death. Health effects related to hypoxia of the cardiovascular system include decrease in exercise time required to produce angina pectoris (chest pain); increase in incidences of myocardosis (degeneration of heart muscle); and, a general increase in the probability of heart failure among susceptible individuals [3].

Population groups at special risk of detrimental effects of CO exposure include fetuses, persons with existing health impairments (especially heart disease), persons under the influence of drugs, and those not adapted to high altitudes who are exposed to both CO and high altitudes.

Proper installation, operation, and maintenance of combustion devices can significantly reduce the health risks associated with these appliances. Manufacturers' instructions regarding the proper size space heater in relation to room size, ventilation conditions, and tuning should be observed. This includes using vented range hoods when operating gas stoves. Studies have indicated reductions in CO, CO_2, and NO_2 levels as high as 60%–87% with the use of range hoods during gas stove operation. Unvented forced draft and unvented range hoods with charcoal filters can be effective for removing grease, odors, and other molecules, but cannot be considered a reliable control for CO and other small molecules. Fireplace flues and chimneys should be inspected and cleaned frequently, and opened completely when in use [2].

14.6 PARTICULATES

Environmental tobacco smoke, ETS (smoke that nonsmokers are exposed to from smokers), has been judged by the Surgeon General, the National Research Council, and the International Agency for Research on Cancer to pose a risk of lung cancer to nonsmokers. Nonsmokers' exposure to ETS is called "passive smoking," "second-hand smoking," and "involuntary smoking." Tobacco smoke contains a number of pollutants, including inorganic gases, heavy metals, particulates, VOCs, and products of incomplete burning, such as polynuclear aromatic hydrocarbons. Smoke can also yield a number of organic compounds. Including both gases and particles, tobacco smoke is a complex mixture of over 4700 compounds [1].

There are two components of tobacco smoke: (1) mainstream smoke, which is the smoke drawn through the tobacco during inhalation, (2) sidestream smoke, which arises from the smoldering tobacco. Sidestream smoke accounts for 96% of gases and particles produced [2].

Studies indicate that exposure to tobacco smoke may increase the risk of lung cancer by an average of 30% in the nonsmoking spouses of smokers. Published risk estimates of lung cancer deaths among nonsmokers exposed to tobacco smoke conclude that ETS is responsible for 3000 deaths each year [4]. It also seriously affects the respiratory health of hundreds of thousands of children. Very young children exposed to smoking at home are more likely to be hospitalized for bronchitis and pneumonia. Recent studies suggest that ETS can also cause other diseases, including other cancers and heart disease in healthy nonsmokers [1].

The best way to reduce exposure to cigarette smoke in the house is to quit smoking and discourage smoking indoors. Ventilation is the most common method of reducing exposure to these pollutants, but it will not eliminate it altogether. Smoking produces such large amounts of pollutants that neither natural nor mechanical methods can remove them from the air as quickly as they build up. In addition, ventilation practices sometimes lead to increased energy costs.

RSP are particles or fibers in the air that are small enough to be inhaled. Particles can exist in either solid or liquid phase or in a combination. Where these particles are deposited and how long they are retained depends on their size, chemical composition, and density. RSP (generally less than $10\,\mu m$ in diameter), can settle on the tissues of the upper respiratory tract, with the smallest particles (those less than $2.5\,\mu m$) penetrating the alveoli, the small air sacs in the lungs.

Particulate matter is a broad class of chemically and physically diverse substances that present risks to health. These effects can be attributed to either the intrinsic toxic chemical or physical characteristics, as in the case of lead and asbestos, or to the particles acting as a carrier of adsorbed toxic substances, as in the case of attachment of radon daughters. Carbon particles, such as those created by combustion processes, are efficient adsorbers of many organic compounds and are able to carry toxic gases such as sulfur dioxide into the lungs.

Asbestos is a mineral fiber used mostly before the mid-1970s in a variety of construction materials. Home exposure to asbestos is usually due to aging, cracking, or physical disruption of insulated pipes or asbestos-containing ceiling tiles and spackling compounds. Apartments and school buildings may have an asbestos compound sprayed on certain structural components as a fire retardant. Exposure occurs when asbestos materials are disturbed and the fibers are released into the air and inhaled. Consumer exposure to asbestos has been reduced considerably since the mid-1970s, when use of asbestos was either prohibited or stopped voluntarily in sprayed-on insulation, fire protection, soundproofing, artificial logs, patching compounds, and hand-held hair dryers. Today, asbestos is most commonly found in older homes in pipe and furnace insulation materials, asbestos shingles, millboard, textured paints and other coating materials, and floor tiles. Elevated concentrations of airborne asbestos can occur after asbestos-containing materials are disturbed by cutting, sanding, or other remodeling activities. Improper attempts to remove these materials can release asbestos fibers into the air in homes, thereby increasing asbestos levels and endangering the people living in those homes. The most dangerous asbestos fibers are too small to be visible. After they are inhaled, they can remain and accumulate in the lungs. Asbestos can cause lung cancer, mesothelioma (a cancer of the chest and abdominal linings), and asbestosis (irreversible lung scarring that can be fatal). Symptoms of these diseases do not show up until many years after exposure began. A more detailed presentation on asbestos can be found in Chapter 28.

Lead has long been recognized as a harmful environmental pollutant. There are many ways in which humans are exposed to lead, including air, drinking water, food, and contaminated soil and dust. Airborne lead enters the body when an individual breathes lead particles or swallows lead dust once it has settled. Until recently, the most important airborne source of lead was automobile exhaust. Lead-based paint has long been recognized as a hazard to children who eat lead-contained paint chips. A 1988 National Institute of Building Sciences Task Force report found that harmful exposures to lead can be created when lead-based paint is removed from surfaces by sanding or open-flame burning. High concentrations of airborne lead panicles in homes can also result from the lead dust from outdoor sources, contaminated soil tracked inside, and use of lead in activities such as soldering, electronics repair, and stained-glass artwork. Lead is toxic to many organs within the body at both low and high concentrations. Lead is capable of causing serious damage to the brain, kidneys, peripheral nervous system (the sense organs and nerves controlling the body), and red blood cells. Even low levels of lead may increase high blood pressure in adults. Fetuses, infants, and children are more vulnerable to lead exposure than are adults because lead is more easily absorbed into growing bodies, and the tissues of small children are more sensitive to the damaging effects of lead. The effects of lead exposure on fetuses and young children include delays in physical and mental development, lower IQ levels, shortened attention spans, and increased behavioral problems. Additional details on lead, as well as other metals, can be found in Chapter 29.

Particles present a risk to health out of proportion to their concentration in the atmosphere because they deliver a high-concentration package of potentially harmful substances. So, while few cells may be affected at any one time, those few that are can be badly damaged. Whereas larger particles deposited in the upper respiratory portion of the respiratory system are continuously cleared away, smaller particles deposited deep in the lung may cause adverse health effects. Particle sizes vary over a broad range, depending on source characteristics.

Major effects of concern attributed to particle exposure are impairment of respiratory mechanics, aggravation of existing respiratory and cardiovascular disease, and reduction in particle clearance and other host defense mechanisms. Respiratory effects can range from mild transient changes of little direct health significance to incapacitating impairment of breathing.

One method of reducing RSP concentrations is to properly design, install, and operate combustion sources. One should make sure there are no existing leaks or cracks in stovepipes, and that these appliances are always vented to the outdoors.

Also available are particulate air cleaners, which can be separated into mechanical filters and electrostatic filters. Mechanical filtration is generally accomplished by passing the air through a fibrous media (wire, hemp, glass, etc.). These filters are capable of removing almost any sized particles. Electrostatic filtration operates on the principle of attraction between opposite electrical charges. Ion generators, electrostatic precipitators, and electric filters use this principle for removing particles from the air.

The ability of these various types of air-cleaning devices to remove respirable particles varies widely. High efficiency particulate air (HEPA) filters can capture over 99% of particles, and are advantageous in that filters only need changing every 3–5 years, but costs can reach \$500–\$800. It is also important to note the location of air-cleaning device inlets in relation to the contaminant sources as an important factor influencing removal efficiencies.

14.7 BIOLOGICAL CONTAMINANTS

Heating, ventilation, and air conditioning systems and humidifiers can be breeding grounds for biological contaminants when they are not properly cleaned and maintained. They can also bring biological contaminants indoors and circulate them. Biological contaminants include bacteria, mold and mildew, viruses, animal dander and cat saliva, mites, cockroaches, and pollen. There are many sources for these pollutants. For example, pollens originate from plants; viruses are transmitted by people and animals; bacteria are carried by people, animals, and soil and plant debris; and, household pets are sources of saliva, hair, and dead skin (known as dander).

Available evidence indicates that a number of viruses that infect humans can be transmitted via the air. Among them are the most common infections of mankind. Airborne contagion is the mechanism of transmission of most acute respiratory infections, and these are the greatest of all causes of morbidity.

The primary source of bacteria indoors is the human body. Although the major source is the respiratory tract, it has been shown that 7 million skin scales are shed per minute per person, with an average of four viable bacteria per scale [3]. Airborne transmission of bacteria is facilitated by the prompt dispersion of particles. Infectious contact requires proximity in time and space between host and contact, and is also related to air filtration and air exchange rate.

Although many important allergens—such as pollen, fungi, insects, and algae—enter buildings from outdoors, several airborne allergens originate predominately in homes and office buildings. House dust mites, one of the most powerful biologicals in triggering allergic reactions, can grow in any damp, warm environment. Allergic reactions can occur on the skin, nose, airways, and alveoli.

The most common respiratory diseases attributable to these allergens are rhinitis, affecting about 15% of the population, and asthma, affecting about 3%–5% [3]. These diseases are most common among children and young adults, but can occur at any age. Research has shown that asthma occurs four times more often among poor, inner-city families than in other families. Among the suspected causes are mouse urine antigens, cockroach feces antigens, and a type of fungus called Alternia.

Hypersensitivity pneumonitis (HP), characterized by shortness of breath, fever, and cough, is a much less common disease, but is dangerous if not diagnosed and treated early. HP is most commonly caused by contaminated forced-air heating systems, humidifiers, and flooding disasters. It can also be caused by inhalation of microbial aerosols from saunas, home tap water, and even automobile air conditioners. Humidifiers with reservoirs containing stagnant water may be important sources of allergens in both residential and public buildings.

Some biological contaminants trigger allergic reactions, while others transmit infectious illnesses, such as influenza, measles, and chicken pox. Certain molds and mildews release disease-causing toxins. Symptoms of health problems caused by biologicals include sneezing, watery eyes, coughing, shortness of breath, dizziness, lethargy, fever, and digestive problems.

Attempts to control airborne viral disease have included quarantine, vaccination, and inactivation or removal of the viral aerosol. Infiltration and ventilation play a large role in the routes of transmission. Because many contaminants originate outdoors, attempts to reduce the ventilation rate might lower indoor pollutant concentrations. However, any reduction in fresh air exchange should be supplemented by a carefully filtered air source.

Central electrostatic filtration (as part of a home's forced-air system) has proven effective in reducing indoor mold problems. Careful cleaning, vacuuming, and air filtration are effective ways to reduce dust levels in a home. Ventilation of attic and crawl spaces help prevent moisture buildup, keeping humidity levels between 30% and 50% [5]. Also, when using cool mist or ultrasonic humidifiers, one should remember to clean and refill water trays often, since these areas often become breeding grounds for biological contaminants.

14.8   MONITORING METHODS

Methods and instrumentation for measuring IAQ vary in their levels of sensitivity (what levels of pollutant they can detect) and accuracy (how close they can come to measuring the true concentration). Instruments that can measure low levels of pollutant very accurately are likely to be expensive and require special expertise to use. Some level of sensitivity and accuracy is required, however, to ensure that data collected are useful in assessing levels of exposure and risk.

In choosing methods for monitoring IAQ, a tradeoff must be made between cost and the levels of sensitivity, accuracy, and precision achieved in a monitoring program. Required levels for each pollutant are based on ranges found in residential buildings. In providing detailed information concerning specific methods or instruments, emphasis is placed on those that are readily available, easy to use, reasonably priced, and that provide the required levels of sensitivity and accuracy.

Methods to monitor indoor air fall into several broad categories. Sampling instruments may be fixed location, portable, or small personal monitors designed to be carried by an individual. These samplers may act in an active or passive mode. Active samplers require a pump to draw in air. Passive samplers rely on diffusion or permeation.

Monitors may be either analytical instruments that provide a direct reading of pollutant concentration, or collectors that must be sent to a laboratory for analysis. Instruments may also be categorized according to the time period over which they sample. These include grab samplers, continuous samplers, and time-integrated samplers, each of which is briefly described below.

1. *Grab sampler*: Collects samples of air in a bag, tube, or bottle, providing a short-term average.
2. *Continuous sampling*: Allows sampling of real-time concentration of pollutants, providing data on peak short-term concentrations and average concentrations over the sampling period.
3. *Time-integrated sampling*: Measures an average air concentration over some period of time (active or passive), using collector monitors that must be sent out for analysis; cannot determine peak concentrations.

More details regarding monitoring methods for specific indoor air pollutants can be found in the *IAQ Handbook* [3].

CosaTron is just one example of a company that produces mechanical air-cleaning devices. The patented CosaTron system has been handling IAQ successfully in thousands of installations for over 25 years. CosaTron is not a filter that ionizes air. It cleans the air electronically, causing the submicron particles of smoke, odor, dirt, and gases to collide and adhere to each other until they become larger and airborne and are easily carried out of the conditioned space by the system air flow to be exhausted or captured in the filter. Mechanical air devices such as this one improve IAQ so much that outside air requirements can be reduced significantly [6].

14.9   FUTURE TRENDS

In recent years, the EPA has increased efforts to address IAQ problems through a building systems approach. EPA hopes to bolster awareness of the importance of prevention and encourage a whole systems perspective to resolve indoor air problems. The EPA Office of Research and Development is also conducting a multidisciplinary IAQ research program that encompasses studies of the health effects associated with indoor air pollution exposure, assessments of indoor air pollution sources and control approaches, building studies and investigation methods, risk assessments of indoor air pollutants, and a recently initiated program on biocontaminants.

Federal research on air quality issues is driven in part by the increasing attention that IAQ has attracted from journalists as well as scientists and engineers. EPA has performed comparative studies that have consistently ranked indoor air pollution among the top five environmental risks to public health. In analyzing over 500 IAQ investigations conducted through the end of 1988, the National Institute for Occupational Safety and Health (NIOSH) categorized its findings into seven broad sources of poor IAQ: inadequate ventilation (53%), inside contamination (15%), outside contamination (13%), microbiological contamination (5%), building materials contamination (4%), and unknown sources (13%) [4]. Since then, ventilation has been the primary focus of most EPA programs.

Requirements for clean air are still changing rapidly and most buildings will need to be refitted with different filters to meet with these new standards and guidelines. EPA's research will continue in these and other areas to try to ensure comfortable and clean air conditions for the indoor environment.

14.10   SUMMARY

1. IAQ is rapidly becoming a major environmental concern since levels of indoor air pollutants are often higher than levels outdoors and a significant amount of people spend the majority of their time indoors.
2. Radon is a naturally occurring, colorless, odorless gas that can be found almost anywhere at very low levels. Radon may be the leading cause of lung cancer among nonsmokers.
3. Formaldehyde is a colorless, water-soluble gas with a pungent odor that can be found in a variety of household products as well as building materials.

4. A wide variety of organic compounds is associated with the use of various household cleaners, pesticides, and painting materials.

5. More than 2000 gaseous compounds have been identified in cigarette smoke. Among them are carbon monoxide and nitrogen oxide, which significantly reduce the oxygen-carrying capacity of the blood.

6. Heating, ventilation, air condition systems, and humidifiers can be breeding grounds for biological contaminants when they are not properly cleaned and maintained.

7. Basic strategies to improve IAQ include source control, ventilation, and mechanical devices.

8. Recent studies have focused on improving ventilation techniques and proper air quality control.

REFERENCES

1. U.S. EPA, Environmental progress and challenges. *EPA's Update*, August 1988.
2. Taylor, J. *Sampling and Calibration for Atmospheric Measurements*. Philadelphia, PA: ASTM Publication, 1987.
3. Mueller Associates, Inc. *Indoor Air Quality Environmental Information Handbook: Building System Characteristics*. Baltimore, MD: Author, 1987.
4. Cox, J. E. and Miro, C. R. EPA, DOE, and NIOSH address IAQ problems. *ASHRAE Journal*, July 1993, 10.
5. Burke, G., Singh, B., and Theodore, L. *Handbook of Environmental Management and Technology*. Hoboken, NJ: John Wiley & Sons, 2000.
6. Four proven solutions for IAQ! *ASHRAE Journal*, March 1993, 2.

15 Vapor Intrusion

CONTENTS

15.1 INTRODUCTION

Vapor intrusion (VI) caused by releases of volatile chemicals from contaminated soil and/or groundwater is rapidly emerging as a serious concern with potentially significant impact on thousands of properties across the nation. Volatile chemicals such as trichloroethylene, perchloroethylene, and benzene may be released from contaminated soil and/or groundwater at properties such as existing or former gas stations, dry cleaners, and industrial facilities. These volatile chemicals may migrate through subsurface soils and into indoor air spaces of overlying structures similar to the way radon gas can seep into homes. The driving force can be convective (bulk flow) or diffusive (concentration) driving force and commonly enters via cracks or openings in the building floors or walls. These vapors can also migrate as a result of pressure differences between the building's interior (lower pressure) and exterior. This condition can create a negative pressure within the structure that effectively draws the vapors into the building. Thus, contaminated groundwater can pose a potential VI threat to inhabitants of nearby buildings and/or structures.

The concern is widespread. According to the U.S. General Accounting Office (GAO, 2002), an estimated 200,000 underground storage tanks currently in operation may be leaking. According to Environmental Protection Agency (EPA), there are more than 36,000 active dry cleaning facilities in operation in the United States, with more than 75% of them estimated to be contaminated with volatile chemical solvents. Tens of thousands of current and former industrial sites across the United States are contaminated with volatile chemicals. The EPA 2002 VI guidance document references a total of 374,000 contaminated sites, the National Research Council reports that the number may be as high as 439,000, and an often cited total in brownfields redevelopment literature is 500,000 sites. The fraction of these contaminated sites with conditions favorable for VI also is not known with certainty, but will depend, in part, on the number of sites that contain volatile organic compounds (VOCs). Volatile contaminants have been reported at approximately one-half of all Superfund and similar cleanup sites. Preliminary estimates suggest that approximately one-half of volatile-contaminated sites have conditions that could be favorable for VI. This suggests, therefore, that VI may be an issue at one-quarter of the total number of contaminated sites in the United States [1].

15.2   HEALTH CONCERNS [2,3]

The health effects from chemical vapor exposures vary based on the individual exposed, the chemical involved, the exposure dose and time. Impacts may include eye and respiratory irritation, headache, and/or nausea. Low-level exposure over long periods of time may also raise a person's lifetime risk for developing cancer.

The three most commonly considered routes for environmental contaminants to enter the human body are ingestion, dermal contact, and inhalation. The VI route of exposure is inhalation.

VI exposures occur indoors and people in the United States spend much of their lives indoors. People who are unhealthy or who are relatively more susceptible to the effects of toxicants—for example, people who are elderly, ill, or immobile; pregnant women and their developing fetuses; newborns, infants, and toddlers—also spend much of their time indoors.

Children are at a higher risk than adults for both physiological and logistical reasons. Physiology-based studies indicate that there is a twofold greater inhalation dose in children than adults. Also, very young children spend substantial amounts of time at floor level, potentially closer to the location of intruding vapors. They presumably could be exposed to higher concentrations than adults since the molecular weight of the vapors of concern is greater than that of air. Their greater density is more likely to produce higher concentrations at ground (floor) level.

Inhalation is not voluntary. The typical adult is assumed to inhale approximately 20,000 L/day of air and consume approximately 2 L/day of drinking water. Individuals may forego drinking tap water and use alternative sources, but they cannot forgo inhaling air. Obviously, the concentration of contaminants in breathing air is important, and measuring or predicting this concentration is a focus of some VI studies.

As expected, the inhalation route of exposure has been observed to lead to higher toxicities than exposures via the oral route or entry. The higher toxicities for inhalation may reflect the fact that the barrier between contaminated air and the human blood system is as small as a single cell and that these cells are membranes whose purpose is the exchange of inhaled gases with the blood.

Pollutant concentrations of $100\,\mu g/m^3$ or greater have been observed in indoor air due to VI. Assuming a 24 h exposure and $20\,m^3/day$ of respiration, the expected adult applied does of $2000\,\mu g/day$ ($2\,mg/day$) of these toxicants could be significant to the health of some individuals. Even higher levels of exposure are possible. Concentrations for a single VOC of $790\,\mu g/m^3$ or even $1700\,\mu g/m^3$ have been observed in indoor air due to VI.

15.3   PROPERTY ENVIRONMENTAL DUE DILIGENCE [3]

There has been considerable confusion and a high degree of inconsistency in the conduct of property environmental due diligence to evaluate the potential for VI resulting from soil and groundwater contamination on the target property or neighboring properties. Moreover, there is even question whether or not this is an appropriate assessment to be included in a Phase I conducted according to ASTM E 1527-05. Notwithstanding, an AAI-compliant Phase I may make it difficult for an environmental professional to ignore VI concerns.

VI can be particularly difficult to assess because vapors tend to migrate along the path of least resistance, without regard to what is up-gradient or down-gradient hydraulically. Considerable concern has already been raised about the significant percentage of false positives in using the screening criteria established by EPA and state regulators. There are questions about the proper application of soil gas surveying, and even more questions about the use of indoor air sampling.

EPA has indicated that the potential costs and liabilities associated with VI impacts may be orders of magnitude greater than those associated with traditional groundwater contamination issues. With the growing trend of federal and state policy and regulations directed at this potential problem, it behooves the industry to provide consistent and reasonable guidance on how VI should be addressed in real estate transaction due diligence.

15.4   CONTROL OPTIONS [2]

For groundwater contamination in general, public health protective and cost-effective exposure controls (e.g., providing permanent alternate water supplies) have been used to successfully avoid a great deal of toxic exposure plus an unknown number of cases of disease over the years. It is possible that similar measures for VI-related exposures would have a similar positive outcome.

Providing alternate water supplies is a recognized technology "standard" for preventing inappropriate tap water exposures from contaminated groundwater. Traditional technology-based equipment, which have typically been used for the control of point-source emissions to air, are recognized as providing some of the most effective and cost-effective improvements in environmental quality in this country. The pollution prevention approach could be another way to help manage the very large uncertainties that exist in the understanding of the VI exposure pathway. Such preventive approaches could help provide environmental managers, property developers, and decision-makers with a more defensible, and likely more cost-effective answer to VI problems.

15.5   EPA, STATE, AND ASTM VI ACTIVITY

To respond to the growing problem of VI, in December 2001, EPA issued its draft *Guidance for Evaluating the Vapor Intrusion into Outdoor Air Pathway*. This document was updated in November 2002 and is currently in the process of being updated again.

States have also been a leading force in VI regulation and policy. New York's comprehensive draft *Guidance for Evaluating Soil Vapor Intrusion in the State of New York* was released for public comment in February 2005. New Jersey's comprehensive draft *Vapor Intrusion Guidance* was published in June 2005. Alaska's *Evaluation of Vapor Intrusion Pathway at Contaminated Sites* was published in September 2005. Today 16 states, including Alaska, California, Colorado, Connecticut, Indiana, Maine, Massachusetts, Michigan, Minnesota, Nebraska, New Hampshire, New Jersey, New York, Pennsylvania, Washington, and Wisconsin, have policy and regulations to address the VI problem.

In October 2005, the real estate, banking, legal, and insurance industries, together with the Phase I industry, seeking to resolve the uncertainty surrounding VI and clarify how it fits in the property environmental due diligence process, approached ASTM with a request for ASTM to develop a national standard for the assessment of VI. On October 28, 2005, ASTM approved the formation of a Vapor Intrusion Task Group (E50.02.06) with responsibility to develop a standard.

15.6   ASTM TASK GROUP [4]

ASTM was selected as the best venue to develop the standard because of ASTM's internationally recognized consensus-based process that has been used so successfully over the years. ASTM is able to bring together stakeholders representing all sides of an issue and work with them to achieve consensus.

More than 200 professionals volunteered to participate on the ASTM Vapor Intrusion Task Group, including representatives from the environmental consultant industry, lenders, lawyers, corporations, real estate investors and developers, and federal and state regulatory agencies. Both EPA and the Interstate Technology and Regulatory Council (ITRC) are represented on the Task Group. ITRC is a state-led, national coalition with representatives from environmental regulatory agencies in 40 states, the District of Columbia and three federal agencies. In addition, representatives from the Aerospace Industries Association, Mortgage Bankers Association, American Petroleum Institute, Halogenated Solvents Industry Alliance and the Environmental Bankers Association participate on the Task Group.

The Task Group's objective was to define good commercial and customary practice for conducting a VI assessment on a property parcel involved in a real estate transaction. The specific intent was to establish a methodology to determine whether or not there is a reasonable probability that VI could present an environmental risk and liability. For commercial real estate transactions, the VI investigation as defined by the standard could be used independently of, or as a supplement to a Phase I environmental site assessment (ESA).

The standard in development prescribes a tiered process designed to quickly screen out properties with a low risk of VI. The standard introduces a number of new terms:

1. *Vapor Intrusion Condition* (VIC), defined as "the presence or likely presence of any *chemicals of concern* in the indoor air environment of existing or planned structures on a property caused by the release of vapor from contaminated soil or groundwater on the property or within close proximity to the property, at a concentration that presents or may present an unacceptable health risk to occupants." The standard only deals with indoor air emissions emanating from contaminated soil or groundwater.
2. *Potential Vapor Intrusion Condition* (pVIC), defined by the standard when screening indicates the possibility of a VIC, but where there is insufficient data to ascertain the presence or likely presence of chemicals of concern (COC) in the indoor air environment.
3. COC, defined as a chemical in the subsurface environment that is known or reasonably expected to be present, that can potentially migrate as a vapor into an existing or planned structure on a property, and that is generally recognized as having the potential for an adverse impact on human health. COC meet specific criteria for volatility and toxicity, and include VOCs, semi-VOCs, and collative inorganic analytes such as mercury. An appendix in the standard lists common COC meeting the criteria.

The process defined in this practice begins with a reasonably conservative screening effort requiring information that would be collected as part of an ASTM E 1527 Phase I ESA. If a pVIC is identified in this initial screening, the process gradually progresses toward a more complex assessment involving increasingly greater use of site-specific data. For those sites unable to be screened out, the process provides alternative methods to determine whether a VIC exists. If a VIC is found to exist, the process describes general mitigation alternatives.

Specifically, the evaluation process consists of four tiers. The first two tiers are screening tiers designed to assess the "potential" for a VI condition (i.e., pVIC) to exist so that properties with a low risk of VI can be screened out quickly and inexpensively as the data justify. If the potential for VI cannot reasonably be eliminated at the Tier 1 and/or Tier 2 levels, the process identifies three options: (1) proceed with a more site-specific and comprehensive investigation (Tier 3), in the hope that this investigation will eliminate VI concerns; (2) proceed directly to mitigation (Tier 4), on the assumption that mitigation conducted preemptively may be more cost effective to address a pVIC; or (3) gain more certainty on the presence of a pVIC through additional investigation. Tier 3 presents a "toolbox" of activities that can accomplish this. Tier 4 addresses mitigation alternatives.

Timeliness may be more important than investigation or mitigation costs during real estate transactions. As such, a user can proceed to any of the tiers in the process. It is not necessary to progress sequentially through each tier. In most cases, however, it is the real estate transaction responsibilities to conduct a Tier 1 screening evaluation, and possibly a Tier 2 screening evaluation before proceeding to a more costly and time consuming Tier 3 investigation or to Tier 4 mitigation.

The VI assessment process described in the standard is designed to complement existing federal or state VI policies or guidance. The flowchart in Figure 15.1

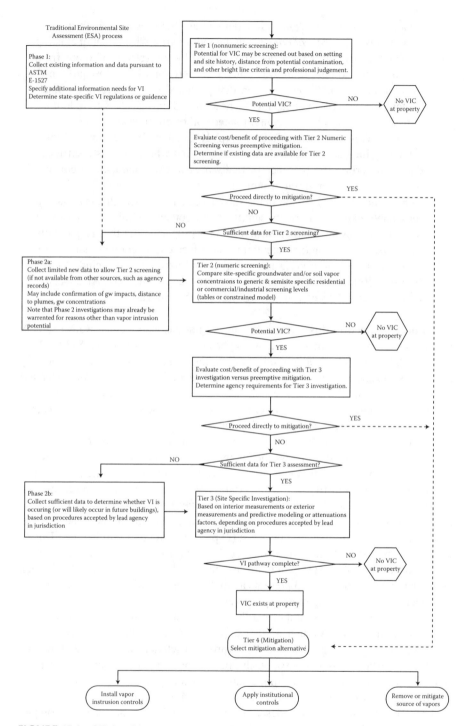

FIGURE 15.1   VI tiered assessment approach.

indicates the four tiers of the VI assessment process when conducted in conjunction with a Phase I. Additional details are available in the literature [5].

15.7 FUTURE TRENDS

In many ways, VI presents a major environmental challenge, but is also provides a tremendous opportunity for regulators and responsible parties to work together toward a better future by helping implement the cost-effective protection of public health by preventing further unnecessary VI exposures today.

Numerous VI information internet links are available to assist the technical community. The two most useful (EPA) are provided below.

1. www.epa.gov/correctiveaction/eis/vapor.html
   OSWER Draft Guidance for Evaluating the Vapor Intrusion to Indoor Air Pathway from Groundwater and Soils (Subsurface Vapor Intrusion Guidance)
2. www.epa.gov/epaoswer/hazwaste/ca/eis/vapor.html#2001
   RCRA Draft Supplemental Guidance for Evaluating the VI to Indoor Air Pathway

Additional sites are provided in the literature [6].

15.8 SUMMARY

1. VI caused by releases of volatile chemicals from contaminated soil and/ or groundwater is rapidly emerging as a serious concern with potentially significant impact on thousands of properties across the nation.
2. Volatile chemicals such as trichloroethylene, perchloroethylene, and benzene may be released from contaminated soil and/or groundwater at properties such as existing or former gas stations, dry cleaners, and industrial facilities; these volatile chemicals may migrate through subsurface soils and into indoor air spaces of overlying structures similar to the way radon gas can seep into homes.
3. The health effects from chemical vapor exposures vary based on the individual exposed, the chemical involved, and the exposure dose and time.
4. The inhalation route of exposure has been observed to lead to higher toxicities than exposures via the oral route or entry.
5. VI exposures occur indoors and people in the United States spend much of their lives indoors. People who are unhealthy or who are relatively more susceptible to the effects of toxicants—for example, people who are elderly, ill, or immobile; pregnant women and their developing fetuses; newborns, infants, and toddlers—also spend much of their time indoors.
6. Children are at a higher risk than adults. Physiology-based studies indicate that there is a twofold greater inhalation dose in children than adults.
7. VI presents a major environmental challenge, but is also provides a tremendous opportunity for regulators and responsible parties to work together toward a better future by helping implement the cost-effective protection of public health by preventing further unnecessary VI exposures today.

REFERENCES

1. H. Schuver, *Vapor Intrusion: Risk and Challenges*, EM, A&WMA, Pittsburgh, February 2007.
2. Adopted from: H. Schuver, *Vapor Intrusion: Risk and Challenges*, EM, A&WMA, Pittsburgh, February 2007.
3. A.J. Buonicore, Private communication, 2006.
4. A.J. Buonicore, Private communications, 2007.
5. A.J. Buonicore, *Upcoming ASTM Standards for Assessment of Vapor Intrusion*, A & WMA Specialty Conference on Vapor Intrusion, September 2007.
6. M. Traister, *Understand the Emerging Issue of Vapor Intrusion*, CEP, New York, 10/2007.

Part III

Water

Part III of this book comprises five chapters and serves as an introduction to water quality. Part III comprises five chapters. Chapter 16 is concerned with the general subject of water or aquatic chemistry. Safe drinking water is reviewed in Chapter 17. Chapter 18 is concerned with municipal water pollution control equipment. A reasonably comprehensive examination of industrial water pollution control equipment is provided in Chapter 19. Chapter 20 addresses the general subject of dispersion modeling in water systems. Part III concludes with material concerned with acid rain (Chapter 21).

16 Water Chemistry

Contributing Author: Richard F. Carbonaro

CONTENTS

16.1 INTRODUCTION

Water chemistry deals with the fundamental chemical properties of water itself, the chemical properties of other constituents that dissolve in water, and the countless chemical reactions that take place in water. The field of natural water chemistry is concerned principally with reactions that occur in relatively dilute solution (low concentrations), although some natural waters have rather high solute concentrations [1]. During a chemical reaction, tiny subatomic particles (e.g., electrons) and atoms (e.g., hydrogen) are transferred, shared, and exchanged. When a chemical reaction occurs in water, these changes require transport through the water medium. Water is not passive in these chemical reactions. Instead, it plays an active role, constantly making and breaking chemical bonds, thereby facilitating chemical change.

16.2 PHYSICAL PROPERTIES OF WATER

A water (H_2O) molecule consists of three atoms: two hydrogen atoms each of which are bonded to a central oxygen atom. Water can exist in three states: solid, liquid, and gas. At room temperature and atmospheric pressure water is a liquid, but below 0°C (32°F) it freezes and turns into ice. Water is present in the gaseous state above 100°C (212°F) and 1.0 atmosphere pressure. This is the boiling point of water, at which water will evaporate.

Water has a number of unusual physical properties that are a consequence of its hydrogen bonding among neighboring water molecules. The hydrogen bond is a weak bond that is the result of the dipolar nature of all water molecules. The hydrogen bonds between water molecules impart water with a relatively large heat capacity, heat of vaporization, and heat of fusion than that expected for a molecule of its size. Hydrogen bonds are also important biologically. Bonding between adjacent base pairs holds double-stranded DNA together. Many proteins also utilize hydrogen bonding to hold their three-dimensional shape and assist in enzymes binding to their substrate.

Hydrogen bonding also causes water to expand upon freezing, which results in water ice being less dense than liquid water. As a result, water collecting in the cracks of rocks will expand upon freezing which is an important mechanism for mechanically breaking rocks apart. This mechanical weathering breaks rocks into smaller fragments, which increases their surface area. This in turn increases the breakdown of the rock by surface chemical reactions, a process known as chemical weathering.

16.3   CHEMICAL PROPERTIES OF WATER

Water is called the "universal solvent" because it is capable of dissolving many substances. This chemical property of water arises from the dipolar nature of water molecules. Water molecules effectively surround positively charged ions (cations) and negatively charged ions (anions) which serve to prevent them from precipitating as solids. This means that wherever water goes, either through the ground or through one's body, it carries with it various solutes such as dissolved minerals, nutrients, organics, and heavy metals.

Even pure water contains some amount of hydrogen ion (H^+) and hydroxide ion (OH^-) as a result of a chemical reaction known as the autoionization of water. The concentration of these ions change as acids and bases are added to water; however, the product of their concentrations is always constant. The relative presence of H^+ and OH^- is measured by the pH, which was defined earlier as the negative logarithm of the hydrogen ion (H^+) concentration in mol/L. A decrease in pH indicates an increase in the H^+ concentration and a corresponding increase in the OH^- concentration.

Pure water has a neutral pH of exactly 7.0. Values of pH less than this are considered acidic, while pH values greater than this are termed alkaline. The typical pH range of water is from 0 to 14, although values outside of this range can be attained under extreme conditions. The pH of water is an important chemical property which controls the distribution of chemical species among various forms. For example, the pH is buffered at precise values in animal cells to maintain functionality of specific enzymes and proteins. Likewise in the environment, the pH controls the distribution of chemicals amongst their various forms and also controls the rate at which many chemical reactions occur. It is therefore often necessary to precisely measure the pH of water to understand its water chemistry.

16.4   CHEMICAL COMPOSITION OF NATURAL WATERS

The composition of natural waters is often described according to its physical qualities, chemical constituents, and/or its biological inhabitants. The focus to follow is primarily concerned with chemical constituents. Water sampling programs are used to obtain information on the chemical characteristics of potential and existing water sources and the performance of water and wastewater treatment plants [2]. Typically, prior knowledge of the type of chemical constituent (i.e., organic vs. inorganic, dissolved vs. suspended) is required to design and implement effective sampling programs. Preservatives are often added to prevent degradation of certain constituents, and holding times have been recommended by the EPA and other agencies so as to maintain proper quality control [3].

16.4.1 DISSOLVED MINERALS

Soil water in the ground reacts with common rock-forming minerals to release ions and form new minerals. Table 16.1 lists the most commonly occurring chemical elements present in the Earth's crust. The most abundant group of minerals in crustal rocks is a family called the feldspars. These minerals are comprised of sodium, potassium, and calcium aluminum silicates. They react with water, thereby producing Na^+, K^+, Ca^{2+}, Al^{3+}, and H_4SiO_4. Magnesium (Mg^{2+}) and iron (Fe^{2+}) are released from other silicate minerals. Carbonate rocks, limestone ($CaCO_3(s)$), and dolomite ($(Ca,Mg)CO_3(s)$) weather to release Ca^{2+}, Mg^{2+}, and HCO_3^-. Phosphorous (PO_4^{3-}) is released by the chemical weathering of apatite, a calcium phosphate mineral. The soluble constituents described above, Na^+, K^+, Ca^{2+}, Mg^{2+}, HCO_3^-, and H_4SiO_4, along with Cl^- and SO_4^{2-} find their way to rivers and streams and eventually reach the ocean. Over time, the concentrations of these constituents have increased to the levels found in the oceans today.

Aluminum (Al) and iron (Fe), mobilized from chemical weathering processes, have low solubility and are not transported over large distances. Under aerobic conditions, iron either stays behind as a hydroxide or oxide coating on the surface of

---

TABLE 16.1
Composition of the Earth's Crust

Element	%	Element	%
O (oxygen)	46.6	Cl (chlorine)	0.1
Si (silicon)	27.7	Cr (chromium)	0.04
Al (aluminum)	8.1	C (carbon)	0.03
Fe (iron)	5.0	V (vanadium)	0.02
Ca (calcium)	3.6	Ni (nickel)	0.008
Na (sodium)	2.8	Cu (copper)	0.007
K (potassium)	2.6	Co (cobalt)	0.002
Mg (magnesium)	2.1	Pb (lead)	0.001
Ti (titanium)	0.6	Sc (scandium)	0.0005
Mn (manganese)	0.1	Zn (zinc)	0.0001

Sources: Lutgens, F.K. and Tarbuck, E.J., *Essentials of Geology*, 7th edn., Prentice Hall, Upper Saddle River, NJ, 2000; and Nicholls, D., *Complexes and First-Row Transition Elements*, American Elsevier, New York, 1975.

---

weathered rocks, or attaches to small particles that remain suspended and are carried with the flowing water. Aluminum precipitates as $Al(OH)_3(s)$, or reacts with H_4SiO_4 to form the mineral kaolinite and other clay minerals.

Hardness is a bulk chemical property that measures the presence of specific dissolved mineral ions. Calcium and magnesium dissolved in water are the two most common minerals that make water "hard." Hard water interferes with washing clothes, dishwashing, and bathing. Clothes laundered in hard water may look dingy and feel harsh or scratchy. Hard water causes a film on surfaces when it evaporates due to the presence of a thin layer of salt that remains. Water flow may be reduced by mineral deposits in pipes. Synthetic detergents are usually less effective in hard water because the active ingredient is partially inactivated by the high levels of calcium and magnesium.

Another bulk measurement of the dissolved ion content of water is total dissolved solids (TDS). TDS is a measure of all of the dissolved ions in solution, and is determined by filtering out any suspended material in the water, evaporating off the water, and weighing the dry residue that remains. TDS levels indicate the potential uses for a water body since the TDS is basically an indicator of the salt content. Freshwater has a TDS of less than 1000 mg/L. Surface waters with significantly high TDS may not be usable for potable water or irrigation purposes. The TDS of sea water is approximately 35,000 mg/L. This means that for every kilogram of seawater there are approximately 35 g of dissolved salt.

Alkalinity is a measure of the buffering capacity of water. It does not refer to pH, but instead refers to the ability of water to resist change in pH upon addition of acid or base. Waters with low alkalinity are very susceptible to changes in pH, while waters with high alkalinity are able to resist major shifts in pH. The buffering chemicals in most natural waters are bicarbonate (HCO_3^-), and carbonate (CO_3^{2-}), although borates, silicates, phosphates, ammonium, sulfides, and organic acids can also contribute to a small degree. Water having a pH below 4.5 contains virtually no alkalinity, because all of the bicarbonate and carbonate have been converted to dissolved carbon dioxide. The amount of alkalinity therefore determines the ability of a water body to neutralize acidic pollution from rainfall or wastewater. Most surface waters typically have alkalinity ranging from 10 to 200 mg/L as $CaCO_3$ [1].

16.4.2 Dissolved Gases

As noted in Part II, the Earth's atmosphere is comprised of a layer of gases that are retained by gravity. It contains roughly 78% nitrogen ($N_2(g)$), 21% oxygen ($O_2(g)$), 0.93% argon ($Ar(g)$), 0.038% carbon dioxide ($CO_2(g)$), trace amounts of other gases, and a variable amount (about 1% on average) of water vapor. The atmosphere protects life on Earth by filtering out harmful ultraviolet solar radiation, and by trapping infrared radiation (heat) from escaping which regulates surface temperatures at habitable levels.

All gases present in the Earth's atmosphere dissolve to some extent into water that is in contact with it. Thus, all surface water has small amounts of N_2, O_2, Ar, CO_2, and other gases dissolved in it. For aquatic life forms, the presence of this small amount of oxygen is essential for survival. Most species of fish, for example, require at least 5 mg/L of dissolved oxygen. CO_2 acts as a weak acid when dissolved in water thereby imparting rainwater with its characteristic slightly acidic pH. N_2 and Ar, although present in all waters, do not engage in chemical reactions to any significant extent and their presence is usually ignored.

16.4.3 HEAVY METALS

The term "heavy metals" is an ambiguous one, and not necessarily associated with any specific set of elements, and therefore does not imply any common set of properties (such as high toxicity, high atomic weight, etc.) [4]. Nonetheless, the term has been used more and more in the literature. A simple, but useful, definition of a heavy metal is any metallic element on the periodic table with an atomic number larger than that of calcium (atomic number = 20). Examples of heavy metals include titanium (Ti), manganese (Mn), chromium (Cr), vanadium (V), copper (Cu), cobalt (Co), lead (Pb), scandium (Sc), and zinc (Zn).

Heavy metals are natural components of the Earth's crust (see Table 16.1). Unlike organic chemicals, they cannot be degraded or destroyed. To a small extent they enter one's body via food, drinking water, and air. Some heavy metals (e.g., copper, selenium, zinc, etc.) are micronutrients, and are essential to maintain the metabolism of the human body. However, at higher concentrations they can cause adverse health effects. Some metals that are in low abundance can have a large environmental impact. Mercury (Hg), arsenic (As), selenium (Se), and silver (Ag), for example, are all widely considered to be environmental stressors. Small amounts of these compounds can be harmful to both human health and aquatic life.

Heavy metals can enter a water supply by industrial and consumer waste, from drinking-water contamination (e.g., lead and copper pipes), or from acidic rain breaking down soils and leaching heavy metals into streams, lakes, rivers, and groundwater. Additional routes of exposure include inhalation of high ambient air concentrations near emission sources, or ingestion of metals via the food chain.

16.4.4 ORGANIC CONSTITUENTS

Organic chemistry is the study of the properties of chemical compounds containing organic carbon. Nearly all chemical compounds that have carbon atoms are considered organic. The exceptions, termed inorganic carbon compounds, include carbon dioxide, bicarbonate and carbonate, cyanide, metal carbides, and a handful of other compounds. On the other hand, the number of organic carbon compounds is impossibly large to count, a quality that arises from the fact that carbon readily bonds with other carbon atoms. This creates countless ways in which carbon atoms can be arranged relative to one another. In many organic carbon molecules, carbon is bound to hydrogen. In addition, carbon atoms are also willing to bind with other elements such as oxygen, nitrogen, phosphorous, sulfur, and chlorine, thereby further increasing the possible number of distinct and different organic carbon constituents that may be present in water.

Organic chemicals of environmental interest are usually classified into various groups or categories based upon similar chemical properties or common origins.

The most basic distinction is made based upon whether an organic chemical is naturally occurring or synthetic. Naturally occurring organic molecules include fossil fuels such as methane (CH_4) gas and the complex mixture of compounds present in petroleum, sugars and starches, and biomolecules such as proteins and enzymes. The G, T, C, and A base pairs of DNA are organic, thereby making DNA an organic molecule.

As plant, animal, and microbial material in soil and water undergo decomposition, a variety of complex organic molecules are produced that are called natural organic matter (NOM). Although NOM is ubiquitous in the environment, the structure and properties of the molecules themselves are not well understood. NOM plays an important role in aquatic toxicology because it interacts with metal ions and minerals to form complexes of a widely differing chemical and biological nature [7]. When NOM binds with metal ions, they become less bioavailable, which lowers the potential toxicity to aquatic life. However, NOM creates problems for the water supply industry, requiring removal to minimize water color and giving rise to potentially harmful chemical by-products after chlorination. Through a process called "biofouling," NOM also degrades the performance of membrane filtration systems used for water purification and desalination.

Synthetic organic chemistry is the science of the design, analysis, and/or construction of organic chemicals for practical purposes. As such, synthetic organic chemicals (SOCs) are chemicals that are produced on a large scale for use by humans. SOCs include several subclasses of chemicals such as pesticides, industrial solvents, chelating agents, and disinfection by-products. Many of these compounds are highly toxic and tested for routinely in public water supply systems.

A pesticide is any substance or mixture of substances intended for preventing, destroying, repelling, or mitigating any pest. Insecticides, herbicides, and fungicides all fall under the pesticide umbrella. Organophosphate pesticides such as malathion and parathion, and carbamate pesticides such as aldicarb and methomyl affect the nervous system of insects by disrupting enzymes that regulate neurotransmitters. Organochlorine insecticides such as DDT and chlordane were common in the past, but many have been banned from use in the United States and many other countries due to their health and environmental effects as well as their persistence. Pyrethroids are a class of pesticides developed as a synthetic version of the naturally occurring pesticide pyrethrin.

Organic solvents are a chemical class of compounds that are used routinely in commercial industries for dissolving other organic compounds. They have found extensive use in dry cleaning (e.g., tetrachloroethylene or PCE), as paint thinners (e.g., toluene, turpentine), as nail polish removers and glue solvents (e.g., acetone), in spot removers (e.g., hexane), in detergents (e.g., limonene), and in perfumes (e.g., ethanol). They are also particularly useful in the synthesis of other organic chemicals. Many organic solvents are recognized by the EPA as known or suspected human carcinogens. Volatile organic compounds (VOCs) are another class of organic chemical compounds that are characterized as having high enough vapor pressures to significantly vaporize and enter the atmosphere. Many of the organic solvents discussed earlier are also VOCs.

Chelating agents, or chelators for short, are a class of SOCs that are used in chemical analysis as water softeners, and are ingredients in many commercial products such as shampoos and food preservatives. The most commonly used synthetic chelating agents are NTA and EDTA [8]. Due to its inability to be broken down in many wastewater treatment plants, significant concentrations of EDTA have been found in the environment. Long-term accumulation of chelating agents is not a concern, however, because they are eventually broken down by bacteria [8]. Their presence in surface waters is more of a concern due to their ability to solubilize heavy metals thereby making them mobile.

Disinfection by-products (DBPs) are a class of chemical compounds that are formed when drinking water supply water is disinfected. Disinfection of drinking water with chlorine has been applied since the 1900s, and has prevented the spread of waterborne diseases such as cholera and typhoid. However, during the 1970s, scientists discovered that chlorination of drinking water containing moderate to high levels of NOM produced a new class of compounds, DBPs, which were later shown to be harmful to human health. Alternative disinfectants such as ozone, chlorine dioxide produce their own characteristic DBPs. Thus, switching from chlorine to ozone or chlorine dioxide is not an exhaustive remedy.

Pharmaceuticals and personal care products (PPCPs) refer to any product used by individuals for personal health or cosmetic reasons or used by agriculture to enhance growth or health of livestock. PPCPs comprise a diverse collection of thousands of individual chemicals, including prescription and over-the-counter therapeutic drugs, veterinary drugs, fragrances, cosmetics, sunscreen products, diagnostic agents, and vitamins. All contribute PPCPs to the environment through excretion, bathing, and disposal of medication to sewers and trash. The various sources of PPCPs include human activity (e.g., bathing, shaving, swimming, etc.), illicit drugs, veterinary drug use—especially antibiotics and steroids, agriculture, and residues from pharmaceutical manufacturing and hospitals. Studies have shown that PPCPs are present in the nation's water bodies. To date, scientists have found no evidence of adverse human health effects from PPCPs in the environment. However, there is strong evidence of ecological harm. PPCPs that can affect the endocrine system in animals, which controls important functions through communication of glands, hormones, and cellular receptors, are known as endocrine-disrupting compounds (EDCs). Many EDCs are associated with developmental, reproductive, and other health problems in fish and wildlife, both in the field and the laboratory.

16.4.5 NUTRIENTS

Nutrients are chemical elements critical to the growth of plant and animal life. In healthy rivers and lakes, nutrients are needed for the growth of algae that form the base of a complex food web that supports the entire aquatic ecosystem. The nutrients that receive the most attention in lakes and streams are nitrogen (as nitrate and ammonia) and phosphorus (as orthophosphate or total phosphate).

If provided an abundance of nutrients, algae and aquatic plants will continue to grow well beyond the amount needed to support the food web. The excess algae and plants will die, and consume dissolved oxygen as microorganisms break down their cellular material. As a result, other aquatic organisms may suffer from lack of oxygen. Other problems associated with excessive algal and plant growth include scum and foam formation, and odor and taste problems if the water is used for drinking.

Eutrophication is the natural process of enrichment of lakes and streams with nutrients, and the associated biological and physical changes that result. Human activity has dramatically increased the rate of eutrophication in many water bodies. Lakes and ponds are particularly vulnerable to eutrophication because the nutrients carried into them continue to build up; in contrast, the nutrients present in rivers and streams can be carried away in moving water.

Phosphorus in the form of phosphates is contributed naturally from soil and dissolution of rocks, while natural sources of nitrogen include leaves and other organic debris from riparian vegetations. The primary anthropogenic sources of these nutrients are wastewater (or sewage) treatment plants, septic systems, suspended sediment resulting from excessive erosion (only P), acid rain (only N), animal manure, and commercial fertilizers. In the past, household detergents brought high loads of phosphorus to treatment plants, which then were discharged with the effluent. In the United States, however, laws restricting the phosphorus content of detergents have produced markedly reduced phosphate levels.

16.5   CHEMICAL REACTIONS

Chemistry is the science of making new substances out of old substances via chemical reactions. All of the chemical constituents described above undergo chemical reactions that result in their being degraded, as is the case for many of the organic chemicals, or transformed to another form, as is the case for many of the inorganic chemicals. When studying a chemical reaction, chemists often pose some basic questions. What is the driving force that makes this reaction occur? What is the equilibrium state of this chemical system? These questions are answered with the help of a subject known as chemical thermodynamics. Why is this reaction so fast? Why is that reaction so slow? These questions are answered with the help of a subject known as chemical kinetics.

In the same way that water will always find its own level, chemical reactions proceed in a way that minimizes the useful energy that is available. Chemical thermodynamic calculations quantify the change in this energy (known as the Gibbs energy) as a reaction proceeds. This allows for one to determine the equilibrium state of this chemical system. The calculations are often relatively simple, and there are many commercially available computer software programs that automate the task. The results from equilibrium calculations are often a reasonable approximation for many systems, and even if the system is not at equilibrium they provide information about the direction and extent in which reactions will proceed [9,10].

The subject of chemical kinetics allows one to quantify how fast chemical reactions occur and answer why certain reactions are faster than others [11,12]. Chemical kinetics is often quantified through the measurement of the rates of change in concentrations of reactants and/or products. The most important factors that influence rates of chemical reactions are the nature and concentration of the reactant(s).

Increasing the temperature of the system imparts more kinetic energy to molecules, thereby serving to increase rates of chemical reactions. The detailed explanation of how a reaction proceeds at a molecular level is called a reaction mechanism. Determination of reaction mechanisms requires a broad and detailed understanding of the properties of reactants and products, and the changes that occur before, during, and after a chemical reaction, and is often difficult if not impossible to confirm unequivocally.

16.6 FUTURE TRENDS

New chemicals or compounds are continually flying onto the radar of environmental chemists and toxicologists. Knowledge gaps in their chemical behavior and toxicological effects must be filled to adequately assess how much of a threat they pose to the environment.

16.7 SUMMARY

1. Water chemistry deals with the fundamental chemical properties of water itself, the chemical properties of other constituents that dissolve in water, and the countless chemical reactions which take place in water.
2. A water (H_2O) molecule consists of three atoms: two hydrogen atoms each of which are bonded to a central oxygen atom. Water can exist in three states: solid, liquid, and gas.
3. Pure water has a neutral pH of exactly 7.0. Values of pH less than this are considered acidic, while pH values greater than this are termed alkaline. The typical pH range of water is from 0 to 14, although values outside of this range can be attained under extreme conditions. The pH of water is an important chemical property which controls the distribution of chemical species among various forms.
4. The composition of natural waters is often described according to its physical qualities, chemical constituents, and/or its biological inhabitants.
5. Nutrients are chemical elements critical to the growth of plant and animal life. In healthy rivers and lakes, nutrients are needed for the growth of algae which form the base of a complex food web that supports the entire aquatic ecosystem.

REFERENCES

1. Hem, J.D. *Study and Interpretation of the Chemical Characteristics of Natural Water*, USGS, Water Supply Paper 2254, 1970.
2. MWH. *Water Treatment: Principles and Design*, John Wiley & Sons, Hoboken, NJ, New York, 2005.
3. Clesceri, L.S., Greenberg, A.E., and Eaton, A.D. *Standard Methods for the Examination of Water and Wastewater*, 20th edition, American Public Health Association Publications, Washington, DC, 1998.
4. Duffus, J.H. "Heavy metals"—A meaningless term? *Pure and Applied Chemistry*, 74, 793–807, 2002.

5. Lutgens, F.K. and Tarbuck, E.J. *Essentials of Geology*, 7th edition, Prentice Hall, Upper Saddle River, NJ, 2000.

6. Nicholls, D. *Complexes and First-Row Transition Elements*, American Elsevier, New York, 1975.

7. Tipping, E. *Cation Binding by Humic Substances*, Cambridge University Press, Cambridge, U.K., 2002.

8. Grundler, O., Hans-Ulrich, J., and Witteler, H. Environmental impact of aminocarboxylate chelating agents, in U. Zoller (Ed.), *Handbook of Detergents, Part B: Environmental Impact (Surfactant Science)*, CRC Press, Boca Raton, FL, 2004.

9. Benjamin, M.M. *Water Chemistry*, McGraw-Hill, New York, 2002.

10. Theodore, L., Ricci, F., and Van Vliet, T. *Thermodynamics for the Practicing Engineer*, John Wiley & Sons, Hoboken, NJ, 2009.

11. Stumm, W. and Morgan, J.J. *Aquatic Chemistry: Chemical Equilibria and Rates in Natural Waters*, John Wiley & Sons, Inc., Hoboken, NJ, 1996.

12. Reynolds, J., Jeris, J., and Theodore, L. *Handbook of Chemical and Enviromental Engineering Calculations*, John Wiley & Sons, Hoboken, NJ, 2004.

17 Safe Drinking Water

Contributing Author: Richard F. Carbonaro

CONTENTS

17.1   INTRODUCTION

Drinking water safety cannot be taken for granted. There are many chemical and physical threats to drinking water supplies. Chemical threats include contamination from improper chemical handling and disposal, animal wastes, pesticides, human wastes, and naturally occurring substances. Drinking water that is not properly treated or disinfected, or that travels through an improperly maintained distribution system, can also become contaminated and pose a health risk. Physical threats include failing water supply infrastructure and threats posed by tampering or terrorist activity.

The reader should note that much of the regulatory material presented below is an extension and/or duplication of text that can be found in Chapter 2.

17.2   THE HYDROLOGIC CYCLE

Water is the original renewable resource. Although the total amount of water on the surface of the Earth remains fairly constant over time, individual water molecules carry with them a rich history. The water molecules contained in the fruit one ate yesterday may have fallen as rain last year in a distant place or could have been used decades, centuries, or even millennia ago by one's ancestors.

Water is always in motion, and the hydrologic cycle describes this movement from place to place. The vast majority (96.5%) of water on the surface of the Earth is contained in the oceans. Solar energy heats the water at the ocean surface and some of it evaporates to form water vapor. Air currents take the vapor up into the atmosphere along with water transpired from plants and evaporated from the soil. The cooler temperatures in the atmosphere cause the vapor to condense into clouds. Clouds move around the world until the moisture capacity of the cloud is exceeded and the water falls as precipitation. Most precipitation in warm climates falls back into the oceans or onto land, where the water flows over the ground as surface run-off. Runoff can enter rivers and streams which transport the water to the oceans, accumulate and be stored as freshwater in lakes, or soak into the ground as infiltration. Some of this water may infiltrate deep into the ground and replenish aquifers which store huge amounts of freshwater for long periods of time. In cold climates, precipitation falls as snow and can accumulate as ice caps and glaciers which can store water for thousands of years.

Throughout this cycle, water picks up contaminants originating from both naturally occurring and anthropogenic sources. Depending upon the type and amount of contaminant present, water present in river, lakes, and streams or beneath the ground may become unsafe for use.

17.3   WATER USAGE

Society uses a significant quantity of water whether it knows it or not. On average, people living in the United States use 110 gal of water per day. Most of this water is used in the bathroom for showers, which uses anywhere from 1.5 to 8.0 gal per minute, and toilet flushing, which uses 10–30 gal per flush [1]. This equates to 1.1 billion gal of water per day for New York City's population of 8 million people. Factoring in other withdrawals of water for irrigation, thermoelectric power, and industry, the United States is estimated to use 408 billion gal per day of freshwater [2].

Natural waters consist of surface waters and groundwater. Surface water refers to the freshwater in rivers, streams, creeks, lakes, and reservoirs, and the saline water present in inland seas and the oceans. The sources of freshwater are vitally important to everyday life. The main uses of surface water include drinking water and other public uses, irrigation uses, and for use by the thermoelectric power industry to cool the electricity-generating equipment. The United States relies heavily on its surface water supplies, accounting for 79% of all the water usage [2].

The remaining 21% of the U.S. water usage is from groundwater. The term aquifer is given to underground soil or fractured rock through which groundwater can move. Groundwater extracted from aquifers provides drinking water for more than 90% of the rural population. Even some major cities rely solely on groundwater for all their needs. Withdrawals of groundwater are expected to rise as the population increases and available sites for surface reservoirs become more limited. Artificial recharge is the practice of increasing the amount of water that enters a groundwater aquifer. This involves the direction of water to the land surface through canals, followed by injection of water into the subsurface through wells. This water can then be called upon when needed by pumping it back to the surface.

Saline water is not directly potable because of its dissolved salt content; however, its use is increasing. In 2000, the United States used about 62 billion gal per day of saline water, which was about 15% of all water used [2]. Currently, the main use is for thermoelectric power-plant cooling, although saline water can be desalinated for use as drinking water with treatment processes to lower the amount of salt. The process traditionally has not been cost effective, but by 2020, desalinized water is predicted to become a major contributor to the water supply of the United States [3]. There are currently over 250 desalination plants in the United States, mostly contained in states with dense populations and arid climates, e.g., California and Texas. Worldwide, there are over 12,000 desalination plants, mostly concentrated in the Middle East where freshwater is in short supply.

17.4   THE SAFE DRINKING WATER ACT

The reader is referred to Chapter 2 for additional details on the Safe Drinking Water Act (SDWA).

The first legislation enacted in the United States to protect the quality of drinking water was the Public Health Service (PHS) Act of 1912. The Public Health Service Act brought together the various federal health authorities and programs, such as the Public Health Service and the Marine Hospital Service, under one statute. The PHS Act authorized scientific studies on the impact of water pollution and human health, and introduced the concept of water quality standards. True national drinking water standards were not established, however, until 60 years later with the SDWA.

The SDWA, originally passed by Congress in 1974, authorizes the EPA to set national health-based standards for drinking water to protect against both naturally occurring and man-made contaminants that may be found in drinking water. Since its enactment, there have been over 10 major revisions and additions, the most substantial changes occurring in the amendments in 1986 and 1996.

The SDWA applies to every public water supply systems (PWS) in the United States, and approximately 87% of all water used in the United States was drawn from PWSs [2]. There are currently more than 160,000 PWS systems currently in the United States. PWS include municipal water companies, homeowner associations, schools, businesses, campgrounds, and shopping malls. The EPA works with PWS systems, along with state and city agencies, to assure that these standards are met. Originally, the SDWA focused primarily on treatment as the means of providing safe drinking water. The 1996 amendments greatly enhanced the existing law which now includes source water protection, protection of wells and collection systems, making certain water is treated by qualified operators, funding for water system improvements, and making information available to the public on the quality of their drinking water.

17.5   DRINKING WATER STANDARDS

The reader is referred to Chapter 2 for additional details on this topic.

Drinking water standards are regulations that EPA has established to control the concentration of contaminants in the U.S. drinking water supply. In most cases, EPA delegates responsibility for implementing drinking water standards to states and tribes. Drinking water standards apply to PWSs, which provide water for human consumption through at least 15 service connections, or regularly serve at least 25 individuals.

The SDWA 1996 Amendments require EPA to identify potential drinking water problems, establish a prioritized list of chemicals of concern, and set standards where appropriate. Peer-reviewed science and data support an intensive technological evaluation which includes many factors such as the occurrence of the chemicals in the environment; human exposure and risks of adverse health effects in the general population and sensitive subpopulations; analytical methods of detection; technical feasibility; and impacts of regulation on water systems, the economy, and public health.

After reviewing health effect studies, EPA sets a maximum contaminant level goal (MCLG). The MCLG is the maximum level of a contaminant in drinking water at which no known or anticipated adverse effect on the health of persons would occur, and which allows an adequate margin of safety. MCLGs are not enforced, but instead are public health goals. Since MCLGs consider only public health and not the limits of detection and treatment technology, they are sometime set at a level which water systems cannot meet. When determining an MCLG, EPA considers the risk to sensitive subpopulations (infants, children, the elderly, and those with compromised immune systems) of experiencing a variety of adverse health effects.

For chemicals that can cause noncancer adverse health effects (noncarcinogens), the MCLG is based on the reference dose. A reference dose (RFD) is an estimate of the amount of a chemical that a person can be exposed to on a daily basis that is not anticipated to cause adverse health effects over a person's lifetime. In RFD calculations, an uncertainty factor is used to account for sensitive subgroups of the population. The RFD is multiplied by typical adult body weight (70 kg) and divided by daily water consumption (2 L/day) to provide a drinking water equivalent level (DWEL). The DWEL is multiplied by a percentage of the total daily exposure contributed by drinking water (usually 20%) to determine the numeric value of the MCLG (in mg/L). Details of these calculations are available in the literature [4].

If there is evidence that a chemical may cause cancer (carcinogens), it is usually assumed that there is no dose below which the chemical is considered safe (i.e., no threshold), and the MCLG is set to zero. For microbial contaminants that may present public health risk, the MCLG is also set at zero because ingestion of one protozoa, virus, or bacterium may cause an adverse health effect.

Once the MCLG is determined, EPA starts the process of setting an enforceable standard. In most cases, the standard is the maximum contaminant level (MCL), the maximum permissible level of a contaminant in water that is delivered to any user of a PWS. The MCL is set as close to the MCLG as feasible, which the SDWA defines as the level that may be achieved with the use of the best available technology, treatment techniques, and other means which EPA finds are available, while taking cost into consideration.

When there is no economically and technically feasible method to measure a contaminant at low concentrations, a treatment technique (TT) is set rather than an MCL. A TT is an enforceable procedure or level of technological performance which a PWS must follow to ensure control of a contaminant. Examples of TT rules are the Surface Water Treatment Rule and the Lead and Copper Rule. The Surface

Water Treatment Rule require systems using surface water or groundwater under the direct influence of surface water to disinfect their water, filter their water, or meet criteria for avoiding filtration to control levels of contamination by bacteria, viruses, and protozoa. The Lead and Copper Rule states that if lead or copper concentrations exceed an action level in more than 10% of customer taps sampled, the system must undertake a number of additional actions to control corrosion.

EPA gathers input from many external groups during the process of establishing standards. For example, the National Drinking Water Advisory Council (NDWAC) is a committee created by the SDWA that advises the EPA Administrator on everything that the EPA does related to drinking water. It is comprised of five members of the general public, five representatives of state and local agencies, and five representations of private organizations and groups that are active in the field of public health and public water supply. Representatives from water utilities, community environmental groups, states, and the general public are encouraged to participate in public meetings.

17.5.1 NATIONAL PRIMARY DRINKING WATER REGULATIONS

National Primary Drinking Water Regulations (NPDWRs or primary standards) are the legally enforceable MCLs and TTs that apply to public water supply. The contaminants are broken up into the following groupings, according to the type of contaminant: inorganic chemicals, organic chemicals, microorganisms, disinfectants, disinfection byproducts, and radionuclides. A list of these contaminants and their respective standard is shown in Table 17.1. Primary standards go into effect 3 years after they are finalized. If capital improvements of PWS are required to meet NPDWRs, the EPA Administrator or state agency may allow this period to be extended up to two additional years.

TABLE 17.1
National Primary Drinking Water Regulations

Inorganic Contaminants

Antimony	0.006	Chromium (total)	0.1	Mercury (inorganic)	0.002
Arsenic	0.010	Copper	TT	Nitrate (as N)	10
Asbestos	7 MFL[a]	Cyanide (free)	0.2	Nitrite (as N)	1
Barium	2	Fluoride	4.0	Selenium	0.05
Beryllium	0.004	Lead	TT	Thallium	0.002
Cadmium	0.005				

Organic Contaminants

Acrylamide	TT	Dichloromethane	0.005	Methoxychlor	0.04
Alachlor	0.002	1,2-Dichloropropane	0.005	Oxamyl (Vydate)	0.2
Atrazine	0.003	Di(ethylhexyl) adipate	0.4	Pentachlrophenol	0.001
Benzene	0.005	Di(ethylhexyl) phthalate	0.006	Picloram	0.5

(continued)

TABLE 17.1 (continued)
National Primary Drinking Water Regulations

Organic Contaminants

Benzo(a)pyrene	0.0002	Dinoseb	0.007	Polychlorinated byphenyls (PCBs)	0.0005
Carbofuran	0.04	Dioxin (2,3,7,8-TCDD)	3×10^{-8}	Simazine	0.004
Carbon tetrachloride	0.005	Diquat	0.02	Styrene	0.1
Chlordane	0.002	Endothall	0.1	Tetrachloroethylene (PCE)	0.005
Chlorobenzene	0.1	Endrin	0.002	Toluene	1
2,4-D	0.07	Epichlorohydrin	TT	Toxaphene	0.003
Dalapon	0.2	Ethylbenzene	0.7	2,4,5-TP (Sivex)	0.05
1,2-Dibromo-3-chloropropane (DBCP)	0.0002	Ethylene dibromide	3×10^{-5}	1,2,4-Trichlorobenzene	0.07
o-Dichlorobenzene	0.6	Glyphosate	0.7	1,1,1-Trichloroethane	0.2
p-Dichlorobenzene	0.075	Heptachlor	0.0004	1,1,2-Trichloroethane	0.005
1,2-Dichloroethane	0.005	Heptachlor epoxide	0.0002	Trichloroethylene (TCE)	0.005
1,1-Dichloroethylene	0.007	Hexachlorobenzene	0.001	Vinyl chloride	0.002
cis-1,2-Dichloroethylene	0.07	Hexachlorocyclopentadiene	0.05	Xylenes (total)	10
trans-1,2-Dichloroethylene	0.1	Lindane	0.0002		

Radionuclides

Radium-226 + radium-228	5 pCi/L	Beta particles and photon emitters	4 mrem/year
Alpha emitters	15 pCi/L	Uranium	0.030

Microbiological Contaminants

Cryptosporidium	TT	Legionella	TT	Enteric viruses	TT
Giardia lamblia	TT	Total coliforms	5%[b]		
Heterotrophic plate count	TT	Turbidity	TT		

Disinfectants/Disinfection By-Products

Bromate	0.010	Chlorine dioxide (as Cl_2)	0.8[c]	Total trihalo-methanes (TTHMs)	0.080
Chloramines (as Cl_2)	4.0[c]	Chlorite	1.0		
Chlorine (as Cl_2)	4.0[c]	Haloacetic acids (HAA5)	0.060		

Note: All concentrations given in mg/L unless otherwise noted.

[a] MFL = million fibers per liter.

[b] No more than 5% of all samples may test positive for coliforms (fecal coliform and *Escherichia coli*) per month.

[c] Referred to as a maximum residual disinfectant level (MRDL).

17.5.2 National Secondary Drinking Water Regulations

National Secondary Drinking Water Regulations (NSDWRs or secondary standards) are nonenforceable guidelines regulating contaminants that may cause cosmetic effects (such as skin or tooth discoloration) or aesthetic effects (such as taste, odor, or color) in drinking water. EPA recommends secondary standards to water systems but does not require systems to comply. State and local agencies may choose to adopt these as enforceable standards. A list of these contaminants and their associated guidelines are shown in Table 17.2 [5].

17.5.3 Unregulated Contaminants

The SDWA includes a process where new contaminants are identified that may require regulation in the future with a primary standard. EPA is required to periodically release a Contaminant Candidate List (CCL) which is used to prioritize research and data collection efforts to help determine whether a specific contaminant should be regulated. On March 2, 1998, the first Drinking Water Contaminant Candidate List (CCL 1) was released which contained 60 contaminants (10 microbiological contaminants and 50 chemical contaminants). After the list was released, research was undertaken to develop analytical methods for detecting the contaminants, to

TABLE 17.2
National Secondary Standards for Drinking Water

Contaminant	Level	Contaminant Effects
Aluminum	0.05–0.02 mg/L	Colored water
Chloride	250 mg/L	Salty taste
Color	15 color units	Visible tint
Copper	1.0 mg/L	Metallic taste; blue-green staining
Corrosivity	Noncorrosive	Metallic taste; corroded pipes/fixtures staining
Fluoride	2.0 mg/L	Tooth discoloration
Foaming agents	0.5 mg/L	Frothy, cloudy; bitter taste; odor
Iron	0.3 mg/L	Rusty color; sediment; metallic taste; reddish or orange staining
Manganese	0.05 mg/L	Black to brown color; black staining; bitter metallic taste
Odor	3 threshold odor number	"Rotten-egg", musty or chemical smell
pH	6.5–8.5	*Low pH*: bitter metallic taste; corrosion; *high pH*: slippery feel; soda taste; deposits
Silver	0.10 mg/L	Skin discoloration; graying of white part of eye
Sulfate	250 mg/L	Salty taste
Total Dissolved Solids (TDS)	500 mg/L	Hardness; deposits; colored water; staining; salty taste
Zinc	5 mg/L	Metallic taste

Source: U.S. EPA Secondary Drinking Water Regulations: Guidance for Nuisance Chemicals. EPA 810/K-92-001, July 1992. http://www.epa.gov/safewater/consumer/2ndstandards.htm.

determine whether they occur in drinking water, to evaluate treatment technologies to remove them from drinking water supplies, and to assess potential health effects resulting from exposure to the contaminants. In July 2003, EPA announced its final determination for nine contaminants from the CCL 1, which was not to regulate the following contaminants: *Acanthamoeba*, aldrin, dieldrin, hexachlorobutadiene, manganese, metribuzin, naphthalene, sodium, and sulfate.

EPA announced the second Drinking Water Contaminant Candidate List (CCL 2) on February 23, 2005. The CCL 2 list included the remaining 51 contaminants from the CCL 1. In February 2008, the EPA announced the draft CCL 3. Approximately 7500 potential chemical and microbial contaminants were considered for inclusion in the CCL 3. Of these, 560 were further evaluated based on the potential for a contaminant to occur in a PWS and the potential for public health concern. Of these, 104 contaminants were included in the CCL 3 based upon more detailed evaluation of occurrence and potential health effects.

17.6   CLEAN WATER ACT

See Chapter 2 for additional details on the Clean Water Act (CWA).

Along with the SDWA, the CWA has played an important role in assuring and maintaining the safety of sources of drinking water. Growing public awareness and concern for controlling water pollution led to enactment of the Federal Water Pollution Control Act Amendments of 1972. As amended in 1977, this law became commonly known as the CWA. The CWA established the basic structure for regulating discharges of pollutants into the waters of the United States. It gave EPA the authority to implement pollution control programs such as setting wastewater standards. The CWA also continued requirements to set water quality standards for all contaminants in surface waters. The CWA made it unlawful for any person to discharge any pollutant from a point source into navigable waters unless a permit was obtained that dictated the terms of the release. It also funded the construction of wastewater treatment plants under the construction grants program. Pollutants regulated under the CWA include biochemical oxygen demand (BOD), total suspended solids (TSS), fecal coliform, oil and grease, and pH (conventional pollutants); toxic chemicals (priority pollutants); and, various contaminants not identified as either conventional or priority (nonconventional pollutants).

The CWA introduced a permit system for regulating point sources of pollution. A "point source" is a single identifiable and localized source of a contaminant. Point source pollution can usually be traced back to a single origin or source. Examples of point sources include industrial facilities (e.g., manufacturing, mining, oil and gas extraction, etc.), municipal and government facilities (e.g., wastewater treatment plants), and some agricultural facilities (e.g., animal feedlots). Point sources are not allowed to be discharged into surface waters without a permit from the National Pollutant Discharge Elimination System (NPDES). This system is managed by the EPA in partnership with state environmental agencies. EPA has authorized 45 states to issue permits directly to the discharging facilities. The EPA regional office issues permits directly in the remaining states and territories.

The CWA employs three general types of standards: technology-based standards, water quality-based standards, and in the case of a small number of toxic compounds, health-based effluent standards. EPA develops technology-based standards for categories of dischargers based on the performance of pollution control technologies without regard to the conditions of a particular receiving water body. The technology-based standard becomes the minimum regulatory requirement in a permit [6]. After application of technology-based standards to a permit, if water quality is still impaired for the particular water body, the permit agency will add water quality-based standards to that permit. The additional limitations are more stringent than the technology-based limitations and require the permit applicant to install additional controls [7].

Water quality standards (WQS) are risk-based requirements which set site-specific allowable pollutant levels for individual water bodies such as rivers, lakes, streams, and wetlands. A water quality standard defines the water quality goals of a water body by designating the use or uses to be made of the water (e.g., recreation, water supply, aquatic life, agriculture), by setting criteria necessary to protect the uses, and by preventing degradation of water quality through antidegradation provisions. The criteria are numeric pollutant concentrations similar to an MCL for drinking water. States adopt water quality standards to protect public health or welfare, enhance the quality of water, and serve the purposes of the CWA.

A total maximum daily load (TMDL) is a calculation of the maximum amount of a pollutant that a water body can receive and still meet WQS. It is the collective sum of the allowable loads of a single pollutant from all contributing point and nonpoint sources. The calculation includes a margin of safety to ensure that the water body can be used for the purposes the State has designated. Since October 1, 1995 over 32,000 TMDLs have been approved by the U.S. EPA. A complete and update listing can be found at http://www.epa.gov/owow/tmdl/.

Section 303(d) of the CWA requires states to identify water bodies that do not meet water quality standards and are not supporting their designated uses. Each state must submit an updated list, called the 303(d) List of Impaired Waterbodies, every even-numbered year. The 303(d) List also identifies the pollutant or stressor causing impairment, and establishes a timeframe for developing a control plan to address the impairment. Placement of a water body on the 303(d) List triggers development of a TMDL for each pollutant listed for that water body.

Nonpoint source (NPS) pollution, unlike pollution from direct discharges, comes from many diffuse sources. NPS pollution is caused by rainfall or snowmelt traveling over the ground surface and through the ground. As the runoff moves, it picks up and carries away natural and human-made pollutants, depositing them into lakes, rivers, wetlands, coastal waters, and groundwater. Pollutants associated with NPS include fertilizers, herbicides, and insecticides from agricultural lands and residential areas, oil and grease, toxic chemicals from urban runoff and energy production, sediment from improperly managed construction sites, crop and forest lands, eroding stream banks, salt from irrigation practices and acid drainage from abandoned mines, and bacteria and nutrients from livestock, pet wastes, and faulty septic systems. Many of the sources of NPS pollution were not subject to the permit program as part of the original 1972 CWA. Stormwater runoff specifically is a significant cause of water quality impairment in many parts of the United States. In the early 1980s, EPA conducted the Nationwide Urban Runoff Program (NURP) to document the extent of the urban stormwater problem.

The results of the NURP report were used to develop the 1987 amendments to the CWA, also known as the Water Quality Act (WQA). The WQA tries to address the stormwater problem by requiring that industrial stormwater dischargers and municipal separate storm sewer systems obtain NPDES permits. A Nonpoint Source Management Program was created where state, territories, and Indian tribes can apply for grant money to support education, training, technology transfer, demonstration projects, and monitoring to assess the success of specific nonpoint source implementation projects.

17.7 WATER SECURITY

As noted in Chapter 2, the Bioterrorism Act of 2002 requires all PWS serving populations of more than 3300 persons to conduct assessments of their vulnerabilities to terrorist attack or other intentional acts, and to defend against adversarial actions that might substantially disrupt the ability of a system to provide a safe and reliable supply of drinking water. In addition to the vulnerabilities assessment, the act requires the PWS to certify and submit a copy of the vulnerability assessment to the EPA Administrator, prepare or revise an emergency response plan based on the results of the vulnerability assessment, and certify that an emergency response plan has been completed or updated within 6 months of completing the assessment.

17.8 FUTURE TRENDS

Improving the security of the nation's drinking water and wastewater infrastructures has become a top priority. Significant actions are underway to assess and reduce vulnerabilities to potential terrorist attacks; to plan for and practice response to emergencies and incidents; and, to develop new security technologies to detect and monitor contaminants and prevent security breaches.

REFERENCES

1. Gleick, P.H. *Water in Crisis: A Guide to the World's Fresh Water Resources*, Oxford University Press, New York, 1993.
2. Hutson, S.S., Barber, N.L., Kenny, J.F., Linsey, K.S., Lumia, D.S., and Maupin, M.A. *Estimated Use of Water in the United States in 2000*, U.S. Department of the Interior, U.S. Geological Survey, Reston, VA, 2004.
3. U.S. Bureau of Reclamation and Sandia National Laboratories. *Desalination and Water Purification Technology Roadmap*, U.S. Bureau of Reclamation and Sandia National Laboratories, Denver, CO, 2003.
4. Reynolds, J., Jeris, J., and Theodore, L. *Handbook of Chemical and Enviromental Engineering Calculations*, John Wiley & Sons, Hoboken, NJ, 2004.
5. Masters, G.M. and Ela, W.P. *Introduction to Environmental Engineering and Science*, 3rd edition, Prentice Hall, Upper Saddle River, NJ, 2008.
6. U.S. EPA. *NPDES Permit Writers' Manual*, U.S. EPA Office of Water, Washington, DC, 1996.
7. Liu, D.H.F. and Liptak, B.G. *Environmental Engineers' Handbook*, CRC Press, Boca Raton, FL, 1997.

18 Municipal Wastewater Treatment

CONTENTS

18.1 INTRODUCTION

The portion of liquid waste produced by the human intervention with the hydrologic cycle is known as wastewater. Such interventions can be the use of water for washing dishes, clothes, and automobiles; the provision of a recreational pool for public use; or, the use of water by a local factory to maintain proper temperatures within their machinery. The use of water is important in each community's daily events in order to function normally and comfortably. In each case, the wastewater must be treated and disposed of, or discharged, into a naturally occurring water source (lakes, rivers, bays, etc.). Therefore, the use and disposal of water can be considered as an artificial water cycle. Wastewater is that which can be generated by the liquid wastes removed from residential, municipal, and industrial areas requiring collection, treatment, and disposal in accordance with local, state, and federal standards. However, municipal wastewater is the general term applied to the liquid collected in sanitary sewers and treated in a municipal plant and will be the focus of this chapter.

In the late 1800s, the United States gave little attention to the treatment and disposal of wastewater from communities. Large, fresh sources of potable water (suitable for human consumption) were available without any threat to human health. The impact upon the public and on the water quality from the discharge of untreated wastewater into adjacent water bodies was considered to be a minor issue. Additionally, large areas of land and water were available; for all waste disposal purposes. However, during the early 1900s, paralleled by a large influx of immigrants, decreasing health conditions were attributed to the concentrated increase of raw wastewater being disposed into surrounding water bodies and the lack of fresh water used to dilute the wastewater before discharge. This led to the demand for a more effective means of wastewater management. Ultimately, the planning, design, construction, and operation of high-level wastewater treatment facilities, sanitary sewer systems, and fresh water collection systems was initiated.

Untreated wastewater is collected in sewer systems and transported underground to a treatment plant prior to disposal. Wastewater has three major characteristics of concern to a community and its surrounding environment: biological, chemical, and physical characteristics. These are discussed in the next three paragraphs.

If wastewater is allowed to accumulate, there are numerous pathogenic, disease-causing microorganisms contained within the waste that can cause outbreaks of intestinal infections within humans. Typical infectious diseases are cholera, typhoid, paratyphoid fever, balantiasis (dysentery), salmonellosis, and shigellosis [1]. During the writing of this chapter, 20,000 people died of cholera in Rwanda, Africa within the time span of 1 week. This was due to the lack of a fresh water supply and mostly the nonexistent treatment for their waste. Shortly after this misfortune, mobile treatment facilities were made available.

The chemical characteristics that are of interest are toxic metals (cadmium, chromium, lead, and mercury) and nutrients (nitrogen, phosphorus, and carbon) being discharged into the water. Toxic metals and/or chemical compounds can cause large fish kills or can lead to the consumption of contaminated fish by humans and other mammals. The discharge of nutrients into a water body at first seems to be beneficial to the local ecology due to the production of algae and its supplement to the food chain. However, too many nutrients will produce gross masses of algae. Eventually the algae will die off and sink to the bottom of the water column forming a layer of biomass. As the algae decays (use of oxygen), the deficit of dissolved oxygen will suffocate the bottom-dwelling fish and shellfish.

The physical characteristics that are of concern to the water environment can basically be described as any organic matter entering a water source. The decomposition (biodegradation) of organic materials (suspended solids [SS], oils, greases, and fats) in local waters occurs by using oxygen. If proper dilution of these waters is not available to accommodate decaying organic matter, the production of offensive odors and gases can occur, indicating a low to zero value for the dissolved oxygen needed for fish to survive in the water body.

For the above reasons, the treatment and disposal of wastewater from each source of generation, is imperative to satisfy those conditions that are beneficial to maintaining a healthy water supply, recreation, harvesting of fish, and future considerations. In order to assure that all oceans, lakes, rivers, bays, harbors, streams, estuaries, and so on are maintained properly, laws have been established with short- and long-term goals that provide standards for every point source of wastewater discharging into a body of water and standards for the overall quality of that water body. This maintains a watch on the quality of the water body and the quality of the treated wastewater being discharged by a municipal wastewater treatment plant. Standards are to be met in order to be given a permit to operate, otherwise each source generating wastewater will have to be upgraded before any permit is approved to operate the facility. Any delay in operation is a potential loss of tax dollars or private funds, depending upon the location.

18.2 REGULATIONS

The first water quality standards were established in 1914 for drinking water. Surface-water standards for the control of wastewater treatment and disposal practices were not introduced until years later. With population growth and dramatic industrial expansion, the untreated wastewater discharges began to exceed the renewal capacity of the natural water body systems. Compounds were often identified with the industrial and municipal waste streams. These compounds could not be removed by simple chemical treatment. As the quality of the drinking water supply became poorer, public complaints forced new legislation in the early 1960s. Initially the surface-water quality standards were established along with drinking-water standards. This was followed by the discharge limits set for those substances that were known to be dangerous to human and aquatic life.

From about the 1900s to the early 1970s, treatment objectives were concerned with

1. The removal of suspended and floatable material
2. The removal of biodegradable organics
3. The removal of pathogenic organisms

Unfortunately, these objectives were not uniformly met throughout the United States. Perhaps the most important piece of wastewater management regulations was the Federal Water Pollution Control Act of 1972, often referred to as the Clean Water Act (CWA). It established levels of treatment, deadlines for meeting these levels, and penalties for violators. It also marked a change in water pollution control philosophy. No longer was the classification of the receiving stream of ultimate importance as it had been before. The quality of the nation's waters was to be improved by the imposition of specific effluent limitations. A National Pollution Discharge Elimination System (NPDES) program was established at that time based on uniform technological minimums with which each point source discharger had to comply [2]. The permit program governs the discharge into navigable waters. The current definition of secondary treatment includes three major effluent parameters: 5 day biological oxygen demand (BOD) (to be discussed later), SS and pH, and is reported in Table 18.1 (40 CFR, Part 133, July 1, 1988 and January 27, 1989). The secondary treatment regulations were amended further in 1989 to clarify the percent removal requirements during dry periods for treatment facilities served by combined sewers.

The CWA of 1977 contains two major provisions for the wastewater solids removed during treatment. It intended to set limits on the quantity and kind of toxic materials reaching the general public. The Resource Conservation and Recovery Act (RCRA)

TABLE 18.1

Minimum National Standards for Secondary Treatment [3,4]

Characteristic of Discharge	Unit of Measurement	Average 30 Day Concentration	Average 7 Day Concentration
BOD_5	μg/L	30	45
SS	μg/L	30	45
pH	pH	6–9	6–9
Fecal coliform bacteria	μg/L	200	400

Sources: Federal Register, 40 *CFR* Part 133, January 27, 1989; Federal Register, 40 *CFR* Part 133, July 1, 1988.

of 1976 requires that solid wastes be utilized or disposed of in a safe and environmentally acceptable manner. The Marine Protection, Research, and Sanctuaries Act 1977 amendments prohibited disposal of sewage sludge by ocean barge dumping after December 31, 1981 [1]. See Chapters 2 and 17 for additional details.

Congress also enacted the Water Quality Act of 1987 (WQA); this was the first major revision to the CWA. Its goals were to eliminate the discharge of pollutants into the nation's waters and to attain water quality capable of supporting recreation and protecting aquatic life and wildlife. Important provisions of the WQA are

1. The strengthening of federal water quality regulations by providing changes in permitting and adding substantial penalties for permit violations.
2. Emphasizing the identification and regulation of toxic pollutants in sludge.
3. Providing funding for EPA and state studies on nonpoint toxic sources of pollution.
4. Establishing new deadlines for compliance of priorities for stormwater.

Within the approach to total maximum daily load (TMDL) rules and subsequent management policy, the emphasis on "Pollutant Trading" or "Water Quality Trading" has evolved. Trading allows sources with responsibility for discharge reductions the flexibility to determine where reduction will occur. Within the trading approach, the economic advantages are emphasized.

The EPA established requirements for facilities with cooling water intake structure to implement best available technologies to minimize adverse environmental impacts. Under section 316 (b) of the CWA, the protection of aquatic organisms from being killed or injured by impingement or entrainment was established. The rule was divided into three phases. Phase I was published in December 2001 and addressed new facilities. Phase II was published in July 2004. The Phase II rule addressed existing electric generating plants withdrawing greater than 50 million gal/day and use at least 25% of their withdrawn water for cooling purposes only. Phase III was prepublished in November 2004. The Phase III rules addresses other electric generating facilities and industrial sectors that withdraw water. It was anticipated that Phase III was to be published within 1–2 years.

This rule provides that the facility may choose one of five compliance alternatives for establishing best technology available for minimizing adverse environmental impact at the site. Under current NPDES program regulations, the 316(b) requirements would occur when an existing NPDES permit is reissued or, when an existing permit is modified or revoked and reissued.

With increasing water sustainability and scarcity issues in the United States, the EPA and the U.S. Agency for International Development (USAID) published the 2004 EPA *Guidelines for Water Reuse* to reflect significant technical advancements and institutional developments since 1992. The guidelines address new areas, including national reclaimed water use trends, groundwater recharge, endocrine disrupters, and approaches to integrated water resources management.

Across the world, reclaimed water is becoming a critical water source, and reuse strategies are recognized in many U.S. states as an integral part of water resources management. The original guidelines, published in 1980, gave many state agencies direction in establishing reuse permits and helped to foster state water reuse regulations. The revised guidelines will include an updated inventory of state reuse regulations.

18.3 CHARACTERISTICS OF MUNICIPAL WASTEWATER

Municipal wastewater is composed of a mixture of dissolved and particulate organic and inorganic materials and infectious disease-causing bacteria. The total amount of each parameter accumulated in wastewater is referred to as the mass loading and is given the units of pounds per day (lbs/day). The concentration, given in pounds per gallon of water (lbs/gal) of any individual component entering a wastewater treatment plant can change as a result of the activities that are producing this waste. The units used to express any concentration, lbs/gal, can also be converted into other nomenclature, such as pounds per liter of water (lbs/L), milligrams per liter of water (mg/L), or even micrograms per cubic meter of water $\mu g/m^3$). The concentration of each individual component while in the treatment plant is usually reduced significantly by the time it reaches the end of the plant, prior to discharge.

Wastewater characteristics depend largely on the mass loading rates flowing from the various sources in the collection system. The flow in sanitary sewers is a composite of domestic and industrial wastewaters, infiltration into the sewer from cracks and leaks in the system, and intercepted flow from combined sewers. During wet weather, the addition of rainfall collected from the combined sewer system and the storm drain collection system (combined sewer overflow systems) can significantly change the characteristics of wastewater and the increased demand of how much water is to be carried by the sewer to the treatment plant. The peak flow rate can be two to three times the average dry (or sunny) weather flow rate. The mass loading rate into the plant also varies cyclically throughout the day. The impact of flow rate is an important determining factor in the design and operation of wastewater treatment plant facilities. The records kept by the treatment plant should include the minimum, average, and maximum flow values (gallons per unit time) on an hourly, daily, weekly, and monthly basis for both wet and dry weather conditions. A moving 7 day daily average flow and mass loading rate entering the plant and at various locations throughout the plant can then be computed from the record. This intricate form of recordkeeping of all factors affecting a wastewater treatment plant must be considered to assess the wastewater flow and variations of wastewater strength in order to operate a facility correctly. The parameters used to indicate the total mass loading in the wastewater entering the treatment plant are the measurements of total suspended solids (TSS), SS, and total dissolved solids (TDS) [1]. The parameters used to indicate the organic and inorganic chemical concentration in the wastewater are the measurements of the BOD and the chemical oxygen demand (COD). Both BOD and COD are discussed in more detail later in this section. Additionally, the total nutrients (carbon, nitrogen, and phosphorus), any toxic chemicals, and trace metals are also characterized prior to the wastewater entering the treatment plant so that the plant operators may adjust their treatment techniques to accommodate the varying waste loads.

Before proceeding to more technical details, some definitions and concerns in the wastewater management field are presented below.

Suspended Solids: Matter that is retained through a filter. SS can lead to the development of sludge deposits and anaerobic conditions (zero dissolved oxygen) when untreated wastewater is discharged in the aquatic environment.

Biodegradable organics: Composed principally of proteins, carbohydrates, and fats. Biodegradable organics are usually measured in terms of BOD and COD. If discharged untreated to the environment, their biological stabilization can lead to the depletion of natural oxygen resources and to the development of septic conditions. As indicated earlier, additional details on BOD and COD are provided later.

Pathogens: Pathogenic organisms that can transmit communicable diseases via wastewater. Typical notified infectious disease reported are cholera, typhoid, paratyphoid fever, salmonellosis, and shigellosis.

Nutrients: Both nitrogen and phosphorus, along with carbon. When discharged to the receiving water, these nutrients can lead to the growth of undesirable aquatic life. When discharged in excessive amounts on land, they can also lead to the pollution of groundwater.

Priority pollutants: Organic and inorganic compounds selected on the basis of their known or suspected carcinogenicity, mutagenicity, or high acute toxicity. Many of these compounds are found in wastewater.

Heavy metals: Heavy metals are usually added to wastewater from commercial and industrial activities and may have to be removed if the wastewater is to be reused.

Municipal wastewater normally contains approximately 99.9% water. The remaining materials (as described earlier) include suspended and dissolved organic and inorganic matter as well as microorganisms. These materials make up the physical, chemical, and biological qualities that are characteristic of residential and industrial waters. Each of these three qualities are briefly described in Sections 18.3.1 through 18.3.3. See Chapter 16 for additional details.

18.3.1 Physical Quality

The physical quality of municipal wastewater is generally reported in terms of temperature, color, odor, and turbidity and is an important parameter because of its effect upon aquatic life and the amount of oxygen available for aquatic respiration. Water temperature varies slightly with the seasons, normally higher than air temperature during most of the year and lower only during the hot summer months. The color of a wastewater is usually indicative of age. Fresh wastewater is usually gray; septic wastes impart a black appearance. Odors in wastewater are caused by the decomposition of organic matter that produces offensive smelling gases such as hydrogen sulfide. Turbidity in wastewater is caused by a wide variety of SS. SS are defined and can be measured as solid matter, which can be removed from water by filtration through a 1-micron pore filter paper. Volatile SS for the most part represent the biodegradable organics. SS may cause undesirable conditions of increased turbidity and silt load in the receiving water. In general, stronger wastewater has a higher turbidity [1].

18.3.2 Chemical Quality

The principal groups of organic substances found in municipal wastewater are proteins (30%–40%), carbohydrates (40%–60%), and fats and oils (15%–25%). Carbohydrates and proteins are easily biodegradable, whereas fats and oils are more stable and require a longer exposure time to be decomposed by microorganisms. In addition, wastewater may also contain small fractions of phenolic compounds, pesticides, PCBs, dioxins, and herbicides. These compounds are usually industrial wastes, depending on their concentration and may create problems such as nonbiodegradability and carcinogenicity.

BOD measurement is very important parameter indicating the organic (e.g., fats and oils) pollution concentration in both wastewater and surface discharge. The 5 day BOD test (BOD_5) measured at 20°C is the most commonly used test for calculating the amount of total organic matter requiring oxygen for its decomposition. The decomposition or biodegradation is accomplished by microorganisms (bacteria and protozoa) that breathe the oxygen in the water while feeding on the amount of organic matter available to them. The BOD_5 value reflects the original organic concentration by observing a depletion of oxygen. The higher BOD content in the wastewater would result in a higher depletion of the oxygen concentration in that wastewater. Dissolved oxygen in the receiving water must be maintained at a level of 4–5 mg/L for the survival of aquatic life. Therefore, it is important to remove the organic matter or to decrease the BOD prior to discharge.

BOD testing is used as the sole basis to determine the efficiency of the treatment plant. There are two basic types of BOD: carbonaceous-BOD (CBOD) from the oxidation of the organic carbon sources, and nitrogenous-BOD (NBOD) from the oxidation of the organic nitrogen (nitrification). The addition of CBOD and NBOD is given the term ultimate BOD (BOD_U) and it usually takes 20–30 days to complete a measurement. BOD_5 is a 5 day measurement that indicates how much CBOD is utilizing a nitrification inhibitor to inhibit the oxidation of NBOD. By inhibiting one type of BOD's ability to oxidize, one may measure the other directly.

The COD test is a measurement of organic matter in wastewater. It is similar to the BOD test in concept but different in the analytical procedure. It is the measurement of the amount of oxygen depleted during the chemical oxidation process, without the use of microorganisms. COD analysis is a more reproducible and less time-consuming test of approximately 3 h. The COD test measures the nonbiodegradable as well as the ultimate biodegradable organics. The COD test and BOD test can be correlated and used as a controlling factor in the treatment of waste. A change in the biodegradable to nonbiodegradable organic ratio affects this correlation, and is therefore waste specific. Calculation details are available in the literature [5].

The most frequently found inorganic compounds in wastewater are chloride salts; acids, hydrogen ions, and alkalinity-causing compounds; bases and heavy metals (cadmium, copper, lead, mercury, and zinc); and, nutrients for the growth of the organism in addition to the, required food substrate, such as ammonia, sulfur, carbon, nitrogen, and phosphorous [1]. A trace amount of metals can be toxic to the organisms in the receiving water. Excessive nutrients of nitrogen and phosphorous discharged to the receiving water can cause eutrophication, causing excessive growth of aquatic plants. As indicated earlier, aged aquatic plants later become the source of particular organic matter that settle to the bottom of the receiving water and indirectly exert an excessive demand of oxygen by their decomposition and deplete the oxygen source for other aquatic life and fish.

Gases commonly found in raw wastewater include nitrogen, oxygen, carbon dioxide, hydrogen sulfide, ammonia, and methane. Of all these gases mentioned, the ones that are most considered in the design of a treatment facility are oxygen and hydrogen sulfide. Oxygen is required for all aerobic life forms either within the treatment facility (microorganisms) or in the receiving water (aquatic life). During the absence of aerobic conditions (extreme low dissolved oxygen levels), oxidation is brought about by the reduction of inorganic salts such as sulfates or through the action of methane-forming bacteria in a treatment process known as sludge thickening. The end products are often very malodorous. To avoid such conditions it is important that an aerobic state be maintained or odor equipment be used. Additional odor control has received major consideration in recent large-sized wastewater treatment facilities. Large capital investments have been made in resolving this offensive smelling issue and complaints from the neighboring residential area of the wastewater treatment plant.

18.3.3 Biological Quality

Within the treatment facility, the wastewater provides the perfect medium for good microbial growth, whether it be aerobic or anaerobic. Bacteria and protozoa are the keys to the biological treatment process used at most treatment facilities. In the presence of sufficient dissolved oxygen, bacteria convert the soluble organic matter into new cells and inorganic elements. This causes a reduction of organic loading through the buildup of more complex organisms [1]. The location of such microbial proliferation is the aeration tank or the activated sludge system. Although the treatment facility utilizes bacteria and protozoa to perform the breakdown of wastewater loads, these are not necessarily the same bacteria, or pathogens, mentioned earlier that cause intestinal (enteric) disease.

Water quality in a receiving body of water is strongly influenced by the biological interactions that take place. The discharged effluent to the receiving waters becomes a normal part of the biological cycle and its effect on aquatic organisms is the ultimate consideration of treatment plant operation. Typically, the species and organisms found in biological examination of the receiving waters include zooplankton, phytoplankton, peryphyton, macroinvertebrates, and fish. The quality and species of micro- and macroscopic plants and animals that make up the biological characteristics in a receiving body of water may be considered as the final test of wastewater treatment effectiveness. Because of the increasing awareness that enteric viruses can be waterborne, attempts have been made to identify and quantify virus contributions to receiving waters via wastewater treatment plants.

18.4   WASTEWATER TREATMENT PROCESSES

Wastewater treatment plants utilize a number of individual or unit operations and processes to achieve the desired degree of treatment. The collective treatment schematic is called a flow scheme, a flow diagram, or a flow sheet. Many different flow schemes can be developed from various processes for the desired level of treatment. Processes are grouped together to provide what is known as primary, secondary, and tertiary (or advanced) treatment. The term primary refers to physical unit operations. Secondary treatment refers to chemical and biological unit processes. Tertiary treatment refers to combinations of all three; this is discussed in the last section.

Treatment methods in which the application of physical process predominate are known as physical unit operations. These were the first methods to be used for wastewater treatment. Screening, mixing, flocculation, sedimentation, flotation, thickening, and filtration are typical processes. Each of these processes removes the initial solid or TSS from the raw sewage entering the facility.

Treatment methods in which the removal or conversion of contaminants is brought about by the addition of chemicals or by other chemical reactions are known as chemical unit processes. Precipitation, adsorption, and disinfection are the most common examples used in wastewater treatment. The first two of these processes will form a solid particle for easier removal and the second is to rid the discharge of any bacteria.

Treatment methods in which the removal of contaminants is brought about by biological activity are known as biological unit processes. Biological treatment is used primarily to remove and convert the biodegradable organic substances, colloidal or dissolved in wastewater, into gases that can escape to the atmosphere. The well-fed organisms are sequentially removed by allowing them to settle in a quiescent pond. Biological treatment can also be utilized to remove the nutrients in wastewater.

18.5   SLUDGE CHARACTERISTICS

Sludge arises when solids in the raw sewage settle prior to treatment. It can also be generated from filtration, aeration treatment, and chemical-addition sedimentation enhancement processes. The characteristics are greatly dependent on the type of treatment to which they have been subjected. Sludge typically consists of 1%–7% of solids with the rest 93%–99% wastewater. There are two basic types of sludge. They are settleable sludges and biological/chemical sludges. These are reviewed in the next two paragraphs.

Settleable sludge is removed during the primary sedimentation in the primary settling tanks. It is fairly easy to manage and can be readily thickened or reduced of its water content by gravity, or can be rapidly dewatered. A higher solids capture and better dry sludge cake can be obtained with primary settleable sludge. Primary sludge production can be estimated by computing the quantity of TSS entering the primary sedimentation tanks assuming a typical 70% efficiency of removal [1]. It is normally within the range of 800–2500 lbs per million gal (100–300 mg/L) of wastewater.

The biological and chemical sludges are produced in the advanced or secondary stages of treatment such as the activated sludge process from the aeration tanks. These sludges are more difficult to thicken; therefore, a portion of the sludge is recycled back into the activated sludge tank (aeration tank) in order to maintain a good population of microorganisms.

The quantity and nature of sludge generated relates to the characteristics of the raw wastewater and the type of process used to settle out the sludge. The operating expenses related to sludge handling can amount to one-third of the total investment of the treatment plant. Due to the high costs related to sludge operations and handling, new innovative technologies have been incorporated into producing fertilizer pellets for farming, addition to compost for the production of rich soils, and the formation of bricks for construction purposes.

18.6 ADVANCED WASTEWATER TREATMENT

Advanced wastewater treatment, known as tertiary treatment, is designed to remove those constituents that may not be adequately removed by secondary treatment. This includes removal of nitrogen, phosphorus, and heavy metals.

Biological nutrient removal of the inorganic constituents in the wastewater has received considerable attention in recent years for the reasons explained above. Excessive nutrients of nitrogen and phosphorus discharged to the receiving water can lead to eutrophication, causing excessive growth of aquatic plants, and indirectly deplete oxygen sources from the aquatic life and fish. There are also other beneficial reasons for biological nutrient removal that include monetary saving through reduced aeration capacity and reduced expense of chemical treatment. This area will see more activity in the future.

The final treatment of a municipal wastewater is disinfection. Currently there are controversial issues as to what type of disinfection techniques should be employed. Historically, chlorine was the choice of many facilities. However, it has come to the point that disinfection byproducts (of chlorides and bromides) that are being formed are toxic to the water environment. Alternate techniques that are being employed are ozonation and ultraviolet processes.

18.7   FUTURE TRENDS

The future outlook for wastewater treatment involves upgrades and retrofits of existing plants to provide increased capacity and better removal through tertiary treatment. Existing plants will need to be updated and retrofitted to handle increased demand from growing populations. New treatment techniques such as vortex separators, membrane bioreactors (MBRs), and ultrafiltration will become more commonplace to provide either better or more efficient removal of wastewater constituents.

Water recycling and reuse has proven to be effective and successful strategy for creating a new and reliable water source. Non-potable reuse is a widely accepted practice that is expected to grow in the years to come. In many parts of the world however, the uses of recycled water are expanding in order to accommodate the needs of the environment and growing water supply demands. Advances in wastewater treatment processes and additional information regarding human health of indirect potable reuse indicate that planned indirect potable reuse will become more acceptable and common in the future.

18.8   SUMMARY

1. The ultimate goal of wastewater treatment is the protection of the environment in a manner commensurate with economic, social, and political concerns.
2. The Federal Water Pollution Control Amendments of 1972 and 1990 require municipalities to prevent, reduce, or eliminate pollution of surface waters and ground water. The planning and design of a wastewater treatment plant must achieve these criteria. Presently, municipal wastewater plants are being upgraded and expanded to meet all government regulations.
3. BOD is the sole basis for determining the efficiency of the treatment plant. Secondary treatment typically utilizes a biological process to further remove the organic content and BOD from the effluent of the primary sedimentation tank. Above 90% BOD removal can be achieved in the aeration biological process tank.
4. Sludge is produced as a waste product during the wastewater treatment process. Sludge management and disposal issues are currently being tackled to provide a safer and cleaner environment. Composting and stabilization of sludge into a recyclable soil conditioning product is widely selected as the final disposal method of the sludge.
5. Primary treatment is used mainly for the removal (approximately 70%) of settleable TSS in the primary sedimentation tank. Paper, rags, sand, coffee grounds, and other solid waste materials are removed at this stage of the process.
6. Advanced wastewater treatment, known as tertiary treatment, is designed to remove those constituents that may not be adequately removed by secondary treatment. This includes removal of nitrogen, phosphorus, and heavy metals.

REFERENCES

1. Burbe, G., Singh, B., and Theodore, L. *Handbook of Environmental Management and Technology*, 2nd editon, John Wiley & Sons, Hoboken, NJ, 2000.
2. Hegewald, M. Setting the water quality agenda: 1988 and beyond, *Journal WPCF*, 60(5), 1988.
3. Federal Register. Amendment to the secondary treatment regulations: Percent removal requirements during dry weather periods for treatment works served by combined sewers, 40 *CFR* Part 133, January 27, 1989.
4. Federal Register. Secondary treatment regulation, 40 *CFR* Part 133, July 1, 1988.
5. Reynolds, J., Jeris, J., and Theodore, L. *Handbook of Chemical and Environmental Engineering Calculations*, John Wiley & Sons, Hoboken, NJ, 2004.

19 Industrial Wastewater Management

CONTENTS

19.1 INTRODUCTION

Clean water is a resource that has been taken for granted. Pure water is often necessary for growing food, manufacturing goods, disposing of wastes, and for consumption. Water conservation is most frequently thought of as a measure to protect against water shortages. While protecting water supplies is an excellent reason to practice conservation, there is another important benefit of water conservation—improved water quality.

The link between water use and water pollution may not be immediately apparent, yet water use is a considerable source of pollution to waste systems. When water is used for household, industrial, agricultural, or other purposes, it is almost always degraded and polluted in the process. Called wastewater, this by-product of human and industrial activities may carry nutrients, biological and chemical contaminants, floating wastes, or other pollutants. Upon discharge, wastewater ultimately finds its way into groundwater or surface waters, contributing to their pollution.

Every day U.S. industries discharge billions of gallons of wastewater generated by industrial processes. This liquid waste stream often contains many toxic metals and organic pollutants. Unfortunately, the discharge point for a large portion of these industries is frequently a municipal sewer system that leads to a publicly owned treatment works (POTW). It is estimated that roughly 60% of the total toxic metals and organics discharged by industry winds up at municipal treatment plants.

This flood of toxic wastewater described above varies from day to day, and from region to region. Its principal pollutants are toxic metals and organic chemicals. Some important toxic metals are lead, zinc, copper, chromium, cadmium, mercury, and nickel. Toxic organics include benzene, toluene, and trichloroethylene. Each of these substances, to a greater or lesser degree, is known to be harmful to human health. Many are toxic to aquatic life as well.

The consequences of these wastewater discharges have been severe. It is estimated that 14,000 mi of streams in 39 states have been polluted by toxic substances. It is also estimated that over half a million acres of lakes in 16 states and nearly 1000 square miles of estuaries in 8 states have been adversely affected.

Industries that send their wastes to POTWs are known as "indirect dischargers," i.e., because their discharges enter America's surface waters by an indirect route via municipal sewage treatment works. Direct dischargers, on the other hand, are industries that release their treated wastewater directly to surface waters.

Another area of concern is groundwater pollution. A few years ago this was almost an unknown problem. Today, groundwater is one of the major environmental areas of concern. Some of the reasons are detailed in the next three paragraphs.

Groundwater is that part of the underground water that is below the water table. Groundwater is in the zone of saturation within which all the pore spaces of rock materials are filled with water. The United States has approximately 15 quadrillion gal of water stored in its groundwater systems within one-half mile of the surface.

Annual groundwater withdrawals in the United States are on the order of 90 billion gal per day, which is only a fraction of the total estimated water in storage. This represents about a threefold increase in American groundwater usage since 1950. Most of this is replenished through and offsets the hydraulic effects of pumpage, except in some heavily pumped arid regions of the Southwest. American groundwater use is expected to rise in the future. Public drinking water accounts for 14% of groundwater use. Agricultural uses, such as irrigation (67%) and water for rural households and livestock (6%), account for 73% of groundwater usage. Self-supplied industrial water accounts for the remaining U.S. groundwater use. Approximately 50% of all Americans obtain all or part of their drinking water from groundwater sources.

The richest reserves of American groundwater are in the mid-Atlantic coastal region, the Gulf Coast states, the Great Plains, and the Great Valley of California. The Ogallala aquifer, which extends from the southern edge of North Dakota southwestward to the Texas and New Mexico border, is the single largest American aquifer in terms of geographical area. The most important American aquifer in agricultural terms is the large unconsolidated aquifer underlying the Great Valley of California. The most important groundwater sources of public drinking water are the aquifers of Long Island, New York, which have the highest per capita usage concentration in the United States.

Six areas of interest need to be addressed in order to obtain a clear picture of industrial wastewater management objectives and solutions. These include:

1. Regulations
2. Sources of industrial wastewater pollution
3. Industrial wastewater characterization
4. Nonpoint-source (NPS) water pollution
5. Wastewater treatment technologies
6. Future trends

These areas of concern will serve as the major focus for this chapter. Details of municipal wastewater management can be found in the previous chapter. Some overlap exists because of the complimentary nature of the two chapters.

19.2 SOURCES OF INDUSTRIAL WASTEWATER POLLUTION

There are literally thousands of industrial sources that contribute to the wastewater pollution problem. Some of the major industrial wastewater contributors are listed as follows:

Textile	Tannery
Laundry	Cannery
Dairy	Brewery, distillery, and winery
Pharmaceutical	Meat packing, rendering, and poultry
Beet sugar	Food processing
Wood fiber	Metal
Liquid material	Chemical
Energy	Nuclear power

The reader should note that a variety of wastes are generated within each industry listed above. For example, the chemical industry produces the following wastes:

Acids	Phosphates
Soaps and detergents	Explosives
Formaldehyde	Pesticides
Plastics and resins	Fertilizers
Toxic chemicals	Mortuary science wastes
Hospital and laboratory wastes	Polychlorinated biphenyls
Chloralkali wastes	Organic chemicals (in general)

Extensive details regarding the types and levels of pollutants discharged from these industries are available in the literature [1].

19.3 INDUSTRIAL WASTEWATER CHARACTERIZATION

The characteristics of wastewater having readily definable effects on water systems and treatment plants can be classified as follows:

1. Biochemical oxygen demand (BOD)
2. Suspended solids
3. Floating and colored materials
4. Volume
5. Other harmful constituents

BOD is defined as the amount of oxygen required by living organisms engaged in the utilization and stabilization of the organic matter present. Standard tests are conducted at 20°C with a 5 day incubation period. BOD is usually exerted by dissolved and colloidal organic matter and imposes a load on the biological units of the treatment plant. Oxygen must be provided so that bacteria can grow and oxidize the organic matter. An added BOD load, caused by an increase in organic waste, requires more bacterial activity, more oxygen, and greater biological-unit capacity for its treatment. Two other tests are generally used to estimate waste organic content: total organic carbon (TOC) and chemical oxygen demand (COD). TOC and COD are primary measures of total organic content, a portion of which may not be removed by biological treatment means.

Suspended solids are found in considerable quantity in many industrial wastes, such as cannery and paper mill effluents. They are screened and/or settled out of the sewage at the disposal plant. Solids removed by settling and separated from the flowing sewage are called sludge. Suspended solids settle to the bottom or wash up on the banks and decompose, causing odors and depleting oxygen in the river water. Fish often die because of a sudden lowering of the oxygen content of a stream, and solids that settle to the bottom will cover their spawning grounds and inhibit propagation. Visible sludge creates unsightly conditions and destroys the use of a river for recreational purposes.

Floating solids and liquids include oils, greases, and other materials that float on the surface; they not only make the river unsightly but also obstruct passage of light through the water, retarding the growth of vital plant food. Color contributed by textile and paper mills, tanneries, slaughterhouses, and other industries is an indicator of pollution. Compounds present in wastewaters absorb certain wavelengths of light and reflect the remainder, a fact generally conceded to account for color development of streams. Color interferes with the transmission of sunlight into the stream and therefore lessens photosynthetic actions [1].

A sewage plant can handle a large volume of flow if its units are sufficiently designed. Unfortunately, most sewage plants are already in operation when a request comes to accept the flow of waste from some new industrial concern.

Finally, other harmful constituents in industrial wastes can cause problems. Some problem areas and corresponding effects are

1. Toxic metal ions that interfere with biological oxidation.
2. Feathers that clog nozzles, overload digesters, and impede proper pump operation.
3. Rags that clog pumps and valves and interfere with proper operation.
4. Acids and alkalis that may corrode pipes, pumps, and treatment units, interfere with settling, upset the biological purification of sewage, release odors, and intensify color.
5. Flammables that cause fires and may lead to explosions.
6. Pieces of fat that clog nozzles and pumps and overload digesters.
7. Noxious gases that present a direct danger to workers.
8. Detergents that cause foaming.
9. Phenols and other toxic organic material.

19.4   NONPOINT SOURCE WATER POLLUTION [2]

In this period of public skepticism over government's ability to solve problems, the results of the Clean Water Act stand as a refreshing counterpoint. By many indicators, this legislation—and the programs it has generated—must be counted as a major success.

Gross pollution of the nation's rivers, lakes, and coastal waters by sewage and industrial wastes is largely a thing of the past. Fish have returned to waters that were once depleted of life-giving oxygen. Swimming and other water-contact sports are again permitted in rivers, in lakes, and at ocean beaches that once were closed by health officials. This success, however, is at best only a partial one. Water pollution remains a serious problem in most parts of the country. Sediment, nutrients, pathogenic organisms, and toxics still find their way into the nation's waters, where they degrade the ecosystem, pose health hazards, and impair the full use of water resources.

It is clear that this success in combatting the gross pollution of yesteryear—however incomplete—is largely the result of tackling the easy things first. This approach has, in large part, brought under control the so-called point sources of pollution. These include municipal and industrial outfalls and other sources that are clearly identified with a well-defined location or place. Government, by requiring permits to operate such facilities, has created a mechanism whereby control technology—such as a waste treatment plant—can be mandated, and the effect of such technology can be monitored.

It is equally clear that to continue the progress made over the past two decades, efforts must now focus on "nonpoint-source" pollution. The task of controlling NPS pollution is in many respects more difficult than controlling pollution from point sources, and requires different control strategies.

NPS pollution—unlike pollution from point sources—is quite diffuse, both in terms of its origin and in the manner in which it enters ground and surface waters. It results from a variety of human activities that take place over a wide geographic area, perhaps many hundreds or even thousands of acres. Unlike pollutants from point sources—which enter the environment at well-defined locations and in a relatively even, continuous discharge—pollutants from NPSs usually find their way into surface and groundwaters in sudden surges, often in large quantities, and are associated with rainfall, thunderstorms, or snowmelt. Seven of the most significant sources of NPS pollution are described below.

1. *Agriculture*: From 50% to 70% of impaired or threatened surface waters is affected by NPS pollution from agricultural activities. Pollutants include sediments from eroded croplands and overgrazed pastures; fertilizers or nutrients, which promote excessive growth of aquatic plants and contamination of groundwater by nitrate; animal waste from confined animal facilities which contains nutrients and bacteria that can cause shellfish bed closures and fish kills; and, pesticides, which can be toxic to aquatic life as well as to humans.

2. *Urban runoff*: Pollutants carried by runoff from such urban artifacts as streets and roadways, commercial and industrial sites, and parking lots

affect between 5% and 15% of surface waters. Urban runoff contains salts and oily residues from road surfaces and may include a variety of nutrients and toxics as well. Elevated temperatures—which are typical of urban runoff—can result in "thermal pollution," contributing to higher-than-normal temperatures in nearby streams, reservoirs, or lakes.

3. *Hydromodification*: Engineering projects, such as reservoir or dam construction, stream channelization, and flood prevention will inevitably result in changes in water flow patterns. When such changes occur, there is often an increase in sediment deposits. By modifying habitats, such projects may adversely affect aquatic life. Between 5% and 15% of surface waters in the United States is estimated to be affected by hydromodification.

4. *Abandoned mines and other past resource-extraction operations*: Up to 10% of surface waters is adversely affected by acid drainage from abandoned mines, pollution from mill tailings and mining waste piles, and pollution from improperly sealed oil and gas wells.

5. *Silviculture*: Pollution associated with commercial timber cutting and other forestry operations affects up to 5% of surface waters. Erosion from deforested lands, and particularly debris from eroded surfaces of logging roads, produces large amounts of sediment that ultimately finds its way into streams and lakes. Habitat altered by logging can adversely affect a wide range of plant and animal species.

6. *Construction*: New building and major land development projects, including highway construction, produce sediment and toxic materials that have been estimated to degrade up to 5% of the nation's surface waters.

7. *Land disposal*: Between 1% and 5% of the nation's surface waters is affected by disposal of waste on land—largely leakage from septic tanks and the spreading of sewage sludge.

19.5 WASTEWATER TREATMENT TECHNOLOGIES

As with municipal wastewater management, numerous technologies exist for treating industrial wastewater. These technologies range from simple clarification in a settling pond to a complex system of advanced technologies requiring sophisticated equipment and skilled operators. Finding the proper technology or combination of technologies to treat a particular wastewater to meet federal and local requirements and still be cost effective can be a challenging task [3,4].

Treatment technologies can be divided into three broad categories: physical, chemical, and biological. Many treatment processes combine two or all three categories to provide the most economical treatment. There are a multitude of treatment technologies for each of these categories. Although the technologies selected for discussion below are among the most widespread employed for industrial wastewater treatment, they represent only a fraction of the available technologies.

Two physical treatment processes are clarification or sedimentation and flotation. When an industrial wastewater containing a suspension of solid particles that have a higher specific gravity than the transporting liquid is in a relatively calm state, the particles will settle out because of the effects of gravity. This process of separating the settleable solids from the liquid is called clarification or sedimentation. In some treatment systems employing two or more stages of treatment and clarification, the terms primary, secondary, and final clarification are used. Primary clarification is the term normally used for the first clarification process in the system. This process is used to remove the readily settleable solids prior to subsequent treatment processes, particularly biological treatment. This treatment step results in significantly lower pollutant loadings to downstream processes and is appropriate for industrial wastewaters containing a high suspended solid content. Flotation, as opposed to clarification, which separates suspended particles from liquids by gravitational forces, accomplishes this operation because of their density difference by the introduction of air into the system. Fine bubbles adhere to, or are absorbed by, the solids, which are then lifted to the surface [3,4].

Two chemical treatment processes include coagulation-precipitation and neutralization. Often the nature of an industrial wastewater is such that the conventional physical treatment methods described in the previous paragraph will not provide an adequate level of treatment. Particularly, ordinary settling or flotation processes will not remove ultrafine colloidal particles and metal ions. Therefore, to adequately treat these particles in industrial wastewaters, coagulation-precipitation may be warranted. Rapid mixing is employed to ensure that the chemicals are thoroughly dispersed throughout the wastewater flow for uniform treatment. The wastewater then undergoes flocculation which provides for particle contact, so that the particles can agglomerate to a size large enough for removal. The final part of this technology involves precipitation. This is effectively the same as settling and thus can be performed in a unit similar to a clarifier. Neutralization is often required because coagulation-precipitation is capable of removing pollutants such as BOD, COD, and total suspended solid (TSS) from industrial wastewater. In addition, depending upon the specifics of the wastewater being treated, coagulation-precipitation can remove additional pollutants such as phosphorus, nitrogen compounds, and metals. This technology is attractive to industry because a high degree of classifiable and toxic pollutants removal can be combined in one treatment process. A disadvantage of this process is the substantial quantity of sludge generated, which presents a sludge disposal problem.

Highly acidic or basic wastewaters are undesirable. They can adversely impact the aquatic life in receiving waters. In addition, they might significantly affect the performance of downstream treatment processes at the plant site or at a POTW. Therefore, in order to rectify these potential problems, one of the most fundamental treatment technologies, the aforementioned neutralization, is employed at industrial facilities. Neutralization involves adding an acid or a base to a wastewater to offset or neutralize the effects of its counterpart in the wastewater flow, namely, adding acids to alkaline wastewaters and bases to acidic wastewaters [4].

The most appropriate industrial treatment technology for removing oxygen-demanding pollutants is biological treatment. Biological treatment processes frequently used in the industrial field include aerobic suspended growth processes (activated sludge), aerobic contact processes, aerated lagoons (stabilization ponds), and anaerobic lagoons. An aerobic suspended growth process (activated sludge) is one in which the biological growth products (microorganisms) are kept in suspension in a turbulent liquid medium consisting of entrapped and suspended colloidal and dissolved organic and inorganic materials. This biological process uses the metabolic reactions of the microorganisms to attain an acceptable effluent quality by removing those substances exerting an oxygen demand. An aerobic attached growth process is one in which the biological growth products (microorganisms) are attached to some type of medium (i.e., rock, plastic sheets, plastic rings, etc.), and where either the wastewater trickles over the surface or the medium is rotated through the wastewater. The process is related to the aerobic suspended growth process in that both depend upon biochemical oxidation of organic matter in the wastewater to carbon dioxide, with a portion oxidized for energy to sustain and promote the growth of microorganisms [4]. Aerobic lagoons (stabilization ponds) are large, shallow earthen basins that are used for wastewater treatment by utilizing natural processes involving both algae and bacteria. The objective is microbial conversion of organic wastes into algae. Aerobic conditions prevail throughout the process. Finally, anaerobic lagoons are earthen ponds built with a small surface area and a deep liquid depth of 8–20 ft. Usually these lagoons are anaerobic throughout their depth, except for an extremely shallow surface zone.

The development of advanced treatment technologies, along with an increasing scarcity of fresh water, has led to marked changes in effluent management. Numerous strategies for purified wastewater reuse are presently being employed in ways appropriate to the particular industrial operation. The combination of scarcity of water with increasingly stiff regulations has made effluent disposal into natural receiving bodies the option of last resort [3].

19.6   FUTURE TRENDS

Today, considerable effort is being expended toward investigating and cleaning up some of the past mistakes, especially those involving hazardous wastes that have led to the contamination of water supplies. These activities, however, must be matched in the future by the equally important effort of preventing water pollution in the first place. Because of the diverse nature of sources of contamination and their widespread occurrence, much of the responsibility for protecting water resources must be left to state and local agencies. This is especially true because programs to protect water quality will not be successful unless they reflect the close relationship of the land, groundwater, and surface water.

Society is still learning more and more each year about the impact that various sources of contamination can have on water. In fact, the emphasis on which source or area to concentrate regulatory efforts has changed drastically over the past two decades. Thus, there is a critical need to give water resource protection the high national priority that it deserves and to encourage federal, state, and local agencies to develop the required strategies and programs to carry out this effort.

19.7   SUMMARY

1. Clean water is a resource that has been taken for granted. Pure water is necessary for growing food, manufacturing goods, disposing of wastes, and for consumption.

2. Congress enacted the Clean Water Act to "restore and maintain the chemical, physical, and biological integrity of the Nation's waters." Waters of the United States protected by the Clean Water Act include rivers, streams, estuaries, the territorial seas, and most ponds, lakes, and wetlands.

3. There are literally thousands of industrial sources that contribute to the wastewater pollution problem.

4. The characteristics of wastewater having readily definable effects on water systems and treatment plants can be classified as follows:
   a. BOD
   b. Suspended solids
   c. Floating and colored materials
   d. Volume
   e. Other harmful constituents

5. Unlike pollutants from point sources—which enter the environment at well-defined locations and in relatively even, continuous discharges—pollutants from NPSs usually find their way into surface and groundwaters in sudden surges, often in large quantities, and are associated with rainfall, thunderstorms, or snowmelt.

6. Treatment technologies can be divided into three broad categories: physical, chemical, and biological. Many treatment processes combine two or all three categories to provide the most economical treatment.

7. Society is still learning more and more each year about the various sources of contamination can have on water.

REFERENCES

1. Nemerow, N. and Dasgupta, A. *Industrial and Hazardous Waste Treatment*, Van Nostrand Reinhold, New York, 1991.
2. U.S. EPA. NPS pollution, EPA Journal, 17(5), November/December, 1991 (22k-1005).
3. Burke, G., Singh, B., and Theodore, L. *Handbook of Environmental Management and Technology*, 2nd edition, John Wiley & Sons, Hoboken, NJ, 2000.
4. Jeris, J. Lecture notes, Manhattan College, Bronx, New York, 1992.

20 Dispersion Modeling in Water Systems

CONTENTS

20.1 INTRODUCTION

Four distinct periods can be distinguished in the development of mathematical models that describe water systems [1]:

1. The precomputer age (1900–1950). During this time, the focus was entirely on water quality with little concern for the environmental aspects.
2. The transition period of the 1950s. Data collection was accelerated but the analysis was slow and costly.
3. The early years of computer use. During the 1960s, the first computer models were developed, and many models were developed during the 1970s due to greater computer access.
4. The mid-1970s to date. Because of the development of inexpensive micro-computers, models can now be used for routine evaluations.

Models are classified by the number of dimensions modeled and by the type of model employed. These can be described as follows [2]:

1. One-dimensional models. The only direction modeled is that of the direction of flow. This is a valid model for flowing streams where the concentration of pollutants is taken to be constant with stream cross section.
2. Two-dimensional models. This is used in wide rivers where concentration may not be uniform across the entire width. The model is a function of both width and flow direction. For deep, narrow rivers, lakes, or estuaries, the

horizontal and vertical dimensions are modeled, while the lateral dimension is held constant.

3. Three-dimensional models. The assumption here is that concentration can vary with length, width, and depth. Of the three models, this is the most accurate. It is also, however, both tedious and time consuming. The potential accuracy is greater, but the development and costs are also greater.

There is also a zero-dimensional model that takes none of the lateral, vertical, or longitudinal motion into consideration. In this model, a segment of stream is treated as a completely mixed reactor [3]. The chemical engineer often refers to this type of model as a continuous stirred tank reactor (CSTR).

The model referred to above is an assembly of concepts in the form of one or more mathematical equations that approximates the behavior of a natural system or phenomenon [4]. Rather than focus on the water systems themselves, however, this chapter will examine the individual pollutants and components that are modeled. These include microorganisms, dissolved oxygen (DO), eutrophication, and toxic chemicals. Within each of these areas, specific types of contaminants as well as the different water systems that are affected will be explored.

20.2   MATHEMATICAL MODELS

There are mass (componential) and flow (overall mass) balance equations for rivers and streams. They are as follows [5]:

Mass balance

$$
\begin{pmatrix} \text{Mass rate of} \\ \text{substance upstream} \end{pmatrix} + \begin{pmatrix} \text{Mass rate added} \\ \text{by outfall} \end{pmatrix} = \begin{pmatrix} \text{Mass rate of substance} \\ \text{immediately downstream} \\ \text{from outfall assuming} \\ \text{complete mixing} \end{pmatrix} \quad (20.1)
$$

Flow balance

$$
\begin{pmatrix} \text{Flow rate} \\ \text{upstream} \end{pmatrix} + \begin{pmatrix} \text{Flow rate added} \\ \text{by outfall} \end{pmatrix} = \begin{pmatrix} \text{Flow rate immediately} \\ \text{downstream from outfall} \end{pmatrix} \quad (20.2)
$$

The physical characteristics of lakes set them apart from other water systems in modeling for microorganisms, as well as for the other contributors to water quality. These characteristics include evaporation due to a large surface area, and temperature stratification due to poor mixing within the lake. Therefore, these differences must be taken into account in any balance equation [6]:

$$\begin{array}{c} \text{Net flow into and} \\ \text{out of the lake due to} \\ \text{river and/or} \\ \text{groundwater flow} \end{array} + \begin{array}{c} \text{Precipitation} \\ \text{directly onto} \\ \text{the lake} \end{array} - \text{Evaporation} = \begin{array}{c} \text{Change in the} \\ \text{lake volume} \\ \text{with time} \end{array}$$

(20.3)

The last and most complicated water system that will be examined is the estuary. Unlike lakes and rivers, there are no simple balance equations that can be written for estuaries. Estuaries are coastal water bodies where freshwater meets the sea. They are traditionally defined as semi-enclosed bodies of water having a free connection with the open sea and within which sea water is measurably diluted with fresh water entering from land drainage [7].

The seaward end of an estuary is easily defined because it is connected to the sea. The landward end, however, is not that well defined. Generally, tidal influence in a river system extends further inward than salt intrusion. That is, the water close to the fall line of the estuary may not be saline, but it may still be tidal. Thus, the estuary is limited by the requirement that both salt and freshwater be measurably present. The exact location of the salt intrusion depends on the freshwater flow rate which can vary substantially from one season to another [8].

The variations in an estuary throughout the year, together with the fact that each estuary is different, make modeling rather difficult. Some simplifications can, however, be made that provide some remarkably useful results in estimating the distribution of estuarine water quality. The simplifications can be summarized through the following assumptions:

1. The estuary is one-dimensional.
2. Water quality is described as a type of average condition over a number of tidal cycles.
3. Area, flow, and reaction rates are constant with distance.
4. The estuary is in a steady-state condition.

A water body is considered to be a one-dimensional estuary when it is subjected to tidal reversals (i.e., reversals in direction of the water velocity) and where only the longitudinal gradient of a particular water quality parameter is dominant [6].

20.3 MICROORGANISMS

The transmission of waterborne diseases (e.g., gastroenteritis, amoebic dysentery, cholera, and typhoid) has been a matter of concern for many years. The impact of high concentrations of disease-producing organisms on water uses can be significant. Bathing beaches may be closed permanently or intermittently during rainfall conditions when high concentrations of pathogenic bacteria are discharged from urban runoff and combined sewer overflows. Diseases associated with drinking water continue to occur [6].

There are four types of organisms that can affect water quality. The first are indicator bacteria which may reflect the presence of pathogens. In the past they were used as a measure of health hazard. Pathogenic bacteria are the cause of such diseases as salmonella, cholera, and dysentery, and continue to be a problem worldwide. Viruses are submicroscopic, inert particles that are unable to replicate or adapt to environmental conditions outside a living host [9], and if ingested, they can cause hepatitis. And finally, pathogenic protozoa are parasitic, but able to reproduce, and are responsible for amoebic dysentery. Table 20.1 lists examples of communicable disease indicators and organisms.

The factors that can affect the survival or extinction of these microorganisms are [6]

1. Sunlight
2. Temperature
3. Salinity
4. Predation
5. Nutrient deficiencies
6. Toxic substances
7. Settling of organism population after discharge
8. Resuspension of particulates
9. Growth of organisms within the body of water

The overall decay rate equation for microorganisms is given as

$$K_B = K_{B1} + K_{BI} \pm K_{Bs} - K_a \tag{20.4}$$

where
K_B is the overall rate of decay
K_{B1} is the death rate due to temperature, salinity, and predation

---

TABLE 20.1

Communicable Disease Indicators and Organisms

Indicators	Viruses
Bacteria	Hepatitis A
Total coliform	Enteroviruses
Fecal coliform	Polioviruses
Fecal streptococci	Echoviruses
Obligate anaerobes	Coxsackieviruses
Bacteriophages (bacterial viruses)	
Pathogenic Bacteria	**Pathogenic Protozoa and Helminths**
Vibrio cholerae	*Giardia lambia*
Salmonella	*Entamoeba histolytica*
Shigella	Facultatively parasitic amoebae
	Nematodes

Source: Thomann, R. and Mueller, J., *Principles of Surface Water Quality Modeling and Control*, Harper & Row, New York, 1987. With permission.

K_{BI} is the death rate due to sunlight
K_{Bs} is the net loss or gain due to settling or resuspension
K_a is the aftergrowth rate [6]

One describing equation for the downstream distribution of bacteria in rivers and streams is given as

$$N = N_0 \exp(-K_B t^*) \tag{20.5}$$

where
N is the concentration of an organism
N_0 is the concentration of the organism at the outfall
K_B is the overall net rate of decay as given previously
t^* is the time it takes to travel a downstream distance x at a water velocity U [6]

As Equation 20.5 states, the organism will decay exponentially with time. (Thus, it takes the form of a classical first-order chemical equation.)

20.4   DISSOLVED OXYGEN (DO)

The problems of DO in surface waters have been recognized for over a century. The impact of low DO concentrations or of anaerobic conditions was reflected in an unbalanced ecosystem, fish mortality, odors, and other aesthetic nuisances. While coliform was a surrogate variable for communicable disease and public health, DO is a surrogate variable for the general health of the aquatic ecosystem [6].

The variations in DO levels are caused by sources and sinks. The sources include reaeration from the atmosphere, which is dependent upon turbulence, temperature, and surface films; photosynthetic oxygen production where plants react with CO_2 and H_2O to form glucose and oxygen; and, incoming DO from tributaries (streams that feed a larger stream or a lake) or effluents. The sinks of DO are oxidation of carbonaceous (CBOD) and nitrogenous (NBOD) waste materials, oxygen demand of the sediments of the water body, and the use of oxygen for respiration by aquatic plants [6].

With these inputs, sources and sinks, the following general mass balance equation for DO in a segmented volume can be written as

$$
\begin{aligned}
&\text{Reaeration} + (\text{Photosynthesis} - \text{Respiration}) - \begin{matrix}\text{Oxidation}\\\text{of CBOD}\\\text{NBOD}\end{matrix} - \begin{matrix}\text{Sediment}\\\text{oxygen}\\\text{demand}\end{matrix}\\
&+ \begin{matrix}\text{Oxygen}\\\text{input}\end{matrix} \pm \begin{matrix}\text{Oxygen transport}\\\text{(into or out}\\\text{of segment)}\end{matrix} = \begin{matrix}\text{Change with time of}\\\text{dissolved oxgen in a}\\\text{specific volume of water}\end{matrix}
\end{aligned} \tag{20.6}
$$

This equation can be applied to a specific water body where the transport, sources, and sinks are unique to that aquatic system [5].

The discharge of municipal and industrial waste, and urban and other nonpoint source runoff will necessitate a continuing effort in understanding the DO resources of surface waters. The DO problem can thus be summarized as the discharge of organic and inorganic oxidizable residues into a body of water, which, during the processes of ultimate stabilization of the oxidizable material (in the water or sediments), and through interaction of aquatic plant life, results in the decrease of DO to concentrations that interfere with desirable water uses [6]. The balance equations that can be applied to the various water systems, namely rivers and lakes, are the same as described in the previous section.

20.5 EUTROPHICATION

Even the most casual observer of water quality has probably had the dubious opportunity of walking along the shores of a lake that has turned into a sickly green pea soup. Or perhaps, one has walked the shores of a slow-moving estuary or bay and had to step gingerly to avoid rows of rotting, matted, stringy aquatic plants. These problems have been grouped under a general term called eutrophication. The unraveling of the causes of eutrophication, the analysis of the impact of human activities on the problem, and the potential engineering controls that can be exercised to alleviate the condition have been a matter of special interest for the past several decades.

Eutrophication is the excessive growth of aquatic plants, both attached and planktonic (those that are free swimming), to levels that are considered to be an interference with desirable water uses. One of the principal stimulants is an excessive level of nutrients such as nitrogen and phosphorus. In recent years, this problem has been increasingly acute due to the discharge of such nutrients by municipal and industrial sources, as well as agricultural and urban runoff. It has often been observed that there is an increasing tendency for some water bodies to exhibit increases in the severity and frequency of phytoplankton blooms and growth of aquatic weeds, apparently as a result of elevated levels of nutrients [6].

The principal variables of importance in the analysis of eutrophication are [6]

1. Solar radiation at the surface and with depth
2. Geometry of water body—surface area, bottom area, depth, volume
3. Flow, velocity, dispersion
4. Water temperature
5. Nutrients
   a. Phosphorus
   b. Nitrogen
   c. Silica
6. Phytoplankton

The nonorganic products that result from oxidation are referred to as nutrients. They include nitrogen found in the form of ammonia, nitrite, and nitrate, and phosphorus, which occurs in the form of phosphates. A third nutrient, silicon in the form of silicate, enters the system through the weathering of soils and rocks. These nutrients, along with carbon dioxide, "feed" the process of photosynthesis, which creates

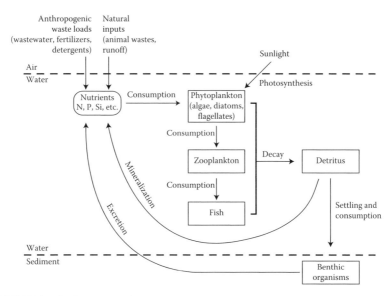

FIGURE 20.1 Anthropogenic inputs.

the beginning components of the biological cycle—phytoplankton, the microscopic plants that drift around in the water, diatoms (which need silicon for their shells), flagellates (organisms possessing one or more whiplike appendages often used for locomotion), and green and blue-green algae among them. Some of these nutrients will go into producing a complementary pool of rooted aquatic plants. Figure 20.1 shows the basic biological cycle in lakes and estuaries [10]. A critical portion of the cycle is the phytoplankton pool. Increased nutrient availability can lead to unsightly plankton blooms and to anoxic conditions as the available oxygen is used up in the plankton decay [10].

Other processes related to algal growth and nutrient recycling are sorption and desorption of inorganic material, settling and deposition of phytoplankton, uptake of nutrients and growth of phytoplankton, death of phytoplankton, mineralization of organic nutrients, and nutrient generation from the sediment [11].

20.6 TOXIC SUBSTANCES

The issue of the release of chemicals into the environment at a level of toxic concentration is an area of intense concern in water quality and ecosystem analyses. Passage of the Toxic Substances Control Act (TSCA) of 1976 in the United States, unprecedented fines, and continual development of data on lethal and sublethal effects attest to the expansion of control on the production and discharge of such substances. However, as illustrated by pesticides, the ever-present potential for insect and pest infestations with attendant effects on humans and livestock results in a continuing demand for product development. As a result of these competing goals, considerable effort has been devoted in recent years to the development of predictive schemes that would permit an a priori judgment of the fate and effects of a chemical in the environment [6].

Table 20.2 summarizes a few specific chemicals that are of special interest in evaluating water quality, with brief descriptions of the problems that they may cause. Note especially the differences in the effects on water quality caused by various chemical forms of the same element, for example, sulfur. This illustrates the

TABLE 20.2
Potential Water Quality Problems That May Be Caused by a Few Selected Chemicals

Chemicals	Potential Problems
Arsenic	Toxicity to humans
	Toxicity to aquatic life
Chlorine	Organic reactions form trihalomethanes
	Toxicity to fish and other aquatic life
Calcium	Causes "hardness" in water
	May result in scale formation in pipes
Iron	Causes stains in laundry and on fixtures
	May kill fish by clogging their gills
Nitrogen: ammonia	May accelerate eutrophication in lakes
	May improve productivity of the water
	May be toxic to aquatic life
Nitrogen: nitrates	May be toxic to babies
	May accelerate eutrophication in lakes
	May improve productivity of the water
Oxygen, dissolved	Low concentrations harmful to fish
	Low concentrations may cause odor problems
	High concentrations accelerate metal corrosion
	Low or zero concentration may allow sulfide formation and concrete corrosion
Phenolics	Tastes and odors in drinking water
	Can cause tainting of fish flesh
	May be toxic to aquatic life
Sulfur: sulfides	Objectionable odors in and near water
	May be toxic to aquatic life
	May corrode concrete through acid formation
	Oxidation of sulfide to sulfate exerts an oxygen demand
Sulfur: sulfites	React with DO and exert oxygen demand
Sulfur: sulfates	Increase water corrosiveness to metals
	Decompose anaerobically to form sulfides
	Salty taste and laxative effects

Source: Rau, J. and Wooten, D., *Environmental Impact Analysis Handbook*, McGraw-Hill, New York, 1980. With permission.

importance, sometimes, of assaying specific ions or molecules instead of merely total content of the element itself [12].

The uniqueness of the toxic substances problem lies in the potential transfer of a chemical to humans with possible attendant public health impacts. This transfer occurs primarily through two principle routes:

1. Ingestion of the chemical from the drinking water supply.
2. Ingestion of the chemical from contaminated aquatic foodstuffs (e.g., fish and shellfish) or from food sources that utilize aquatic foodstuffs as a feed.

The toxic substances water quality problem can therefore be summarized as the discharge of chemicals into the aquatic environment. This results in concentrations in the water or aquatic food chain at levels that are determined to be toxic, in a public health sense or to the aquatic ecosystem itself, and thus may interfere with the use of the water body for water supply or fishing or contribute to ecosystem instability [6].

20.7 FUTURE TRENDS

The future predicts more sophisticated models.

20.8 SUMMARY

1. A model is an assembly of concepts in the form of one or more mathematical equations that approximates the behavior of a natural system or phenomena.
2. There exist mass balance equations for both rivers and lakes, but a simple equation for estuaries does not exist due to their complexity.
3. The levels of microorganisms in a water system depend on sunlight, temperature, salinity, predation, nutrients, toxic substances, settling of the organic population, resuspension of particulates, and growth within the body of water.
4. The impact of DO concentrations or of anaerobic conditions is reflected in an unbalanced ecosystem.
5. Eutrophication is the excessive growth of aquatic plants, both attached and planktonic, to levels that are considered to interfere with desirable water uses.
6. With toxic substances, it is as important to look at the effects of individual ions or molecules as it is to look at the total content of the element itself.

REFERENCES

1. Novotny, V. Agricultural nonpoint source pollution, model selection and application, *Dev. Environ. Model*, 10, 1986.
2. Rau, J. and Wooten, D. *Environmental Impact Analysis Handbook*, McGraw-Hill, New York, 1980.
3. Dortch, M. and Martin, J. *Alternatives in Regulated Flow Management*, CRC Press, Boca Raton, FL, 1988.

4. American Society of Testing and Materials (ASTM). *Standard Practice for Evaluating Environmental Fate Models for Chemicals*, proposed standard, Subcommittee E-47.06 on Environmental Fate, Committee E-47 on Biological Effects and Environmental Fate, 1983.
5. Burke, G., Singh, B., and Theodore L. *Handbook of Environmental Management and Technology*, John Wiley & Sons, Hoboken, NJ, 2008.
6. Thomann, R. and Mueller, J. *Principles of Surface Water Quality Modeling and Control*, Harper & Row, New York, 1987.
7. Pritchard, D. What is an estuary? *Estuaries*, American Association for the Advancement or Sciences, 83(2), 1967.
8. Mills, W., Porcella, D., Ungs, M., Gherini, S., Summers, K., Mok, L., Rupp, G., and Bowie, G. *Water Quality Assessment: A Screening Procedure for Toxic and Conventional Pollutants*, Part II, EPA/600/6-85/002b, 1985.
9. National Academy of Sciences. *Drinking Water and Health*, Safe Drinking Water Committee, Natural Resources Council, Washington, DC, 1977.
10. Officer, C. and Page, J. *Tales of the Earth*, Oxford University Press, New York, 1993.
11. Lung, W. Application to estuaries. *Water Quality Modeling*, Vol. III, CRC Press, Boca Raton, FL, 1993.
12. Lamb, J., III. *Water Quality and Its Control*, John Wiley & Sons, Hoboken, NJ, 1989.

21 Acid Rain

CONTENTS

21.1   INTRODUCTION

Acid deposition, popularly known as acid rain, has long been suspected of damaging lakes, streams, forests, and soils, decreasing visibility, corroding monuments and tombstones, and potentially threatening human health in North America and Europe. The National Academy of Sciences and other leading scientific bodies first gave credence to these concerns in the early 1980s when they suggested that emissions of sulfur dioxide from electric power plants were being carried hundreds of miles by prevailing winds, being converted in the atmosphere into sulfuric acid, falling into pristine lakes, and killing off aquatic life. The process of acid deposition also begins with emissions of nitrogen oxides (primarily from motor vehicles and coal-burning power plants). These pollutants interact with sunlight and water vapor in the upper atmosphere to form acidic compounds. During a storm, these compounds fall to earth as acid rain or snow; the compounds also may join dust or other dry airborne particles and fall as "dry deposition" [1]. Regulations have been passed concerning the amount of SO_2 and NO_x (oxides of nitrogen) emitted in the air. These regulations have caused the power industries to find ways to cut their emissions. The three ways of lowering emissions—before combustion, during combustion, and after combustion—will be discussed.

Sulfur dioxide, the most important of the two gaseous acid pollutants, is created when the sulfur in coal is released during combustion and reacts with oxygen in the air. The amount of sulfur dioxide created depends on the amount of sulfur in the coal. All coal contains some sulfur, but the amount varies significantly depending on where the coal is mined. Over 80% of sulfur dioxide emissions in the United States originate in the 31 states east of or bordering the Mississippi River [1]. Most emissions come from the states in or adjacent to the Ohio River Valley.

The extent of damage caused by acid rain depends on the total acidity deposited in a particular area and the sensitivity of the area receiving it. Areas with acid-neutralizing compounds in the soil, for example, can experience years of acid deposition without problems. Such soils are common in much of the United States. But the thin soils of the mountainous and glaciated northeast have very little acid-buffering capacity, making them vulnerable to damage from acid rain. Surface waters, soils, and bedrock that have a relatively low buffering capacity are unable to neutralize the acid effectively. Under such conditions, the deposition may increase the acidity of water, reducing much or all of its ability to sustain aquatic life. Forests and agriculture may be vulnerable because acid deposition can leach nutrients from the ground, kill nitrogen-fixing microorganisms that nourish plants, and release toxic metals.

Acid rain with a pH below 5.6 is formed when certain anthropogenic air pollutants travel into the atmosphere and react with moisture and sunlight to produce acidic compounds. Sulfur and nitrogen compounds released into the atmosphere from different sources are believed to play the biggest role in the formation of acid rain. The natural processes which contribute to acid rain include lightning, ocean spray, decaying plant and bacterial activity in the soil, and volcanic eruptions. Anthropogenic sources include those utilities, industries, businesses, and homes that burn fossils fuels, plus motor vehicle emission. Sulfuric acid is the type of acid most commonly formed in areas that burn coal for electricity, while nitric acid is more common in areas that have a high density of automobiles and other internal combustion engines.

There are several ways that acid rain affects the environment:

1. Contact with plants can harm plants by damaging outer leaf surfaces and by changing the root environment.
2. Contact with soil and water resources. Due to the acid in the rain, fish kills in ponds, lakes and oceans, as well as effects on aquatic organisms, are common. Acid rain can cause minerals in the soil to dissolve and be leached away. Many of these minerals are nutrients for both plants and animals.
3. Acid rain mobilizes trace metals, such as lead and mercury. When significant levels of these metals dissolve from surface soils they may accumulate elsewhere, leading to poisoning.
4. Acid rain may damage building structures and automobiles due to accelerated corrosion rates.

The general chemical formulae for the formation of acid rain are as follows:

$$SO_x + O_2 \rightarrow SO_2 + H_2O \rightarrow H_2SO_4 \tag{21.1}$$

$$NO_x + O_2 \rightarrow NO_2 + HNO_3 \tag{21.2}$$

$$CO_2 + H_2O \rightarrow H_2CO_3 \tag{21.3}$$

21.2   EMISSIONS REDUCTION—BEFORE COMBUSTION

In an attempt to mitigate the effects of acid rain several proposals have appeared before the U.S. Congress. Reductions in sulfur dioxide emissions, particularly from utility and industrial coal-fired boilers, is the primary target for combatting acid rain.

Several means exist to limit the amount of sulfur in the fuel prior to combustion and include the use of lower-sulfur coals and coal cleaning. Reduction of nitrogen oxides emissions cannot be accomplished at this point because it is formed after (and following) incomplete combustion. The two major reduction procedures described below include coal switching and coal cleaning. Although this development is directed toward coal (the fuel of primary concern in this nation), it may also be applied, in some instances to other fossil fuels.

21.2.1   COAL SWITCHING

For many power plants, the most economical strategy for reducing sulfur dioxide emissions tends to be switching from higher-sulfur to lower-sulfur coals. Because the sulfur content in coal varies across regions, this move to consume lower-sulfur coals would result in a major shift in regional coal production from higher-sulfur supply regions, to lower-sulfur supply regions. It could also generate regional hostility by causing shifts in existing coal markets.

Existing coal-fired power plants now burning higher-sulfur coals without scrubbers would be faced with the most stringent requirements for reducing emissions. Most of these plants were initially designed to burn bituminous coals.

Under most circumstances, higher-sulfur plants would face relatively little technical difficulty in shifting to lower-sulfur coals, but there may be some additional costs, especially for upgrading electrostatic precipitators. A few plants, such as cyclone-fired boilers, are not technically well-suited for burning lower-sulfur coals because of the difference in ash-fusion temperatures.

Also, there is some question as to the type of lower-sulfur coals likely to be in demand by power plants shifting to these coals. Since most existing boilers were initially designed to burn bituminous coals, it is not clear whether these units can economically shift to lower-sulfur subbituminous coals. Some of the questions are technical, such as the potential for slagging or fouling when an off-design coal is burned in these boilers.

Consumption of subbituminous coals will entail higher heat rates, higher handling costs, and capacity derates. Together, these economic and technical considerations tend to make the use of subbituminous coals in bituminous boilers a very site-specific issue that does not provide clear economic advantages.

21.2.2 COAL CLEANING

Another way to control sulfur emission is to clean the coal before burning. This process can reduce sulfur dioxide emission by 20%–90%. Coal cleaning can be accomplished in three ways: physical (gravity separation), chemical (reaction or bioremediation), or electrical. Details on the principal cleaning method gravity are provided below. Details on other separation techniques are available in the literature [2].

Gravity separation is used by industry. This method depends on the size, shape, density, and surface properties of the coal. The process first crushes the coal into small particles and then allows gravity to separate the pyritic sulfur from the coal. The gravity separation process is usually accomplished in a water medium. The coal containing impurities sink to the bottom and the usable coal stays on the top. The benefits of this process is that the coal is easy to handle and the coal can burn better. Although there are some benefits to this process, there are limitations as well.

One of the limitations of gravity separation is that it is ineffective in reducing sulfur content when the coal particles are very fine. When the particles are very fine the metals will not be attached to the sulfur enriched coal and separation will not occur. Another problem is with the medium. The coal must be dried after using water to separate the burning coal from its impurities; drying the coal is expensive. There is also a major loss of energy release with the purified coal; this means more coal must be burned to generate the same amount of energy.

21.3 EMISSIONS REDUCTION—DURING COMBUSTION

The second method that can be used to reduce the emissions of the precursors of acid rain is during combustion. Both NO_x and SO_2 emissions can be reduced during this stage of the combustion process.

The reduction of NO_x emissions is accomplished by primarily three methods: low-NO_x burners, overfire air, and fuel staging or reburning. These methods have the ability to reduce emissions by 80% and are cost-effective. The three processes are based on using combustion with stoichiometric air and controlling the temperature. The fuel staging process seems to work the best in reducing the emissions.

In a coal-fired boiler, reburning is accomplished by substituting 15%–20% of the coal with natural gas or low-sulfur oil and burning it at a location downstream of the primary combustion zone of the boiler. Oxides of nitrogen formed in the primary zone are reduced to nitrogen and water vapor as they pass through the reburn zone. Additional air is injected downstream of the reburn zone to complete the combustion process at a lower temperature. In general, NO_x reductions of 50% or more are achievable by reburning. When combined with other low-NO_x technologies (such as low-NO_x burners), NO_x reductions of up to 90% may be achievable.

Reduction of sulfur dioxide emissions cannot be accomplished easily because many problems occur and these methods are expensive. Two methods that are frequently used in industry are limestone injection multistage burner (LIMB) and fluidized bed combustion (FBC).

LIMB is an emerging control process that can be retrofitted on a large number of existing coal-fired boilers. In a LIMB system, an SO_2 sorbent (limestone) is injected into a boiler equipped with low-NO_x burners. The sorbent absorbs the SO_2 and the low-NO_x burners limit the amount of NO_x formed. LIMB is capable of reducing both SO_2 and NO_x by about 50%–60%.

The benefit of using the LIMB process is that it is one of the least expensive processes for reducing SO_2 emissions during combustion, but the sorbent injected into the boiler tends to increase slagging and fouling, which in turn increase operation and maintenance costs. Because boilers retrofitted with LIMB tend to produce more particulates of smaller sizes, particulate control becomes more difficult. Technical questions remain as to what sorbents are most effective in a LIMB system, and how and where to inject the sorbents.

In an FBC boiler, pulverized coal is burned while suspended over a turbulent cushion of injected air. This technique allows improved combustion efficiencies and reduced boiler fouling and corrosion. Such boilers also are capable of burning different kinds of low-grade fuels like refuse, wood bark, and sewage sludge. In addition, if the coal is mixed with limestone or some other sorbent material during combustion, the SO_2 is captured and retained in the ash.

FBC boilers have the potential to control NO_x as well as SO_2. FBC boilers must operate within a narrow temperature range (1500°F–1600°F) and lower combustion temperatures inherently limit the formation of NO_x. FBC boilers may be able to control NO_x by 50%–75% at the same time as they control SO_2 by up to 90%. An FBC system does have one major flaw: it requires the construction of a new boiler. The FBC system is more of a replacement technology than a retrofit.

21.4   EMISSIONS REDUCTION—AFTER COMBUSTION

The final method of reducing emissions that cause acid rain is after combustion has occurred. This method is most frequently used by industry, although it is not preferred by environmentalists because it creates other wastes while reducing emissions.

The most popular process for reducing NO_x after combustion is selective catalyst reduction (SCR). It is mainly used in Japan. In the SCR system, a mixture of ammonia gas and air is injected upstream of a catalytic reactor chamber. The flue–gas mixture then travels in a vertical, downward-flow direction through a catalytic reactor chamber, where the ammonia gas disassociates NO_x to nitrogen gas and water vapor [1]. This process benefits from its high removal rate (80%–90%) and no retrofitting on the unit is necessary. Catalyst selection is very important for this process. Catalyst selection is based on the following criteria: resistance to toxic materials, abrasion resistance, mechanical strength, resistance to thermal cycling, resistance to the oxidation of SO_2, and resistance to plugging [3].

Industry has mainly chosen the flue–gas desulfurization process (FGD) to combat SO_2 emissions. FGD uses sorbents such as limestone to soak up (or scrub) SO_2 from exhaust gases. This technology, which is capable of reducing SO_2 emissions by up to 95%, can be added to existing coal-fired boilers.

FGD has several drawbacks. The control equipment is very expensive and bulky. Smaller facilities do not always have the capital or the space needed for FGD equipment. If, however, the sorbent could be injected into existing ductwork, the cost of the reaction vessel could be eliminated, and it would be much easier to retrofit controls on a wider range of sources.

ETS International, Inc. (Roanoke, VA) developed a Limestone Emission Control (LEC) system. The first ever full-scale LEC system was applied to the control of acid gas emissions from a metal alloy production process located in Taiwan, ROC. The LEC system is a proprietary acid gas control system that has demonstrated high levels of acid gas removals (99%) at very competitive costs. A 10 year R&D program included scrutiny by the Ohio Coal Development Office, the U.S. EPA, and DOE as well as major U.S. industries. The system holds great promise for acid gas control from both the chemical and utility industries [2].

21.5   NATIONAL ACID PRECIPITATION ASSESSMENT PROGRAM

In addition to enforcement and monitoring under the provisions of the Clean Air Act, the EPA is actively pursuing a major research effort with other federal agencies under the National Acid Precipitation Assessment Program (NAPAP). This ongoing research project is designed to resolve the critical uncertainties surrounding the causes and effects of acid rain. About $300 million has been spent for federal research since NAPAP was initiated in 1980. In September 1987, NAPAP published an interim assessment on the causes and effects of acid deposition.

21.5.1   Aquatic Effects

One of the most important acid rain research projects being conducted by EPA is the National Surface Water Survey. This survey is designed to provide data on the present and future status of lakes and streams within regions of the United States believed to be susceptible to change as a result of acid deposition. Phase I of the Eastern and Western Lakes Surveys showed that there are essentially no lakes or reservoirs in the mountainous West, northeastern Minnesota, and the Southern Blue Ridge of the Southeast that are considered acidic. The four subregions with the highest percentages of acidic lakes are the Adirondacks of New York, where 10% of the lakes were found to be acidic; the Upper Peninsula of Michigan, where 10% of the lakes were also found to be acidic; the Okefenokee Swamp in Florida, which is naturally acidic; and, the lakes in the Florida Panhandle where the cause of acidity is unknown.

The 1988 Stream Survey determined that approximately 2.7% of the total stream sampled in the mid-Atlantic and Southeast were acidic. About 10% of head waters in the forested ridges of Pennsylvania, Virginia, and West Virginia were found to be acidic. Streams in Florida found to have a low pH are naturally acidic. The study indicated that atmospheric deposition is the major cause of sulfates in streams. Atmospheric deposition was also found to be a major cause of sulfates in the lakes surveyed as part of the National Surface Water Survey.

21.5.2   FOREST EFFECTS

The NAPAP interim assessment reviewed research concerning the effects of acid deposition on forests. It focused on the effects of precursor pollutants (sulfur dioxide and nitrogen oxides) and volatile organic compounds (VOCs) and their oxidants (including ozone and hydrogen peroxides) on eastern spruce-fir, southern pine, eastern hardwood, and western conifer. The assessment found that air pollution is a factor in the decline of both managed and natural forests. The San Bernardino National Forest in California and some types of white pine throughout the eastern United States are seriously affected by ozone.

Forests found to have unknown causes of damage included northeastern spruce-fir, northeastern sugar maple, southeastern yellow pine, and species in the New Jersey Pine Barrens. The high-elevation forests such as the spruce-fir in the eastern United States were found to be exposed to severe natural stresses as well as being frequently immersed in clouds containing pollutants at higher concentrations than those observed in rain. Research has shown no direct impacts to seedlings by acidic precipitation or gaseous sulfur dioxide and nitrogen oxides at ambient levels in the United States. Ozone is the leading suspected pollutant that may stress regional forests and reduce growth. Research is underway to resolve the relative importance of physical and natural stresses.

21.5.3   CROP EFFECTS

The NAPAP assessment indicated that there are no measurable consistent effects on crop yield from the direct effects of simulated acidic rain at ambient levels of acidity. This finding was based on yield measurements of grains, forage, vegetable, and fruit crops exposed to a range of simulated rain acidity levels in controlled exposure studies [1]. Continuing research efforts will examine whether stress agents such as drought or insect pests cause crops to be more sensitive to rainfall acidity.

Average ambient concentrations of sulfur dioxide and nitrogen oxides over most agricultural areas in the United States are not high enough or elevated frequently enough to affect crop production on a regional scale. However, crops may be affected locally in areas close to emission sources. Controlled studies also indicate that ambient levels of ozone in the United States are sufficient to reduce the yield of many crops.

21.5.4   MATERIALS EFFECTS

The NAPAP Interim Report indicated that many uncertainties need to be reduced before a reliable economic assessment could be made of the effects of acid deposition on materials, such as building materials, statues, monuments, and car paint. Major areas of uncertainty include inventories of materials at risk, variability of urban air quality, effects on structures, and cost estimates for repair and replacement.

21.5.5   HUMAN HEALTH EFFECTS

The NAPAP interim assessment reported that there are also many uncertainties associated with assessing the influence of ambient levels of atmospheric pollutants on human health. The primary factors involved are a lack of information on the levels of exposure to acidic aerosols for various population groups across North America; chronic health problems caused by short-term changes in respiratory symptoms and decrease in lung function; and, the effects of repetitive or long-term exposures to air pollutants. Studies on toxicity of drinking water have linked rain acidity to unhealthy levels of toxic metals in drinking water and fish.

21.6 FUTURE TRENDS

The EPA, in coordination with other federal agencies, is conducting wide-ranging research on the causes and effects of acid deposition. Major research efforts include determining effects on aquatic and forest ecosystems, building materials and human health. In the area of human health, EPA is conducting exposure studies on acid aerosols. EPA is also conducting ongoing aquatics research projects that will continue into the future. As part of the National Surface Water Survey, seasonal variability of lakes in the Northeast will be studied.

Over the next several years, major research results are anticipated for improving the basis of decision making on acid rain issues. EPA also expects that Congress and other groups will continue to propose options to reduce acid deposition. As proposals are offered, EPA will provide analyses of costs, consequences, and the feasibility of implementation.

EPA's greatest challenge is to continue to reduce emissions of sulfur dioxide and nitrogen oxides. The Agency must also continue research to reduce the level of scientific and economic uncertainties about acid deposition and work to resolve the regional conflicts related to this problem. In addition to the research efforts, major federal research programs are being funded by the Department of Energy, the Tennessee Valley Authority, and the Argonne, Brookhaven, Lawrence Berkley, and Oak Ridge national laboratories.

21.7 SUMMARY

1. The process of acid deposition begins with emissions of nitrogen oxides and sulfur dioxide from motor vehicles and coal-burning power plants.
2. Title IV of the 1990 Clean Air Act Amendments calls for historic reductions in sulfur dioxide and nitrogen oxides emissions. EPA is implementing a market-based allowance-trading system that will provide power plants with maximum flexibility in reducing emissions.
3. Means to limit the amount of sulfur in the fuel prior to combustion include the use of lower-sulfur coals and coal cleaning. Reduction of nitrogen oxides emissions cannot be accomplished at this point because it is formed due to incomplete combustion.
4. Reduction of NO_x emissions during combustion can be accomplished by three methods: low-NO_x burners, overfire air, and fuel staging or reburning. Two methods that are frequently used in industry to reduce sulfur dioxide emissions during combustion are limestone injection multistage burning and FBC.

5. One of the more popular processes for reducing NO_x after combustion is SCR. Industry has mainly employed FGD processes to combat against SO_2 emissions after combustion.

6. The NAPAP is an ongoing research project designed to resolve the critical uncertainties surrounding the causes and effects of acid rain.

7. EPA, along with other federal agencies, is conducting wide-ranging research on the causes and effects of acid deposition. EPA's greatest challenge is to continue to reduce emissions of sulfur dioxide and nitrogen oxides.

REFERENCES

1. U.S. EPA. Acid deposition, National Acid Precipitation Assessment Program Interim Report, 1987.
2. Burke, G., Singh, B., and Theodore, L. *Handbook of Environmental Management and Technology*, John Wiley & Sons, Hoboken, NJ, 2000.
3. Frankel, K. Acid rain, Manhattan College term paper (submitted to L. Theodore), April 13, 1992.

Part IV

Solid Waste

Part IV of this book, comprised of eight chapters, serves as an introduction to solid waste, since today's industries are faced with the major technological challenge of identifying ways to manage solid waste effectively. Chapter 22 is concerned with municipal waste management. Chapter 23 is specifically concerned with industrial waste management, while comprehensive examination of the hospital waste problem and the various waste management options available is provided in Chapter 24. Chapter 25 focuses on the highly sensitive issue of nuclear wastes. The problems arising with underground storage tanks are presented in Chapter 26. Chapter 27 addresses Superfund and all of its ramifications. Part IV concludes with chapters concerned with asbestos and metals; the general subject of metals is treated in Chapter 29, while the questionable concerns associated with asbestos are reviewed in Chapter 28.

22 Municipal Solid Waste Management*

CONTENTS

22.1 INTRODUCTION [1,2]

It is not news that many communities in America are faced with a garbage disposal problem. In 1990, Americans generated over 195 million tons of municipal solid waste. In 2006, EPA reported that this number had increased slightly. Since over two-thirds of municipal solid waste is sent to landfills, this chapter will primarily key on landfilling—the solid waste management option that simply will not go away. However, some landfills are closing and the siting of new landfills has become increasingly difficult because of public opposition. Past problems sometimes associated with older landfills might have contributed to this situation. Landfills that were poorly designed, that were located in geologically unsound areas, or that might have accepted toxic materials without proper safeguards have contaminated some groundwater sources. Many communities use groundwater for drinking, and people living where contamination has occurred understandably worry about its threat to their health and the cost of cleaning it up. Communities where new landfills are needed share these concerns. Consequently, at a time when more are needed, there is increasing resistance to building new landfills.

---

* See EPA [1].

To ease these worries and to make waste management work better, federal, state, Native American tribal, and local governments have adopted an integrated approach to waste management. This approach involves a mix of three waste management techniques:

1. Decreasing the amount and/or toxicity of waste that must be disposed of by producing less waste to begin with (source reduction)
2. Increasing recycling of materials such as paper, glass, steel, plastics, and aluminum, thus recovering these materials rather than discarding them
3. Providing safer disposal capacity by improving the design and management of incinerators and landfills

The EPA has defined a number of activities that need to be carried out in order to help solve the municipal solid waste problem. These management options include increased source reduction, increased recycle/reuse, and improving the design and operation of both incinerators and landfills. The remainder of the chapter addresses the following topic areas: regulations, source reduction and recycle/reuse, incineration and landfilling.

22.2   REGULATIONS [1,3]

In a general sense, the regulations attempt to establish a cost-effective and practical system for managing the nation's waste by:

1. Encouraging source reduction and recycling to maximize landfill life.
2. Specifying safe design and management practices that will prevent releases of contaminants into groundwater.
3. Specifying operating practices that will protect human health.
4. Protecting future generations by requiring careful closure procedures, including monitoring of landfill conditions and the effects of landfills on the surrounding environment.

The federal government sets minimum national standards applicable to municipal solid waste disposal, but state, tribal, and local governments are responsible for actually implementing and enforcing waste programs. States are required to develop their own programs based on the federal regulations. The EPA is offering the same opportunity to tribes. The EPA's role is to evaluate states' and tribes' programs and decide if they are adequate to ensure safe disposal of municipal solid waste.

States and tribes that apply for and receive EPA approval of their programs have the opportunity to provide significant flexibility in implementing the regulations. This added flexibility allows states and tribes to take local conditions and needs into account, and can make the costs of municipal solid waste management more affordable. States and tribes also may establish requirements that are more stringent than those set by the federal government.

Private citizens have a role, too. Individuals can help ensure that adequate landfill capacity exists for their wastes by supporting the siting and development of facilities that comply with the regulations. Individuals can exercise their responsibilities through grassroots activities, such as participating in public meetings regarding landfill or incinerator siting, by taking part in permitting processes, and by working closely with the responsible state or tribal officials. Citizens also have the right to sue landfill owners/operators who are not in compliance with federal regulations.

Under the regulations, a municipal solid waste landfill (MSWLF) is defined as a discrete area of land or an excavation that receives household waste, and is not a land application unit, surface impoundment, injection well, or waste pile, as those terms are defined in the law. Household waste includes any solid waste, including garbage, trash, and septic tank waste, derived from houses, apartments, hotels, motels, campgrounds, and picnic grounds. An MSWLF unit may also receive other types of wastes as defined under Subtitle D of the Resource Conservation and Recovery Act (RCRA), such as commercial solid waste, nonhazardous sludge, small quantity generator waste, and industrial solid waste. Such a landfill may be publicly or privately owned. An MSWLF unit can be a new unit, and existing unit, or a lateral expansion. An existing unit is defined as an MSWLF unit that received solid waste as of October 9, 1993. Waste placement in existing units must be consistent with past operating practices or modified practices to ensure good management. A new unit is any MSWLF unit that did not receive waste prior to October 9, 1993. A landfill serving a community that disposes of less than 20 tons of municipal solid waste per day, averaged yearly, is referred to as a small landfill. These regulations were still in place at the time of the preparation of this chapter in 2008.

22.3 SOURCE REDUCTION AND RECYCLE/REUSE [2–4]

The general subject area of source reduction and recycle/reuse is treated extensively in the pollution prevention sections (Parts V through VII) of this book. Specific details on waste reduction in the home, office, and other areas are examined in Chapters 32, 33, and 34, respectively. The interested reader should review these chapters to obtain a better understanding of the problems associated with all the management options available for municipal solid waste.

To increase recycling nationwide, the EPA has undertaken a number of efforts to stimulate markets for secondary materials and to promote increased separation, collection, processing, and recycling of waste. The EPA also funded the establishment of a National Recycling Institute, composed of high-level representatives from business and industry, to identify and resolve issues in recycling.

Composting is another process commonly associated with recycling. Composting is the microbiological decay of organic materials in an aerobic environment. Materials that potentially could be composted include agricultural waste, grass clippings, leaves and other yard waste, food waste, and paper products. Many municipalities have implemented leaf composting programs.

One of the problems with the implementation of any recycling program is the public perception of associated costs. Many people believe that recycling is free or, at the very least, inexpensive. However, in most instances, that is not the case. Costs are associated with every aspect of the program, including collection of the materials, processing of the materials, and disposing of any residues. Purchase of new equipment or the retrofitting of existing equipment that is used to separate or recycle materials, or incorporating recycled materials into a process, is often very expensive. Direct operational costs include labor and utilities.

With the exceptions of glass and aluminum, it is unfortunately usually more cost effective to use virgin materials rather than recycled materials in manufacturing processes. Many times markets are not available for sorted materials. The plastics industry is representative of this problem. Although much research has gone into plastics recycling in the past few years, markets for both the segregated material and the end products are very limited. The public needs to realize that recycling is not cheap; and, many times the cost of recycling is only offset by the avoided cost of disposal rather than by any profits generated [2,3].

22.4   INCINERATION [3–5]

Incineration is not a new technology and has been commonly used for treating wastes for many years in Europe and the United States. The major benefits of incineration are that the process actually destroys most of the waste rather than just disposing of or storing it; it can be used for a variety of specific wastes; and, it is reasonably competitive in cost compared to other disposal methods.

Municipal solid waste incineration involves the application of combustion processes under controlled conditions to convert wastes containing hazardous materials to inert mineral residues and gases. Four parameters influence the mechanisms of incineration:

1. Adequate free oxygen must always be available in the combustion zone.
2. Turbulence, the constant mixing of waste and oxygen, must exist.
3. Combustion temperatures must be maintained; the combustion process must provide enough heat to raise the burning mixture to a sufficient temperature to destroy all organic components.
4. Elapsed time of exposure to combustion temperatures must be adequately long in duration to ensure that even the slowest combustion reaction has gone to completion. In other words, transport of the burning mixture through the high temperature region must occur over a sufficient period of time.

Municipal solid waste can be combusted in bulk form or in reduced form. Shredding, pulverizing, or any other size reduction method that can be used before incineration decreases the amount of residual ash due to better contact of the waste material with oxygen during the combustion process [6]. Shredded waste used as fuel is generally referred to as refuse-derived fuel (RDF) and is sometimes combined with other fuel types. Table 22.1 lists the American Society of Testing and Materials (ASTM) classification for RDF.

The types of incinerators used in municipal waste combustion include fluidized bed incinerators, rotary waterwall combustors, reciprocating grate systems, and modular incinerators. The basic variations in the design of these systems are related

TABLE 22.1
ASTM Classifications for RDF

ASTM	RDF Classification	Nomenclature Description
RDF1	Raw	Solid waste used as a fuel as discarded form, without oversize bulky waste.
RDF2	Coarse	Solid waste processed to a coarse particle size, with or without ferrous metal extraction, such that 95% by weight passes through a 6 in., 2-mesh screen.
RDF3	Fine or fluff	Solid waste processed to a particle size such that 95% by weight passes through a 2 in., 2-mesh screen, and from which the majority of metals, glass, and other inorganics have been extracted.
RDF4	Powder	Solid waste processed into a powdered form such that 95% by weight passes through a 10-mesh screen and from which most metals, glass, and other inorganics have been extracted.
RDF5	Densified	Solid waste that has been processed and densified into the form of pellets, slugs, cubettes, or briquettes.
RDF6	Liquefied	Solid waste that has been processed into a liquid fuel.
RDF7	Gaseous	Solid waste that has been processed into a gaseous fuel.

to the waste feed system, the air delivery system, and the movement of the material through the system. Specific details are available in the literature [4,5].

22.5   LANDFILLING [1,3]

As indicated earlier, approximately two-thirds of the nation's municipal solid waste is landfilled. This is due to the fact that it is not possible to reuse, recycle, or incinerate the entire solid waste stream; therefore, a significant portion of the waste must be landfilled.

Landfills have been a common means of waste disposal for centuries. A process that originally was nothing more than open piles of waste has now evolved into sophisticated facilities. Perhaps the best approach to both describe and discuss the solid waste management option is to examine the federal regulations pertaining to landfills. The federal regulations for MSWLFs cover the following six basic areas:

1. Location
2. Operation
3. Design
4. Groundwater monitoring and corrective action
5. Closure and postclosure care
6. Financial assurance

The following material presents the applicable regulations in some detail. However, states and tribes with EPA-approved programs have the opportunity to exercise flexibility in implementing these regulations. Some of the exceptions described below are only available in states and tribes with EPA-approved programs.

22.5.1 LOCATION

Because landfills can attract birds that can interfere with aircraft operation, owners/operators of sites near airports must show that birds are not a danger to aircraft. This restriction applies to new, existing, and laterally expanding landfills. Landfills may not be located in areas that are prone to flooding unless the owner/operator can prove the landfill is designed to withstand flooding and prevent the waste from washing out. This restriction also applies to new, existing, and laterally expanding landfills. Since wetlands are important ecological resources, new landfills and laterally expanding ones may not be built in wetlands unless the landfill is in a state or on tribal lands with an EPA-approved program, and the owner/operator can show that it will not pollute the area. The owner/operator must also show that no alternative site is available. This restriction does not apply to existing landfills. To prevent pollution that could be caused by earthquakes or other kinds of earth movement, new and laterally expanding landfills may not be built in areas prone to them. This restriction does not apply to existing landfills. Finally, landfills cannot be located in areas that are subject to landslides, mudslides, or sinkholes; this restriction applies to new, existing, and laterally expanding landfills.

22.5.2 OPERATION

The EPA and the states have developed regulations specifically covering the disposal of hazardous wastes in special landfills. Owners/operators of municipal landfills must develop programs to keep these regulated hazardous wastes out of their units. In general, each day's waste must be covered to prevent the spread of disease by rats, flies, mosquitoes, birds, and other animals that are naturally attracted to landfills. Methane gas, which occurs naturally at landfills, must be monitored routinely. If emission levels at the landfill exceed a certain limit, the proper authorities must be notified and a plan must be developed to solve the problem.

Owners/operators must restrict access to their landfills to prevent illegal dumping and other unauthorized activities. So that no pollutants are swept into lakes, rivers, or streams, landfills must be built with ditches and levees to keep storm water from flooding their active areas and to collect and control stormwater runoff. Landfills cannot accept liquid waste from tank trucks or in 55 gal drums. This restriction helps reduce both the amount of leachate (liquids that have passed through the landfill) and the concentrations of contaminants in the leachate. Finally, landfills must be operated so they do not violate state and federal clean air laws and regulations. This means, among other things, that the burning of waste is prohibited at landfills, except under certain conditions.

22.5.3 DESIGN

New and expanding landfills must be designed for groundwater protection by making sure that levels of contaminants do not exceed federal limits for safe drinking water. In states and tribes with EPA-approved programs, landfill owners/operators have flexibility in designing their units to suit local circumstances, providing the state or tribal program director approves the design. This allows owners/operators to ensure environmental protection at the lowest possible cost to citizens served by the landfill. This flexibility means, for example, that the use of a liner, and the nature and thickness of the liner system, may vary from state to state, and perhaps from site to site. In states and tribal areas without EPA-approved programs, owners/operators must build their landfills according to a design developed by EPA, or seek a waiver.

The EPA design lays out specific requirements for liners and leachate collection systems. Liners must be composite, that is, a synthetic material over a 2-feet layer of clay. This system forms a barrier that prevents leachate from escaping from the landfill into groundwater. The design also requires leachate collection systems that allow the leachate to be captured and treated.

22.5.4 GROUNDWATER MONITORING AND CORRECTIVE ACTION

Generally, landfill owners/operators must install monitoring systems to detect groundwater contamination. Sampling and analysis must be conducted twice a year. States and tribes with EPA-approved programs have the flexibility to tailor facility requirements to specific local conditions. For example, they may specify different frequencies for sampling ground water for contaminants, or phase in the deadline for complying with the federal groundwater monitoring requirements.

If the groundwater becomes contaminated, owners/operators in approved states and tribal areas must clean it up to levels specified by the state or tribal director. In states and tribes without EPA-approved programs, the federal regulations specify that contaminants must be reduced below the federal limits for safe drinking water.

22.5.5 CLOSURE AND POST CLOSURE CARE

When a landfill owner/operator stops accepting waste, the landfill must be closed in a way that will prevent problems later. The final cover must be designed to keep liquid away from the buried waste. For 30 years after closure, the owner/operator must continue to maintain the final cover, monitor ground-water to ensure the unit is not leaking, collect and monitor landfill gas, and perform other maintenance activities. (States and tribes with approved programs may vary this period based on local conditions.)

22.5.6 FINANCIAL ASSURANCE

To ensure that monies are available to correct possible environmental problems, landfill owners/operators are now required to show that they have the financial means to cover expenses for site closure, postclosure maintenance, and cleanups. The regulations spell out ways to meet this requirement, including (but not limited to) surety bonds, insurance, and letters of credit.

22.6 FUTURE TRENDS

As described earlier, there is significant public opposition to the siting of any type of municipal solid waste management facility. In the future, the public needs to be educated and informed so that these facilities can be properly located. Most of these facilities are found in commercial and/or industrial zones, and away from restricted areas.

The effects of the facilities on health and safety have not been measured at this time. However, even with proper management, wastes containing contaminated materials and dangerous chemicals are potential hazards to millions of people. The health of an entire community can be jeopardized if these wastes are temporarily inadequately and or improperly managed. The whole health risk assessment area needs to be addressed in the future.

The future is also certain to bring a reduced dependence on landfilling of municipal solid waste. Source reduction and recycle/reuse options will be emphasized. And, although incineration has come under pressure recently with environmentalist, it too may very well gain favor if the authorities and the public are educated as to the inherent advantages of this solid waste management option [3].

22.7 SUMMARY

1. In 1990, Americans generated over 195 million tons of municipal solid waste; this annual amount is expected to increase in the future.
2. The federal government sets minimum national standards applicable to municipal solid waste disposal, but state, tribal, and local governments are responsible for actually implementing and enforcing waste programs. States are required to develop their own programs based on the federal regulations.
3. To increase recycling nationwide, the EPA has undertaken a number of efforts to stimulate markets for secondary materials and to promote increased separation, collection, processing, and recycling of waste.
4. Incineration is not a new technology and has been commonly used for treating wastes for many years in Europe and the United States. The major benefits of incineration are that the process actually destroys most of the waste rather than just disposing of or storing it; it can be used on a variety of specific wastes and is reasonably competitive in cost compared to other disposal methods.
5. The federal regulations for MSWLFs cover the following six basic areas: location, operation, design, groundwater monitoring and corrective action, closure and post-closure care, and financial assurance.
6. There is significant public opposition to the siting of any type of municipal solid waste management facility. In the future, the public needs to be educated and informed so that these facilities can be properly located.

REFERENCES

1. Adapted from: U.S. EPA. *Safer Disposal for Solid Waste*, Document EPA/530SW91092, March 1993.
2. Burke, G., Singh, B., and Theodore, L. *Handbook of Environmental Management and Technology*, 2nd edition, John Wiley & Sons, Hoboken, NJ, 2000.

3. Theodore, L. Personal lecture notes, 2003.
4. Theodore, L. and McGuinn, Y. *Pollution Prevention*, Van Nostrand Reinhold, New York, 1992.
5. Santoleri, J., Reynolds, J., and Theodore, L. *Introduction to Hazardous Waste Incineration*, 2nd edition, John Wiley & Sons, Hoboken, NJ, 2002.
6. Geiger, G. Incineration of municipal and hazardous waste. *Natl. Environ. J.* 1(2), November/December, 1991.

23 Industrial Waste Management

CONTENTS

23.1 INTRODUCTION

Pollution has grown to proportions where a reasonable solution to the total problem is almost unfathomable, but not necessarily unattainable. One of the human problems has always been the proper disposal of refuse. Upon the discovery that diseases and illnesses develop as the result of inadequate and unsanitary disposal of wastes, demand for sanitary systems grew. Industrial pollution then started to become intolerable to society.

It is established that the chemical process industries contribute only a small part of the total pollution problem; however, they now expend and will continue to expend in the future resources for corrective methods [1]. Sometimes, pollution from process industries poses problems more complex than pollution from other areas. The magnitude of air pollution from automotive exhaust and central power stations, and nonindustrial water pollution from raw sewage dumped into streams, lakes, and oceans surpass the pollution generated by the chemical industry.

Large processing plants may have a multitude of different individual waste problems, each requiring a separate solution to meet air, water, and solid waste pollution standards. The solutions to these problems may range from the very simple to the very complex. In developing solutions, it is of primary importance to know the characteristics of the waste. Learning as much as possible about the various management tools that are available is often secondary.

The composition of industrial wastes varies not only with the type of industry but with the processes used within the same industry. They may be classified according to composition in many ways, but in general, the wastes may be classified as wastes that may be utilized, and as wastes that require treatment. Industry continues to recognize the importance of saving and making use of all available resources from certain wastes. The wastes that require treatment should attract more attention. These wastes are usually of a form that contain waste materials in a more or less dilute solution or suspension. Since any materials of value are present in small quantities in a dilute solution, they cannot be economically recovered by most processes. In turn, these wastes that require treatment can be divided into three classes of wastes. There are those where organic compounds predominate and constitute the undesirable components, those that contain poisonous substances, and those that contain certain inert materials in such concentrations as to have undesirable features.

This chapter will review a wide range of industries and their respective wastes. In discussing these wastes, both liquid and solid wastes are treated together since it is difficult, if not impossible, to compartmentalize each phase/class of waste in any presentation. Each particular waste requires a different method of handling and treatment. In general, the predominating compounds in a waste usually determine the treatment process that will be required for that distinctive waste.

23.2   FOOD PROCESSING

Food-processing industries are industries whose main concern is the production of edible goods for human or animal consumption. The production processes usually consist of the cleaning, the removal of inedible portions, the preparation, and the packaging of the final product. The generated wastes are the spoiled raw material or the spoiled manufactured product, the liquid or water used (rinsing, washing, condensing, cooling, transporting, and processing), the cleaning liquids of the equipment, the drainage of the product, the overflow from tanks, and the unused portions of the product.

The wastes that result from food processing usually contain varying degrees of concentrations of organic matter. To provide the proper environmental conditions for the microorganisms upon which biological treatment depends, additional adjustments such as continuous feeding, temperature control, pH adjustment, mixing, supplementary nutrients, and microorganism population, adaptation are necessary.

The major and more effective methods of aerobic or anaerobic biological treatments make use of activated sludge, biological filtration, anaerobic digestion, oxidation ponds, and spray irrigation (see Chapters 18 and 19 for additional details about their treatment methods). Since many of the wastes contain high concentrations of organic matter, the loadings of the biological units must be maintained with care. Most often, long periods of aeration or high-rate two-stage biofiltration (a biological control process) is required to produce an acceptable effluent.

The selection of the type of treatment depends on the degree of treatment required, the nature and phase of the organic waste, the concentration of organic matter, the variation in waste flow (if applicable), the volume of the waste, and the capital and operating costs.

23.3 CANNERY WASTES

Cannery wastes are classified according to the product being processed, its growth season, and its geographic location. Many canneries are designed to process more than one product because vegetables, fruits, and citrus fruits have short harvesting and processing periods. The wastes from these plants are primarily organic. These wastes are the result of the trimming, juicing, blanching, and pasteurizing of raw materials; the cleaning of the processing equipment; and, the cooling of the finished product. The most common and effective methods of treatment for the bulk of these wastes are discharging to a municipal treatment plant, lagooning with the addition of chemical stabilizers, soil absorption or spray irrigation, and anaerobic digestion.

The vegetables that produce "strong" wastes when processed for canning are peas, beets, carrots, corn, squash, pumpkins, and beans. The origin of all vegetable wastes is analogous since the canning procedures are alike even though the processing preparations differ for each vegetable. The wastes that result from food processing consist of the wash liquid; the solids from sorting, peeling, and coring operations; the spillage from filling and sealing the machines; and, the wash liquid from cleaning the facilities.

The fruits that present the most common problems in the discharge of waste after processing are peaches, tomatoes, cherries, apples, pears, and grapes. Their wastes come from lye peeling, spray washing, sorting, grading, slicing and canning, removing condensates, cooling of cans, and plant cleanup.

The main citrus fruits (oranges, lemons, and grapefruit) are usually processed in one plant to make canned citrus juices, concentrates, citrus oils, dried meal, molasses, and other by-products. The wastes come from cooling waters, pectin wastes, pulp-press liquors, processing-plant wastes, and floor washings. The canning solid waste is a mixture of peel, rag, and seeds of the fruits, surplus juices, and blemished fruits.

The selection of the most suitable type of treatment of cannery wastes involves the review of the volume, phase, and treatment involved in the process and the unique conditions of the packaging. Cannery wastes are most efficiently treated by screening, chemical precipitation, lagooning, and spray irrigation (digestion and biological filtration are also used, but to a lesser extent).

The preliminary step of screening is designed to remove large solids prior to the final treatment or discharge of the waste to a receiving stream or municipal wastewater system. Only slight reductions in biological oxygen demand (BOD) are accomplished by screening. The machines either rotate or vibrate, and have loads ranging from 40 to 50 lb per 1000 gal of waste water. The wastes retained on the screens are disposed of by being spread on the ground, used as sanitary fill, dried and burned, or used as animal food supplement.

To reduce the concentration of solids in the wastes, chemical precipitation is used to adjust the pH. This method is quite effective for treating apple, tomato, and cherry wastes. Ferric salts or aluminate and lime have produced 40%–50% BOD reductions [2]. The product of this procedure is normally dried on sand beds without producing an odor for a week.

Treatment in lagoons involves biological action, sedimentation, soil absorption, evaporation, and dilution. When adequate land is available, lagooning may be the only practical and economical treatment of cannery wastes, $NaNO_3$ (sodium nitrite) is used to eliminate odors produced by lagoons with unmaintained aerobic conditions. However, the use of these treated lagoons for complete treatment may be costly because of the large volumes of wastes involved. Surface sprays are used to reduce the flies and other insect nuisances that breed around these lagoons.

Whenever the cannery waste is nonpathogenic and nontoxic to plants, spray irrigation is the preferred economical method to use. Ridge-and-furrow irrigation beads are used on soils of relatively high water-absorbing capacity. In general, wastes should be screened before spraying, although comminution alone has been used successfully in conjunction with spray irrigation.

Oxygen-demanding materials in cannery wastes can be removed by biological oxidation. When the operation is limited by seasonal conditions, it is difficult to justify capital investment for bio-oxidation facilities. However, in many instances cannery wastes can be combined with domestic sewage, and then, bio-oxidation processes provide a practical and economic solution.

23.4 DAIRY WASTES

Most dairy wastes are made up of various dilutions of whole milk, separated milk, butter-milk, and whey. They result from accidental or intentional spills, drippings, and washings. Dairy wastes are largely neutral or slightly alkaline, but have a tendency to become acid quite rapidly because of the fermentation of milk sugar to lactic acid. Lactose in milk wastes may be converted to lactic acid when streams become lacking of oxygen, and the resulting lowered pH may cause precipitation of casein. Because of the presence of whey, cheese-plant waste is decidedly acid. Milk wastes have very little suspended material and their pollution effects are almost entirely due to the oxygen demand that they impose on the receiving stream. Decomposing casein causes heavy black sludge and strong butyric-acid odors that characterize milk-waste pollution.

There is a considerable variation in the size of the dairy plants and in the type of products they manufacture. The disposal or treatment of milk waste may be accomplished through irrigation on land, hauling, biological filtration on either the standard or the recirculating filter, biochemical treatment, or the oxidized sludge process. Milk-plant wastes have a tendency to ferment and become anaerobic and odorous because they are composed mostly of soluble organic materials. This characteristic enables them to respond ideally to treatment by biological methods. The selection of a treatment method hinges on the location and size of the plant. The most effective conventional methods of treatment are aeration, trickling filtration, activated sludge, irrigation, lagooning, and anaerobic digestion.

There is a wide variation in the flow rates and "strength" of milk wastes, and through holding and equalization, a desirable uniform waste could be achieved. Aeration for one day often results in 50% BOD reduction and eliminates odors during conversion of the lactose to lactic acid. Some two-stage filters yield greater than 90% BOD reduction, while single-stage filters yield about 75%–80% BOD reduction [2].

A successful method for the complete treatment of milk wastes is the activated-sludge process. It uses aeration to cause the accumulation of an adapted sludge. When supplied with sufficient air, the flora and fauna in the active sludge oxidize the dissolved organic solids in the waste. Excess sludge is settled out and subsequently returned to the aeration units. Properly designed plants that provide ample air for handling the raw waste and returned sludge are not easily upset, nor is the control procedure difficult.

The amount of milk and milk products lost in waste water from factories depends very much on the degree of control and attention to detail in the operation of the plants. The first and most important step in reducing pollution from milk factories is to make sure that whole whey and buttermilk are never discharged with the waste-water. Also, churns in which the milk is delivered should be adequately drained. The effects of whey and buttermilk on the environment are intense if neglected; besides, they have high food values and can be used as food or in the preparation of foods.

23.5 FERMENTATION AND PHARMACEUTICAL INDUSTRIES

The fermentation industries range from breweries and distilleries to some parts of the pharmaceutical industry (the producers of antibiotics); the pharmaceutical industry is treated later in this section. To produce alcohol or alcoholic products, starchy materials (barley, oats, rye, wheat, corn, rice, potatoes) and materials containing sugars (blackstrap and high-sugar molasses, fruits, sugar beets) are used. The process of converting these raw materials to alcohol depends upon the desired alcoholic product. Beer manufacturers focus on taste, while distillers are concerned about alcohol yield.

The brewing of beer has two stages. The first stage involves the malting of the barley and the second involves the brewing the beer from the malt. Both these operations occur at the same plant. The two major wastes produced by the malting process come after grain has been removed, and those remaining in the germinating drum after the green malt has been removed. A considerable amount of water is required for cooling purposes in the actual brewing process. Brewery wastes are composed mainly of liquor pressed from the wet grain, liquor from yeast recovery, and wash water from the various departments. The residue remaining after the distillation process is referred to as "distillery slops," "beer slops," or "still bottoms."

In a distillery, there are several sources of wastes. The dealcoholized still residue and evaporator condensate are major concerns. Minor wastes include redistillation residue and equipment washes. In the manufacture of compressed yeast seed, yeast is planted in a nutrient solution and allowed to grow under aerobic conditions until maximum cell multiplication is attained. The yeast is then separated from the spent nutrient solution, compressed, and finally packaged. The yeast-plant effluent consists of filter residues resulting from the preparation of the nutrient solutions, spent nutrients, wash water, filter-press effluent, and cooling and condenser waters or liquid.

Pharmaceutical wastes arise primarily from spent liquors from the fermentation process, with the addition of the floor washings and laboratory wastes. Wastes from pharmaceutical plants producing antibiotics and biologicals can be categorized as strong fermentation beers, inorganic solids, washing of floors and equipment, chemical waste, and barometric condenser water from evaporation. The wastes from pharmaceutical plants that produce penicillin and similar antibiotics are strong and generally should not be treated with domestic sewage, unless the extra load is considered in the design and operation of the treatment plant.

Stillage is the principal pollution load from a distillery; it is the residual grain mash from distillation columns. Industry attempts to recover as much of this as possible as a by-product to manufacture animal feed or for conversion to chemical products. Centrifuging has also been used to concentrate distillery slops.

23.6 MEAT INDUSTRY

The three main sources of waste in the meat industry are stockyards, slaughterhouses, and packinghouses. The stockyard is where the animals are kept until they are killed. The actual killing, dressing, and some by-product processing are carried out in the slaughterhouse. Packinghouse operations include the manufacture of sausages, canning of meat, rendering of edible fats into lard and edible tallow, cleaning of casings, drying of hog's hair, and some rendering of inedible fats into grease and inedible tallow.

Packinghouse wastes are generated from various operations on the killing floor, during carcass dressing, rendering, bag-hair removal and processing, casing, and cleaning. Stockyard wastes contain both liquid and solid excretions. The amount and strength of the wastes vary widely, depending on the presence or absence of cattle horns, the thoroughness and frequency of manure removal, the frequency of washing, and so on.

Blood should be recovered as completely as possible, even in small plants. Blood is a rich source of protein and is more economical to recover for large plants. Small plants do not have the equipment nor the conditions necessary to profit from the sales of the blood. Paunch manure should be recovered and used for fertilizer purposes. There is little reason for this material to enter the waste system except as washings from the floor. Grease recovery or removal should be common practice in all packing houses and even in smaller slaughterhouses. Grease removal is accomplished through the use of baffled tanks or grease traps. Cleanup by water from high-pressure hoses has been and continues to be the general practice in the meat-packing industry. The use of dry cleanup prior to wet cleanup reduces pollution loads substantially; although it reduces wastewater volume, it does increase solid waste volume.

Slaughterhouse processes are centered about the killing floor. Meat plant wastes are similar to domestic sewage in regard to their composition and effects on receiving bodies of water. The total organic contents of these wastes are considerably higher than those of domestic sewage. Without adequate dilution, the principal detrimental effects of meat plant wastes are oxygen depletion, sludge deposits, discoloration, and general nuisance conditions. The total liquid waste from the poultry-dressing process contains varying amounts of blood, feathers, fleshings, fats, washings from evisceration, digested and undigested foods, manure, and dirt. The largest amount of pollution from the process is contributed by the manure from receiving and feeding stations and blood from the killing and sticking operations.

The treatment processes adapted to slaughterhouse and packing plant wastes depend on the size of the industry. The most common methods used for treatment are fine screening, sedimentation, chemical precipitation, trickling filters, and activated sludge. Biological filtration is perhaps the most dependable process for the medium- and larger-sized plants.

Poultry-plant wastes should and do respond readily to biological treatment; it is attainable if troublesome materials such as feathers, feet, heads, etc., are removed beforehand. Treatment facilities include stationary screens in pits, septic tanks, and lagoons.

The small packinghouse or slaughterhouse requires a process of treatment that is dependable and simple to operate. Small plants operate sporadically resulting in an undesirable operation conditions for biological processes. Biological processes are much more easily upset by careless treatment or large variations in waste content than are chemical processes.

23.7 TEXTILE INDUSTRY

The textile industry has been one of the largest of users and polluters of water, and unfortunately, there has been little success in the development of the low-cost treatment methods needed by the industry to lessen the pollution that is discharged into streams. The operations of textile mills consist of waving, dyeing, printing, and finishing. Many processes involve several steps, each contributing a particular type of waste, like sizing of the fibers, kiering (alkaline cooking at elevated temperature), desizing the woven cloth, bleaching, mercerizing, dyeing, and printing. Textile wastes are generally colored, highly alkaline, high in BOD and suspended solids, and high in temperature. Manufacturing synthetic fiber generates wastes that resemble chemical-manufacturing wastes, and their treatment depends on the chemical process used. Equalization and holding are generally preliminary steps to the treatment of those wastes because of their varying compositions. Additional methods are chemical precipitation, trickling, filtration, and, more recently, biological treatment and aeration.

23.8 FUTURE TRENDS [3]

Often it is not necessary for the producer of the waste to reprocess the material internally. Many companies, through the ingenious application of sound engineering practices, have been able to sell waste products, particularly solid waste products, as raw materials to other processors. Numerous small companies have geared their business toward the sale of reprocessed waste materials. In the evaluation of any waste disposal program, the possible reuse or sale of the waste or its components is certain to receive more attention in the future. Successful strategies will take into account both short-term waste disposal costs and long-term site treatment liabilities. The total cost of waste management consists of the disposal costs, which include taxes and fees, transportation costs, administration costs, and present value for future liability costs; disposal and transportation costs are usually the only ones considered. The time spent by personnel in handling the wastes on-site to off-site approvals, to conduct analytical testing, and to fill out any state, federal, or industry association reports are costly. Long-term costs can include the costs to clean up misused sites. This obligation is due to the presence of substances identified as hazardous under the Comprehensive Environmental Response, Compensation, and Liability Act (CERCLA, which established Superfund), which is the basis for most site remediation actions [4].

An important factor for waste management planning in the future is the rise in the costs of disposal. Costs vary depending on the type of the waste, the quantity (bulk), its containment, and its method of disposal. Landfills close to the location of the waste generation are rapidly shrinking in capacity. The trend toward incineration, led extensively by land disposal restrictions, may also increase incinerations costs in the near future.

As older sites close, the newer containment sites will refuse to take wastes or at least seriously cut back levels [5]. The ideal option will be the development of successful waste minimization programs so that there is no longer the need for disposal. Unfortunately, this attractive situation will take some time. Another option is having the companies take on the responsibility of waste treatment themselves, but not every company wants, or is able, to treat its own wastes on site.

For some companies, hiring a contractor to do the treatment is a better option than building their own treatment plant. The amount of waste generated may not even justify the amount spent for the facility. Naturally, there are additional costs for running the plant and for the personnel to run the plant. Also, if the amount the products being produced changes from time to time, then the amount of waste treatment/management would change accordingly. There could be a large disparity between the quantities generated, and the plant may not be able to comply with that range, thereby making contracting a preferred option.

Waste management issues must be combined with realistic company factors to develop practical, attainable strategies in the future. Capital planning will influence the decision of whether to invest in new production that can use existing waste-handling equipment or to add new equipment that may not be justified. The business plan of the company is essential to estimate the future generation of waste.

The solution to industrial waste problems normally do not present themselves directly, but rather, some ingenuity and practicality must be employed so that the management of these wastes can be carried out safely and efficiently. The methods, strategies, and equipment to be employed in the future will vary according to the situation and the type of waste. Regarding industry and manufacturers, the major factors that determine the way wastes are dealt with will continue to be cost and necessity.

23.9 SUMMARY

1. The predominating compounds in a waste usually determine the treatment processes that will be required for that particular waste.
2. The wastes that result from food processing usually contain varying degrees of concentration of organic matter. To provide the proper environmental conditions for the microorganisms upon which biological treatment depends,

additional adjustments such as continuous feeding, temperature control, pH adjustment, mixing, supplementary nutrients, and microorganism population adaptation are necessary.

3. The selection of the most suitable type of treatment for cannery waste is concerned with a review of the volume, character, and treatment involved in the process and the unique conditions of the packaging periods.

4. Milk wastes have very little suspended material and their pollution effects are almost entirely due to the oxygen demand that they impose on the receiving stream.

5. Pharmaceutical wastes come primarily from spent liquors from the fermentation process, with the addition of the floor washings and laboratory wastes.

6. The three main sources of waste in the meat industry are stockyards, slaughterhouses, and packinghouses.

7. The textile industry has been one of the largest of users and pollutants of water, and unfortunately, there has been little success in the development of the low-cost treatment methods that the industry needs in order to lessen the pollution that is discharged into the environment.

8. An important factor for waste management planning in the future is the rise in the costs of disposal. Costs vary depending on the type of the waste, the quantity (bulk), its containment, and its method of disposal.

REFERENCES

1. Ross, R. *Industrial Waste Disposal*, Reinhold Book Corporation, New York, 1968.
2. Nemerow, N. *Liquid Waste of Industry: Theories and Treatment*, Syracuse University, New York, 1971.
3. Burke, G., Singh, B., and Theodore, L. *Handbook of Environmental Management and Technology*, 2nd edition, John Wiley & Sons, Hoboher, NJ, 2000.
4. Rice, S. Waste management: The long view, *Chemical Technology*, 543–546, September 1991.
5. Brown, S. Washing your hands of waste water disposal. *Process Engineering*, 34–35, July 1991.

24 Hospital Waste Management

CONTENTS

24.1 INTRODUCTION [1,2]

Virtually all of the approximately 6600 hospitals in the United States house x-ray equipment, laboratories, kitchens, pharmacies, and waste disposal stations. More than half also have diagnostic radioisotope facilities, CT scanners, and ultrasound equipment. The environmental impact of the waste is considerable. All of these substances are subject either to the Environmental Protection Agency (EPA) and/or the Occupational Safety and Health Administration (OSHA) rules and regulations on the environment and worker exposure (see Chapter 2). OSHA has numerous regulations pertaining specifically to workers in health care settings.

Medical wastes are not only generated by hospitals but also by laboratories, animal research facilities, and by other institutional sources. The term "biomedical waste" is coming into usage to replace what had been referred to as pathological waste or infectious wastes. Hospitals, however, are generating more and more medical waste with their increasing use of disposable products as well as their increasing service to the community. Hence, the focus of this chapter will be on the issue of hospital waste management.

Progress has been made in methods and equipment for the care of hospital patients. Hundreds of single-service items have been marketed to reduce the possibility of hospital-acquired infections. Yet, hospitals generally have been slow to improve their techniques for the handling and disposing of the waste materials which are increasing in quantity as a result of more patients and higher per-patient waste loads.

Medical waste comes in a wide variety of forms. These forms include packaging, such as wrappers from bandages and catheters; disposable items, such as tongue depressors and thermometer covers; and, infectious wastes, such as blood, tissue, sharps, cultures and stocks of infectious agents.

The location of these wastes includes laboratories, x-ray facilities, surgical departments, pharmacies, emergency rooms, offices, and service areas.

There is an equally wide variety of sources. While hospitals, clinics, and health care facilities may generate the vast majority of medical waste, both infectious and noninfectious waste is also generated by private practices, home health care, veterinary clinics, and blood banks. In New York and New Jersey alone, there are approximately 150,000 sources producing nearly 250 million lb a year.

The beach closures along coastal New Jersey and along the south shore of Long Island have focused attention on medical wastes. Their volume is relatively small (probably less than 1% of the total), but as with sewage wastes, concern centers around the issue of public health. Why these wastes are appearing more frequently is not certain. However, there are several possible contributing factors. The three major factors include:

1. A marked increase in disposable medical care materials.
2. An increase in the use of medically associated equipment on the streets as drug paraphernalia.
3. An increase in illegal disposal of medical wastes as a consequence of the increased costs of disposal.

24.2   MEDICAL WASTE REGULATIONS AND DEFINITIONS

On March 24, 1989, the EPA published regulations in the Federal Register as required under the Medical Waste Tracking Act of 1988. The term "medical waste" was defined as any solid waste that is generated in the diagnosis, treatment, or immunization of human beings or animals, in research pertaining thereto, or in the production or testing of biologicals. Medical waste can be either infectious or noninfectious. The term "medical waste" does not include any hazardous or household waste, as defined in regulations under Subtitle C of the Act.

Infectious waste is waste that contains pathogenic microorganisms. In order for a disease to be transmitted, the waste must contain sufficient quantity of the pathogen that causes the disease. There must also be a method of transmitting the disease from the waste material to the recipient.

Medical waste that has not been specifically excluded in the EPA provisions (for example, household waste) and is either a listed medical waste or a mixture of a listed medical waste and a solid under the demonstration program of the act is known as "regulated medical waste." Seven classes of listed wastes are defined by the EPA as regulated medical waste. Details on these seven classes are provided below.

1. Cultures and stocks. Cultures and stocks of infectious agents and associated biologicals, including: cultures and stocks of infectious agents from research and industrial laboratories; wastes from the production of biologicals; discarded live and attenuated vaccines; and culture dishes and devices used to transfer, inoculate, and mix cultures.

2. Pathological waste. Human pathological wastes, including (a) tissues, organs, body parts, and body fluids that are removed during surgery, autopsy, or other medical procedures, and (b) specimens of body fluids and their containers.

3. Human blood and blood products. Products here include: liquid waste human blood; products of blood; items saturated and/or dripping with human blood that are now caked with dried human blood including serum, plasma, and other components; and, containers that were used or intended for use in either patient care, testing, laboratory analysis, or the development of pharmaceuticals (intravenous bags are also included in this category).

4. Sharps. The category includes sharps that have been used in animal or human patient care or treatment, in medical research, or in industrial laboratories, including hypodermic needles, syringes (with or without the attached needles), Pasteur pipettes, scalpel blades, blood vials, needles with attached tubing, and culture dishes (regardless of presence of infectious agents).

5. Animal waste. Contaminated animal carcasses, body parts, and bedding of animals that were known to have been exposed to infectious agents during research, production of biologicals, or testing in pharmaceuticals.

6. Isolation wastes. Biological waste and discarded materials contaminated with blood, excretions, or secretions from humans known to be infected with certain highly communicable diseases.

7. Unused sharps. These included hypodermic needles, cuture needles, syringes, and scalpel blades.

For additional information about the regulations and the background and implementation documents the reader should go to the EPA website at: http://www.epa.gov/ttn/atw/129/hmiwi/rihmiwi.html.

24.3   WASTE STORAGE AND HANDLING

Hospital wastes are stored in many kinds of receptacles: wastepaper baskets, garbage cans, empty oil drums, laundry hampers, carts, buckets, and even on the floor. Plastic containers are coming into widespread use; they are easier to lift and clean than metal containers, and the bases and sides are impermeable to insects since they do not rust, bend, or dent.

Most hospitals segregate their medical wastes prior to treatment and disposal. However, most hospitals do not segregate all medical waste categories from one another, although certain wastes, most often sharps, cultures and stock, are segregated from other medical wastes prior to treatment or disposal. Most hospitals carefully segregate sharps in rigid plastic sharps containers. Medical wastes are usually segregated from the general trash (e.g., office and cafeteria wastes). Medical waste is almost always separated into red, orange, or biohazard marked bags, and general waste is usually placed in clear, white, or brown bags.

In some hospitals a sharps container is mounted on the wall of every patient room. In other hospitals, the sharps containers are placed in central collection areas on the patient floors, in other areas as necessary (operating room [OR], emergency room [ER], laboratory), and on the carts themselves.

Red bags are almost always used in the laboratory, OR, ER, and isolation rooms. In some cases, red bags also appear in patient rooms. As an alternative to redbagging all waste from patient rooms, some hospitals place red bags and sharps containers on patient care carts. In this fashion, medical wastes are segregated from other discarded wastes.

Medical waste from the "floors" (as patient wings are called) are sometimes stored in "soiled utility rooms" on the patient floors until carried to central storage rooms for incineration or transport. The housekeeping staff is often responsible for gathering the waste and carting it to the storage area. In some cases, general trash is collected in the same cart with red bag wastes.

Suctioned fluids are more commonly discharged into a sanitary sewer rather than containerized or incinerated. However, some hospitals use disposable suction containers and place the entire container in red bags. Other fluids—from the laboratory, for example—are often contained and then redbagged. In some cases, fluids are poured into red bags.

Sharps, other than needles and syringes (such as discarded slides and test tubes), are often placed in the red plastic bags without first being placed in a punctureproof container. Some of these are then placed in cardboard boxes to avoid punctures. The boxes also provide support for heavier sharps such as glassware, slides, and tubes of blood.

An unusual feature of hospital waste management is that wastes are generated continuously around the clock, but they are collected at fixed intervals during the day shift. The housekeeping department usually has the primary responsibility for collection within the hospitals, although a number of other departments have regular responsibility for other facets of waste collection. Generally, only minimal qualifications are required for individuals collecting wastes.

Hospitals often use manually propelled carts of some variety to collect waste materials. Hospital carts are frequently constructed in such a way that sanitizing them is impossible, thus providing surfaces where bacteria can multiply. The routing of carts into and through areas where freedom from contamination is critical increases the probability of contamination from wastes. In addition, individuals collecting wastes are repeatedly exposed to chemical and microbiological contamination and other hazards, but usually have minimal knowledge, skill, or equipment to protect themselves.

Gravity chutes are a simple and inexpensive means of transferring wastes vertically. However, the chutes are seldom constructed with mechanical exhausts, interlocking charging doors, or other systems for preventing the spread of microbiological contamination. In several instances, linen chutes are reserved for conveying solid wastes during certain times of the day thus providing another potential way of spreading contamination. Chute usage has additional drawbacks: fire hazards, spilling of wastes during loading, blockages, difficulties in cleaning, and odors. Proper design and construction can help to prevent some of these, especially the fire hazard and cleaning problems. Others can be avoided by excluding certain wastes, especially grossly contaminated articles, and by exercising more care in the use of chutes.

24.4 WASTE PROCESSING AND DISPOSAL

Hospital wastes are disposed of in a number of ways, usually by the hospital's maintenance or engineering department. Eventually, almost two-thirds of the wastes leave the hospitals and go out into the community for disposal. Approximately 35% by weight, principally combustible rubbish and biological materials, are disposed of in hospital incinerators. Noncombustibles are usually separated and, along with the incinerator residue, leave the hospital to be disposed of on land.

Waste management methods include incineration, autoclaving, sanitary landfilling, sewer systems, chemical disinfection, thermal inactivation, ionizing radiation, gas vapor sterilization, segregation, and bagging [3]. Table 24.1 lists typical treatment/disposal methods for each waste type. This table reveals that only 55% of hospitals that segregate infectious from noninfectious waste incinerate their infectious waste. Eighteen percent treat infectious waste by steam sterilization and then incinerate or landfill the waste. Three percent of hospitals dispose of infectious waste in sanitary landfills without prior treatment [4].

24.5 WASTE MANAGEMENT PROGRAMS

The large amounts of potentially contaminated wastes generated by hospitals raise the possibility that they are a concentrated source of environment health problems. Many hospital solid wastes are indeed contributing to occupational injuries, environmental pollution, and insect and rodent infestation. Some remedial steps that can be taken include the following:

1. Seal as many wastes as possible in disposable bags at the point of generation, or enclose them in such a way as to prevent or minimize contamination of the hospital environment.
2. Construct carts and other equipment used to handle waste so they are easy to keep in sanitary condition.

TABLE 24.1
Approximate Percent of Hospitals Using Treatment/Disposal Methods for Each Waste Type

Type of Waste	Incineration (%)	Sanitary Landfill (%)	Steam Sterilization (%)	Sewer (%)
Blood and blood products	58	12	25	23
Body fluids and wastes	58	32	11	6
Lab wastes	61	16	33	3
Pathological wastes	92	4	4	2
Sharps	79	16	14	0
Animal wastes	81	2	2	0
Disposable materials	29	54	4	6

3. Construct and operate chutes in such a way as to prevent or minimize microbiological contamination of air, linen, and various areas of the hospital.
4. Reduce the danger to personnel handling wastes. Provide preventive health services such as immunization, as well as protective equipment such as gloves and uniforms. In the future, introduce equipment and systems that require less manpower.
5. Require higher qualifications for those handling wastes. Provide them with training on the hazards associated with hospital wastes and the means of protecting not only themselves but others in the hospital and the community.
6. Improve operation of incinerators by training operators to keep loads within incinerator capacity and to maintain temperatures high enough for proper combustion.
7. Provide for safe management of hazardous wastes within the hospital so that they cannot pose a danger to the community.

Most hospitals have comprehensive and sound policies on solid waste management, including specific directives on segregation and special handling of hazardous materials. But, in practice, the policies break down. Employees fail to make the right judgments consistently, and stricter supervision is needed to ensure that employees maintain proper handling and disposal of pathological and sharp wastes, separate disposable wastes from reusable wastes such as dinnerware and linens, bag materials properly, and deposit chute materials promptly. In addition, storage, processing, and disposal areas should be supervised closely and security maintained so that unauthorized personnel cannot gain access.

24.6   INFECTIOUS WASTE MANAGEMENT PROGRAMS

A waste management plan for an institution should be a comprehensive written plan that includes all aspects of management for different types of waste, including infectious, radioactive, chemical, and general wastes as well as wastes with multiple hazards (e.g., infectious and radioactive, infectious and toxic, infectious and radioactive and carcinogenic). In addition, it is appropriate for each laboratory or department to have specific detailed written instructions for the management of the types of waste that are generated in that unit. The waste management section would probably constitute one part of a general, more comprehensive document that also addresses other policies and procedures. Many such documents that include sections on the management of infectious waste have been prepared by various institutions and government agencies [5].

An infectious waste management system should include the following elements:

1. Designation of infectious wastes
2. Handling of infectious wastes, including:
   a. Segregation
   b. Packaging
   c. Storage

    d. Transport and handling
    e. Treatment techniques
    f. Disposal of treated waste
3. Contingency planning
4. Staff training

Various options are available for the development of an infectious waste management system. Management options for an individual facility should be selected on the basis of what is most appropriate for the particular facility. Factors such as location, size, and budget should be taken into consideration. The selected options should be incorporated into a documented infectious waste management plan. An infectious waste management system cannot be effective unless it is fully implemented. Therefore, a specific individual at the generating facility should be responsible for implementation of the plan. This person should have the responsibility as well as the authority to make sure that the provisions of the management plan are being followed.

There are a number of areas in which alternative options are available in an infectious waste management system, (e.g., treatment techniques for the various types of infectious waste, types of treatment equipment, treatment sites, and various waste handling practices). The selection of available options at a facility depends upon a number of factors, such as the nature of the infectious waste, the quantity of infectious waste generated, the availability of equipment for treatment on-site and off-site, regulatory constraints, and cost considerations. These factors are presented here in order to provide assistance in the development of an infectious waste management program.

Since treatment methods vary with waste type, the waste must be evaluated and categorized with regard to its potential to cause disease. Such characteristics as chemical content, density, water content, bulk, etc., are known to influence waste treatment decisions. For example, many facilities use a combination of treatment techniques for the different components of the infectious waste stream, for example, steam sterilization for laboratory cultures and incineration for pathological waste.

The quantity of each category of infectious waste generated at the facility may also influence the method of treatment. Decisions should be made on the basis of the major components of the infectious waste stream. Generally, it would be desirable and efficient to handle all infectious waste in the same manner. However, if a selected option is not suitable for treatment of all wastes, then other options must be included in the infectious waste management plan.

Another important factor in the selection of options for infectious waste management is the availability of on-site and off-site treatment. On-site treatment of infectious waste provides the advantage of a single facility or generator maintaining control of the waste. For some facilities, however, off-site treatment may offer the most cost-effective option. Off-site treatment alternatives include such options as morticians (for pathological wastes), a shared treatment unit at another institution, and commercial or community treatment facilities. With off-site treatment, precautions should be taken in packaging and transporting to ensure containment of the infectious waste. In addition, generators should comply with all state and local regulations pertaining to the transport of regulated medical waste, and ensure that the waste is being handled and treated properly at the off-site treatment facility.

It is also important to consider prevailing community attitudes in such matters as site selection for off-site treatment facilities. These include local laws, ordinances, and zoning restrictions as well as unofficial public attitudes that may result in changes in local laws.

Cost considerations are also important in the selection of infectious waste management options. Cost factors include personnel, equipment cost (capital expense, annual operating, and maintenance expenses—see Chapter 47 for more details), hauling costs (for infectious waste and the residue from treatment), and, if applicable, service fees for the offsite treatment option.

As indicated earlier, the EPA recommends that each facility establish an infectious waste management plan. A responsible individual at the facility should prepare a comprehensive document that outlines policies and procedures for the management of infectious waste (including infectious wastes with multiple hazards). This recommendation is consistent with the standard of the Joint Commission on Accreditation of Hospitals (JCAH), which specifies a system "to safely manage hazardous materials and wastes" [3].

24.7   FUTURE TRENDS

Hopefully, the infectious waste management plans in the future that deal with health and safety will include a contingency plan to provide for emergency situations. It is important that these measures be selected in a timely manner so that they can be implemented quickly when needed. This plan should include, but not be limited to, procedures to be used under the following circumstances:

1. Spills of liquid infectious waste, cleanup procedures, protection of personnel, and disposal of spill residue.
2. Rupture of plastic bags (or other loss of containment), cleanup procedures, protection of personnel, and repackaging of waste.
3. Equipment failure, alternative arrangements for waste storage and treatment (e.g., off-site treatment).

Facilities that generate waste should provide employees with waste management training. This training should include an explanation of the waste management plan and assignment of roles and responsibilities for implementation of the plan. Such education is important for all employees who generate or handle wastes regardless of the employee's role (i.e., supervisor or supervised) or type of work (i.e., technical/scientific or housekeeping/maintenance).

Training programs should be implemented when:

1. The infectious waste management plans are first developed and instituted.
2. New employees are hired.
3. Waste management practices are changed.

Continuing education is also an important part of staff training, including refresher training aids in maintaining personnel awareness of the potential hazards posed by wastes. Training also serves to reinforce waste management policies and procedures that are detailed in the waste management plan.

Many hospitals are beginning to address the issues raised in this chapter on a comprehensive basis. Developing environmental management health and safety programs does more than meeting the letter of the law. They can cut costs, reduce liability, and ensure that the hospital's primary mission of delivering health care is not jeopardized by a fine or an incident that requires shutting down a facility. On average, though, most hospitals still have attained only partial compliance. That will surely change in the future [6].

24.8 SUMMARY

1. Medical wastes are not only generated by hospitals, but also by laboratories, animal research facilities, and by other institutional sources.
2. On March 24, 1989, the EPA published regulations in the Federal Register as required under the Medical Waste Tracking Act of 1988. The term "medical waste" was defined as any solid waste that is generated in the diagnosis, treatment, or immunization of human beings or animals, in research pertaining thereto, or in the production of testing of biologicals.
3. Most hospitals segregate their medical wastes prior to treatment and disposal. However, most hospitals do not segregate all medical waste categories from one another, although certain wastes, most often sharps, cultures and stock, are segregated from other medical wastes prior to treatment or disposal.
4. Hospital wastes are disposed of in a number of ways, usually by the hospital's maintenance or engineering department. Eventually, almost two-thirds of the wastes leave the hospitals and go out into the community for disposal.
5. Most hospitals have comprehensive and sound policies on solid waste management, including specific directives on segregation and special handling of hazardous materials.
6. A waste management plan for an institution should be a comprehensive written plan that includes all aspects of management for different types of waste, including infectious, radioactive, chemical, and general wastes as well as wastes with multiple hazards (e.g., infectious and radioactive, infectious and toxic, infectious and radioactive and carcinogenic).
7. Future trends in hospital waste management are certain to more carefully address accident/emergency situations.

REFERENCES

1. Adapted from: Burke, G., Singh, B., and Theodore, L. *Handbook of Environment Management and Technology*, 2nd edition, John Wiley & Sons, Hoboken, NJ, 2000.
2. Adopted from: Theodore, L. *Air Pollution Control for Hospital and Other Medical Facilities*, Garland STPM Press, New York, 1988 (Copyright owned by Theodore, L.).
3. Doucet, L. *Update of Alternative and Emerging Medical Waste Treatment Technologies*, AHA Technical Document Series, 1991.

4. U.S. EPA, Proceedings of the Meeting on Medical Wastes, Annapolis, MD, November 14–16, 1988.
5. U.S. Department of the Army, U.S. Army Medical Research Institute of Infectious Diseases (USAMRIID), *"Hot" Suite Operations: Standard Operating Procedure*, USAMRIID, Fort Detrick, Frederick, MD, November 30, 1978.
6. Lundy, K. It's enough to make you sick, *Resources*, February 1994.

25 Nuclear Waste Management

CONTENTS

25.1   INTRODUCTION

As with many other types of waste disposal, radioactive waste disposal is no longer a function of technical feasibility but rather a question of social or political acceptability. The placement of facilities for the permanent disposal of municipal solid waste, hazardous chemical waste, and nuclear wastes alike has become an increasingly large part of waste management. Today, a large percentage of the money required to build a radioactive waste facility will be spent on the siting and licensing of the facility.

Nuclear or radioactive waste can be loosely defined as something that is no longer useful and that contains radioactive isotopes in varying concentrations and forms. Radioactive waste is then further broken down into categories that classify the waste by activity, by generation process, by molecular weight, and by volume.

Radioactive isotopes emit energy as they decay to more stable elements. The energy is emitted in the form of alpha particles, beta particles, neutrons, and gamma rays. The amount of energy that a particular radioactive isotope emits, the time frame over which it emits that energy, and the type of contact with humans, all help determine the hazard it poses to the environment. The major categories of radioactive waste that exist are high-level waste (HLW), low-level waste (LLW), transuranic

(TRU) waste, uranium mine and mill tailings, mixed wastes, and naturally occurring radioactive materials (NORMs).

25.2 CURRENT STATUS OF NUCLEAR WASTE MANAGEMENT

Nuclear or radioactive materials are used in many applications throughout today's society. Radioactive materials are used to generate power in nuclear power stations, and are used to treat patients in hospitals (see Chapter 24). The generators of radioactive waste in today's society are primarily the federal government, electrical utilities, private industry, hospitals, and universities. Although, each of these generators uses radioactive materials, the waste that is generated by each of them may be very different and must be handled accordingly. Any material that contains radioactive isotopes in measurable quantities is considered nuclear or radioactive waste. For the purposes of this chapter, the terms nuclear waste and radioactive waste will be considered synonymous.

Waste management is a field that involves the reduction, stabilization, and ultimate disposal of waste. Waste reduction is the practice of minimizing the amount of material that requires disposal. Some of the common ways in which waste reduction is accomplished are incineration, compaction, and dewatering. The object of waste disposal is to isolate the material from the biosphere, and in the case of radioactive waste allow it time to decay to sufficiently safe levels. Table 25.1 is a chronology

---

TABLE 25.1
Major Events Affecting Nuclear Waste Management

Year	Event
1954	The Atomic Energy Act is passed
1963	First commercial disposal of LLW
1967	DOE facilities begin to store TRU wastes retrievably
1970	National Environmental Policy Act becomes effective; Environmental Protection Agency is formed
1974	Atomic Energy Commission divides into the Nuclear Regulatory Commission (NRC) and the Energy Research and Development Administration (ERDA)
1975	WIPP proposed as unlicensed defense TRU waste disposal facility; West Valley, New York low-level disposal facility closed
1977	President Carter deferred reprocessing, pending the review of the proliferation implications of alternative fuel cycles
1979	Three Mile Island, Unit #2 accident; report to the President of the Interagency Review Group on Radioactive Waste Management
1980	Low Level Waste Policy Act is passed; all commercial disposal of TRU wastes ends
1982	Nuclear Waste Policy Act is passed; 10 CFR Part 61 issued as final regulation for LLW
1985	Low Level Radioactive Waste Policy Act Amendments
1986	The reactor explosion at Chernobyl
1987	Nuclear Waste Policy Act Amendments provide for the characterization of the proposed HLW repository at Yucca Mountain, Nevada

Source: Berlin, R. and Stanton, C. *Radioactive Waste Management*, John Wiley & Sons, Hoboken, NJ, 1989. With permission.

---

of the laws that have affected radioactive waste management practices over the last 50 years.

The federal government has mandated that individual states or interstate compacts, which are formed and dissolved by Congress, be responsible for the disposal of the LLW generated within their boundaries. Originally, these states were to bring the disposal capacity online by 1993. Although access to the few remaining facilities is drawing to an end, none of the states or compacts have a facility available to accept waste. Some states are making progress, but none of the proposed facilities is currently in the construction phase.

Both the HLW and the TRU waste programs have sites defined for their respective facilities at Yucca Mountain, and at the Waste Isolation Pilot Plant (WIPP) in Carlsbad, New Mexico. The WIPP facility is a Department of Energy (DOE) research and development facility that has been designed to accept 6 million ft^3 of contact-handled TRU waste, as well as 25,000 ft^3 of remote-handled TRU waste. The facility will accept defense-generated waste and place it into a retrievable geologic repository. A geologic repository is in this instance the salt formations located near Carlsbad. The facility has a design-based lifetime of 25 years.

At the time of the submission of this manuscript in 2009, President Obama's proposed budget would cut off most the money for the Yucca Mountain nuclear waste project, a decision that fulfills a campaign promise and reportedly wins the president political points in Nevada. This raises new questions about what to do with radioactive waste from the nation's nuclear power plants, since the decision could cost the federal government additional billions in payments to the utility industry. If the cut stands, it would mean that most of the $10.4 billion spent since 1983 to find a place to put nuclear waste was wasted. In addition, a final decision to abandon the repository would leave the nation with no solution to a problem it has struggled with for a half century. Lawyers are understandably predicting tens of billions of dollars in damage suits from utilities that must pay to store their wastes instead of having the government bury them.

Additional details are available at:

1. http://www.nytimes.com/2009/03/06/science/earth/06yucca.html
2. http://www.greenpeace.org/raw/content/international/press/reports/nuclear-waste-crisis-france.pdf.

25.3 RAMIFICATIONS OF NUCLEAR ACCIDENTS

The three largest radiological accidents of the last 20 years are the explosion at Chernobyl, the partial core meltdown at Three Mile Island Unit #2, and the mishandling of a radioactive source in Brazil. The least publicized, but perhaps the most appropriate of these accidents, with respect to waste management, was the situation in Brazil.

The uncontrolled radiotherapy source was overlooked in an abandoned medical clinic, and was eventually discarded as scrap. The stainless steel jacket and the platinum capsule surrounding the radioactive cesium were compromised by scavengers in a junkyard. The cesium was distributed among the people for use as "carnival glitter," because of its luminescent properties. The material was spread directly onto

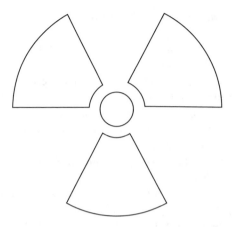

FIGURE 25.1   Radioactive materials' warning sign.

individuals' skin and face, as well as their clothing. Severe illness was immediately evident to most of the exposed victims. Four people died from exposure by the spring of 1988, and it was estimated that additional people would die in the future. Over 40 tons of material, including clothing, shoes, and housing materials, were contaminated from the release of less than 1 g of radioactive cesium.

25.3.1   Biological Effects of Radiation

Although much still remains to be learned about the interaction between ionizing radiation and living matter, more is known about the mechanism of radiation damage on the molecular, cellular, and organ system level than most other environmental hazards. The radioactive materials warning sign is shown in Figure 25.1. A vast amount of quantitative dose–response data has been accumulated throughout years of studying the different applications of radionuclides. This information has allowed the nuclear technology industry to continue at risks that are no greater than any other technology. The following subsections will provide a brief description of the different types of ionizing radiation and the effects that may occur upon overexposure to radioactive materials.

25.3.2   Radioactive Transformations

Radioactive transformations are accomplished by several different mechanisms, most importantly alpha particle, beta particle, and gamma ray emissions. Each of these mechanisms is a spontaneous nuclear transformation. The result of these transformations is the formation of different more stable elements. The kind of transformation that will take place for any given radioactive element is a function of the type of nuclear instability as well as the mass/energy relationship. The nuclear instability is dependent on the ratio of neutrons to protons; a different type of decay will occur to allow for a more stable daughter product. The mass/energy relationship states that for any radioactive transformations, the laws of conservation of mass and the conservation of energy must be followed.

An alpha particle is an energetic helium nucleus. The alpha particle is released from a radioactive element with a neutron to proton ratio that is low. The helium nucleus consists of two protons and two neutrons. The alpha particle differs from a helium atom in that it is emitted without any electrons. The resulting daughter product from this type of transformation has an atomic number that is 2 less than its parent and an atomic mass number that is 4 less. Below is an example of alpha decay using polonium (Po); polonium has an atomic mass number (protons and neutrons) and atomic number of 210 and 84, respectively.

$$^{210}_{84}Po \rightarrow {}^{4}_{2}He + {}^{206}_{82}Pb \qquad (25.1)$$

The terms "He" and "Pb" represent helium and lead, respectively. This is a useful example because the lead daughter product is stable and will not decay further. The neutron to proton ratio changed from 1.5 to 1.51, just enough to result in a stable element. Alpha particles are known as having a high linear energy transfer (LET). The alphas will only travel a short distance while releasing energy. A piece of paper or the top layer of skin will stop an alpha particle. So, alpha particles are not external hazards, but can be extremely hazardous if inhaled or ingested.

Beta particle emission occurs when an ordinary electron is ejected from the nucleus of an atom. The electron (e) appears when a neutron (n) is transformed into a proton within the nucleus.

$$^{1}_{0}n \rightarrow {}^{1}_{1}H + {}^{0}_{(-1)}e \qquad (25.2)$$

Note that the proton is shown as a hydrogen (H) nucleus. This transformation must conserve the overall charge of each of the resulting particles. Contrary to alpha emission, beta emission occurs in elements that contain a surplus of neutrons. The daughter product of a beta emitter remains at the same atomic mass number, but is one atomic number higher than its parent. Many elements that decay by beta emission also release a gamma ray at the same instant. These elements are known as betagamma emitters. Strong beta radiation is an external hazard, because of its ability to penetrate body tissue.

Similar to beta decay is positron emission, where the parent emits a positively charged electron. Positron emission is commonly called betapositive decay. This decay scheme occurs when the neutron to proton ratio is too low and alpha emission is not energetically possible. The positively charged electron, or positron, will travel at high speeds until it interacts with an electron. Upon contact, each of the particles will disappear and two gamma rays will result. When two gamma rays are formed in this manner it is called annihilation radiation.

Unlike alpha and beta radiation, gamma radiation is an electromagnetic wave with a specified range of wavelengths. Gamma rays cannot be completely shielded against, but can only be reduced in intensity with increased shielding. Gamma rays typically interact with matter through the photoelectric effect, Compton scattering, pair production, or direct interactions with the nucleus.

25.3.3   DOSE–RESPONSE

The response of humans to varying doses of radiation is a field that has been widely studied. The observed radiation effects can be categorized as stochastic or nonstochastic effects, depending upon the dose received and the time period over which such dose was received. Contrary to most biological effects, effects from radiation usually fall under the category of stochastic effects. The nonstochastic effects can be noted as having three qualities: A minimum dose or threshold dose must be received before the particular effect is observed; the magnitude of the effect increases as the size of the dose increases; and a clear causal relationship can be determined between the dose and the subsequent effects. Cember [2] uses the analogy between drinking an alcoholic beverage and exposure to a noxious agent. For example, a person must exceed a certain amount of alcohol before he or she shows signs of drinking. After that, the effect of the alcohol will increase as the person continues to drink. Finally, if he or she exhibits drunken behavior, there is no doubt that this is a result of his or her drinking.

Stochastic effects, on the other hand, occur by chance. Stochastic effects will be present in a fraction of the exposed population as well as in a fraction of the unexposed population. Therefore, stochastic effects are not unequivocally related to a noxious agent as the above example implies. Stochastic effects have no threshold; any exposure will increase the risk of an effect but will not wholly determine if any effect will arise. Cancer and genetic effects are the two most common effects linked with exposure to radiation. Cancer can be caused by the damaging of a somatic cell, while genetic effects are caused when damage occurs to a germ cell that results in a pregnancy.

25.4   SOURCES OF NUCLEAR WASTE

25.4.1   NATURALLY OCCURRING RADIOACTIVE MATERIALS [3,4]

NORMs are present in the earth's crust in varying concentrations. The major naturally occurring radionuclides of concern are radon, radium, and uranium. These radionuclides have been found to concentrate in water treatment plant sludges, petroleum scale, and phosphate fertilizers.

In the United States, an estimated 40 billion gal of water is distributed daily through public water supplies. Since water comes from different sources, streams, lakes, reservoirs, and aquifers, it contains varying levels of naturally occurring radioactivity. Radioactivity is leached into ground or surface water while in contact with uranium- and thorium-bearing geologic materials. The predominant radionuclides found in water are radium, uranium, and radon, as well as their decay products.

For reasons of public health, water is generally treated to ensure its quality before consumption by the public. Water treatment includes passing the water through various fitters and devices that rely on chemicals to remove any impurities and organisms. If water with elevated radioactivity is treated by one or more of these systems, there exists the possibility of generating waste sludges or brines with elevated levels of radioactive materials. These wastes may be generated even if the original intention of the treatment process was not to remove radionuclides.

Mining of phosphate rock (phosphorite) is the fifth largest mining industry in the United States in terms of quantity of material mined. The southeastern United States is the center of the domestic phosphate rock industry, with Florida, North Carolina, and Tennessee having over 90% of the domestic rock production capacity.

Phosphate rock is processed to produce phosphoric acid and elemental phosphorus. These two products are then combined with other materials to produce phosphate fertilizers, detergents, animal feeds, other food products, and phosphorus-containing materials. The most important use of phosphate rock is the production of fertilizer, which accounts for 80% of the phosphorite in the United States.

Uranium in phosphate ores found in the United States ranges from 20 to 300 parts per million (ppm), or about 7–100 pCi/g. Thorium occurs at a lower concentration between 1 and 5 ppm, or about 0.1–0.6 pCi/g. The unit picocuries per gram (pCi/g) represents a concentration of each radionuclide based on the activity of that radionuclide. The units of curies represent a fixed number of radioactive transformations in a second. Phosphogypsum is the principal waste byproduct generated during the phosphoric acid production process. Phosphate slag is the principle waste byproduct generated from the production of elemental phoshorous. Elevated levels of both uranium and thorium as well as their decay products are known to exist at elevated levels in these wastes. Since large quantities of phosphate industry wastes are produced, there is a concern that these materials may present a potential radiological risk to individuals that are exposed to these materials if distributed in the environment.

Fertilizers are spread over large areas of agricultural land. The major crops that are routinely treated with phosphate-based fertilizer include coarse grains, wheat, corn, soybeans, and cotton. Since large quantities of fertilizer are used in agricultural applications, phosphate fertilizers are included as a NORM material. The continued use of phosphate fertilizers could eventually lead to an increase in radioactivity in the environment and in the food chain.

Currently, there are no federal regulations pertaining directly to NORM-containing wastes. The volume of wastes produced is sufficiently large that disposal in a LLW facility is generally not feasible. A cost-effective solution must be implemented to both guard industry against large disposal costs and ensure the safety and health of the public.

25.4.2 Low-Level Radioactive Waste

Low-level radioactive waste (LLRW) is produced by a number of processes and is the broadest category of radioactive waste. LLW is frequently defined for what it is not rather than for what it is. According to the Low Level Waste Policy Act of 1980, LLRW is defined as "radioactive waste not classified as high-level radioactive waste (HLRW), TRU waste, spent nuclear fuel, or byproduct material as defined in Section 11(e) [2] of the Atomic Energy Act of 1954."

This definition excludes HLW and spent nuclear fuel because of its extremely high activity. TRU wastes (those containing elements heavier than uranium) are excluded because of the amount of time needed for them to decay to acceptable levels. Finally, byproduct material or mill tailings are excluded because of the very low concentrations of radioactivity in comparison to the extreme volume of waste that is present.

TABLE 25.2

Typical Waste Streams by Generator Category

Waste Stream	Power Reactors	Medical and Academic	Industrial	Government
Compacted trash or solids	X	X	X	X
Dry active waste	X			
Dewatered ion exchange resins	X			
Contaminated bulk	X		X	X
Contaminated plant hardware	X		X	X
Liquid scintillation fluids		X	X	X
Biological wastes		X		
Absorbed liquids		X	X	X
Animal carcasses		X		
Depleted uranium MgF_2			X	

Source: EG&G Idaho, Inc. *The State by State Assessment of Lowlevel Radioactive Wastes Shipped to Commercial Disposal Sites*, DOE/LLW50T, December 1985.

The generators of LLW include nuclear power plants, medical and academic institutions, industry, and the government. LLW can be generated from any process in which radionuclides are used. A list of the different waste streams and the possible generators of each is presented in Table 25.2.

Each of the aforementioned generators produces wastes that fall into the category of LLW. The waste streams identified in Table 25.2 are categorized by generation process, but may also, in some instances, be identified by the type of generating facility.

The disposal of LLW is accomplished through shallow land burial. This process usually involves the packaging of individual waste containers in large concrete overpacks. The overpack is designed to reduce the amount of water that may come into contact with the waste. Another function of the overpack is to guard against intruders coming into contact with the waste once institutional control of the facility is lost. When waste is delivered to the facility in drums, boxes, or in HDPE liners, they are placed in an overpack and sealed with cement before being buried in the landfill.

25.4.3 High-Level Radioactive Waste

HLW consists of spent nuclear fuel, liquid wastes resulting from the reprocessing of irradiated reactor fuel, and solid waste that results from the solidification of liquid HLW. Spent reactor fuel is the fuel that has been used to generate power in a reactor. The spent fuel may be owned by a government reactor, a public utility reactor, or a commercial reactor. The wastes resulting from fuel reprocessing are either governmentally or commercially generated. Only a small fraction of the liquid HLW has been generated commercially. Approximately 600,000 gal of waste was produced in the nation's only commercial fuel reprocessing facility in West Valley, New York. The remainder of the HLW present in the United States today has been generated by the government in weapon facilities.

Spent nuclear fuel is removed from a reactor and stored in a pool of water on the site. The water in the spent fuel storage pools shields the workers and the environment from the fission products, as well as provides cooling to the fuel. The residual heat from a fuel assembly is quantified as approximately 6% of the operating power level of the reactor. Failure to provide additional cooling after the fission reaction has stopped was the reason for the fuel damage at Three Mile Island. Once a geologic repository is constructed, the spent fuel assemblies will be placed in a sealed canister and disposed of.

Most of the liquid HLW is stored in underground storage tanks. Many of these tanks are getting old and the availability of a geologic repository in the near future is doubtful. Many methods of solidifying the wastes for transport and ultimate disposal have been investigated. Plans are under way to store HLW in one central location in the United States. The chosen location is Yucca Mountain, Nevada.

25.4.4 TRANSURANIC WASTE

TRU wastes are those wastes containing isotopes that are heavier than uranium, U. Generally, transuranic isotopes are not found in nature. These isotopes are man-made, produced by the irradiation of heavy elements, such as uranium and thorium. TRU wastes are

$$^{238}_{92}U + {}^{1}_{0}n \rightarrow {}^{239}_{93}Np + {}^{0}_{(-1)}e \leftarrow {}^{239}_{93}Np \rightarrow {}^{239}_{94}Pu + {}^{0}_{(-1)}e \qquad (25.3)$$

where Np and Pu represent neptunium and plutonium, respectively. They are normally generated by the government, particularly from weapons testing. The TRU waste is now being stored at a number of DOE facilities across the country, awaiting permanent disposal at WIPP in Carlsbad, New Mexico.

25.5 RADIOACTIVE WASTE TREATMENT AND DISPOSAL

Many treatment processes can be employed to reduce the volume, or increase the stability, of waste that must ultimately be permanently disposed. Landfill fees for radioactive waste is assessed largely on the volume of the waste to be disposed. Current trends in the rising cost of waste disposal have led to the generators' implementing one or a number of waste minimization techniques. The physical form of the waste is a critical factor in determining the probability that the waste will remain isolated from the biosphere.

Compacting is a method of directly reducing the volume and increasing the specific weight of the resulting waste. Materials such as glass vials, protective clothing, and filter media can be compacted to reduce the volume. Compacting does not reduce the environmental hazard of the waste stream—its purpose is purely waste minimization.

Incineration of waste both reduces the volume and provides a more stable waste stream. Many biological wastes, including animal carcasses, are incinerated. The storage of animal carcasses in drums is generally not cost effective because of the gas generation of the materials as they decay biologically. A drum packed with animal carcasses must be filled with absorbent material so that the pressure inside the drum does not rise to unsafe levels. Incineration is a very cost-effective waste reduction technique for large generators of combustible materials.

Dewatering or evaporation is another waste minimization and stabilization technique that is practiced by waste generators. Evaporating sludges or slurries can greatly reduce the volume of the waste stream and stabilize the waste prior to disposal.

25.6   FUTURE TRENDS

Current regulations call for each individual state or interstate compact to store and dispose of all of the LLW generated within its boundaries. An interstate compact consists of a group of states that have joined together to dispose of LLW. Interstate compacts can only be formed and dissolved by Congress. Many regulatory mile-stones have passed, leaving most states with restricted access to Barnwell, South Carolina. Barnwell is the only remaining LLW disposal facility for such wastes. It is most certain that the Barnwell facility will close before most states have centralized storage capacity on line. Some states, like New York, have unsuccessfully attempted to sue the federal government, arguing that it is unconstitutional to mandate that states dispose of radioactive waste within their boundaries. Without the individual states or compacts taking immediate action to site and construct a permanent disposal facility or temporary centralized storage facility, generators of waste will be forced to either store radioactive materials on-site or stop generating radioactive wastes by ceasing all operations that utilize radioactive materials. While these two options may seem appropriate, neither of them will solve the problem of waste disposal for any extended period of time. Many radioactive waste generators, like hospitals, do not have the storage space allocated to handle on-site storage for periods exceeding 1 or 2 years. Much of the waste generated at hospitals is directly related to patient care, and it is unacceptable to assume that all processes, like chemotherapy, that produce radioactive waste will be stopped.

Both the HLW and TRU waste programs are limping along because of public concern for the areas surrounding the proposed facilities. The WIPP facility has per-formed some waste emplacement, but this has only been accomplished as a research and development activity. The HLW program has met drastic public opposition because of the amount of time that the waste will remain extremely hazardous. This time period is in the order of thousands of years. Opponents to this facility are argu-ing that the ability to properly label the disposal facility and guard against future intruders is lacking. Many symbols, such as thorns or unhappy faces, have been proposed.

The public at large will continue to oppose most activities involving nuclear waste until they are made aware of the unwanted characteristics of current more acceptable technologies as well as the extreme benefits that radioactive materials have made to society.

25.7 SUMMARY

1. After an individual state or an interstate compact is denied access to the current nationwide disposal facilities, the generators of the state will be forced to store LLW on site. The only other alternative is to stop generating waste until such time as the state or compact develops and constructs an appropriate disposal or storage facility. Neither of these options constitutes an appropriate choice for generators such as hospitals that offer nuclear medical services.

2. The interaction between ionizing radiation and living matter is one of the most understood environmental hazards. Radioactive isotopes are transformed into more stable elements through the mechanisms of alpha, beta, gamma, and neutron emission.

3. NORMs may be concentrated by many industrial and municipal processes. The individual states and interstate compacts now have the responsibility to site and construct facilities to dispose of LLW. Both HLW and TRU waste are being stored on the site of generation until the respective geologic repositories begin to accept waste for disposal.

4. Waste disposal fees are assessed primarily on the volume of waste. Generators have invested in treatment technologies because of the rising cost of disposal. Many of the treatment technologies also improve the stability of the waste.

5. Generators of radioactive wastes will be forced to store all generated materials on-site until the next generation of disposal facilities comes on line.

REFERENCES

1. Berlin, R. and Stanton, C. *Radioactive Waste Management*, John Wiley & Sons, Hoboken, NJ, 1989.
2. Cember, H. *Introduction to Health Physics*, McGraw Hill, New York, 1992.
3. Burke, G., Singh, G., and Theodore, L. *Handbook of Environmental Management and Technology*, 2nd edition, John Wiley & Sons, Hoboken, NJ, 2000.
4. U.S. EPA. *Draft: Diffuse NORM Wastes Waste Characterization and Risk Assessment*, May 1991.
5. EG&G Idaho, Inc. *The State by State Assessment of Lowlevel Radioactive Wastes Shipped to Commercial Disposal Sites*, DOE/LLW50T, December 1985.

26 Underground Storage Tanks

CONTENTS

26.1 INTRODUCTION

Leaking underground storage tanks (LUSTs) contribute to the contamination of the environment and pose a great risk to human health. The acronym LUST has apparently been dropped by industry and the government probably because the term is politically/socially incorrect or unacceptable. There are an estimated 5–6 million underground storage tanks (USTs) in the United States that contain either a hazardous substance or petroleum. Of those, approximately 400,000 are believed to be leaking. Many more will begin to leak in the near future.

Under the Resource Conservation and Recovery Act (RCRA), an UST is defined as a tank with 10% or more of its volume underground. This 10% includes piping. Only about 16,000 of the 5–6 million UST systems are protected against corrosion. Another 200,000 are made of fiberglass, which breaks easily. Almost half of the tanks to be regulated by the Environmental Protection Agency (EPA) are petroleum storage tanks owned by gas stations. Another 47% store petroleum for other industries such as factories, farms, police and fire departments, and individuals. The remaining 3% are used by a variety of industries for chemical storage.

Many of the petroleum tanks were installed during the oil boom of the 1950s and 1960s. Two 1985 studies of tank age distribution indicate that approximately one-third of the existing motor fuel storage tanks are over 20 years old or of unknown age. Most of these tanks are constructed of bare steel and not protected against corrosion. With steel tanks, corrosion is the leading cause of leakage, accounting for 92% of the leaks in the tank, and 64% of leaks in the pipes.

Substances released from leaking tanks can poison crops, damage sewer lines and buried cables, and lead to fires and explosions. The most serious concern, however, is groundwater contamination. One gallon of gasoline is enough to render 1 million gallons of groundwater unusable based on federal drinking water standards. Groundwater represents two-thirds of the freshwater on the planet, and if you eliminate unavailable freshwater such as glaciers and the ice caps, groundwater makes up 95% of available freshwater. More than half of the United States relies on groundwater as a source of drinking water. Groundwater drawn for large-scale agricultural and industrial can also be adversely affected by contamination from leaking USTs.

26.2   EARLY REGULATIONS

It had become apparent that regulations for leaking USTs were needed. In 1984 federal laws were enacted in response to increasing problems resulting from leaking USTs. These laws were provided in the RCRA amendments that formed regulations on USTs. Certain regulations required all owners and operators to register their tanks with state agencies giving tank age, location, and substance stored. They also set up design requirements for new tanks installed after May 1985. Owners were now responsible for detecting leaks and the cleanup of releases. Tank owners also had to demonstrate that they were financially capable of cleaning up leaks and compensating third parties for damages. There were some exclusions to these regulations. There were statutory exclusions, which were tank systems that Congress specifically exempted from regulations, and regulatory exclusions, which were tanks exempted after the EPA determined they did not pose a danger to human health or the environment. A listing of each group is provided below [1].

Statutory exclusions

1. Farm or residential tanks of 1100 gal capacity or smaller used for storing motor fuel for noncommercial purposes.
2. Tanks that store heating oil for use on the premises where they are stored.
3. Septic tank systems.
4. Pipeline facilities regulated under specific federal or state laws.
5. Any surface impoundments, pits, ponds, or lagoons.
6. Storm water or wastewater collection systems.
7. Flow-through process tanks.
8. Liquid traps or associated gathering lines directly related to oil or gas production and gathering operations.
9. Tanks in underground areas.

EPA regulatory exclusions

1. Tanks that store (a) hazardous wastes listed or identified under Subtitle C of the Solid Waste Disposal Act, or (b) a mixture of such hazardous wastes and regulated substances.
2. Wastewater treatment tank systems regulated under Section 402 or 307(b) of the Clean Water Act.

3. Equipment or machinery containing regulated substances for operational purposes, such as hydraulic lift tanks.
4. USTs whose capacity is 110 gal or less.
5. USTs containing de minimis concentrations of regulated substances.
6. Any emergency spill or overflow containment system that is expeditiously emptied after use.

A later amendment of the Comprehensive Environmental Response, Compensation and Liability Act (CERCLA) in 1986 provided $500 million dollars over the next 5 years for a Leaking Underground Storage Tank Trust Fund. This revenue was collected primarily through a tax on motor fuels.

26.3   FEDERAL REGULATIONS

There are three main sets of federal regulations: the technical regulations, which address technical standards for corrective action requirements for owners and operators of USTs; the UST State Program Approval Regulations, which set regulations for approval of states to run UST programs in lieu of the federal program; and, the financial responsibility requirements in which owners and operators must show financial responsibility [2]. All three are addressed in the EPA's Federal Register proposed on April 17, 1988. Subpart D has the greatest impact of any part of the proposal and will be discussed below in more depth [3].

Adherence to the performance and operating standards for USTs will reduce both new and existing incidents of leaks from UST systems. The regulations also include extensive requirements for release detection. EPA views release detection as an essential backup measure combined with prevention techniques. Release detection monitoring on a frequent and consistent basis is the best-known method for quickly detecting a release from a UST and reducing the potential environmental damages and liability. Thus, these requirements are in keeping with the overall goal of the UST regulations.

Seven general categories of release detection methods are acceptable. These are tank tightness or precision tests, manual tank gauging systems, automatic tank gauging systems, inventory control methods, groundwater monitoring, vapor monitoring, and interstitial monitoring. The EPA believes that any of these methods can be successful if proper procedures are followed; therefore, no one method is preferred [2].

Release detection is required for all UST systems. The deadline for providing detection for existing tank systems varies with the year the tank was installed. Older tanks require release detection sooner while new tanks must include release detection upon installation. The detection system must be able to detect leaks from any part of the tank or piping that routinely contains product. The system must be installed, operated, and maintained in accordance with the instructions set forth by the manufacturer. The regulations also address performance standards for release detection. Four of the seven methods—inventory, manual and automatic tank gauging, and tank tightness tests—have specific volume or leak rate limits above which the method must be able to detect a leak with a probability of detection of at least 0.95 and a probability of a false alarm less than 0.05. The other three methods—vapor monitoring, groundwater, and interstitial monitoring—have no numerical standards to be met, only general standards.

The release detection regulations are divided into petroleum UST systems and hazardous substance UST systems. Generally, UST systems that store petroleum must conduct release detection every 30 days; however, several exceptions apply. New or upgraded USTs may use monthly inventory controls in conjunction with a tank tightness test every 5 years (until December 22, 1998) or until 10 years after installation (for new tanks). Existing USTs that are not upgraded can use monthly inventory controls with annual tank tightness tests until December 22, 1998, by which time the tank must be upgraded or closed. Tanks of less than 550 gal capacity may use weekly tank gauging.

The piping of petroleum UST systems must meet different standards. Pressurized piping must be equipped with an automatic line leak detector and have an annual line tightness test or monthly monitoring. Suction piping does not require release detection, provided certain conditions are met. These conditions include that the below-grade piping operates at a negative pressure and the piping is sloped so the products will flow back into the tank. Only one check valve per line is included, and that valve is located as close as practical to the suction pump. Suction piping that does not meet these requirements must have line tightness tests at least every 3 years or monthly monitoring.

UST systems that store hazardous substances are covered by separate regulations. All existing systems must meet the petroleum release detection requirements described above, and must have been upgraded by December 22, 1998, to meet the requirements for new hazardous substance systems. Release detection for new systems must include secondary containment systems that are able to contain all substances released from the tank system until the substances are detected and removed. They also must prevent the release of any regulated substance to the environment throughout the operational life of the UST system and must be checked for leakage at least every 30 days. If a double tank is used, the outer wall must be strong enough to contain a release from the inner tank. The inner tank leak must also be detected. If the entire system is surrounded by a liner, that liner must be strong enough to contain a leak. These requirements also pertain to piping for hazardous substance systems. In addition, pressurized piping must be equipped with an automatic line leak detector, similar to petroleum systems [2].

A performance standard is also specified for each of the seven specific types of release detection that can be used to meet the requirements of these regulations. For example, if product inventory control is used, that method must be able to detect a release of 1% of monthly flow-through, plus 130 gal, on a monthly basis. If inventory control cannot provide this level of accuracy, it would not be considered as an acceptable release detection method.

The use of manual tank gauging is restricted to tanks of 2000 gal or less. This method's performance standards are specified in terms of a weekly standard and a monthly standard. For tanks of 550 gal or less the required precision is 10 gal weekly and 5 gal monthly. For tanks of 550–1000 gal, the standards are 13 gal weekly and 7 gal monthly. Tanks of 1000–2000 gal capacity are allowed standards of 26 weekly and 13 gal monthly, respectively [2].

A tank tightness test must be able to detect a leak rate of 0.1 gal/h. These tests must also take into account the effects of thermal expansion or contraction of the product. In addition, vapor pockets, tank deformation, evaporation or condensation, and a location of the water table need to be considered.

Automatic tank gauging systems may be used if they can detect a leak rate of 0.2 gal/h, and if they are integrated with inventory control.

Vapor monitoring systems are a fifth available release detection method. Prior to their installation, the site must be assessed to ensure that their use is appropriate. These systems must be able to detect any significant increase in vapor level above the background concentration.

Groundwater monitoring may be used as a release detection method but again requires a site assessment. Such an assessment should show that the stored substance is immiscible in water and will float; also, that the water table is usually 20 m or less below the ground surface, that the soil's hydraulic conductivity is at least 0.01 cm/s, and that the monitoring wells are designed and placed properly.

Interstitial monitoring may be used for UST systems with secondary containment. Several standard requirements apply, such as the assurance that any leak from the inner tank of a double-walled tank be detected.

Any other method not listed in the regulations may be used if it can detect a leak rate of 0.2 gal/h with a probability of 0.95, a probability of a false alarm less than 0.05, and if such a method is approved by the implementing agency [2].

26.4 RELEASE RESPONSE AND CORRECTIVE ACTION

Following the confirmation of a release, the leak must be investigated and remedied. Emergency and corrective actions will then be implemented. Emergency actions will identify and reduce any immediate health threats, such as explosions or fire. Corrective actions will be taken to mitigate long-term threats to human health and to the environment. Immediate actions might include pumping the remaining product from a leaking tank or dispersing explosive vapors, while long-term correction could include groundwater cleanup plans using air stripping or other such measures.

Subpart F of the Federal Register details seven actions to be taken after a leak is discovered. These are provided below.

1. The initial response of any release should include three steps: notification of the release to the appropriate agency; prevention of any further release; and, mitigation of any immediate fire, explosion, or vapor hazards. These steps should occur within 24 h of the confirmation of the release.

2. Following the initial response, the owner or operator should immediately begin initial abatement measures. First, the regulated substance to prevent further leakage should be removed. Then, any exposed portion of the release must be inspected to stop its spreading. The response should attempt to stop any fire or explosion hazards. Problems caused by polluted soil should be investigated. Finally, the removal of free product, if it is present, is to be initiated.

3. A site characterization must also be performed and a report must be submitted to the implementing agency within 45 days of release confirmation. This report should contain information about the nature and quantity of the release, surrounding populations, location and use of nearby wells, land use, climatological conditions, and similar factors.

4. If free product is found on the site, steps must be taken to remove the free product. Not only will this prevent spreading, but the product can still be used, saving money. Extra care should be taken when dealing with a flammable product. Another report must be submitted within 45 days detailing the removal, treatment, and disposal of the product.

5. Investigations for soil and groundwater cleanup must be conducted if needed. This is warranted in the case of contaminated groundwater, wells, or soil.

6. The owner or operator may be required to submit a corrective action plan for soil and groundwater cleanup. The plan will only be accepted if the agency finds that it adequately protects human health and the environment.

7. For all confirmed releases that require a corrective action plan, the implementing agency must take a number of steps to assure public participation and notice of corrective action procedures. These steps could include public notice in a newspaper or letters to affected parties.

26.5 CLEANUP PROCEDURES AND ECONOMIC CONSIDERATIONS

Only a limited number of technologies to clean soil, air, and water of the contaminants principally associated with gasoline are available that have demonstrated performance records and have progressed to full-scale applications. One reason for this is there are different types of spills and each should be treated using an appropriate procedure. Whether groundwater is in jeopardy will mostly determine what method should be used for cleanup. The final cost of the project is often determined more by what method is used rather than the size of the spill.

The first type of spill is petroleum that has not yet infiltrated the water table. Removal here is essential because the toxin in the unsaturated soil can eventually enter the groundwater. The methods available for this type of removal are excavation and disposal, enhanced volatilization, incineration, venting, vitrification, and microbial degradation.

Excavation and disposal can be 100% effective in this situation. This method entails actually removing the dirt, many times with the use of a backhoe, and carrying it off-site for disposal. The drawbacks of this procedure include difficulty for deep excavation and increased risk of exposure for the workers. Removal-off site can also be dangerous. The cost for this method runs about $200–$300 per cubic yard if the soil is considered hazardous. This is relatively expensive and therefore only used for small spills.

Enhanced volatilization has not been widely used but has an effectiveness near 99.99%. This method can be enhanced through rototilling, pneumatic conveyer systems, and low-temperature thermal stripping. Only low-temperature thermal stripping is effective for petroleum spills. The limitations of enhanced volatility are soil characteristics, contaminant concentrations, and the need to control dust and organic vapors. The cost is $250–$300 per cubic yard.

Incineration is widely used and is very reliable [1,4]. It can achieve 99.99% effectiveness depending on what is burned. However, to burn the contaminants, they must be brought to the surface. Many substances are not safe to be burned at all and have regulations against incineration. Even when approved, location for incineration plants are hard to come by since residential areas do not want them "in their backyard." The cost will usually be between $250 and $640 per cubic yard.

Venting is not widely used but can have an effectiveness of 99%. An advantage to venting is that it is relatively easy to implement and causes minimal damage to structures and pavement. An example of venting is vapor stripping. Calculational details on vapor stripping are available in the literature [5–7]. The drawbacks include critical design requirements that often remain undefined, soil characteristics, and the possibility of explosion. The cost is only $15–$20 per cubic yard.

Two relatively new methods are microbial degradation of contaminants and vitrification. The advantage of microbial degradation is that the soil is treated on site and contaminants are completely destroyed. For its effectiveness, this technique depends on oxygen levels, nutrient levels, temperature, and moisture content of the soil. The cost is between $66 and $123 per cubic yard. A combination of soil venting and microbial degradation is often one of the least costly and most effective corrective actions.

The next type of spill to consider is one where the products have reached the water level but have not yet dissolved into the groundwater. There are two ways often used to capture the free-floating spill—the trench method and the pumping well method. The trench method is most effective when the water table is no more then 15 ft deep. Excavation of the trench is easy to undertake and with this method, the entire edge of the gasoline plume can be captured. It is similar to digging a moat around a castle that is the source of the spill, and waiting for it to fill up. This method does not reverse water flow and should not be used if a drinking water well is immediately threatened. This method can cost about $100 per cubic yard of soil excavated.

For deep spills, a pumping well system is normally used. This method can reverse the direction of groundwater flow. The cost is $100–$200 per foot of depth. Dual-pump systems and oil/water separators are typically used for deeper releases.

The final situation to examine is the most dangerous, and is the most expensive to clean up. This occurs when the contaminants get into the water table and dissolve. There are three technologies widely used with relative success: air stripping, filtration through granular activated carbon (GAC), and biorestoration [8].

Air stripping is a proven effective means of removing organic chemicals from the groundwater. It works by providing intimate contact of air and water, thus allowing diffusion of volatile substances from the liquid phase to the gas phase. There are three types of air stripping: diffused aeration, which has a 70%–90% effectiveness; tray aerators, which have an effectiveness of 40%–60%; and packed towers, which can be almost 100% effective [5–7]. The limitations of air stripping are the types of chemicals that can be effectively removed and the possible air pollution impact. Also, possible high noise levels and zoning laws may restrict the maximum height of the tower. The cost will run anywhere from $50,000 to $100,000.

Next there is the use of GAC which is excellent for removing organic compounds dissolved in water, but is very costly. It works by adsorbing the contaminants onto the activating carbon. The design of a GAC system is very complex and requires more complete pilot testing. More judgment is needed for GAC than other methods, since a system can vary from site to site. The limitations include high-solubility components that do not adsorb welt, the presence of iron, manganese, and hard water will decrease effectiveness, and the dangers of fire increase due to the fact that gasoline-soaked carbon can self-ignite. The cost can run about $300,000–$400,000 for a typical GAC unit. A combination of a GAC system and air stripping is usually the best alternative.

Biorestoration, unlike the other two methods shown, is a destructive technique. This is a distinct advantage since the pollution is not simply transformed into another media; it is completely destroyed. The end products are carbon dioxide and water. However, the degree of cleanup is highly dependent on specific environmental conditions affecting microbial growth. It does not promise to be a cost-effective alternative at this time. The system cannot be used where quick startup is needed and must remain running 24 h a day, 7 days a week. The cost is about $30–$40 per cubic yard.

Underground storage tank cleanup is now being taken seriously. In 2007, the Michigan Department of Environmental Quality issued a series of demand letters to BP Products of North America Incorporated, a subsidiary of BP P.L.C., formerly British Petroleum and Amoco Oil Company, for failing to submit required reports related to contamination from historical releases from leaking underground storage tank systems at eight formerly-owned gasoline stations across Michigan. The letters notified BP that their failure to properly address these issues resulted in $869,150 in penalties being issued against the company. In 2008, the Arizona Department of Environmental Quality completed cleanups of leaking underground storage tanks (USTs) at sites in 11 school districts around the state. The cleanup was part of ADEQ's innovative School Assitance Initiative, which was launched in 2007 to help schools across Arizona clean up contamination from USTs on school property and to prevent future leaks from the tanks.

Leak detection and tank level details are available in the literature [9].

26.6 FUTURE TRENDS

The problems caused by leakage from USTs has become too big for the EPA to handle alone and a new approach needs to be taken. Traditionally, the EPA has controlled all areas of the tank program until states demonstrated it could operate independently. Under the new "franchise concept," the EPA will help states and counties succeed in implementing and enforcing their own tank programs. The EPA initially will focus on assisting the states to establish basic tank programs and then provide a range of services to help them improve their performance. This includes providing special expertise, develop training videos, publishing handbooks, and job training. In this sense, the EPA can be viewed as a franchiser while states and counties are owners of franchises. Through education and training, the EPA, states, and local communities will do a better job of communicating the dangers of leaking tanks to their owners [1].

26.7 SUMMARY

1. Leakage from USTs contribute to the contamination of the environment and pose a great risk to human health.
2. It had become apparent that regulations for USTs were needed. Federal laws were enacted in 1984 in response to increasing problems resulting from leaking USTs. The RCRA amendments formed regulations for USTs but left many exceptions.
3. The problems caused by leakage from USTs has become too big for the EPA to handle alone. Under the new "franchise concept" the EPA will help states and counties succeed in implementing and enforcing their own tank programs.
4. The federal regulations consist of three major sets of regulations; these technical regulations are set forth in EPA's Federal Register 40 CFR Part 280.
5. Following the confirmation of a release the leak must be investigated and remedied. Emergency and corrective actions must also be implemented.
6. Only a limited number of technologies to clean soil, air, and water of the contaminants principally associated with gasoline are available that have demonstrated performance records and have progressed to full-scale applications. One reason for this is there are different types of spills and each should be treated using an appropriate procedure.

REFERENCES

1. Burke, G., Singh, B., and Theodore, L. *Handbook of Environmental Management and Technology*, 2nd edition, John Wiley & Sons, Hoboken, NJ, 2000.
2. Noonan, D. and Curtis, J. *Groundwater Remediation and Petroleum*, Lewis Publishers, Boca Raton, FL, 1990.
3. Rules and Regulations, Federal Register, 52(185), pp. 27197–37207, Friday, September 23, 1988.
4. Santoleri, J., Reynolds, J., and Theodore, L. *Introduction to Hazardous Waste Incineration*, 2nd edition, John Wiley & Sons, Hoboken, NJ, 2002.
5. Theodore, L. and Allen, R. *ETS Theodore Tutorial, Air Pollution Control Equipment*, ETS International, Inc., Roanoke, VA, 1994.
6. Theodore, L. and Barden, J. *ETS Theodore Tutorial, Mass Transfer Operations* (in preparation), ETS International, Inc., Roanoke, VA, 1995.
7. McKenna, J., Mycoch, J., and Theodore, L. *Handbook of Air Pollution Control Technology* (in publication), Lewis Publishers, Boca Raton, FL, 1995.
8. Clarke, J., *In Situ Treatment of Contaminated Soil and Rock*, 1989.
9. Russell, D.L. and Hart, S., *Underground Storage Tanks: Potanlead for Disaster, Chemical Engineering*, New York, March 16, 1987.

27 Superfund

CONTENTS

27.1 INTRODUCTION

During the 1970s, people started to realize that the planet Earth and its environment had reached a critical point and pollution could cause potential health risks. If one examines the environmental laws passed during these times, one can see how the need for Superfund arose out of the public's concern for the environment.

The environmental problem became a major national issue of public concern because of a number of reasons. People were unable to fish or swim in some of the waterways and the air quality was poor. The health effects of smog and industrial air pollution alarmed the people and environmental concern initially moved toward clean air. In order to improve the nation's air quality, Congress passed the Clean Air Act in 1970, which was amended in 1990. This act reduced the pollutants being released into the air by forcing emission standards and regulations on individuals and private industry. Congress then responded in 1977 to the public's concern for clean water by passing the Clean Water Act. This act regulated safe drinking water by requiring secondary treatment on all wastewater facilities. The second provision of the Clean Water Act required previously polluted natural water bodies to become suitable for animal life. In 1976, Congress passed the Resources Conservation and Recovery Act (RCRA), which was one of the first laws to regulate solid and hazardous waste. Landfills had become a big problem since swampland, which was otherwise useless, was utilized for dumping waste. This led to groundwater contamination and other problems. RCRA addressed the issue of landfill sites and led to the issuance of permits for dumping hazardous waste.

During the late 1970s, the press and the American people's attention focused on hazardous waste sites like Love Canal and Times Beach. Love Canal is a hazardous waste site located in upstate New York, which at one time, was merely an unfinished canal. As it was never completed, this large hole remained until a chemical company bought the property, used it as a landfill, and buried tons of hazardous waste chemicals. Once the dumping stopped, the land was covered with fill. The land was later used as a residential area where houses and a school were built. The people living in the houses became ill, and they soon realized that their illness was caused by the dumping that had taken place 25 years earlier. The media began to publicize the story and stir up social concern as the public began to see the health risks of pollution. Because of this concern about hazardous waste sites, Congress passed the first law in 1980 to deal with the nation's hazardous waste sites. This law is called the Comprehensive Environmental Response, Compensation, and Liability Act (CERCLA), now commonly known as Superfund. In 1980, Superfund was given $1.6 billion by Congress to clean up the nation's highest risk hazardous waste sites. A method was sought to determine the worst sites. The act required every individual state to compile a list of the worst sites in their state and submit it to the Environmental Protection Agency (EPA). From this list, each site was evaluated and ranked on its risk to public health and to the environment. The sites were then placed in risk order from highest to lowest. This method of ranking sites was known as the National Priorities List (NPL). Congress initially had no idea how big the hazardous waste site problem was, and Superfund was not given enough money to clean up many of the sites now on the NPL. Congress extended Superfund in 1986 for five more years by passing the Superfund Amendments and Reauthorization Act (SARA). This gave Superfund $8.5 billion more in order to clean up hazardous waste sites. SARA also set up an infrastructure to run daily transactions and provided other means to obtain money for Superfund.

Recent updates and additional details are available at:

1. http://www.epa.gov/superfund/programs/er/nrsworks.htm
2. http://www.epa.gov/superfund/programs/er/hazsubs/lauths.htm
3. http://www.epa.gov/tri/whatis.htm

This chapter will examine the following topics:

1. Funding and legal considerations
2. The ranking systems
3. The cleanup process
4. The role of the private sector
5. The progress up to date
6. Future trends

27.2 THE FUNDING OF SUPERFUND AND LEGAL CONSIDERATIONS

When CERCLA was first passed in 1980, the law set up a trust fund of $1.6 billion, commonly known as Superfund. Congress initially obtained the money to fund the trust from taxes on crude oil and some commercial chemicals. Once this money ran out, Superfund was reauthorized by SARA in 1986, and Congress was again faced with the problem of how to fund the law. Congress decided that "these monies are to be made available to the Superfund directly from excise taxes on petroleum and feedstock chemicals, a tax on certain imported chemical derivatives, environmental tax on corporations, appropriations made by Congress from general tax revenues, and any monies recovered or collected from parties responsible for site contamination" [1].

CERCLA has three concepts that make it an unusual law. These are ex post facto, innocent landowner liability, and joint and several liability. These are discussed below.

Ex post facto means that after the fact, a party can be liable for what was once legal, but now is illegal. For example, a party could have legally disposed of waste at the time of disposal. However, they could later be found liable under CERCLA for whatever that waste was and legally responsible for its cleanup. Since it is necessary to obtain money for the cleanup of a sight, it is very important that the EPA find the parties responsible for the hazardous site. The potentially responsible party (PRP) under Section 107(a) of CERCLA is defined as

1. The current owner or operator of the site that contains hazardous substances.
2. Any person who owned or operated the site at the time when hazardous substances were disposed.
3. Any person who arranged for the treatment, storage, or disposal of the hazardous substances at the site.
4. Any generator who disposed of hazardous substances at the site.
5. Any transporter who transported hazardous substances to the site.

The persons listed above are liable for

1. All costs of removal or remedial action (RA) incurred by the government.
2. Any other necessary costs of response incurred by any other person consistent with the National Contingency Plan (NCP).
3. Damages for injury to, destruction of, or loss of natural resources, including the reasonable costs for assessing them.
4. The costs of any health assessment or health effects study carried out under Section 104(i).

The second unique part of CERCLA is the innocent landowner liability. This states that anyone who buys property that is contaminated with a hazardous substance may be liable for the cost of cleanup even if they did not know the site was contaminated. The only way they might avoid liability is if they made an "all appropriate inquiry" before purchase and found nothing. The following factors are to be examined to see if an "all appropriate inquiry" was made:

1. Any specialized knowledge or experience on the part of the defendant.
2. Commonly known or reasonably ascertainable information about the property.

3. Relationship of the purchase price to the value of the property if uncontaminated.
4. Obviousness of the presence or likely presence of contamination at the property.
5. Ability to detect such contamination by appropriate inspection [2].

A third unique part of the law is the joint and several liability clause. This simply means that liability for a site can be shared between several PRPs or just one. "Each party could be liable for the same amount or one party may be liable for the entire amount even though the parties did not dispose of equal amounts" [2]. Joint and several liability makes the enforcement side of CERCLA easier because the EPA can sue only one PRP and get all the money. In turn, that PRP can then sue the other contributors for their part. This saves the EPA money in legal fees. Under Section 107(b) of CERCLA, there are only four legal defenses to avoid liability for hazardous site contamination. They are

1. An act of God
2. An act of war
3. An act of omission of a third party
4. Any combination of the foregoing [2]

Enforcement and liability go hand-in-hand when cleaning up a Superfund site. Some of Superfund's goals are to encourage potentially responsible parties to finance and conduct the necessary response action and to recover the costs for response action(s) that were financed using the fund's money. The EPA has several enforcement options. Of the several options, the EPA usually seeks voluntary compliance. An enforcement agreement could be one of two options. The first is a judicial consent decree, which "is a legal document that specifies an entity's obligations when that entity enters into a settlement with the government" [2]. The second is an administrative order which is a mutual agreement between the PRP and the EPA outside of court. If the PRP does not chose to reach an agreement with the EPA, then the EPA can issue a unilateral administrative order forcing the PRP to take charge. If the PRP still refuses, then the EPA can file a lawsuit. If the EPA wins, they may recover treble (triple) damages, which means that an uncooperative PRP could be charged three times what it cost the government to clean up the site. This is done to encourage the PRP to take responsibility early on in the cleanup process.

27.3   RANKING OF HAZARDOUS WASTE SITES

When Superfund first began, every state was told to compile a list of their worst hazardous waste sites to be evaluated by the EPA for cleanup. It was soon realized that the number of sites were too large for federal action, so the government decided to rank the sites in order from the highest to the lowest risk to human health and the environment. This ranking system list is known as the NPL. If a site makes the NPL, it is then eligible for federal money through the Superfund program. In order to be placed on the NPL, the site must meet at least one of the following three criteria:

1. Receive a health advisory from the Agency for Toxic Substances and Disease Registry (ATSDR) recommending that people be relocated away from the site.
2. Score 28.5 or higher in the hazard ranking system (HRS), which is the method that the EPA uses to assess the relative threat from a release, or potential release of hazardous substances; HRS is the scoring system used to enhance the process for identifying the most hazardous and threatening sites for Superfund cleanup.
3. Be selected as the state top priority [1].

As risk is a very difficult quantity to measure, the HRS score was used to reflect the potential harm to human health and the environment from the migration of hazardous substances. In order to understand how waste can be ranked, it is necessary to look at the different types of hazardous wastes, and how they end up in the environment. The three media in which hazardous wastes can enter the environment are air, water, and soil. Hazardous wastes may leach, percolate, wash into ground or surface water, evaporate, explode, or they may get carried with the wind or rain into any media. Hazardous waste can bioaccumulate and end up in the food chain or water supply. The risk assessment looks at the waste quantity, where it is, who or what is near it, and relates these to its potential effects on the public health and the environment.

In 1990, the NCP was created to implement the response authorities and responsibilities created by Superfund and the Clean Water Act. The NCP outlines the steps that the federal government must follow in responding to situations in which hazardous substances are released or are likely to be released into the environment. The four basic components of the hazardous substance response provisions of the NCP are:

1. Methods for discovering sites at which hazardous substances have been disposed.
2. Methods for evaluating and remedying releases that pose substantial danger to public health and the environment.
3. Methods and criteria for determining the appropriate extent of cleanup.
4. Means of assuring that RA measures are cost effective [2].

In order to understand how Superfund works it is necessary to look at the structure of the Superfund program. The EPA was given responsibility as the designated manager of the trust fund by CERCLA. The policies Superfund follows comes from the EPA headquarters. However, the EPA has ten offices in different regional cities throughout the country. They have more control of the day-to-day program decisions and operations. This makes it easier to keep a closer eye on what is going on at any particular site. Figure 27.1 shows the political structure of Superfund.

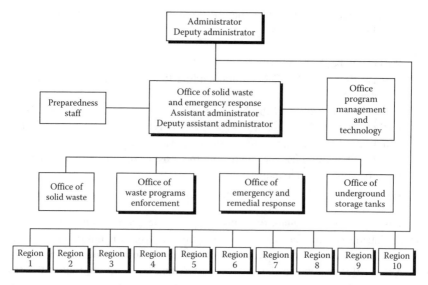

FIGURE 27.1   EPA Superfund organization.

27.4   THE CLEANUP PROCESS

The purpose of Superfund is to eliminate the short- and long-term effects of a hazardous waste. There are ten basic steps in the cleanup process, and they are

1. Site discovery
2. Preliminary assessment
3. Site inspection
4. Hazard ranking analysis (HRS)
5. NPL determination
6. Remedial investigation and feasibility study (RI/FS)
7. Remedy selection/record of decision (ROD)
8. Remedial design (RD)
9. Remedial action
10. Project closeout

The first five steps have been discussed in previous sections. Steps six through ten will be discussed here. Once a site has been officially placed on the NPL, it is eligible for money from the fund. It first becomes necessary to determine which is the lead agency. If there is a PRP, then fund money is not used but the EPA follows up on the progress of the site. If a PRP cannot be found, or cannot pay, then the site becomes a State Lead, Federal Agency Lead, or a Fund Lead. Regardless of who the lead is, the EPA's regional office will provide a remedial project manager (RPM) to coordinate between the EPA and the lead agency. The RPM oversees the technical, enforcement, and financial aspects at the site until completion.

Once a site is placed on the NPL a RI/FS report is performed. In order to choose a remedy that would best protect the public health and the environment, a detailed study of the site is done. The RI consists of sampling to determine a risk assessment. The risk assessment for a Superfund site has three different parts. The first is the human health and environmental evaluation that examines baseline risks. This part measures levels of chemicals and helps in the evaluation of the site characteristics and the selection of possible response alternatives. The second part is the health assessment, which is conducted by the Agency for Toxic Substances and Disease Registry (ATSDR). ATSDR looks at risk in a qualitative way, and its effect on the neighboring people. The third part is the endangerment assessment, which is a legal determination of risk and the requirements to satisfy the RI/FS process.

The overall RI is made up of two phases: (1) site characterization, which involves field sampling and laboratory analyses, and (2) treatability investigations that examine how treatable the wastes are and the possible treatment technology alternatives.

The FS works with the RI by taking the data from the RI and developing and screening different treatment alternatives for a given site. Once the FS is started, the RI continues as more sampling may be needed to make a determination. It then becomes necessary to look at the RI/FS and choose a remedy. This is done by a ROD, which is a written report of the alternatives found in the RI/FS and reasons for the selection of a treatment technology. In the ROD, a remedy is proposed for the site and then it goes out for public evaluation. The comments are studied and then a final selection is made. In the ROD, there is a decision summary that explains the site characteristics and the determination of the method chosen for that site.

The next step in the treatment process is the RD phase. This is when the remedy selected in the ROD is engineered to meet the specifications and cleanup levels specified by law and the ROD. Detailed engineering plans are drawn up to implement the selected remedy and then the site enters the next phase, which is RA. The RA phase is when the construction on the site begins and the treatment, removal, and other tasks are undertaken.

The last stage of cleanup is project closeout and deletion from the NPL. Project closeout is divided into three phases: NPL deletion, operation and maintenance, and final project closeout. For a site to be eligible for deletion, at least one of the following three criteria must be met:

1. EPA, in consultation with the state, must have determined that responsible or other parties have implemented all appropriate response actions.
2. All appropriate fund-financed responses must have been implemented, and EPA in consultation with the state, must have determined that no further response is appropriate.
3. Based on a RI, EPA, in consultation with the state, must have determined that no further response is appropriate [2].

The EPA has recently created a new way to declare a site complete even if all the cleanup standards have not been met. It is called construction completion of a superfund site and it would occur after the RD/RA and before the actual deletion of the site from the NPL. A site may be declared construction complete if the entire RD/RA process is finished and the remedy has taken effect and has been proven to be working. The site may, however, need operations and maintenance (O&M) of the equipment, and an extended period of time to reach the cleanup levels.

For example, a site that has contaminated groundwater could take years of pumping and treatment to actually clean it up to the standards specified in the ROD. All other procedures may be met and the treatment plant may be functioning, but the site is not eligible for deletion from the NPL until it reaches the specified cleanup levels. Declaring the site construction complete makes it possible for the site to be counted as complete as far as the public progress reports are concerned.

27.5   THE ROLE OF THE PRIVATE SECTOR

Superfund follows the policy that the public has a right to know what happens at a site. The EPA needs the public's help for many aspects of Superfund cleanups. Often, it is the private citizens of a community that report hazardous waste dumping. Superfund is mandated by law to involve the public in all aspects of a site except PRP legal activities. The public is given an information and comment period whenever a site makes it on to the NPL. Public concerns are analyzed and sometimes influence the decision toward an alternative remedy. Also, people in a town where a site is located are often helpful in locating a PRP. In this way, the public can help with liability and enforcement issues by assisting the EPA in finding a PRP.

In order to inform the public the EPA creates a community relations plan where they outline activities that they will use to inform the public of what is going on at a site. Once a plan is made, the EPA has a general informational meeting to explain to the public what will happen at the site. The EPA collects any comments and includes them in the ROD as part of the considerations for selecting a remedy. The EPA then establishes an information repository which contains the site updates, news releases, and phone numbers to call for questions or concerns about the site. Here are some of the things citizens can do:

1. Report hazardous waste dumping: call the National Response Center at 1-(800)-424-8802.
2. Individual or organizations that suspect they are or may be affected by a hazardous waste release may petition the EPA to preform a preliminary assessment. Contact ATSDR at 1600 Clifton Road NE, Atlanta, GA 30333.
3. Find out when cleanup investigators will arrive and share information with them.
4. Get information from the EPA or state Superfund office.
5. Learn about the EPA's Community Involvement Programs.
6. Write the EPA for information on the status of any site [1].

27.6   PROGRESS TO DATE

Superfund often deals with dangerous contaminated sites that can have serious effects on the public health and the environment. There are nearly 1300 sites now on the NPL and of those sites over 200 have made it through the entire cleanup process discussed early in this chapter. One of the common questions often asked is what kind of progress is the EPA making on every site to deal with such a serious problem. First, the Superfund program is required to evaluate, stabilize, treat, or otherwise take actions to make dangerous sites safe [1]. This is accomplished through emergency response action tailored to specific sites. After any emergency responses have been completed a site goes through the ten phases of cleanup discussed in the earlier sections of this chapter.

Regarding progress, the EPA has reported:

> The net results of the work done at the NPL sites has been to reduce the potential risks from hazardous waste for an estimated 23.5 million people who live within 4 miles of these sites. Other results are to bring technology to bear by increased use of permanent treatment remedies at NPL sites, to remove contamination from the environment, and control the sources of contamination [1].

The reader is referred to Chapter 42, The EPA Dilemma, for an alternative view of the progress to date.

27.7 FUTURE TRENDS

The Superfund program has come under fire in recent years. Critics feel that Superfund has not been efficient in cleaning up the nation's hazardous waste sites. A large portion of the money in Superfund goes to "transaction costs," which are lawyer and consultant fees. Since enforcement is a very large part of the Superfund program, significant money gets spent on legal considerations instead of on actual site cleanup. The future of Superfund lies in improving the existing system. Some argue that Superfund cannot be abolished because it is still needed. Many hazardous sites that pose threats to public health and the environment still exist. The administration has suggested two ideas that may improve Superfund. The first would cause remedies for cleaning toxic sites to be based partly on "probable future use," and the second would reduce wasteful transaction costs [3]. The idea of "probable future use" means that a hazardous waste site will be cleaned up to a certain level depending on what the site will be used for after the cleanup. For example, sites that will be used as parking lots will be cleaned to different specifications than will sites that are to become hospitals. The second suggestion for reducing wasteful transaction costs involves engaging a "neutral professional," such as a judge, to be an arbitrator between the PRP and government. This idea would reduce the legal costs incurred by Superfund because it expedites a decision in payment responsibility. These two changes in the future of Superfund may improve the program. However, based on past history, the EPA continues to move toward a legally based rather than a technology-driven agency.

27.8 SUMMARY

1. Because of concern about hazardous waste sites, Congress passed a law in 1980 to deal with the nation's hazardous waste sites. This law is called the CERCLA, now commonly known as Superfund.
2. CERCLA has three concepts that make it an unusual law. These are ex post facto, innocent landowner liability, and joint and several liability.

3. The NPL is a government list that ranks hazardous sites in order from the highest to the lowest risk to human health and the environment.
4. There are ten basic steps in the cleanup process: site discovery, preliminary assessment, site inspection, HRS, NPL determination, RI/FS, ROD, RD, RA, and project closeout.
5. Superfund maintains a policy that the public has a right to know what happens at a site.
6. There are over 1300 sites now on the NPL and of those sites, over 200 have made it through the entire cleanup process.
7. Two ideas to improve Superfund are to base remedies for cleaning toxic sites on "probable future use," and to reduce wasteful transaction costs. However, based on past history, the EPA continues to move toward a legally based rather than a technology-driven agency.

REFERENCES

1. U.S. EPA, Focusing on the Nation at Large, *EPA's Update*, EPA/540/8-91/016, September 1991.
2. Wagner, T. *The Complete Guide to the Hazardous Waste Regulation*, 2nd edition, Van Nostrand Reinhold, New York, 1991.
3. Cushman, J., Jr., Not so Superfund. *New York Times*, National section, Monday, February 7, 1994, p. Al, A15.

28 Asbestos

CONTENTS

28.1 INTRODUCTION [1]

Asbestos fibers can cause serious health problems. If inhaled, they can cause diseases that disrupt the normal functioning of the lungs. Three specific diseases—asbestosis (a fibrous scarring of the lungs), lung cancer, and mesothelioma (a cancer of the lining of the chest or abdominal cavity)—have been linked to asbestos exposure. These diseases do not develop immediately after inhalation of asbestos fibers; it may be 20 years or more before symptoms appear.

In general, as with cigarette smoking and the inhalation of tobacco smoke, the more asbestos fibers a person inhales, the greater the risk of developing an asbestos-related disease. Most of the cases of severe health problems resulting from asbestos exposure have been experienced by workers who held jobs in industries such as shipbuilding, mining, milling and fabricating where they were exposed to very high levels of asbestos in the air without benefit of the worker protections now afforded by law. Many of these same workers were also smokers. These employees worked directly with asbestos materials on a regular basis, and generally for long periods of time as part of their jobs. Additionally, there is an increasing concern for the health and safety of construction, renovation, and building maintenance personnel because of possible periodic exposure to elevated levels of asbestos fibers while performing their jobs.

Whenever one discusses the risk posed by asbestos, one must keep in mind that asbestos fibers can be found nearly everywhere in the environment (usually at very low levels). There is, at this time, insufficient information concerning health effects resulting from low-level asbestos exposure, either from exposures in buildings or from the environment. This makes it difficult to accurately assess the magnitude of cancer risk for building occupants, tenants, and building maintenance and custodial workers. Although, in general, the risk is likely to be negligible for occupants, health concerns remain, particularly for the building's custodial and maintenance workers. Their jobs are likely to bring them into close proximity to ACM (asbestos-containing materials), and may sometimes require them to disturb the ACM in the performance of maintenance activities. For these workers in particular, a complete and effective operation and maintenance (O&M) program can greatly reduce asbestos exposure. This kind of O&M program can also minimize asbestos exposure for other building occupants as well.

The term "asbestos" describes six naturally occurring fibrous minerals found in certain types of rock formations. Of that general group, the minerals chrysotile, amosite, and crocidolite have been most commonly used in building products. When mined and processed, asbestos is typically separated into very thin fibers. When these fibers are present in the air, they are normally invisible to the naked eye. Asbestos fibers are commonly mixed during processing with a material which binds them together so that they can be used in many different products. Because these fibers are so small and light, they may remain in the air for many hours if they are released from ACM in a building. When fibers are released into the air they may be inhaled by people in the building.

Asbestos became a popular commercial product because it is strong, will not burn, resists corrosion, and insulates well. In the United States, its commercial use began in the early 1900s and peaked in the period from World War II into the 1970s. It has been downhill ever since. The movie "Libby Montana" added further to asbestos's decline. The PBS movie was shown in 2007. Libby is a tale of how an entire town in Montana was exposed to asbestos for decades without its knowledge, resulting at last count in an estimated 1,500 cases of lung abnormalities in a population of about 8,000, and of how W. R. Grace & Company supposedly knew the asbestos was there.

Additional details on asbestos are available at:

1. http://www.nytimes.com/2007/08/28/arts/28mont.html
2. http://yosemite.epa.gov/opa/admpress.nsf/0/5CBEBAC73CDA2046852575
AC00622D03

28.2 REGULATORY CONCERNS

Over the last 15 years, the U.S. Environmental Protection Agency (EPA) and several other federal agencies have acted to prevent unnecessary exposure to asbestos by prohibiting some uses and by setting exposure standards in the workplace. Now, the government is also acting to limit exposure to the public at large [2]. Five agencies have major authority to regulate asbestos [2].

1. The Occupational Safety and Health Administration (OSHA) sets limits for worker exposure on the job.
2. The Food and Drug Administration (FDA) is responsible for preventing asbestos contamination in food, drugs, and cosmetics.
3. The Consumer Product Safety Commission (CPSC) regulates asbestos in consumer products. It already has banned the use of asbestos in drywall patching compounds, ceramic logs, and clothing. The CPSC is now

studying the extent of asbestos use in consumer products generally, and is considering a ban on all nonessential product uses that can result in the release of asbestos fibers.

4. The Mine Safety and Health Administration (MSHA) regulates mining and milling of asbestos.

5. The EPA regulates the use and disposal of toxic substances in air, water, and land, and has banned all uses of sprayed asbestos materials. The effects of cumulative exposure to asbestos have been established by dozens of epidemiological studies. In addition, EPA has issued standards for handling and disposing of asbestos-containing wastes.

EPA has a program to help abate asbestos exposure in schools. Since 1982, when EPA issued the Asbestos-in-Schools Identification and Notification Rule, the agency has required all local education agencies to inspect for friable asbestos materials; to notify parents and teachers if such materials are found; to place warning signs in schools where asbestos is found; and, to keep accurate records of their actions to eliminate the problem.

Congress passed the Asbestos School Hazard Abatement Act of 1984 to help those schools with the most serious hazards and the greatest financial need. The Act gives EPA the responsibility for providing both financial and technical assistance to local education agencies.

EPA offers technical assistance and guidance on asbestos. Under the Technical Assistance Program (TAP), each of the agency's 10 regions has a Regional Asbestos Coordinator backed up by a staff of technical experts. These are listed in Table 28.1 [3].

The EPA has also published several documents that provide state-of-the-art guidance on how to identify and control friable asbestos-containing materials. In addition, the Agency is beginning the operation of several new programs. These include

1. Contractor certification
2. Pilot information centers
3. Rules to provide worker protection during asbestos abatement activities
4. Expanded technical assistance materials

In July 1989, EPA promulgated the Asbestos Ban and Phasedown Rule. The rule applies to new product manufacture, importation, and processing, and essentially bans almost all asbestos-containing products in the United States by 1997. Interestingly, this rule does not require removal of ACM currently in place in buildings much of this is still applicable as of 2009.

28.3 SOURCES

In February 1988, the EPA released a report titled "EPA Study of Asbestos-Containing Materials in Public Buildings: A Report to Congress." EPA found that "friable" (easily crumbled) ACM can be found in an estimated 700,000 public and

TABLE 28.1

Regional Asbestos Coordinators (TAP)

Region	Address and Phone	Jurisdiction
EPA Region 1	JFK Federal Building Boston, MA 02203 (617) 565-3835	Connecticut, Massachusetts, Maine, New Hampshire, Rhode Island, Vermont
EPA Region 2	Woodbridge Avenue Edison, NJ 08837 (201)321-6671	New Jersey, New York, Puerto Rico, Virgin Islands
EPA Region 3	841 Chestnut Street Philadelphia, PA 19107 (215) 597-3160	Delaware, District of Columbia, Maryland, Pennsylvania, Virginia, West Virginia
EPA Region 4	345 Corland Street, NE Atlanta, GA 30365 (404) 347-4727	Alabama, Florida, Georgia, Kentucky, Mississippi, North Carolina, South Carolina, Tennessee
EPA Region 5	230 S. Dearborn Street Chicago, IL 60604 (312) 886-6003	Illinois, Indiana, Michigan, Minnesota, Ohio, Wisconsin
EPA Region 6	Allied Bank Tower 1445 Ross Avenue Dallas, TX 75202-2733 (214) 655-7244	Arkansas, Louisiana, New Mexico, Oklahoma, Texas
EPA Region 7	726 Minnesota Avenue Kansas City, KS 66101 (913) 236-2835	Iowa, Kansas, Missouri, Nebraska
EPA Region 8	One Denver Place 999-18th Street Suite 500 Denver, CO 80202-2413 (303) 293-1744	Colorado, Montana, North Dakota, South Dakota, Utah, Wyoming
EPA Region 9	215 Fremont Street San Francisco, CA 94105 (415) 974-7290	Arizona, California, Hawaii, Nevada, American Samoa, Guam
EPA Region 10	1200 6th Avenue Seattle, WA 98101 (20) 442-4762	Alaska, Idaho, Oregon, Washington, DC

commercial buildings. About 500,000 of those buildings are believed to contain at least some damaged asbestos, and some areas of significantly damaged ACM can be found in over half of them.

According to the EPA study, significantly damaged ACM is found primarily in building areas not generally accessible to the public, such as boiler and machinery rooms, where asbestos exposures generally would be limited to service and maintenance workers. Friable ACM, if present in air plenums, can lead to distribution of the material throughout the building, thereby possibly exposing building occupants. ACM can also be found in other building locations.

Asbestos in buildings has been commonly used for thermal insulation, fireproofing, and in various building materials, such as floor coverings and ceiling tile, cement pipe wrap, and acoustical and decorative treatment for ceilings and walls. Typically, it is found in pipe and boiler insulation and in spray-applied uses such as fireproofing or sound-deadening applications.

The amount of asbestos in these products varies widely (from approximately 1% to nearly 100%). The precise amount of asbestos in a product cannot always be accurately determined from labels or by asking the manufacturer; nor can positive identification of asbestos be ascertained merely by visual examination. Instead, a qualified laboratory must analyze representative samples of the suspect material.

28.4   HEALTH CONCERNS [4]

Intact and undisturbed asbestos materials do not pose a health risk. The mere presence of asbestos in a building does not mean that the health of building occupants is endangered. ACM that is in good condition, and is not somehow damaged or disturbed, is not likely to release asbestos fibers into the air. When ACM is properly managed, release of asbestos fibers into the air is prevented or minimized, and the risk of asbestos-related disease can be reduced to a negligible level.

However, asbestos materials can become hazardous when, due to damage, disturbance, or deterioration over time, they release fibers into building air. Under these conditions, when ACM is damaged or disturbed, e.g., by maintenance repairs conducted without proper controls, elevated airborne asbestos concentrations can create a potential hazard for workers and other building occupants.

As described above, the potential for an asbestos-containing material to release fibers depends primarily on its condition. As described earlier, if the material, when dry, can be crumbled by hand pressure—a condition known as "friable"—it is more likely to release fibers, particularly when damaged. The fluffy spray-applied asbestos fireproofing material is generally considered friable. Pipe and boiler insulation materials can also be friable, but they often are enclosed in a protective casing that prevents fiber release unless the casing is damaged. Some materials that are considered "nonfriable," such as vinyl-asbestos floor tile, can also release fibers when sanded, sawed, or otherwise disturbed. Materials such as asbestos cement pipe can release asbestos fibers if broken or crushed when buildings are demolished, renovated, or repaired.

28.5   CONTROL MEASURES [4]

Most asbestos-containing material can be properly managed where it is. In fact, asbestos that is managed properly and maintained in good condition appears to pose relatively little risk.

Proper asbestos management begins with a comprehensive inspection by qualified, trained, and experienced inspectors, accredited through an EPA or state-approved training course. Inspecting the condition of asbestos materials initially with accredited inspectors and at least semiannually visits with trained custodial or maintenance staff is extremely important so that changes in the material's condition, such as damage or deterioration, can be detected and corrected before the condition worsens. Normal activities can sometime damage asbestos material and cause fiber release, particularly if the material is friable. A thorough initial inspection and regular surveillance can prevent accidental exposure to high levels of asbestos fibers.

The proper methods for dealing with asbestos are

1. Developing and carrying out a special maintenance plan to insure that asbestos-containing materials are kept in good condition. This is the most common method when the materials are in good condition at the time of initial inspection.
2. Repairing damaged pipe or boiler covering which is known as thermal system insulation.
3. Spraying the material with a sealant to prevent fiber release—a process called encapsulation.
4. Placing a barrier around the materials, known as an enclosure.
5. Removing asbestos under special procedures.

The last three methods of response actions—encapsulation, enclosure, and removal—and sometimes the second method—repair—must be performed by accredited asbestos professionals.

The final response action, asbestos removal, is generally necessary only when the material damage is extensive and severe, and other actions will not control fiber release. Removal decisions should not be made lightly. An ill-conceived or poorly conducted removal can actually increase rather than eliminate risk. Consequently, all removal projects must be designed, supervised, and conducted by accredited professionals and should be performed in accordance with state-of-the-art procedures. In addition, one may wish to hire an experienced and qualified project monitor to oversee the asbestos contractor's work to make sure the removal is conducted safely.

28.6   FUTURE TRENDS

Training of custodial and maintenance workers is one of the major approaches that can be employed in a successful asbestos control program. This is the key to future activities. If building owners do not emphasize the importance of well-trained custodial and maintenance personnel, asbestos O&M tasks may not be performed properly. This could result in higher levels of asbestos fibers in the building air and an increased risk faced by both building workers and occupants.

With proper training, custodial and maintenance staffs in the future will successfully deal with ACM in place, and greatly reduce the release of asbestos fibers. Training sessions should provide basic information on how to deal with all types of maintenance activities involving ACM. However, building owners should also recognize that O&M workers in the field often encounter unusual, "nontextbook" situations. As a result, training should provide key concepts of asbestos hazard control. If these concepts are clearly understood by workers and their supervisors, workers can develop techniques to address a specific problem in the field. Building owners who need to provide O&M training to their custodial and maintenance staff should contact an EPA environmental assistance center (listed earlier under Regulatory Concerns) or an equally qualified training organization for more information.

28.7 SUMMARY

1. Asbestos fibers can cause serious health problems. If inhaled, they can cause diseases that disrupt the normal functioning of the lungs. Three specific diseases—asbestosis (a fibrous scarring of the lungs), lung cancer, and mesothelioma (a cancer of the lining of the chest or abdominal cavity)—have been linked to asbestos exposure.
2. Over the last 25 years, the EPA and several other federal agencies have acted to prevent unnecessary exposure to asbestos by prohibiting some uses and by setting exposure standards in the workplace.
3. According to an EPA study, significantly damaged ACM is found primarily in building areas not generally accessible to the public, such as boiler and machinery rooms, where asbestos exposures generally would be limited to service and maintenance workers.
4. Intact and undisturbed asbestos materials do not pose a health risk. The mere presence of asbestos in a building does not mean that the health of building occupants is endangered. ACM that is in good condition, and is not somehow damaged or disturbed, is not likely to release asbestos fibers into the air. When ACM is properly managed, release of asbestos fibers into the air is prevented or minimized, and the risk of asbestos-related disease can be reduced to a negligible level.
5. The proper methods for dealing with asbestos are maintenance, repairing, encapsulation, enclosure, and removal.
6. Training of custodial and maintenance workers is one of the major approaches that can be employed in a successful asbestos control program. This is the key to future activities.

REFERENCES

1. U.S. EPA, 2003, Managing Asbestos in Place, June 1990.
2. U.S. EPA, Office of Public Affairs, *Asbestos Fact Book*, February 1985.
3. U.S. EPA, *The ABCs of Asbestos in Schools*, TS799, June 1989.
4. Burke, G., Singh, B., and Theodore, L. *Handbook of Environmental Management and Technology*, 2nd edition, John Wiley & Sons, Hoboken, NJ, 2000.

29 Metals

CONTENTS

29.1 INTRODUCTION

In recent years a great amount of media attention has focused on the effects of industry on the environment. While a large portion of the attention has focused on "politically correct" issues such as waste incineration, nuclear waste disposal, and rainforest destruction, the environmental and biological damage due to metal contamination has largely been ignored by the media (although not by the scientific community). Lead poisoning, long known to occur in children through the ingestion of lead-based paint chips, is also virtually ignored by the media. In addition, the disposal of large quantities of nickel–cadmium batteries and lead–acid batteries in landfills, a source of groundwater contamination, receives scant attention.

Since heavy metals are strongly attracted to biological tissues and to the environment, and remain in them for long periods of time, metal pollution is a serious issue. Overall, metals are quite abundant and persistent in the environment [1]. This chapter will discuss the metals posing the greatest threat to health and the environment; i.e., the most toxic metals, including metals used in large quantities by industry, and metals found in common everyday products such as batteries and paint. The four major metals to be considered are lead, mercury, cadmium, and arsenic.

The scientific community has long known the dangers of metal contamination. Exposure to mercury occurred in nineteenth-century hatmakers. They developed a shaking and slurring of speech due to exposure to large quantities of inorganic mercury during the manufacturing process [2]. In the 1930s, reports of wastes containing chromium were reported in the United Kingdom. In addition, a chromium discharge onto a filter bed resulted in the termination of a biological process. These findings were reported by H.E. Monk and J.H. Spencer in the *Proceedings of the Institute of Sewage Purification* in the 1930s [3]. Since the 1950s, the determination of health hazards posed by chemical and other environmental agents has received more attention. However, man's total exposure to a chemical or metal from various sources such as water, air, food, home, and work only has been considered since the 1970s [4].

See also: http://www.abc.net.au/news/stories/2008/10/03/2381869.htm

29.2   LEAD

Lead occurs naturally in soil, air, water, and plants. However, industrial and technological uses of lead contribute the most damage to man's health. The highest exposure occurs during mining, smelting, and other manufacturing operations that produce lead. While the air concentration in large cities having dense automobile traffic (e.g., New York City) is about $2-4 \, g/m^3$, the lead concentration in smelting and storage battery facilities usually exceeds $1000 \, g/m^3$. Children also are exposed to lead via ingestion of lead-based paint chips and other lead-containing objects [5].

Lead normally localizes near the points of discharge and any amount (normally about 20%) that does not localize usually is widely dispersed in the atmosphere. Discharges of lead do occur into soil and water but air discharges are of the most concern. Lead ore smelters, for instance, create emission pollution problems in localized areas. The height of the stack and its air pollution control devices, the topography of the surrounding area, and numerous local characteristics determine the exact amount of pollution caused by a lead ore smelter. Lead emissions from a lead ore smelter contribute to soil and water pollution in addition to air pollution. As emission control devices become more effective, air pollution from lead emissions should start to decrease [5].

Drinking water becomes contaminated with lead due to corrosion of plumbing material in homes and water distribution systems. Most public water systems deliver water to households containing lead solder; in addition, materials used in faucets contribute a certain amount of lead to drinking water [2]. Homes containing lead-lined water storage tanks and lead pipes have the highest concentrations of lead. This occurs primarily in areas where the water is soft; i.e., the water contains low amounts of both calcium and magnesium [5].

Health effects of high lead ingestion are significant and may lead to brain damage, high blood pressure, premature birth, low birth weight, and nervous system disorders. Children are most affected by lead in drinking water. It is imperative to test for lead in drinking water if it is suspected the water may contain lead [2].

Amounts of lead in gasoline have been drastically reduced due to efforts by the Environmental Protection Agency (EPA) that began in the early 1970s. The EPA's overall automotive emission control program required the use of unleaded gasoline in many cars beginning in 1975. By 1986, the lead content of leaded gasoline had been reduced to 0.1 g/gal.

During the past decade, many communities have begun to implement recycling programs to reduce the amount of waste in the municipal solid waste (MSW). While recycling has had an impact on the amount of solid waste in landfills, an abundant amount of lead still is present in MSW streams. Greater amounts of lead occur in

TABLE 29.1
Lead in Products Discarded in MSW, 1970–2000 (in Short Tons)

Products	1970	1986	2000 (Est)	Tonnage	Percentage
Lead–acid batteries	83,825	138,043	181,546	Increasing	Variable
Consumer electronics	12,233	58,536	85,032	Increasing	Increasing
Glass and ceramics	3,465	7,956	8,910	Increasing	Increasing; stable after 1986
Plastics	1,613	3,577	3,288	Increasing; decreasing after 1986	Fairly stable
Soldered cans	24,117	2,052	787	Decreasing	Decreasing
Pigments	27,020	1,131	682	Decreasing	Decreasing
All others	12,567	2,537	1,701	Decreasing	Decreasing
Totals	164,840	213,832	281,886		

MSW streams than amounts of cadmium; lead in municipal solid waste has grown in several areas (see Table 29.1). No additional data was available as of 2005. Lead–acid batteries contribute the greatest amount of lead to the municipal solid waste stream along with consumer electronics, glass, ceramics, and plastics. In contrast, the amount of lead contributed by soldered cans and pigments has dropped considerably.

Lead–acid batteries contributed 65% of the lead in municipal solid waste in 1986 although the percentage ranged between 50% and 85% during the 1970–1986 period. However, lead–acid batteries are recycled to a large extent compared with the other categories in Table 29.1. The other main contributor of lead to the municipal solid waste stream is consumer electronics, which account for approximately 27% of lead discards. This includes soldered circuit boards, leaded glass in television sets, and plated-steel [2]. No updated information on the 21st century was available at the time of the preperation of this manuscript.

29.3 MERCURY

In March 1970, fish from Lake Erie and Lake St. Clair outside of Detroit were determined to contain mercury. This was the first of a series of events that has since raised public consciousness about mercury pollution. The alarming discovery that mercury could be transported so easily throughout the aqueous environment led to a high degree of concern from both the scientific community and the general public [5].

How is mercury distributed in the environment? Globally, land sources emit mercury vapor, which circulates throughout the atmosphere and enters the oceans. Since the ocean's mercury content is so large (approximately 70 million tons), it is difficult to determine yearly increases in the mercury content of the world's oceans [5]. Mercury pollution in the aqueous environment occurs when organic molecules from dead organisms and sewage react with mercury to form soluble organic complexes. Anaerobic bacteria in river and lake bottoms convert mercury into the organic compound methyl mercury, a highly poisonous substance. Methyl mercury dissolves in water; this results in the transport of mercury into the aquatic food chain, which ultimately leads to human consumption of mercury [5].

Although the aquatic environment and food chain contains a surprising amount of mercury contamination (particularly in fish), the exposure to elemental mercury vapor in the workplace still poses the greatest threat to human health. Diseases caused by mercury and its toxic compounds are numerous and in most countries qualify for worker's compensation. However, a lack of reporting of mercury poisoning occurs in most developing countries; evidence suggests that a large number of workers are exposed to high mercury concentrations in these countries. Most people exposed to mercury in industry work in the mining industry or in chloralkali plants. Mercury levels in the atmosphere in these industrial settings may attain levels as high as $5\,mg/m^3$ [4].

The natural degassing of the earth's crust, the major source of mercury, contributes between 20,000 and 125,000 tons per year of mercury to the atmosphere. Industrial production of mercury via mining and smelting was about 10,000 tons per year in 1973 and has been increasing 2% annually. Total mercury releases into the environment by man amounted to 20,000 tons per year in 1975; this includes the burning of fossil fuels, steel, cement and phosphate production, and metal smelting from sulfide ores [4]. No additional data is available at this time.

29.4   MERCURY REMOVAL FROM COAL-FIRED POWER PLANTS

Coal-fired power plants are a major source of mercury emissions in the United States. The EPA proposed to further reduce these emissions in 2004. Previously, the mercury emissions from these plants were controlled by fabric filters and wet flue gas desulfurization (FGD). However, the amount of mercury removed can be enhanced by the introduction of a catalyst or a sorbent.

Mercury is found in coal in trace amounts (about 0.1 ppm). During combustion, the mercury is released into the flue gas as elemental mercury vapor (Hg^0). This mercury can then be oxidized to Hg^{2+} by homogeneous and heterogeneous reactions. The oxidized mercury is much easier to capture than its elemental counterpart. The main homogeneous reaction is with gas-phase chlorine. This is a slow process with oxidized mercury yields ranging from a few percent to 90%. The heterogeneous reactions involve surfaces with electrophilic groups that attract the elemental mercury. These usually occur on fly ash or boiler surfaces, especially if the fly ash contains unburned carbon.

The heterogeneous reaction of oxidizing the mercury emissions can be enhanced by catalysts. Studies show that the use of selective catalytic reduction (SCR) [3] promotes the oxidation of elemental mercury to Hg^{2+}. The SCR catalysts provided a 90% oxidation of mercury in bituminous coal. However, studies showed that when these catalysts were used for coals, the reaction was equilibrium limited and not kinetic limited. This basically means that the catalytic oxidation of mercury is not effective for low-rank coals such as subbituminous.

Another method in the development for mercury control is the use sorbent injection. The injection of dry sorbents such as powdered activated carbon (PAC) has been used to control mercury emissions from waste combustors. The sorbent is usually injected in the ductwork upstream in a fabric filter or electrostatic precipitator. Other sorbents have also been tested and have shown greater results, such as enhanced PAC and silica-based sorbents. This is due to the fact that the effectiveness of PAC is reduced greatly above 350°F, while enhanced PAC or silica-based sorbents can operate at much higher temperatures.

The removal of mercury emissions from coal-fired power plants is an evolving field. As more studies are done on the use of catalyzed oxidation and sorbent injection, the removal of mercury from these plants will increase.

29.5 CADMIUM

Cadmium, a rare but toxic metal, is most commonly found in rechargeable nickel–cadmium batteries. Its color is silvery-white and is soft and ductile; in addition, it possesses good electrical and thermal conductivity. When cadmium is exposed to moist air, it slowly oxidizes to form a thin layer of cadmium oxide, thereby protecting itself from further corrosion.

Cadmium occurs in nature most often as the mineral greenockite (CdS). Normally it is mined with zinc, but occasionally it is mined with lead and copper ores. The end uses of cadmium include batteries, pigments, and plastic stabilizers. However, the consumption of cadmium in 1986 was a small 4800 tons (compared to 1.2 million tons of lead). The consumption in the United States decreased until 1983, but then started to increase once again.

The most prevalent use of cadmium today is in nickel–cadmium rechargeable batteries; their popularity is growing due to their rechargeable nature. Although they were invented in the early 1900s, nickel–cadmium batteries were not widely used until the mid-1940s when they came into use in industrial and military applications. Currently, military and industrial uses of cadmium include satellites, missile guidance systems, naval signaling, computer memories, television and camera lighting, and portable hospital equipment. Consumer use of nickel–cadmium batteries began in the early 1960s; however, their popularity grew more rapidly in the early 1970s. Their uses are endless toys, hand-held tools, flashlights, hedge trimmers, VCRs, cameras, electric shavers, and alarm systems.

Although it occurs in much smaller quantities in municipal solid waste than does lead, the amounts of cadmium in MSW have increased due to the disposal of nickel–cadmium batteries (indicated in Table 29.2). No additional data was available as of 2005. As of 1980, nickel–cadmium batteries were the largest contributor to cadmium in MSW [2].

29.6 ARSENIC

Arsenic is likely familiar to most people as a poison used by villains in mystery novels and movies to kill their innocent victims. Many movie buffs can relate this statement to the Oscar award-winning movie in the early 1940s titled *Arsenic and Old Lace* starring Cary Grant. But, although arsenic is a poison, its more detrimental effect is its ability to cause cancer. Lead arsenate was used as a pesticide in farms

TABLE 29.2
Cadmium in Products Discarded in MSW, 1970–2000 (in Short Tons)

Products	1970	1986	2000 (Est)	Tonnage	Percentage
Household batteries	53	930	2035	Increasing	Increasing
Plastics	342	520	380	Variable	Variable; decreasing after 1986
Consumer electronics	571	161	67	Decreasing	Decreasing
Appliances	107	88	57	Decreasing	Decreasing
Pigments	79	70	93	Variable	Variable
Glass and ceramics	32	29	37	Variable	Variable
All others	12	8	11	Variable	Variable
Totals	1196	1806	2680		

and gardens but now has been replaced by synthetic pesticides. Arsenic compounds are found in the home; typical products containing arsenic are rat poison and plant killers. However, most products now contain little or no arsenic [2].

29.7 FUTURE TRENDS

This chapter has highlighted those metals posing the greatest threat to human health and the environment. Since increasing metal contamination is occurring through the disposal of products such as lead–acid and nickel–cadmium batteries in the municipal solid waste stream, the use of prevention and recycling methods in the future would alleviate metal contamination. In addition, as environmental regulations become more stringent, industry must focus on perhaps other less damaging materials to replace those metals causing contamination, or must attempt to integrate both pollution prevention and recycling methods into their processes.

29.8 SUMMARY

1. Recent media attention has focused on environmental issues such as nuclear waste disposal, waste incineration, and rainforest destruction. Unfortunately, only a small amount of media coverage is placed on the effect of metal contamination on health and the environment through lead poisoning and metal products in municipal solid waste.
2. Lead has virtually been eliminated from gasoline and the paint industry. However, the disposal of lead–acid batteries and other products containing lead in municipal solid waste has increased. Health effects of lead ingestion include brain damage, increased blood pressure, nervous system disorders, and premature birth.
3. The dangers of mercury poisoning were first discovered during the nineteenth century. Human exposure to mercury occurs through the food chain

(particularly through fish) and to a greater extent in the workplace. Mercury poisoning leads to kidney damage, birth defects, and even death.

4. Cadmium is best known for its use in nickel–cadmium rechargeable batteries. Although it is a rare metal and is not used to the same extent as lead, cadmium consumption is increasing and is present in the municipal waste stream in significant amounts.

5. Although arsenic is best known as a poison, its main characteristic is its ability to cause cancer. Lead arsenate, a pesticide, is no longer used; however, some household products such as rat poison and plant killers still contain arsenic. Overall, the use of arsenic is decreasing.

6. Since metal contamination is steadily increasing through the disposal of lead–acid and nickel–cadmium batteries and the industrial use of mercury, a greater emphasis in the future will be placed on preventing health and environmental damage from metals.

REFERENCES

1. Oehme, F. *Toxicity of Heavy Metals in the Environment, Part 1*, Marcel Dekker, Inc., New York, 1978.
2. Burkes, G., Singh, B., and Theodore, L. *Handbook of Environmental Management and Technology*, 2nd edition, John Wiley & Sons, Hoboken, NJ, 2000.
3. Lester, J., Ed. *Heavy Metals in Wastewater and Sludge Treatment Processes: Treatment and Disposal* (*Foreword*), Vol. II, CRC Press, Boca Raton, FL, 1987.
4. World Health Organization. *Environmental Health Criteria 1: Mercury*, World Health Organization, Geneva, 1976.
5. World Health Organization. *Environmental Health Criteria 3: Lead*, World Health Organization, Geneva, 1977.
6. Jones, H. *Mercury Pollution Control.* Noyes Data Corp., Park Ridge, NJ, 1971.
7. EPA. Control of mercury emissions from coal fired electric utility boilers: An update, EPA, 2005.
8. Pseudo, R., term paper submitted to Theodore, L., 2007.

Part V

Pollution Prevention

Part V serves to introduce the reader to the general subject of pollution prevention and the three major elements in the pollution prevention field. Chapter 30 introduces the general concept of waste reduction, particularly at the industrial level. Numerous companies have already established formal pollution prevention programs and reported successes in reducing the amount of waste they produce. Chapter 31 is concerned solely with industrial applications where it is demonstrated that the best alternative today for companies usually is to produce less waste in the first place. Chapters 32 and 33 serve as an introduction to health, safety, and accident management issues. Development of plans for handling accidents and emergencies must precede the actual occurrence of these events. The latter chapter is concerned with industrial safety applications, since incidents related to the chemical, petrochemical, and refinery industries have caused particular concern to safety in recent years. The pollution prevention sequence continues here with two chapters devoted to energy conservation. Chapter 34 serves as an introduction to energy conservation while Chapter 35 is concerned solely with industrial applications. Part V concludes with Chapter 36, which examines architectural environmental considerations.

30 The Pollution Prevention Concept

CONTENTS

30.1   INTRODUCTION

The amount of waste generated in the United States has reached staggering proportions; according to the United States Environmental Protection Agency (EPA), nearly 300 million tons of solid waste alone are generated annually. Although both the Resource Conservation and Recovery Act (RCRA) and the Hazardous and Solid Waste Act (HSWA) encourage businesses to minimize the wastes they generate, the majority of current environmental protection efforts are centered around treatment and pollution cleanup.

The passage of the Pollution Prevention Act of 1990 has redirected industry's approach to environmental management; pollution prevention has now become the environmental option of that decade and the twenty-first century. Whereas typical waste management strategies concentrate on "end-of-pipe" pollution control, pollution prevention attempts to handle waste at the source (i.e., source reduction). As waste handling and disposal costs increase, the application of pollution prevention measures is becoming more attractive than ever before. Industry is currently exploring the advantages of multimedia waste reduction and developing agendas to "strengthen" environmental design while "lessening" production costs.

There are profound opportunities for both the individual and industry to prevent the generation of waste; indeed, pollution prevention is today primarily stimulated by economics, legislation, liability concerns, and the enhanced environmental benefit of managing waste at the source. The EPA's Pollution Prevention Act of 1990 has established pollution prevention as a national policy declaring "waste should be prevented or reduced at the source wherever feasible, while pollution that cannot be prevented should be recycled in an environmentally safe manner" [1]. The EPA's policy establishes the following hierarchy of waste management:

1. Source reduction
2. Recycling/reuse
3. Treatment
4. Ultimate disposal

The hierarchy's categories are prioritized so as to promote the examination of each individual alternative prior to the investigation of subsequent options (i.e., the most preferable alternative should be thoroughly evaluated before consideration is given to a less accepted option.) Practices that decrease, avoid, or eliminate the generation of waste are considered source reduction and can include the implementation of procedures as simple and economical as good housekeeping. Recycling is the use, reuse, or reclamation of wastes and/or materials and may involve the incorporation of waste recovery techniques (e.g., distillation, filtration). Recycling can be performed at the facility (i.e., on-site), or at an off-site reclamation facility. Treatment involves the destruction or detoxification of wastes into nontoxic or less toxic materials by chemical, biological or physical methods, or any combination of these methods. Disposal has been included in the hierarchy because it is recognized that residual wastes will exist; the EPA's so-called ultimate disposal options in the past included landfilling, land farming, ocean dumping, and deep-well injection. However, the term "ultimate disposal" is a misnomer, but is included here because of its earlier adaptation by the EPA. Table 30.1 provides a rough timetable demonstrating the national approach to waste management. Note how waste management has begun to shift from pollution "control" to pollution prevention.

One of the authors [2] has developed a popular pollution prevention calendar for home or office use. Each of the two calendars contains 365 one-line suggestions (one for each day) dealing with waste reduction, energy conservation, and health safety and accident prevention. Each topic receives 4 months of coverage. The calendar is available in either hard copy or electronic format.

Some of key EPA literature and websites on pollution prevention are provided below,

1. The Pollution Prevention Information Clearinghouse is located at http://www.epa.gov/opptintr/ppic.
2. Introduction to Pollution Prevention Training Manual is located at http://www.epa.gov/opptintr/ppic/pubs/intropollutionprevention.pdf.
3. The Guide to Industrial Assessments for Pollution Prevention and Energy Efficiency is located at http://www.epa.gov/Pubs/2001/energy/complete.pdf.

TABLE 30.1
Waste Management Timetable

Timeframe	Control
Prior to 1945	No control
1945–1960	Little control
1960–1970	Some control
1970–1975	Greater control (EPA founded)
1975–1980	More sophisticated control
1980–1985	Beginning of waste reduction management
1985–1990	Waste reduction management
1990–1995	Pollution Prevention Act
1995–2000	Pollution prevention activities
2000–	???

4. The final discussion of the EPA program concerning the 33-50 program can be found at http://www.epa.gov/opptintr/3350/with the final report at http://www.epa.gov/opptintr/3350/3350-fnl.pdf.

5. A discussion of the EPA program concerning "Persistent Bioaccumulative and Toxic Chemical Program" (PBT) is located at http://www.epa.gov/pbt/. The Waste Minimization Program discussed at http://www.epa.gov/epaoswer/hazwaste/minimize. Priority chemicals are listed at http://www.epa.gov/epaoswer/hazwaste/minimize/chemlist.htm.

6. Compliance assistance for a variety of industrial sectors can be viewed at the EPA Website concerning compliance center notebooks at http://www.epa.gov/compliance/resources/publications/assistance/sectors/notebooks/

7. General information:
    a. http://www.epa.gov/glossary
    b. http://www.epa.gov//P2/pubs/basic.htm.

30.2   POLLUTION PREVENTION HIERARCHY

As discussed in Section 30.1, the hierarchy set forth by the EPA in the Pollution Prevention Act establishes an order in which waste management activities should be employed to reduce the quantity of waste generated. The preferred method is source reduction, as indicated in Figure 30.1. This approach actually precedes traditional waste management by addressing the source of the problem prior to its occurrence.

Although the EPA's policy does not consider recycling or treatment as actual pollution prevention methods per se, these methods present an opportunity to reduce the amount of waste that might otherwise be discharged into the environment. Clearly, the definition of pollution prevention and its synonyms (e.g., waste minimization) must be understood to fully appreciate and apply these techniques.

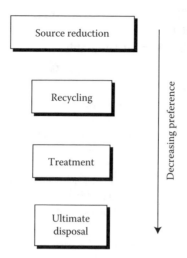

FIGURE 30.1   Pollution prevention hierarchy.

Waste minimization generally considers all of the methods in the EPA hierarchy (except for disposal) appropriate to reduce the volume or quantity of waste requiring disposal (e.g., source reduction). The definition of source reduction as applied in the Pollution Prevention Act, however, is "any practice which reduces the amount of any hazardous substance, pollutant, or contaminant entering any waste stream or otherwise released into the environment … prior to recycling, treatment, or disposal" [1]. Source reduction reduces the amount of waste generated; it is therefore considered true pollution prevention and has the highest priority in the EPA hierarchy.

Recycling (reuse, reclamation) refers to the use or reuse of materials that would otherwise be disposed of or treated as a waste product. Wastes that cannot be directly reused may often be recovered on-site through methods such as distillation. When on-site recovery or reuse is not feasible due to quality specifications or the inability to perform recovery on-site, off-site recovery at a permitted commercial recovery facility is often a possibility. Such management techniques are considered secondary to source reduction and should only be used when pollution cannot be prevented.

The treatment of waste is the third element of the hierarchy and should be utilized only in the absence of feasible source reduction or recycling opportunities. Waste treatment involves the use of chemical, biological, or physical processes to reduce or eliminate waste material. The incineration of wastes is included in this category and is considered "preferable to other treatment methods (i.e., chemical, biological, and physical) because incineration can permanently destroy the hazardous components in waste materials" [3].

Of course, many of these pollution prevention elements are used by industry in combination to achieve the greatest waste reduction. Residual wastes that cannot be prevented or otherwise managed are then disposed of only as a last resort.

30.3 MULTIMEDIA ANALYSIS AND LIFECYCLE COST ANALYSIS

30.3.1 MULTIMEDIA ANALYSIS

In order to properly design and then implement a pollution prevention program, sources of all wastes must be fully understood and evaluated. A multimedia analysis involves a multifaceted approach. It must not only consider one waste stream but all potentially contaminant media (e.g., air, water, and land). Past waste management practices have been concerned primarily with treatment. All too often, such methods solve one waste problem by transferring a contaminant from one medium to another (e.g., air-stripping); such waste shifting is "not" pollution prevention or waste reduction.

Pollution prevention techniques must be evaluated through a thorough consideration of all media, hence the term multimedia. This approach is a clear departure from previous pollution treatment or control techniques where it was acceptable to transfer a pollutant from one source to another in order to solve a waste problem. Such strategies merely provide short-term solutions to an ever-increasing problem. As an example, air pollution control equipment prevents or reduces the discharge of waste into the air but at the same time can produce a solid hazardous waste problem. (See Chapter 5 for additional details on multimedia analyses.)

30.3.2 LIFECYCLE ANALYSIS

The aforementioned multimedia approach to evaluating a product's waste stream(s) aims to ensure that the treatment of one waste stream does not result in the generation or increase in an additional waste output. Clearly, impacts resulting during the production of a product must be evaluated over its entire history or lifecycle.

A lifecycle analysis, or "Total Systems Approach," [4] is crucial to identifying opportunities for improvement. This type of evaluation identifies "energy use, material inputs, and wastes generated during a product's life: from extraction and processing of raw materials, to manufacture and transport of a product to the marketplace, and, finally, to use and dispose of the product" [5].

During a forum convened by the World Wildlife Fund and the Conservation Foundation in May 1990, various steering committees recommended that a three-part lifecycle model be adopted. This model consists of the following:

1. An inventory of materials and energy used, and environmental releases from all stages in the life of a product or process.
2. An analysis of potential environmental effects related to energy use and material resources and environmental releases.
3. An analysis of the changes needed to bring about environmental improvements for the product or process under evaluation.

Traditional cost analysis often fails to include factors relevant to future damage claims resulting from litigation, the depletion of natural resources, the effects of energy use, and so on. Therefore, waste management options such as treatment and disposal may appear preferential if an overall lifecycle cost analysis is not performed. It is evident that environmental costs from "cradle-to-grave" have to be evaluated together with more conventional production costs to accurately ascertain genuine production costs. In the future, a total systems approach will most likely involve a more careful evaluation of pollution, energy, and safety issues. For example, if one was to compare the benefits of coal versus oil as a fuel source for an electric power plant, the use of coal might be considered economically favorable. In addition to the cost issues, however, one must be concerned with the environmental effects of coal mining, transportation, and storage prior to use as a fuel. Many have a tendency to overlook the fact that there are serious health and safety matters (e.g., miner exposure) that must be considered, along with the effects of fugitive emissions. When these effects are weighed alongside of standard economic factors, the cost benefits of coal usage may no longer appear valid. Thus, many of the economic benefits associated with pollution prevention are often unrecognized due to inappropriate cost accounting methods. For this reason, economic considerations are detailed in a later chapter.

30.4   POLLUTION PREVENTION ASSESSMENT PROCEDURES

The first step in establishing a pollution prevention program is the obtainment of management commitment. Management commitment is necessary given the inherent need for project structure and control. Management will determine the amount of funding allotted for the program as well as specific program goals. The data collected during the actual evaluation is then used to develop options for reducing the types and amounts of waste generated. Figure 30.2 depicts a systematic approach that can be used during the procedure. After a particular waste stream or area of concern is identified, feasibility studies are performed involving both economic and technical considerations. Finally, preferred alternatives are implemented.

The four phases of the assessment (i.e., planning and organization, assessment, feasibility, and implementation) are introduced in the following subsections. Sources of additional information as well as information on industrial programs are also provided in this section.

30.4.1   PLANNING AND ORGANIZATION

The purpose of this phase is to obtain management commitment, define and develop program goals, and to assemble a project team. Proper planning and organization are crucial to the successful performance of the pollution prevention assessment. Both managers and facility staff play important roles in the assessment procedure by providing the necessary commitment and familiarity with the facility, its process(es), and current waste management operations. The benefits of the program, including economic advantages, liability reduction, regulatory compliance, and improved public image, often lead to management support.

Once management has made a commitment to the program and goals have been set, a program task force is established. The selection of a team leader will be dependent upon many factors including his or her ability to effectively interface with both the assessment team and management staff.

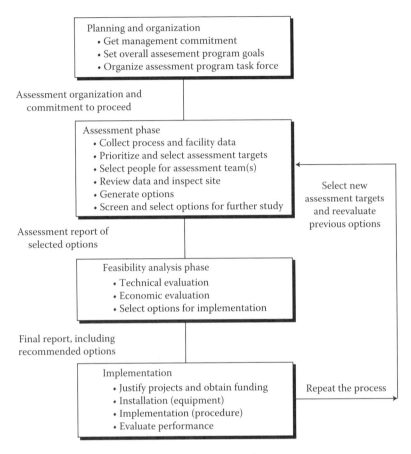

FIGURE 30.2 Pollution prevention assessment procedures.

The task force must be capable of identifying pollution reduction alternatives, as well as be cognizant of inherent obstacles to the process. Barriers frequently arise from the anxiety associated with the belief that the program will negatively affect product quality or result in production losses. According to an EPA survey, 30% of industry comments responded that they were concerned that product quality would decline if waste minimization techniques were implemented [6]. Thus, the assessment team, and the team leader in particular, must be prepared to react to these and other concerns [3].

30.4.2 ASSESSMENT PHASE

The assessment phase aims to collect data required to identify and analyze pollution prevention opportunities. Assessment of the facility's waste reduction needs includes the examination of hazardous waste streams, process operations, and the identification of techniques that often promise the reduction of waste generation.

Information is also derived from observations made during a facility walk-through, interviews with employees (e.g., operators, line workers), and review of site or regulatory records. The American Society of Testing and Materials (ASTM) suggests the following information sources be reviewed, as available [7].

1. Product design criteria.
2. Process flow diagrams for all solid waste, wastewater, and air emissions sources.
3. Site maps showing the location of all pertinent units (e.g., pollution control devices, points of discharge).
4. Environmental documentation, including: Material Safety Data Sheets (MSDS), military specification data, permits (e.g., NPDES, POTW, RCRA), SARA Title III reports, waste manifests, and any pending permits or application information.
5. Economic data, including: cost of raw material management; cost of air, wastewater, and hazardous waste treatment; waste management operating and maintenance costs; and, waste disposal costs.
6. Managerial information: environmental policies and procedures; prioritization of waste management concerns; automated or computerized waste management systems; inventory and distribution procedures; maintenance scheduling practices; planned modifications or revisions to existing operations that would impact waste generation activities; and, the basis of source reduction decisions and policies.

The use of process flow diagrams and material balances are worthwhile methods to "quantify losses or emissions, and provide essential data to estimate the size and cost of additional equipment, data to evaluate economic performance, and a baseline for tracking the progress of minimization efforts" [5]. Material balances should be applied to individual waste streams or processes, and then utilized to construct an overall balance for the facility. Details on these calculations are available in the literature [6].

The data collected is then used to prioritize waste streams and operations for assessment. Each waste stream is assigned a priority based on corporate pollution prevention goals and objectives. Once waste origins are identified and ranked, potential methods to reduce the waste stream are evaluated. The identification of alternatives is generally based on discussions with the facility staff, review of technical literature, and contacts with suppliers, trade organizations, and regulatory agencies.

Alternatives identified during this phase of the assessment are evaluated using screening procedures so as to reduce the number of alternatives requiring further exploration during the feasibility analysis phase. The criteria used during this screening procedure include: cost-effectiveness; implementation time; economic, compliance, safety, and liability concerns; waste reduction potential; and, whether the technology is proven [3,6]. Options which meet established criteria are then examined further during the feasibility analysis.

30.4.3 FEASIBILITY ANALYSIS

Preferred alternative selection is performed by an evaluation of technical and economic considerations. The technical evaluation determines whether a given option will work as planned. Some typical considerations follow:

1. Safety concerns
2. Product quality impacts or production delays during implementation
3. Labor and/or training requirements
4. Creation of new environmental concerns
5. Waste reduction potential
6. Utility and budget requirements
7. Space and compatibility concerns

If an option proves to be technically ineffective or inappropriate, it is deleted from the list of potential alternatives. Either following or concurrent with the technical evaluation, an economic study is performed weighing standard measures of profitability such as payback period, investment returns, and net present value. Many of these costs (or more appropriately, cost savings) may be substantial yet are difficult to quantify [4].

30.4.4 IMPLEMENTATION

The findings of the overall assessment are used to demonstrate the technical and economic worthiness of program implementation. Once appropriate funding is obtained, the program is implemented, not unlike any other project requiring new procedures or equipment. When preferred waste pollution prevention techniques are identified, they are implemented, and should become part of the facility's day-to-day management and operation. Subsequent to the program's execution, its performance should be evaluated in order to demonstrate effectiveness, generate data to further refine and augment waste reduction procedures, and maintain management support.

It should be noted that waste reduction, energy conservation, and safety issues are interrelated and often complementary to each other. For example, the reduction in the amount of energy a facility consumes results in reduced emissions associated with the generation of power. Energy expenditures associated with the treatment and transport of waste are similarly reduced when the amount of waste generated is lessened; at the same time worker safety is elevated due to reduced exposure to hazardous materials.

30.5 SOURCES OF INFORMATION

The successful development and implementation of any pollution prevention program is not only dependent on a thorough understanding of the facility's operations but also requires an intimate knowledge of current opportunities and advances in the field. In fact, 32% of industry respondents to an EPA survey identified the lack of technical information as a major factor delaying or preventing the implementation of a waste minimization program [1]. Fortunately, the EPA has developed a national Pollution Prevention Information Clearinghouse (PPIC) and the Pollution Prevention Information Exchange System (PIES) to facilitate the exchange of information needed to promote pollution prevention through efficient information transfer [1].

PPIC is operated by the EPA's Office of Research and Development and the Office of Pollution Prevention. The clearinghouse is comprised of four elements:

1. *Repository*: Including a hard copy reference library and collection center, and an online information retrieval and ordering system.
2. *PIES*: A computerized conduit to databases and document ordering, accessible via modem and personal computer—(703) 506-1025.
3. *Hotline*: PPIC uses the RCRA/Superfund and Small Business Ombudsman Hotlines as well as a PPIC technical assistance line to answer pollution prevention questions, access information in the PPIC, and assist in document ordering and searches. To access PPIC by telephone, call:
   RCRA/Superfund Hotline (800) 242-9346
   Small Business Ombudsman Hotline (800) 368-5888
   PPIC Technical Assistance (703) 821-4800
4. *Networking and outreach*: PPIC compiles and disseminates information packets and bulletins, and initiates networking efforts with other national and international organizations.

Additionally, the EPA publishes a newsletter entitled *Pollution Prevention News*, which contains information including EPA news, technologies, program updates, and case studies. The EPA's Risk Reduction Engineering Laboratory and the Center for Environmental Research Information has published several guidance documents, developed in cooperation with the California Department of Health Services. The manuals supplement generic waste reduction information presented in the EPA's *Waste Minimization Opportunity Assessment Manual* [6]. Additional information is available through PPIC.

Pollution prevention or waste minimization programs have been established at the State level and as such are good sources of information. Both Federal and State agencies are working with universities and research centers and may also provide assistance. For example, the American Institute of Chemical Engineers has established the Center for Waste Reduction Technologies (CWRT), a program based on targeted research, technology transfer, and enhanced education.

30.5.1 INDUSTRY PROGRAMS

A significant pollution prevention resource may very well be found with the "competition." Several large companies have established well-known programs that have successfully incorporated pollution prevention practices into their manufacturing processes. These include, but are not limited to: 3M—Pollution Prevention Pays (3P); Dow Chemical—Waste Reduction Always Pays (WRAP); Chevron—Save Money and Reduce Toxics (SMART); and General Dynamics—Zero Discharge Program.

Smaller companies can benefit by the assistance offered by these larger corporations. It is clear that access to information is of major importance when implementing efficient pollution prevention programs. By adopting such programs, industry is affirming pollution prevention as a good business practice and not simply a noble effort.

30.6 FUTURE TRENDS

The development of waste management practices in the United States has recently moved toward securing a new pollution prevention ethic. The performance of pollution prevention assessments and their subsequent implementation will encourage increased research into methods that will further aid in the reduction of wastes and pollution.

It is evident that the majority of the present day obstacles to pollution prevention are based on either a lack of information or an anxiety associated with economic concerns. By strengthening the exchange of information among businesses, a better understanding of the unique benefits of pollution prevention will be realized in the future.

30.7 SUMMARY

1. The passage of the Pollution Prevention Act of 1990 has redirected industry's approach to environmental management; pollution prevention has now become the environmental option of that decade and the twenty-first century. Whereas typical waste management strategies concentrate on "end-of-pipe" pollution control, pollution prevention attempts to handle waste at the source (i.e., source reduction).
2. The EPA's policy establishes the following hierarchy of waste management:
   a. Source reduction
   b. Recycling/reuse
   c. Treatment
   d. Ultimate disposal
3. In order to properly design and then implement a pollution prevention program, sources of all wastes must be fully understood and evaluated. A multimedia analysis involves a multifaceted approach. It must not only consider one waste stream but all potentially contaminant media (e.g., air, water, and land).
4. The four phases of a pollution prevention assessment procedure are: planning and organization, assessment, feasibility analysis, and implementation.
5. EPA has developed a national PPIC and the PIES to facilitate the exchange of information needed to promote pollution prevention through efficient information transfer.
6. It is evident that the majority of the present-day obstacles to pollution prevention are based on either a lack of information or an anxiety associated with economic concerns. By strengthening the exchange of information among businesses, a better understanding of the unique benefits of pollution prevention will be realized in the future.

REFERENCES

1. U.S. EPA. *Pollution Prevention Fact Sheet*, Author, Washington, DC, March 1991.
2. Theodore, M.K. *Pollution Prevention Calendar*, East Williston, New York, 2000 (and subsequent years).
3. Dupont, R., Theodore, L., and Haneson, K. *Pollution Prevention*, CRC Press, Boca Raton, FL, 2002.
4. Theodore, L., Personal notes, 2000.
5. World Wildlife Fund. *Getting at the Source*, New York, 1991.
6. U.S. EPA. *The EPA Manual for Waste Minimization Opportunity Assessments*, EPA: Cincinnati, OH, August 1988.
7. ASTM. *Standard Guide for Industrial Source Reduction* (Draft Copy), American Society of Testing and Materials, New York, June 16, 1992.

31 Pollution Prevention Applications

CONTENTS

31.1 INTRODUCTION [1,2]

One of the key elements of the assessment phase of a pollution prevention program involves mass balance equations. These calculations are often referred to as material balances; the calculations are performed via the conservation law for mass. The details of this often-used law are described below.

The conservation law for mass can be applied to any process or system. The general form of the law follows:

$$\text{Mass in} - \text{mass out} + \text{mass generated} = \text{mass accumulated} \quad (31.1)$$

This equation can be applied to the total mass involved in a process or to a particular species, on either a mole or mass basis. The conservation law for mass can be applied to steady-state or unsteady-state processes and to batch or continuous systems. A steady-state system is one in which there is no change in conditions (e.g., temperature, pressure, etc.) or rates of flow with time at any given point in the system; the accumulation term then becomes zero. If there is no chemical reaction, the generation term is zero. All other processes are classified as unsteady-state.

To isolate a system for study, the system is separated from the surroundings by a boundary or envelope that may either be real (e.g., a reactor vessel) or imaginary. Mass crossing the boundary and entering the system is part of the mass-in term. The equation may be used for any compound whose quantity does not change by chemical reaction, or for any chemical element, regardless of whether it has participated in a chemical reaction. Furthermore, it may be written for one piece of equipment, several pieces of equipment, or around an entire process (i.e., a total material balance).

The conservation of mass law finds a major application during the performance of pollution prevention assessments. As described in the previous chapter, a pollution prevention assessment is a systematic, planned procedure with the objective of identifying methods to reduce or eliminate waste. The assessment process should characterize the selected waste streams and processes [3]—a necessary ingredient if a material balance is to be performed. Some of the data required for the material balance calculation may be collected during the first review of site-specific data; however, in some instances, the information may not be collected until an actual site walk-through is performed.

Simplified mass balances should be developed for each of the important waste-generating operations to identify sources and gain a better understanding of the origins of each waste stream. Since a mass balance is essentially a check to make sure that what goes into a process (i.e., the total mass of all raw materials), what leaves the process (i.e., the total mass of the product(s) and byproducts), the material balance should be written individually for all components that enter and leave the process. When chemical reactions take place in a system, there is an advantage to doing "elemental balances" for specific chemical elements in a system. Material balances can assist in determining concentrations of waste constituents where analytical test data are limited. They are particularly useful when there are points in the production process where it is difficult or uneconomical to collect analytical data.

Mass balance calculations are particularly useful for quantifying fugitive emissions, such as evaporative losses. Waste stream data and mass balances will enable one to track flow and characteristics of the waste streams over time. Since in most cases the accumulation equals zero (steady-state operation), it can then be assumed that any buildup is actually leaving the process through fugitive emissions or other means. This can be useful in identifying trends in waste/pollutant generation and can also be critical in the task of measuring the performance of implemented pollution prevention options.

The result of these activities is a catalog of waste streams that provides a description of each waste, including quantities, frequency of discharge, composition, and other important information useful for material balance. Of course, some assumptions or educated estimates will be needed when it is impossible to obtain specific information.

By performing a material balance in conjunction with a pollution prevention assessment, the amount of waste generated becomes known. The success of the pollution prevention program can therefore be measured by using this information on baseline generation rates (i.e., that rate at which waste is generated without pollution prevention considerations).

31.2   BARRIERS TO POLLUTION PREVENTION [4]

As discussed previously, industry is beginning to realize that there are profound benefits associated with pollution prevention including cost effectiveness, reduced liability, enhanced public image, and regulatory compliance (some details are discussed in the next section). Nevertheless, there are barriers or disincentives identified with pollution prevention. This section will briefly outline barriers that may need to be confronted or considered during the evaluation of a pollution prevention program.

There are numerous reasons why more businesses are not reducing the wastes they generate. The following "dirty dozen" are common disincentives:

1. *Technical limitations.* Given the complexity of present manufacturing processes, waste streams exist that cannot be reduced with current technology. The need for continued research and development is evident.
2. *Lack of information.* In some instances, the information needed to make a pollution prevention decision may be confidential or is difficult to obtain. In addition, many decision makers are simply unaware of the potential opportunities available regarding information to aid in the implementation of a pollution prevention program.
3. *Consumer preference obstacles.* Consumer preference strongly affects the manner in which a product is produced, packaged, and marketed. If the implementation of a pollution prevention program results in the increase in the cost of a product, or decreased convenience or availability, consumers might be reluctant to use it.
4. *Concern over product quality decline.* The use of a less hazardous material in a product's manufacturing process may result in decreased life, durability, or competitiveness.
5. *Economic concerns.* Many companies are unaware of the economic advantages associated with pollution prevention. Legitimate concerns may include decreased profit margins or the lack of funds required for the initial capital investment.
6. *Resistance to change.* The unwillingness of many businesses to change is rooted in their reluctance to try technologies that may be unproven, or based on a combination of the barriers discussed in this section.
7. *Regulatory barriers.* Existing regulations that have created incentives for the control and containment of wastes, are at the same time discouraging the exploration of pollution prevention alternatives. Moreover, since regulatory enforcement is often intermittent, current legislation can weaken waste reduction incentives.
8. *Lack of markets.* The implementation of pollution prevention processes and the production of environmentally friendly products will be of no avail if markets do not exist for such goods. As an example, the recycling of newspaper in the United States has resulted in an overabundance of waste paper without markets prepared to take advantage of this "raw" material.
9. *Management apathy.* Many managers capable of making decisions to begin pollution prevention activities, do not realize the potential benefits of pollution prevention and may therefore take on an attitude of passiveness.
10. *Institutional barriers.* In an organization without a strong infrastructure to support pollution prevention plans, waste reduction programs will be difficult to implement. Similarly, if there is no mechanism in place to hold

individuals accountable for their actions, the successful implementation of a pollution prevention program will be limited.

11. *Lack of awareness of pollution prevention advantages.* As mentioned in *economic concerns*, decision makers may be uninformed of the benefits associated with pollution reduction.

12. *Concern over the dissemination of confidential product information.* If a pollution prevention assessment reveals confidential data pertinent to a company's product, fear may exist that the organization will lose a competitive edge with other businesses in the industry.

31.3 POLLUTION PREVENTION ADVANTAGES [4]

Various means exist to encourage pollution prevention through regulatory measures, economic incentives, and technical assistance programs. Since the benefits of pollution prevention undoubtedly surpass prevention barriers, a baker's dozen incentives is presented below:

1. *Economic benefits.* The most obvious economic benefits associated with pollution prevention are the savings that result from the elimination of waste storage, treatment, handling, transport, and disposal. Additionally, less tangible economic benefits are realized in terms of decreased liability, regulatory compliance costs (e.g., permits), legal and insurance costs, and improved process efficiency. Pollution prevention almost always pays for itself, particularly when the time investment required to comply with regulatory standards is considered. Several of these economic benefits are discussed separately below.

2. *Regulatory compliance.* Quite simply, when wastes are not generated, compliance issues are not a concern. Waste management costs associated with recordkeeping, reporting, and laboratory analysis are reduced or eliminated. Pollution prevention's proactive approach to waste management will better prepare industry for the future regulation of many hazardous substances and wastes that are currently unregulated. Regulations have, and will continue to be, a moving target.

3. *Liability reduction.* Facilities are responsible for their wastes from "cradle-to-grave." By eliminating or reducing waste generation, future liabilities can also be decreased. Additionally, the need for expensive pollution liability insurance requirements may be abated.

4. *Enhanced public image.* Consumers are interested in purchasing goods that are safer for the environment and this demand, depending on how they respond, can mean success or failure for many companies. Business should therefore be sensitive to consumer demands and use pollution prevention efforts to their utmost advantage by producing goods that are environmentally friendly.

5. *Federal and state grants.* Federal and State grant programs have been developed to strengthen pollution prevention programs initiated by states and

private entities. The EPA's Pollution Prevention By and For Small Business Grant Program awards grants to small businesses to assist their development and demonstration of new pollution prevention technologies.

6. *Market incentives*. Public demand for environmentally preferred products has generated a market for recycled goods and related products; products can be designed with these environmental characteristics in mind, offering a competitive advantage. In addition, many private and public agencies are beginning to stimulate the market for recycled goods by writing contracts and specifications that call for the use of recycled materials.

7. *Reduced waste treatment costs*. As discussed in "economic benefits," the increasing costs of traditional end-of-pipe waste management practices are avoided or reduced through the implementation of pollution prevention programs.

8. *Potential tax incentives*. As an effort to promote pollution prevention, taxes may eventually need to be levied to encourage waste generators to consider reduction programs. Conversely, tax breaks to corporations that utilize pollution prevention methods could similarly be developed to foster pollution prevention.

9. *Decreased worker exposure*. By reducing or eliminating chemical exposures, businesses benefit by lessening the potential for chronic workplace exposure, and serious accidents and emergencies. The burden of medical monitoring programs, personal exposure monitoring, and potential damage claims are also reduced.

10. *Decreased energy consumption*. As mentioned previously, energy conservation strategies are often interrelated and complementary to each other. Energy expenditures associated with the treatment and transport of waste are reduced when the amount of waste generated is lessened, while at the same time the pollution associated with energy consumed by these activities is abated.

11. *Increased operating efficiencies*. A potential beneficial side effect of pollution prevention activities is a concurrent increase in operating efficiency. Through a pollution prevention assessment, the assessment team can identify sources of waste that results in hazardous waste generation and loss in process performance. The implementation of a waste reduction program will often rectify such problems through modernization, innovation, and the implementation of good operating practices.

12. *Competitive advantages*. By taking advantage of the many benefits associated with pollution prevention, businesses can gain a competitive edge.

13. *Reduced negative environmental impacts*. Through an evaluation of pollution prevention alternatives which consider a total systems approach, consideration is given to the negative impact of environmental damage to natural resources and species that occur during raw material procurement and waste disposal. The performance of pollution prevention endeavors will therefore result in enhanced environmental protection.

The development of new markets by means of regulatory and economic incentives will further assist the effective implementation of waste reduction. Various combinations of the pollution prevention barriers provided earlier have appeared on numerous occasions in the literature, and in many different forms. However, there is one other concern that both industry and the taxpayer should be aware of. EPA Administrators have repeatedly claimed that pollution prevention is the organization's top priority. "Nothing could be further from the truth." Despite near unlimited resources, the EPA has contributed little to furthering the pollution prevention effort. The EPA offices in Washington, Research Park Triangle, and Region II have exhibited a level of bureaucratic indifference that has surpassed even the traditional attitudes of many EPA employees. It is virtually impossible to contact any responsible pollution prevention individual at the EPA. Calls are rarely returned. Letters are rarely returned. On the rare occasion when contact is made, the regulatory individual typically passes the caller onto someone else who "really is in a better position to help you," and the cycle starts all over again [5].

This standard bureaucratic phenomena has been experienced by others in both industry and the EPA [3,4]. Two letters of complaint to the EPA Region II Administrator resulted in a response that was somewhat cynical and suggestive of a reprimand. The Administrator chose not to reply to the complaints [5]. Notwithstanding some of the above comments, pollution prevention efforts have been successful in industry because these programs have often either produced profits or reduced costs, or both. The driving force for these successes has primarily been economics and "not" the EPA.

A more detailed presentation on economic considerations is considered in Section 31.4.

31.4 ECONOMIC CONSIDERATIONS ASSOCIATED WITH POLLUTION PREVENTION PROGRAMS

The purpose of this section is to outline the basic elements of a pollution prevention cost accounting system that incorporates both traditional and less tangible economic variables. The intent is not to present a detailed discussion of economic analysis but to help identify the more important elements that must be considered to properly quantify pollution prevention options.

Pollution prevention is now recognized as one of the lowest-cost options for waste/pollutant management. The greatest driving force behind any pollution prevention plan is the promise of economic opportunities and cost savings over the long term. Hence, an understanding of the economics involved in pollution prevention programs/options is important in making decisions at both the engineering and management levels. Every organization should be able to execute an economic evaluation of a proposed project. If the project cannot be justified economically after "all" factors and considerations have been taken into account, it should obviously not be pursued. The earlier such a project is identified, the fewer resources will be wasted.

Before the true cost or profit of a pollution prevention program can be evaluated, the factors contributing to the economics must be recognized. There are two traditional contributing factors—capital costs and operating costs—but there are other important costs and benefits associated with pollution prevention that need to be quantified if a meaningful economic analysis, is to be performed.

The economic evaluation referred to above is usually carried out using standard measures of profitability. Each company and organization has its own economic criteria for selecting projects for implementation. In performing an economic evaluation, various costs and savings must be considered. The economic analysis presented in this section represents a preliminary, rather than a detailed, analysis. For smaller facilities with only a few (and perhaps simple) processes, the entire pollution prevention assessment procedure will tend to be much less formal. In this situation, several obvious pollution prevention options, such as the installation of flow controls and good operating practices, may be implemented with little or no economic evaluation. In these instances, no complicated analyses are necessary to demonstrate the advantages of adopting the selected pollution prevention option. A proper perspective must also be maintained between the magnitude of savings that a potential option may offer and the amount of manpower required to perform the technical and economic feasibility analyses. Details on economics are provided in Chapter 47.

31.5 FUTURE TRENDS

The main problem with the traditional type of economic analysis is that it is difficult—nay, in some cases, impossible—to quantify some of the not-so-obvious economic merits of a pollution prevention program. Several considerations, in addition to those provided in the previous sections, have just recently surfaced as factors that need to be taken into account in any meaningful economic analysis of a pollution prevention effort. These factors are certain to become an integral part of any pollution prevention analysis in the future. What follows is a listing of these considerations (see also Chapter 47):

1. Decreased long-term liabilities
2. Regulatory compliance
3. Regulatory recordkeeping
4. Dealings with the EPA
5. Dealings with state and local regulatory bodies
6. Elimination or reduction of fines and penalties
7. Potential tax benefits
8. Customer relations
9. Stockholder support (corporate image)
10. Improved public image
11. Reduced technical support
12. Potential insurance costs and claims
13. Effect on borrowing power
14. Improved mental and physical well-being of employees
15. Reduced health maintenance costs
16. Employee morale
17. Other process benefits
18. Improved worker safety
19. Avoidance of rising costs of waste treatment and/or disposal
20. Reduced training costs
21. Reduced emergency response planning

Many proposed pollution prevention programs have been quenched in their early stages because a comprehensive analysis was not performed. Until the effects described above are included, the true merits of a pollution prevention program may be clouded by incorrect and/or incomplete economic data. Can something be done by industry to remedy this problem? One approach [5] is to use a modified version of the standard Delphi panel that the authors have modestly defined as the WTA (an acronym for the Wainwright–Theodore Approach). In order to estimate these "other" factors and/or economic benefits of pollution prevention, several knowledgeable individuals within and perhaps outside the organization are asked to independently provide estimates, with explanatory details, on these benefits. Each individual in the panel is then allowed to independently review all responses. The cycle is then repeated until the group's responses approach convergence.

Finally, pollution prevention measures can provide a company with the opportunity of looking their neighbors in the eye and truthfully saying that all that can reasonably be done to prevent pollution is being done ... in effect, the company is doing right by the environment. Is there an advantage to this? It is not only a difficult question to answer quantitatively but also a difficult one to answer qualitatively. The reader is left with pondering the answer to this question in terms of future activities [6].

31.6 SUMMARY

1. One of the key elements of the assessment phase of a pollution prevention program involves mass balance equations. These calculations are often referred to as material balances; the calculations are performed via the conservation law for mass.
2. Industry is beginning to realize that there are profound benefits associated with pollution prevention including cost effectiveness, reduced liability, enhanced public image, and regulatory compliance. Nevertheless, there are barriers or disincentives identified with pollution prevention.
3. Various means exist to encourage pollution prevention through regulatory measures, economic incentives, and technical assistance programs.
4. The greatest driving force behind any pollution prevention plan is the promise of economic opportunities and cost savings over the long term.
5. The main problem with the traditional type of economic analysis is that it is difficult, or in some cases, impossible, to quantify some of the not-so-obvious economic merits of a pollution prevention program.

REFERENCES

1. Santoleri, J., Reynolds, J., and Theodore, L. *Introduction to Hazardous Waste Incineration*, 2nd edition, John Wiley & Sons, Hoboken, NJ, 2002.
2. Reynolds, J., Jeris, J., and Theodore, L. *Handbook of Chemical and Environmental Engineering Calculations*, John Wiley & Sons, Hoboken, NJ, 2004.
3. ICF Technology Incorporated. *New York State Waste Reduction Guidance Manual*, Author, Alexandria, VA 1989.
4. Perry, R. and Green, D. *Perry's Chemical Engineers' Handbook*, 8th edition, McGraw-Hill, New York, 2008.
5. Theodore, L. and Wainwright, B. Personal notes, 1995.
6. Theodore, L. Personal notes, 2000.

32 Introduction to Health, Safety, and Accident Management*

CONTENTS

32.1 INTRODUCTION

Accidents have occurred since the birth of civilization and were just as damaging in early times as they are today. Anyone who crosses a street, skis, or swims in a pool runs the risk of injury through carelessness, poor judgment, ignorance, or other circumstances. This has not changed through history. This introductory section examines a number of accidents and disasters that took place before the advances of modern technology.

Catastrophic explosions have been reported as early as 1769, when one-sixth of the city of Frescia, Italy, was destroyed by the explosion of 100 tons of gunpowder stored in the state arsenal. More than 3000 people were killed in this, the second deadliest explosion in history [1].

The worst explosion in history occurred in 1856 on the Greek Island of Rhodes. A church, which had gunpowder stored in its vaults, was struck by lightning. The resulting blast is estimated to have killed 4000 people. This remains the highest death toll for a single explosion [2].

One of the most legendary disasters occurred in Chicago in October 1871. The "Great Chicago Fire," as it is now known, is alleged to have started in a barn owned by Patrick O'Leary, when one of his cows overturned a lantern. The O'Leary house escaped unharmed, since it was upwind of the blaze, but the barn was destroyed, as well as 2124 acres of Chicago real estate.

---

* This chapter is a condensed, revised, and updated version of an unpublished (but copyrighted) 1992 text prepared by M. K. Theodore and L. Theodore titled *A Citizens Guide to Pollution Prevention.*

355

Four persons died and eight others were injured on March 26, 1976, when two gondola cars fell more than 100 feet down the slopes of Vail Mountain in Vail, Colorado. The accident occurred because an automatic shutoff mechanism failed to respond to the partial derailment of a car ahead of the two that crashed. The cause was traced to five strands of steel sheath encasing the cable. The strands had begun to unravel at a point about two-thirds of the way up the mountain. The frayed cable caused the cars bearing the victims to derail and jam up, which should have activated an electrical overload switch designed to shut down the gondola. There is no explanation of why this safety device did not function [1,3].

In Caracas, Venezuela on April 9, 1952, a large crowd gathered at a church at the beginning of Holy Week. Apparently a pickpocket, wishing to create confusion, shouted "Fire!", whereupon the worshippers rushed toward exits at the rear of the church. Many people fell and were trampled by their fellow parishioners who were rushing to escape the imaginary fire. Fifty-three people were killed, nearly half of whom were small children and infants ("Be a Firesafe Neighbor," NFPA, 1988).

On January 2, 1971, at the Ibrox soccer stadium in Glasgow, Scotland, 66 persons were killed and 145 injured when a reinforced steel barrier collapsed under the weight of a surging crowd. The tragedy came after 8000 spectators had thronged the exits near the end of a hotly contested game between the Glasgow Rangers and the Glasgow Celtics, traditional rivals. A group of Rangers supporters at an exit stairway reportedly attempted to reenter the stadium when they heard that their team had scored to tie the game at 1–1. A massive human pileup was created, causing the loss of many lives [3].

On June 27, 1978, a freak accident at a fountain pool on Hilton Head Island, South Carolina, claimed four victims. A young man and woman apparently broke one of the lights illuminating the fountain as they jumped into it at 8:00 PM, sending several hundred volts through the water. The woman's roommate entered the pool, either unaware of the danger or attempting to save the pair. A neighbor then jumped in to attempt a rescue. All four were electrocuted [3].

On March 20, 1980, a regional seismic network operated by the U.S. Geological Survey and University of Washington recorded an earthquake of Richter magnitude 4.0 from a point north of the summit of Mount St. Helens, a dormant volcano that had last erupted in 1857. Two days later, the intensity of the seismic activity increased. Geologists suspected that magma, or melted rock, was moving up inside the mountain. The activity continued to increase. On March 27, a plume of steam and ash was emitted from Mount St. Helens and rose about 66,000 ft above the mountain. At a point one mile north of the summit crater, a large bulge was observed to be forming in the mountain's side. By early May, this bulge had grown to a length of one mile and a width of six-tenths of a mile. Volcanologists watched this bulge for signs that it might split open, extruding magma. On May 18, without any warning, Mount St. Helens suddenly exploded.

Apparently triggered by an earthquake of Richter magnitude 5.0, the entire north slope burst open along the upper edge of the bulge, releasing the bottled up gases and magma. Up to 3 km^3 of rock and ash were blown away from the mountain laterally. The blast spewed over one and a third billion cubic yards of material into the atmosphere.

Almost everything within 5 miles of the volcano was destroyed. Tons of choking ash and dust were dropped on central and eastern Washington, northern Oregon, and Idaho, and even parts of western Montana. The volcano continued to erupt during the remainder of 1980 and throughout the summer and fall of 1981. The death toll was confirmed in 1981 as 34, and 27 remain missing. Wildlife officials estimate that approximately 10,000 wild and domestic animals may also have been killed.

Eight "less significant" incidents are described below [3]:

1. Kandy, Ceylon, August 19, 1959. An elephant ran amok at a religious festival, killing 14 persons and injuring many others.
2. Bombay, India, September 20, 1959. A crush created at the scene of a religious "miracle" is reported to have killed 75 persons.
3. Kumaon Hills, India, February 13, 1970. A man-eating tiger, roaming a hilly area 50 miles northeast of New Delhi, was reported to have killed 48 persons.
4. Baltimore, Maryland, August 2, 1970. State health authorities reported that an outbreak of salmonella food poisoning at a city nursing home caused the deaths of 12 elderly patients; 60 others who were stricken recovered.
5. Sallen, France, May 15, 1971. The floorboards of a rented hall gave way at the close of a wedding reception, plunging the guests into a well beneath the floor; 13 persons perished.
6. Mozambique, November 1973. A large quantity of methyl alcohol washed ashore in drums and was mistakenly consumed as whiskey; 58 deaths were confirmed, but hundreds were believed to have died.
7. Near Jaipur, India, September 7, 1977. The roof of a village class room collapsed under the weight of a troop of baboons; 15 schoolgirls were killed instantly.
8. Harrisburg, Pennsylvania, June 13, 1978. An attempt by 2200 students and teachers to set a world record for tug-of-war ended with 70 persons injured when the 2000-ft nylon rope they were using broke. Four persons had parts of their hands and fingers ripped-off.

32.2   RISK CONCERNS [1]

Accidents can occur in many ways. There may be a chemical spill, an explosion, or a nuclear plant out of control. There are often accidents in transport: trucks overturning, trains derailing, or ships capsizing. There are "acts of God" such as earthquakes and storms. It is painfully clear that accidents are a fact of life. The one common thread through all of these situations is that accidents are rarely expected and, unfortunately, they are frequently mismanaged.

Development of plans for handling accidents and emergencies must precede the actual occurrence of these events. In recent years incidents related to the chemical, petrochemical, and refinery industries have caused particular concern. Since the products of these industries are essential in a modern society, every attempt must be made to identify and reduce the risk of accidents or emergencies in these areas.

Whether a careless mishap at home (to be discussed shortly), an unavoidable collision on the freeway, or a miscalculation at a chemical plant, accidents are a fact of life. Even in prehistoric times, long before the advent of technology, a club-wielding caveman swings at his prey and inadvertently topples his friend in what can only be classified as an "accident." As humanity progressed, so did the severity of our misfortunes. The "modern era" has brought about assembly lines, chemical manufacturers, nuclear power plants, and other technological complexes, all carrying the capability of disaster. To keep pace with the changing times, safety precautions must constantly be upgraded. It is no longer sufficient, as with the caveman, to shout a warning "Hey, watch out with that thing!" Today's problems require more elaborate systems of warnings and controls to minimize the chances of serious accidents. A crucial part of any design project is the inclusion of safety controls. Whether the plans involve a chemical plant, a nuclear reactor, or a thruway, steps must be taken to minimize the likelihood, or consequences, of accidents. It is also important to realize how accident planning has improved in order to monitor today's advanced technologies.

The word "home" refers to the living quarters where a person or family dwells, be it a house, apartment, or room. Contrary to popular belief, the home is not the safest place in the world. In fact, most industrial (including chemical) plants have records that indicate it is not safer to be at home. Top management of many chemical companies now require that new and/or proposed plants, or changes to existing plants, require a work environment that is "safer" than that at home.

Accidents at home take the lives of more than 20,000 Americans each year. These occurrences are the number one cause of the death of young children; two-thirds of these accidents involve boys. Accidents claim the lives of more children aged 1–14 than do the leading diseases combined [2].

Often after an accident has occurred, one sadly realizes that if simple safety practices had been followed in a timely manner, the accident could have been prevented. Yet each year, more accidents and injuries take place in the home than anywhere else. Injuries and deaths from fires, burns, and falls lead the list of home accidents. Many accidents are automobile-related. In competition for this infamous list is the gun-related accident category, which now has the potential to surpass automobile-related fatalities. There are also an estimated 5 million plus home fires in the United States each year. Building fires claim over 5000 lives a year, and most of these victims die in their own homes [2].

32.3 EXPOSURE TO TOXIC SUBSTANCES

It has only been in the latter half of this century that there has been a recognition of the threats posed by toxic chemicals. In fact, until 1962, the ever-increasing production and use of chemical substances went unchallenged. The general public made the naive assumption that these chemicals were safe based on the facts that "billions of dollars were being expended in manufacturing facilities and that governments did not disapprove of chemical use." Today, the public, now far more mistrusting, demands fact-based information on what occurs inside the industrial plants and what substances are being released into the atmosphere. This attitude is reflected in such legislation as the federal right-to-know standard.

Recently, there has been growing concern over the toxicity of new chemicals. This concern has been escalated due to a number of well-publicized incidents in which toxic substances caused injury to or loss of human life. At the incidence in Bhopal, India, a tremendous gas leak of methyl isocyanate (MIC) from a Union Carbide factory sent a toxic cloud into the atmosphere and the city surrounding the factory. The effect was devastating—thousands of people were sent fleeing blindly through the streets, choking and vomiting. In the end, it is estimated that more than 2500 people lost their lives on that December night in 1984, and many more, possibly 200,000, have been physically scarred from the release of the toxin.

Toxic and chemically active substances, including radioactive materials and biological agents, present a special concern because they can be readily inhaled or ingested, or can be absorbed through the skin. This fact makes any toxic gas leak hard to contain, because once the toxin becomes airborne, it can spread rapidly and affect an untold number of people, as well as present adverse effects to the atmosphere and water supply. Liquids or sludges present problems, not only because they can be splashed on the skin and cause damage, but because they can leak into the soil and affect vegetation and the water supply. The severity of danger of these substances can vary significantly. If ingested or inhaled, a substance may cause no apparent illness, or the results can be fatal. When spilled on the skin, one chemical may incur a slight rash or no reaction at all, or, as in the case of liquid mercury, may be absorbed through it, leading to systemic toxic effects.

32.4 ADVANCES IN SAFETY FEATURES [1]

Today's sophisticated equipment and technologies require equally sophisticated means of accident prevention. Unfortunately, the existing methods of detection and prevention are often assumed to be adequate until proven otherwise. This approach to determining a technology's effectiveness sometimes is costly and often leads to loss of life. Chemical manufacturers and power plants are businesses, and thus are not as likely to "unnecessarily" update their present controls.

Before the advent of technology, there was still a need for safety features and warnings; yet these did not exist. Many accidents occurred because of a lack of knowledge of the system, process, or substance being dealt with. Many of the pioneers of modern science were sent to an early grave by their experiments. Karl Wilhelm Scheele, the Swedish chemist who discovered many chemical elements and compounds, often sniffed or tasted his finds. He died of mercury poisoning. As noted earlier, Marie Curie died of leukemia contracted from overexposure to radioactive elements. Had either of these brilliant scientists an accurate idea of the properties of their materials, their methods certainly would have been significantly different. In those days, safety precautions often were devised by trial and error; if inhaling a certain gas was found to make someone sick, the prescribed precaution was not to smell it. Today, since the physical properties of most known compounds are readily found in handbooks, proper care can be exercised when working with these chemicals.

Laboratories are equipped with exhaust hoods and fans to minimize a buildup of gases; in addition, safety glasses and eye-wash stations are required, and gloves and smocks must be worn.

Many natural disasters are now accurately predicted, buying precious time in which warnings can be made and possible evacuation plans implemented. Radar equipment commonly track storms, and seismographs detect slight rumblings in the earth, which can provide early warning of potential earthquakes. Volcanic eruptions can be predicted by using seismic event counters and aerial scanning of anomalies detected in the infrared region. Where natural disasters often occurred unexpectedly in the past, similar occurrences today are more predictable. Thus there is more time for preparation, and less likelihood of loss of life.

The use of computers and modern instrumentation has greatly enhanced plant safety. System overloads, uncontrollable reactions, and unusual changes in temperature or pressure can be detected, with the information being relayed to a computer. The computer can then shutdown the system or take the steps necessary to minimize the danger. Industry has come a long way from sniffing and tasting its way to safety.

The subject of accident prevention and plant safety is addressed in more detail in the literature [1,4].

32.5   FUTURE TRENDS

For the most part, future trends will be found in hazard accident prevention in both the home and at work. To help promote hazard accident prevention, companies should start employee training programs. These programs should be designed to alert the technical staff and employees about the hazards they are exposed to on the job. Training should also cover company safety policies and the proper procedures to follow in case an accident does occur. A major avenue to reducing risk will involve source reduction of hazardous materials. Awareness issues for the home will also continue to increase.

32.6   SUMMARY

1. Accidents often resulted in the tightening of safety controls.
2. Advances in technology have brought about new problems. Nuclear power plant accidents (Three Mile Island and Chernobyl) have been the most frightening, perhaps because no one really knows what to expect from them.
3. Not all accidents involve the loss of human life. An oil spill at Ashland left more than a million people with limited water supplies. In addition, large numbers of fish and waterfowl were killed.
4. There have been numerous "less publicized" accidents, many under unusual or unlikely circumstances. It is important to remember that an accident can occur at any time.

5. Along with the rise of technology, industry has improved its accident prevention measures. Unfortunately, many improvements are not made until after an accident has occurred.

REFERENCES

1. Theodore, L., Reynolds, J., and Taylor, F. *Accident and Emergency Management*, John Wiley & Sons, Hobohen, NJ, 1989.
2. Burke, G., Singh, R., and Theodore, L. *Handbook of Environmental Management and Technology*, 2nd edition, John Wiley & Sons, Hobohen, NJ, 2000.
3. *79 Tips to Make Your House Safer*, Long Island Lighting Company, Nassau County, New York, undated.
4. Flynn, A.M. and Theodore, L. *Accident and Emergency Management in the Chemical Process Industries*, CRC Press, Boca Raton, FL, 2004.

33 Health, Safety, and Accident Management Applications

CONTENTS

33.1 INTRODUCTION

This chapter deals not only with the dangers posed by hazardous substances but also examines the general subject of health, safety, and accident prevention. In addition, the laws and legislation passed to protect workers, the public, and the environment from the effects of these chemicals are also reviewed. The chapter discusses the regulations with particular emphasis on emergency planning and the training of personnel.

33.2 EXPOSURE CONCERNS

Two general types of potential health, safety and accident exposures exist. These are classified as

1. *Chronic*: Continuous exposure occurs over longer periods of time, generally several months to years. Concentrations of inhaled contaminants are usually relatively low, direct skin contact by immersion, by splash, or by contaminated air involves contact with substances exhibiting low dermal activity. (See Chapter 38 for more details.)

2. *Acute*: Exposures occur for relatively short periods of time, generally minutes to 1–2 days. Concentration of air contaminants is usually high relative to their protection criteria. In addition to inhalation, airborne substances might directly contact the skin, or liquids and sludges may be splashed on the skin or into the eyes, leading to toxic effects. (See Chapter 39 for more details.)

In general, acute exposures to chemicals in air are more typical in either transportation accidents and fires, or releases at chemical manufacturing or storage facilities. High concentrations of contaminants in air usually do not persist for long periods of time. Acute skin exposure may occur when workers come in close contact with the substances in order to control a release—for example, while patching a tank car, offloading a corrosive material, uprighting a drum, or while containing and treating a spilled material.

Chronic exposures, on the other hand, are usually associated with longer-term removal and remedial operations. Contaminated soil and debris from emergency operations may be involved, soil and groundwater may be polluted, or temporary impoundment systems may contain diluted chemicals. Abandoned waste sites typically represent chronic exposure problems. As activities start at these sites, personnel engaged in certain operations such as sampling, handling containers, or bulking compatible liquids, face an increased risk of acute exposures. These exposures stem from splashes of liquids, or from the release of vapors, gases, or particulates that might be generated.

In any specific incident, the hazardous properties of the materials may only represent a potential risk. For example, if a tank car containing liquefied natural gas is involved in an accident but remains intact, the risk from fire and explosion is low. In other incidents, the risks to response personnel are high, e.g., when toxic or flammable vapors are released from a ruptured tank truck. The continued health and safety of response personnel requires that the risks, both real and potential, at an accident be assessed, and appropriate measures instituted to reduce or eliminate the threat to response personnel.

Specific chemicals and chemical groups affect different parts of the body. One chemical, such as an acid or base, may affect the skin, whereas another, such as carbon tetrachloride, attacks the liver. Some chemicals will affect more than one organ or system. When this occurs, the organ or system being attacked is referred to as the "target organ." The damage done to a target organ can differ in severity depending on chemical composition, length of exposure, and the concentration of the chemical.

When two different chemicals simultaneously enter the body, the result can be intensified or compounded. The "synergistic effect," as it is referred to, results when one substance intensifies the damage done by the other. Synergism complicates almost any exposure due to a lack of toxological information. For just one chemical, it may typically take a toxological research facility approximately 2 years of studies to generate valid data. The data produced in that 2 year timeframe applies only to the effect of that one chemical acting alone. With the addition of another chemical, the original chemical may have a totally different effect on the body. This fact results in a great many unknowns when dealing with toxic substances, and therefore increases risk due to lack of dependable information.

The National Institute of Occupational Safety and Health (NIOSH) recommends standards for industrial exposure that the Occupational Safety Hazard Administration (OSHA) uses in its regulations. The NIOSH pocket guide to chemical hazards contains a wealth of information on specific chemicals such as:

1. Chemical name, formula, and structure
2. Trade names and synonyms
3. Chemical and physical properties
4. Time-weighted average threshold limit values (TLVs)
5. Exposure limits
6. Lower explosive limit (LEL)
7. "Immediately dangerous to life and health" (IDLH) concentrations
8. Measurement methods
9. Personal protection and sanitation guidelines
10. Health hazards information

As discussed above, there are many different dangers resulting from the toxicity of reactive substances. It therefore becomes evident that a knowledge of the specific hazards is important, as are detailed regulations on how to handle situations arising from the use of these chemicals.

33.3 SAFETY AND ACCIDENTS

In the chemical industry, there is a high risk of accidents due to the nature of the processes and the materials used. Although precautions are taken to ensure that all processes run smoothly, there is always (unfortunately) room for error, and accidents will occur. This is especially true for highly technical and complicated operations, as well as processes under extreme conditions such as high temperatures and pressures. In general, accidents occur due to one or more of the following primary causes:

1. Equipment breakdown
2. Human error
3. Fire exposure and explosions
4. Control system failure
5. Natural causes
6. Utilities and ancillary system outage
7. Faulty siting and plant layout

These causes are usually at the root of most industrial accidents. Although there is no way to guarantee that these problems will not arise, steps can be taken to minimize the number, as well as the severity, of incidents. In an effort to reduce occupational accidents, measures should be taken in the following areas [1]:

1. *Training*: All personnel should be properly trained in the use of equipment and made to understand the consequences of misuse. In addition, operators should be rehearsed in the procedures to follow should something go wrong.
2. *Design*: Equipment should only be used for the purposes for which it was designed. All equipment should be periodically checked for damage or errors inherent in the design.
3. *Human performance*: Personnel should be closely monitored to ensure that proper procedures are followed. Also, working conditions should be such that the performance of workers is improved, thereby simultaneously reducing the chance of accidents. Periodic medical examinations should be provided to assure that workers are in good health, and that the environment of the workplace is not causing undue physical stress. Finally, under certain conditions, it may be advisable to test for the use of alcohol or drugs—conditions that severely handicap judgment, and therefore make workers accident-prone.

Each day, an average of 9,000 U.S. workers sustain disabling injuries on the job, 16 workers die from an injury suffered at work, and 137 workers die from work-related diseases. The Liberty Mutual 2005 Workplace Safety Index estimated that employes spent $50.8 billion in 2003 on wage payments and medical care for workers hurt on the job.

33.4 EMERGENCY PLANNING AND RESPONSE

The extent of the need for emergency planning is significant, and continues to expand as new regulations on safety are introduced. Planning for emergency must begin at the very start, when the plant itself is still being planned. The new plant will have to pass all safety measures and OSHA standards. This is emphasized by Piero Armenante, author of *Contingency Planning for Industrial Emergencies* [2], "The first line of defense against industrial accidents begins at the design stage. It should be obvious that it is much easier to prevent an accident rather than to try to rectify the situation once an accident has occurred."

Successful emergency planning begins with a thorough understanding of the event or potential disaster being planned for. The impacts on public health and the environment must also be estimated. Some of the types of emergencies that should be included in the plan are

1. Natural disasters such as earthquakes, tornados, hurricanes, and floods
2. Explosions and fires
3. Hazardous chemical leaks
4. Power or utility failures
5. Radiation accidents
6. Transportation accidents

In order to estimate the impact on the public or the environment, the affected area or emergency zone must be studied in depth. A hazardous gas leak, fire, or explosion may cause a toxic cloud to spread over a great distance, as it did in Bhopal. An estimate of the minimum affected area, and thus the area to be evacuated, should be performed based on an atmospheric dispersion model. There are various models that can be used. While the more difficult models produce the most realistic results, simpler models are faster to use and usually still provide adequate data and information for planning purposes.

The main objective for any plan should be to prepare a procedure to make maximum use of the combined resources of the community in order to accomplish the following:

1. Safeguard people during emergencies
2. Minimize damage to property and the environment
3. Initially contain and ultimately bring the incident under control
4. Effect the rescue and treatment of casualties
5. Provide authoritative information to the news media who will communicate the facts to the public
6. Secure the safe rehabilitation of the affected area

33.5 REGULATIONS

Each company must develop a health and safety program for its workers. OSHA has regulations governing employee health and safety at hazardous waste operations and during emergency responses to hazardous substance releases. These regulations (29 CFR 1910.120) contain general requirements for:

1. Safety and health programs
2. Training and informational programs
3. Work practices along with personal protective equipment
4. Site characterization and analysis
5. Site control and evacuation
6. Engineering controls
7. Exposure monitoring and medical surveillance
8. Material handling and decontamination
9. Emergency procedures
10. Illumination
11. Sanitation

The EPA's Standard Operating Safety Guides supplement these regulations. However, for specific legal requirements for industry, OSHA's regulations must be used. Other OSHA regulations pertain to employees working with hazardous materials or working at hazardous waste sites. These, as well as state and local regulations, must also be considered when developing worker health and safety programs [3].

The OSHA Hazard Communication Standard was first promulgated on November 25, 1983, and can be found in 29 CFR Part 1910.120. The standard was developed to inform workers who are exposed to hazardous chemicals of the risk associated with those specific chemicals. The purpose of the standard is to ensure that the hazards of all chemicals produced or imported are evaluated, and information concerning chemical hazards is conveyed to employers and employees.

Information on chemical hazards must be dispatched from manufacturers to employers via material safety data sheets (MSDSs) and container labels. This data must then be communicated to employees by means of comprehensive hazard communication programs, which include training programs, as well as the MSDSs and container labels.

The basic requirements of the Hazardous Communication Standard are as follows:

1. There must be an MSDS on file for every hazardous chemical present or used in the workplace.
2. MSDSs must be readily available during each work shift, and all employees must be informed how to obtain the information. If employees "travel" on the shift, the MSDSs may be kept in a central location at the primary job site, as long as the necessary information is immediately available in the event of an emergency.
3. It must be ensured that every container holding hazardous chemicals in the workplace is clearly and properly labeled and includes appropriate hazard warnings.
4. Labels of incoming hazardous chemical containers must not be removed or defaced in any way.
5. Prior to initial assignments, employees must be informed of the requirements of the standard operations in their work area where hazardous chemicals are present.
6. Employers must train employees how to identify and protect themselves from chemical hazards in the work area, as well as how to obtain the employer's written hazard communication program and hazard information.
7. Employers must develop, implement, and maintain a written communication program for each workplace that describes how MSDSs, labeling, and employee information and training requirements will be met. This written program must also include a list of hazardous chemicals present in the workplace and the methods that will be used to inform employees of the hazards associated performing nonroutine tasks.

Companies with multiemployer workplaces must include with the MSDS methods the employer will use for contractors at the facility. These employers must also describe how they will inform the subcontractors' employees about the precautions which must be followed and the specific labeling system used in the workplace.

The Superfund Amendments and Reauthorization Act (SARA) of 1986 renewed the national commitment to correcting problems arising from previous mismanagement of hazardous wastes. Title III of SARA, specifically known as the Emergency Planning and Community Right-to-Know Act, forever changed the concept of environmental management. Planning for emergencies became law. While SARA was similar in many respects to the original law, it also contained new approaches to the program's operation. The 1986 Superfund legislation accomplished the following:

1. Reauthorization of the original program for five more years, dramatically increasing the cleanup fund from $1.6 to $8.5 billion.
2. Setting of specific goals and standards, stressing permanent solutions.
3. Expansion of state and local involvement in decision-making policies.
4. Provision for new enforcement authorities and responsibilities.
5. Strengthened the focus on human health issues caused by hazardous waste sites.

This law was more specific than the original statute with regard to such things as remedies to be used at Superfund sites, public participation, and accomplishment of cleanup activities.

The Emergency Planning and Community Right-to-Know Act is undeniably the most important part of SARA when it comes to public acceptance and support. Title III addresses the most important issues regarding community awareness and participation in the event of a chemical release. Title III establishes requirements for emergency planning, hazardous emissions reporting, emergency notification, and "community right-to-know." For instance, it is now law that companies release any data that a local or community planning committee needs in order to develop and implement its emergency plan [4]. The objectives of Title III are to improve local chemical emergency response capabilities, primarily through improved emergency planning and notification, and to provide citizens and local governments with access to information about chemicals in their area. Title III has four major sections that aid in the development of contingency plans. They are as follows:

1. Emergency Planning (Sections 301–03)
2. Emergency Notification (Section 304)
3. Community Right-to-Know Reporting Requirements (Sections 311–12)
4. Toxic Chemicals Release Reporting—Emissions Inventory (Section 313)

Title III has also developed timeframes for the implementation of the Emergency Planning and Community Right-to-Know Act of 1986. Although the material discussed above was first published in 1986, much of this was still applicable at the time of the preparation of this chapter in 2008.

33.6 TRAINING

Safety and health training must be an integral part of a total health, safety, and accident prevention program. Safety training must be frequent and up-to-date for response personnel to maintain their proficiency in the use of equipment and their knowledge of safety requirements. Personnel must also be familiar with the substances they are dealing with in order to respond appropriately. The consequences of improper response can be devastating. For example, in Canning, Nova Scotia in 1986, a fire broke out in a warehouse. The firefighters responding to the blaze assumed it was a "normal" fire, and went in with fire hoses to spray it down. However, the building housed pesticides and other agricultural products. The chemicals mixed with the tons of water and escaped through run-off streams into the streets and finally into a nearby river. All vegetation and animals in the path of the deadly streams were killed. Although, fortunately, no human lives were lost, the community had to be evacuated, and could not safely return for weeks. If the firefighters had known about the chemicals, they would have been better off to let the warehouse burn, and merely contained the blaze to keep it from spreading. This would have prevented the escape of the toxins to a greater extent. Fighting the blaze in the normal manner only compounded the problem [5].

All personnel involved in responding to environmental incidents, and who could be exposed to hazardous substances, health hazards, or safety hazards, must receive safety training prior to carrying out their response functions. Health and safety training must, as a minimum, include:

1. Use of personal protective equipment (i.e., respiratory protective apparatus and protective clothing)
2. Safe work practices, engineering controls, and standard operating safety procedures
3. Hazard recognition and evaluation
4. Medical surveillance requirements, symptoms that might indicate medical problems, and first aid
5. Site safety plans and plan development
6. Site control and decontamination
7. Use of monitoring equipment, if applicable

Training must be as practical as possible and include hands-on use of equipment and exercises designed to demonstrate and practice classroom instruction. Formal training should be followed by at least 3 days of on-the-job experience working under the guidance of an experienced, trained supervisor. All employers should, as a minimum, complete an 8 hour safety refresher course annually. Health and safety training must comply with OSHA's training requirements as defined in 29 CFR 1910.120.

The personnel at an industrial plant, particularly the operators, are trained in the operation of the plant. These people are critical to proper emergency response. They must be taught to recognize abnormalities in operations and report them immediately. Plant operators should also be taught how to respond to various types of accidents. Emergency squads at plants can also be trained to contain an emergency until outside help arrives, or, if possible, to terminate the emergency. Shutdown and evacuation procedures are especially important when training plant personnel.

Training is important for the emergency teams to ensure that their roles are clearly understood, and that accidents can be reacted to safely and properly without delay. The emergency teams include police, firefighters, medical people, and volunteers who will be required to take action during an emergency. These people must be knowledgeable about the potential hazards. For example, specific antidotes for different types of medical problems must be known by medical personnel. The entire emergency team must also be taught the use of personal protective equipment.

33.7 FUTURE TRENDS

As evident in the lessons from past accidents, it is essential for industry to abide by stringent safety procedures. The more knowledgeable the personnel, from the management to the operators of a plant, and the more information that is available to them, the less likely a serious incident will occur. The new regulations, and especially Title III of 1986, help to insure that safety practices are up to standard. However, these regulations should only provide a minimum standard. It should be up to the companies, and specifically the plants, to see that every possible measure is taken to insure the safety and well-being of the community and the environment in the surrounding area. It is also up to the community itself, under Title III, to be aware of what goes on inside local industry, and to prepare for any problems that might arise.

The future promises to bring more attention to the topics discussed in the above paragraph. In addition, it appears that there will be more research in the area of risk assessment, including fault-tree, event-tree, and cause-consequence analysis. Details on these topics are beyond the scope of this book, but are available in the literature [1].

33.8 SUMMARY

1. Toxic and chemically active substances present special concern because they can be dangerous when inhaled, ingested, or absorbed through the skin.
2. Although accidents cannot be completely prevented, careful planning and stringent safety procedures can significantly lower the potential risk that an accident will occur.
3. Emergency planning is essential in preventing a potential disaster, and in foreseeing what possible incidents might occur.
4. The OSHA has guidelines and regulations for the safe operation of industrial plants and handling of emergencies.
5. Safety and health training for personnel is essential in preventing accidents. Workers must know what they are dealing with, and understand the consequences.
6. In the future, more stringent regulations and hopefully better safety techniques will help minimize industrial accidents.

REFERENCES

1. Flynn, A.M. and Theodore, L.F. *Accident and Emergency Management in the Chemical Process Industries*, CRC Press, Boca Raton, FL, 2004.
2. Armenante, P. *Contingency Planning for Industrial Emergencies*, Van Nostrand Reinhold, New York, 1991.
3. EPA Office of Emergency and Remedial Response Division. *Standard Operating Safety Guides*, July 1988.
4. Lees, F. *Safety Cases within the Control of Industrial Major Accidents Hazards Regulations 1984*, Butterworths and Co., Ltd., London, 1989.
5. Côté, R. *Controlling Chemical Hazards*, Unwin Hyman, London, 1991.

34 Introduction to Energy Conservation*

CONTENTS

34.1 INTRODUCTION

Energy is the keystone of American life and prosperity as well as a vital component of environmental rehabilitation. The environment must be protected and the quality of life improved but, at the same time, economic stability must also be maintained. These two objectives will be prime factors in determining domestic and foreign policies for years to come. Since energy consumption is a major contributor to environmental pollution, decisions regarding energy policy alternatives require comprehensive environmental analysis. Environmental impact data must be developed for all aspects of an energy system and/or energy conservation program and must not be limited to separate components.

Because energy has been relatively cheap and plentiful in the past, many energy-wasting practices have been allowed to develop and continue in all sectors of the economy. Industries have wasted energy by discharging hot process water instead of recovering the heat, and by wasting the energy discharged from power plant stacks. Waste hydrocarbons have been discharged or combusted with little consideration for recovering their energy value. There are many more examples, too numerous to mention. Elimination of these practices will, at least temporarily, partially reduce the rate of increase in energy demand. If conservation can reduce energy demand, it can reduce the associated pollution.

The most dramatic environmental improvements can be developed by energy conservation in the industrial sector of the economy. Industry accounts for approximately 40% of the energy consumed in this country. Also, industry might be

---

* See Dupont et al. [1] and Burke et al. [2].

considered more dynamic, progressive, and strongly motivated by the economic incentives offered by conservation than the other energy-user sectors (residential, commercial, and transportation).

The environmental impacts of energy conservation and consumption are far-reaching, affecting air, water, and land quality as well as public health. Combustion of coal, oil, and natural gas is responsible for air pollution in urban areas; acid rain that is damaging lakes and forests; and, some of the nitrogen pollution that is harming estuaries. Although data show that for the period from 1977 to 1989 annual average ambient levels of all criteria air pollutants were down nationwide, 96 major metropolitan areas still exceeded the national health-based standard for ozone, and 41 metropolitan areas exceeded the standard for carbon monoxide.

Energy consumption also appears to be the primary man-made contribution to global warming, often referred to as the greenhouse effect (see Chapter 12 for more details). The Environmental Protection Agency (EPA) has concluded that energy use—through the formation of carbon dioxide during combustion processes—has contributed approximately 50% to the global warming that has occurred in the last 15 years. Although the scientific community is not unanimous in regard to the causes of global warming, most individuals and groups have indicated that a "reasonable" chance of climatic change exists and have already begun to define the potential implications of such changes, many of which are catastrophic. In light of this situation, the Alliance to Save Energy has challenged Congress to pass meaningful legislation to promote and achieve energy efficiency.

34.2   CONSERVATION OF ENERGY [3–5]

The concept of energy developed slowly over a period of several hundred years and culminated in the establishment of the general principle of conservation of energy about 1850. The germ of this principle as it applies to mechanics was present in the work of Galileo (1564–1642) and Isaac Newton (1642–1726).

Joule's experiments cleared the way for the enunciation of the "first law of thermodynamics: when a closed system is taken through a cyclic process, the work done on the surroundings equals the heat absorbed from the surroundings." Mathematically this statement, in a very broad reuse, introduced the conservation law of energy.

A presentation of the conservation law for energy would be incomplete without a brief review of some introductory thermodynamic principles. Thermodynamics is defined as that science that deals with the relationships among the various forms of energy. A system may possess energy due to its

1. Temperature
2. Velocity
3. Position
4. Molecular structure
5. Surface and so on

The energies corresponding to these states are, for example,

1. Internal
2. Kinetic
3. Potential
4. Chemical
5. Surface

This law, in steady-state equation form for batch and flow processes, is presented here.
For batch processes:

$$\Delta E = Q - W \tag{34.1}$$

For flow processes:

$$\Delta H = Q - W_s \tag{34.2}$$

where
   potential, kinetic, and other energy effects have been neglected
   Q is the energy in the form of heat transferred across the boundaries of the system
   W is the energy in the form of work transferred across the boundaries of the system
   W_s is the energy in the form of mechanical work, transferred across the boundaries of the system
   E is the internal energy of the system
   H is the enthalpy of the system
   ΔE, ΔH is the changes in the internal energy and enthalpy, respectively, during the process

The internal energy and enthalpy in Equations 34.1 and 34.2, may be on a *mass* basis (i.e., for 1 gal or 1 lb of material), on a *mole* basis (i.e., for 1 gmol or 1 lbmol of material), or represent the total internal energy and enthalpy of the entire system. It makes no difference as long as these equations are dimensionally consistent.

Perhaps the most important thermodynamic function the engineer works with is the aforementioned enthalpy. The enthalpy is defined by

$$H = E + PV \tag{34.3}$$

where
   P is the pressure of the system
   V is the volume of the system

The terms E and H are *state* or *point* functions. By fixing a certain number of variables on which the function depends, this automatically fixes the numerical value of the function; i.e., it is single valued. For example, fixing the temperature and pressure of a one-component single-phase system immediately specifies the enthalpy and internal energy.

The change in enthalpy as it undergoes a change in state from (T_1, P_1) to (T_2, P_2) is given by

$$\Delta H = H_2 - H_1 \tag{34.4}$$

Note that H and ΔH are independent of the path. This is a characteristic of all state or point functions, i.e., the state of the system is independent of the path by which the state is reached. The terms Q, W, and W_s in Equations 34.1 and 34.2 are "path" functions; their values depend on the path used between the two states. Unless a process or change of state is occurring, path functions have no value.

34.3 ENERGY TERMS [3–5]

It is known from experience that a hot object brought in contact with a cold object becomes cooler, whereas the cold object becomes warmer. It is reasonable to adopt the view that something is transferred from the hot object to the cold one, and one calls that something heat, Q.

The heat added to a system that causes no change in temperature but a change in phase is termed "latent heat." For example, water at one atmosphere total pressure can be heated to 212°F; further addition of heat will cause boiling, e.g., a change of phase from liquid to gas. If both phases are kept in contact, a change in temperature will not result until all the liquid has evaporated. The heat added to evaporate a liquid is termed "latent heat of vaporization." When a gas condenses, the heat of vaporization is released. When a solid is heated at its melting point, the heat required to melt it is termed "latent heat of fusion."

When a chemical reaction takes place there is usually an evolution or absorption of heat because the absolute enthalpy of the products of a reaction is usually quite different from that of the reactants, even when both are at the same temperature. The amount of heat absorbed or evolved is called the "heat of reaction" and is equal to the enthalpy change of the reaction. In order to evaluate and understand these heat effects which take place during chemical reactions several of the previously derived energy relationships should be reviewed by the reader.

Work W is done whenever a force acts through a distance. The quantity of work done is defined by the equation:

$$dW = F \, dl \tag{34.5}$$

where F is the component of the force acting in the direction of the displacement dl. This equation must be integrated if the work for a finite process is required. In engineering thermodynamics an important type of work is that which accompanies a change in volume of a fluid.

"Power" is defined as the time rate of doing work, or

$$\text{Power}, P = \frac{\text{Work}}{\text{Time}} \tag{34.6}$$

The most common unit for power is horsepower (hp), defined as work being done at the rate of $550\,\text{ft}\cdot\text{lb f/s}$. Most continuously operating pieces of equipment such as electrical motors or internal combustion engines are rated in terms of horsepower and the "efficiency" of energy conversion of such units is defined as

$$\text{Efficiency} = \frac{\text{Power output}}{\text{Power input}} \tag{34.7}$$

For most engineering work the following approximate conversion factors are adequate:

$$1(\text{Btu}) = 1055(\text{J}) = 252(\text{cal}) = 778(\text{ft}\cdot\text{lb}_\text{f})$$

Another useful conversion factor is given to a close approximation by

$$1(\text{cal})/(\text{g}) = 1.8 \text{ Btu/lb}$$

An extensive table of conversion factors is available in the literature [3,4].

34.4 CONSERVATION OF ENERGY AT HOME [6]

Some key suggestions regarding cooking include:

1. A microwave oven is an energy efficient alternative to a conventional oven. It cooks food more quickly and it uses 70%–80% less electricity than a regular oven.
2. When cooking on top of the range, use pots and pans that are properly sized to fit the burners. A small pot on a large burner wastes energy.
3. When using a conventional oven, avoid "peeking" by opening the oven door. Each "peek" can lower the oven temperature by 25°.

Some key suggestions regarding lighting include:

1. If one prefers incandescent bulbs, try to use "energy saver" bulbs. These bulbs use halogen gases that allow the filament to burn brighter while consuming less electricity.
2. Lighting controls or "timers" can help save energy dollars.
3. Consider using task lighting (lighting directed at a specific area) instead of overhead or general lighting, which may light unused areas of the room.

Some key suggestions regarding new appliances include:

1. When shopping for a new appliance, check for the yellow Energy Guide label that indicates the unit's energy efficiency.
2. For refrigerators and other appliances, the Energy Guide label provides the estimated yearly energy cost for operating the appliance based on an average national utility rate.
3. With any appliance, it is helpful to compare units in the same size range when trying to determine which model has the lowest annual operating cost.

34.5 FUTURE TRENDS

As will be noted in Chapter 35, nuclear fusion, solar energy, wind energy, biomass, energy produced as a result of thermal gradients in the earth or the oceans, tidal energy, and advanced chemical energy systems show promise as potential power sources in the future with minimum environmental damage. Energy conservation program will reduce the environmental damage from the various energy systems. Finally, energy systems in industry must be evaluated in light of their impact on the total environment in all aspects of their respective production methods and uses.

34.6 SUMMARY

1. Energy is the keystone of American life and prosperity as well as a vital component of environmental rehabilitation.
2. Thermodynamics is defined as that science that deals with the relationships among the various forms of energy.
3. It is known from experience that a hot object brought in contact with a cold object becomes cooler, whereas the cold object becomes warmer. It is reasonable to adopt the view that something is transferred from the hot object to the cold one, and one calls that something heat, Q.
4. When a chemical reaction takes place there is usually an evolution or absorption of heat because the absolute enthalpy of the products of a reaction is usually quite different from that of the reactants, even when both are at the same temperature.
5. Work W is done whenever a force acts through a distance.
6. Power is defined as the time rate of doing work.
7. Nuclear fusion, solar energy, wind energy, biomass, energy produced as a result of thermal gradients in the earth or the oceans, tidal energy, and advanced chemical energy systems show promise as potential power sources in the future with minimum environmental damage.

REFERENCES

1. Adapted from: Dupont, R., Ganesan, K., and Theodore, L. *Pollution Prevention*, CRC Press (Taylor & Francis Group), Boca Raton, FL, 1996.
2. Adapted from: Burke, G., Singh, B., and Theodore, L. *Handbook of Environmental Management and Technology*, 2nd edition, John Wiley & Sons, Hoboken, NJ, 2000.

3. Reynolds, J., Jeris, J., and Theodore, L. *Handbook of Chemical and Environmental Engineering Calculations*, John Wiley & Sons, Hoboken, NJ, 2004.
4. Santoleri, J., Reynolds, J., and Theodore, L. *Introduction to Hazardous Waste Incineration*, 2nd edition, John Wiley & Sons, Hoboken, NJ, 2000.
5. Adopted from: Theodore, L., Ricci, F., and VanVliet, T. *Thermodynamics for the Practicing Engineer*, John Wiley & Sons, Hoboken, NJ, 2009.
6. Theodore, M.K. *Environmental Calendar*, Theodore Tutorials, East Williston, NY, 2000 (one annually thereafter).

35 Energy Conservation Applications

CONTENTS

35.1 INTRODUCTION

At present, there are two major industrial energy sources in use: fossil fuel and nuclear energy. Fossil fuel may be subdivided according to the various raw fuels such as coal, oil, natural gas, liquefied petroleum gas, wood, smoke, coke, refining gas, blast furnace gas, and byproduct fuels. This section covers only the three major fossil fuels: coal, oil, and natural gas.

The fossil fuel energy picture has changed drastically since the 1973 oil crisis. The current so-called national energy policy aims toward energy independence through conservation and promotes increased coal utilization, due to increasingly severe shortages of domestic oil and natural gas. However, an increase of coal consumption would produce more pollutants that are carried through the environment affecting individual organisms and ecosystems. Therefore, emphasis is placed on the impacts of fossil fuel cycles on the physical environment.

35.2 ENVIRONMENTAL IMPLICATIONS OF ALTERNATE ENERGY SOURCES

There has been considerable controversy in recent years concerning the roles that various forms of energy production should assume in the nation's energy supply. As discussed earlier the air pollution impacts of coal combustion, the fears of radiation from nuclear accidents and/or wasters, and the dependence on foreign oil have numerous questions about conventional supplies. A shift toward alternate energy sources (i.e., those that do not currently contribute significantly to the nation's energy supply) has long been suggested.

Many of the alternate sources have been termed "soft" or "benign," and their relative costs, risks, and other points have been debated in the literature. The risk to human health from nonconventional sources can be as high as, or even higher than, that of conventional sources. One of the principal problems in comparing conventional and alternate energy systems, is that alternate sources will not be independent of conventional systems, at least in the short term. Analyses must therefore proceed by comparing various mixes of source types, rather than direct comparisons. This limitation holds for environmental comparisons as well.

While the balance may eventually favor alternate sources, the very real environmental impacts of these systems should not be ignored. Some of the alternate energy sources are listed below.

1. Oil shale
2. Geothermal systems
3. Solar energy
4. Hydrogen
5. Wind power
6. Tar sands
7. Ocean thermal energy

Extensive details regarding environmental impacts are available in the literature [1].

The American people appear to be overwhelmingly in favor of alternate energy sources. They favor cooperation between industry and the government to develop alternate fuel sources. However, public support for the environment is also high. In any event, the overwhelmingly popular support for alternate sources should not be allowed to mask the public's support for environmental protection; the environmental impacts outlined earlier must still be taken into account.

35.3 GENERAL CONSERVATION PRACTICES IN INDUSTRY

There are numerous general energy conservation practices that can be instituted at plants. Ten of the simpler ones are detailed below.

1. Lubricate fans
2. Lubricate pumps
3. Lubricate compressors
4. Repair steam and compressed air leaks
5. Insulate bare steam lines
6. Inspect and repair steam traps
7. Increase condensate return
8. Minimize boiler blowdown
9. Maintain and inspect temperature measuring devices
10. Maintain and inspect pressure measuring devices

Providing details on fans, pumps, compressors, and steam lines is beyond the scope of this book. Descriptive information [2,3] and calculational procedures [4,5] are available in the literature.

Some energy conservation practices applicable to specific chemical operations are also provided below.

1. Recover energy from hot gases
2. Recover energy from hot liquids
3. Reduce reflux ratios in distillation columns
4. Reuse hot wash water
5. Add effects to existing evaporators
6. Use liquefied gases as refrigerants
7. Recompress vapor for low-pressure steam
8. Generate low-pressure steam from flash operations
9. Use waste heat for absorption
10. Cover tanks of heated liquid to reduce heat loss

Providing details on distillation columns, evaporators, and refrigerators is beyond the scope of this book. Descriptive information [2,3] and calculational procedures [4,5] are available in the literature.

For the purposes of implementing an energy conservation strategy, process changes and/or design can be divided into four phases, each presenting different opportunities for implementing energy conservation measures. These include:

1. Product conception
2. Laboratory research
3. Process development (pilot plant)
4. Mechanical (physical) design

Energy conservation training measures that can be taken in the chemical process and other industries include:

1. Implementing a sound operation, maintenance, and inspection (OM&I) program
2. Implementing a pollution prevention program (see Chapter 31)
3. Instituting a formal training program for all employees

It should be obvious to the reader that a multimedia approach that includes energy conservation considerations requires a total systems approach (see Chapter 5). Much of the environmental engineering work in future years will focus on this area, since it appears to be the most cost-effective way of solving many energy problems. This is discussed in more detail in the last section.

Energy efficiency is a cornerstone of Environmental Protection Agency's (EPA's) pollution prevention strategy. If less electricity is used to deliver an energy service—such as lighting—the power plant that produces the electricity burns less fuel and thus generates less pollution.

Lighting accounts for 20%–25% of all electricity sold in the United States. Lighting for industry, stores, offices, and warehouses represents 80%–90% of total lighting electricity use, so the use of energy-efficient lighting has a direct effect on pollution prevention. Every kilowatt-hour of lighting electricity not used prevents emissions of approximately 1.5 lbs of carbon dioxide, 5.8 g of sulfur dioxide, and 2.5 g of nitrogen oxides. If energy-efficient lighting were used where profitable, the nation's demand for electricity would be cut by more than 10%. This would result in annual reductions of 200 million metric tons of carbon dioxide—the equivalent of taking 44 million cars off the road; 1.3×10^6 metric tons of sulfur dioxide; and 600,000 metric tons of nitrogen oxides. These reductions represent 12% of U.S. utility emissions. These goals may not be fully achievable, but EPA's Green Lights program (detailed below) seeks to capture as much of the efficiency "bonus" as possible.

Lighting is not typically a high priority for the vast majority of U.S. institutions. Often the responsibility of facility management, lighting is viewed as an overhead item. Because of this, most facilities are equipped with the lowest first-cost (rather than the lowest lifecycle-cost) lighting systems, and profitable opportunities to upgrade the systems are ignored or passed over in favor of higher-visibility projects. As a result, institutions pay needless overhead every year, reducing their own competitiveness and that of the country. And, wasteful electricity use becomes a particularly senseless source of pollution.

By signing the Green Lights Memorandum of Understanding (part of program), senior management makes it clear that energy-efficient lighting is now one of the organization's highest priorities. Authority is granted, budgets are approved, procedures are streamlined, and staff is assigned to make the upgrades happen. The commitment to maximize energy savings by upgrading an organization's facilities often requires a change in the way an organization does business. Management will have to take a fresh look at how the organization maintains and upgrades its facilities, ensures environmental responsibility, and plans for maximum workforce production. For some organizations, this change will require significant planning and coordination among several different sectors of the organization. In addition, partners and allies agree to provide annual documentation of the lighting upgrades they complete. To simplify this process, EPA asks them to submit a one-page form for each facility—the Green Lights Implementation Report—to report their progress.

The Green Lights approach to lighting upgrades defines as "profitable" those projects that—in combination and on a facility aggregate basis—maximize energy savings while providing an annualized internal rate of return (IRR) that is at least equivalent to the prime interest rate plus 6% points. Projects that maximize energy savings while providing internal rates of return higher than the prime interest rate plus 6% points meet the Green Lights profitability criterion. The typical Green Lights upgrade yields a posttax IRR of 20%–40%.

35.4 FUTURE TRENDS

Nuclear fusion, solar energy, wind energy, biomass, energy produced as a result of thermal gradients in the earth or the oceans, tidal energy, and advanced chemical energy systems show promise as potential power sources in the future with minimum environmental damage. Controlled thermonuclear fusion is receiving increasing research and development funds. It would make use of light-element fuels that are sufficiently abundant to supply power needs almost indefinitely. Solar energy is also receiving increasing attention as a virtually pollution-free and inexhaustible source of energy. The renewed interest in tidal power systems stems from their environmental advantages. They produce no harmful wastes, cause minor scenic and ecological disturbances, and are inexhaustible. Although not economically feasible at this time, tidal energy is a future source that could reduce the environmental consequences of power generation.

Energy conservation will reduce the environmental damage from the various energy systems. Conservation will also enhance the reliability of future energy supplies. By slowing the rate of growth of energy demand, the longevity of energy supplies may be extended, allowing more flexibility in developing systems for meeting long-term needs. For too long a time, energy has been considered a limitless commodity. Energy was continuously wasted because it was abundant and cheap. This situation is now reversed. No longer will industry refuse and other solid wastes that are potential sources of energy be discarded. Instead, they will be used to supplement fuel supplies. No longer will reusable items be discarded. Recycling, which inherently will extend the lifetime of many natural resources will be found profitable in many instances and compatible with environmental goals.

There is also tremendous potential for conservation in the energy production and consumption stages. On the average, only 30% of the oil in a reservoir is being extracted from onshore wells; offshore extraction is somewhat more efficient. As the price of crude oil rises, more extensive use of secondary recovery techniques, such as water flooding and thermal stimulation, will become evident. In the deep mining of coal, less than 60% of the resource in place is recovered, and over 10% of the energy in coal can be lost in cleaning. The pillar method of mining coal limits primary coal recovery to 30%–60%. Secondary coal recovery techniques, such as the "robbing the pillars" method, will become economical and increase the amount of recoverable coal from a mine.

For electric power systems, a major source of inefficiency is the power plant itself. Thermionic or magnetohydrodynamic topping of electric power plants and the use of combined cycles show promise in increasing power plant efficiency. This will significantly reduce the thermal discharge to the environment and conserve fuel resources, and the constructive use of the waste heat will benefit the environment. Waste heat from power plants is a rich source of energy for plant growth. Already there have been very successful applications of warm water irrigation to increase yields. There is a great deal to be learned about aquaculture, but it appears that clams, shrimp, and scallops are adaptable to this procedure.

Energy systems in industry must be evaluated in light of their impact on the total environment in all aspects of their respective production methods and uses. As more and more air and water pollution control devices are being employed, air and water emissions have been reduced considerably, but increasing amounts of solid waste have been generated. More land is needed for the disposal of this waste and this will reduce the net effects of reclamation of mined-out areas. Although the damages from air and water pollution are much less severe with controls, the need to avoid unintentionally shifting environmental problems from one medium or location to another must be recognized (see Chapter 5).

35.5 SUMMARY

1. Environmental consumption is a major contributor to environmental pollution; thus, decisions regarding energy policy alternatives require comprehensive environmental analysis. Environmental impact data must be developed for all aspects of an energy system and/or conservation program and must not be limited to their separate components.

2. At present, there are two major industrial energy sources in use: fossil fuel and nuclear energy. Fossil fuels may be subdivided according to the various raw fuels such as coal, oil, natural gas, liquefied petroleum gas, wood, coke refining gas, blast furnace gas, and byproduct fuels.

3. The American people appear to be overwhelmingly in favor of alternate energy sources. They favor cooperation between industry and the government to develop alternate fuel sources. However, public support for the environment is also high.

4. For the purposes of implementing an energy conservation strategy, process changes and/or design can be divided into four phases, each presenting different opportunities for implementing energy conservation measures. These include product conception, laboratory research, process development, and mechanical design.

5. Nuclear fusion, solar energy, wind energy, biomass, energy produced as a result of thermal gradients in the earth or the oceans, tidal energy, and advanced chemical energy systems show promise as potential power sources in the future with minimum environmental damage.

REFERENCES

1. Raufer, R. and Yates, J. Alternate energy sources: An environmental perspective, *Proceedings of the Fifth Conference on Energy and the Environment*, Cincinnati, OH, 1979.

2. Reynolds, J., Santoleri, J., and Theodore, L. *Introduction to Hazardous Waste Incineration*, 2nd edition, John Wiley & Sons, Hoboken, NJ, 2002.

3. Reynolds, J., Jeres, J., and Theodore, L. *Handbook of Chemical and Environmental Engineering Calculations*, John Wiley & Sons, Hoboken, NJ, 2004.

4. Kauffman, D. *Process and Plant Design, an ETS Theodore Tutorial*, ETS International Inc., Roanoke, VA, 1992.

5. Abulencia, P. and Theodore, L. *Fluid Flow for the Practicing Engineer*, John Wiley & Sons, Hoboken, NJ, 2009.

36 Architecture in the Environment: History, Practice, and Change

CONTENTS

36.1   INTRODUCTION

As environmental concerns present some of the most pressing issues to the world, both professional and academic architects have begun to address how planning and built form affect the environment. Although the term *build environment* has come to mean different things to different people, one may state in general terms that it is the result of human activities that impact the environment. It essentially includes everything that is constructed or built, i.e., all types of buildings, chemical plants, roads, railways, parks, farms, gardens, bridges, etc. Thus, the built environment includes everything that can be described as a structure or "green" space. Generally the built environment is organized into six interrelated components:

1. Products
2. Interiors
3. Structures
4. Landscapes
5. Cities
6. Regions

While "architecture" may appear to be one of the many contributors to the current environmental state, in reality, the energy consumption and pollution affiliated with the materials, the construction, and the use of buildings contributes to most major environmental crises. In fact, architectural planning, design, and building significantly contribute to the destruction of the rain forest, the extinction of plant and animal species, the depletion of nonrenewable energy sources, the reduction of the ozone layer, the proliferation of chlorofluorocarbons (CFCs), and exposure to carcinogens and other hazardous materials. Where one chooses to build, which construction materials are selected, how a comfortable temperature is maintained, or what type of transportation is needed to reach it—each issue, decided by both architect and user, significantly impacts the overall environmental condition. Sadly, despite these opportunities to shape a healthier future, an analysis of American planning and building describes an assault on the existing ecological conditions.

Most architects have committed to build green. New buildings will incorporate a range of green elements including: radiant ceiling panels that heat and cool, saving energy and improving occupant comfort; a cogeneration plant that utilizes, waste heat; a green roof that is irrigated exclusively with rainwater and mitigates the heat island effect; materials that are rapidly renewable and regionally manufactured, etc. Additionally, buildings are being designed to maximize day-lighting and air circulation. For example, a bird nest's design has been employed that is efficient, withstanding wind loads and wind shear while simultaneously enabling light and air to move through it. Throughout the building process, construction and demolition waste is recycled. Measurement and verification plans are also being employed to track utility usage for sustainability purposes.

Urban planners are employing designs that operate like a wall of morning glories—adjusting to sunlight throughout the day, both regulating light and gathering solar energy. In effect, the design can create an energy surplus that can be employed elsewhere.

36.2 HISTORICAL CONCERNS

A schematic review of the United States' architectural expansion reveals a strict adherence to the grid. While facilitating the organization of a new country, the gridding of land parcels and urban plans made few allowances for existing conditions. In fact, "slapped down anywhere," the grid imposed a man-made order on nature. From the New England town, to the first cities of Philadelphia, New York, and Washington DC, the grid etched an order atop the country with little acknowledgment of or regard for the natural landscape. Instead, in the case of the earliest urban examples, the grid contained nature in the form of the village or town "green." Unfortunately, the green did not retain a reserve of the natural landscape. Rather, as nature controlled, it set the precedent for the simulation and subjugation of nature. Today, many housing developments raze forests only to turf and replant the area with something else. The simulation of landscapes, rather than reserving or using the existing landscapes, increases net land usage, energy consumption, and pollution. The retention of untouched and undeveloped land protects more than just trees. Each area—forest, wetland, prairie, coastal plain—sustains a complete ecosystem of plant and animal life. In examining the clearing of a forest, not only are the trees lost, but also the birds that used to live in and off of them, the plants that needed those trees' shade to survive, the animals that ate those understory plants, and so on. These losses diagram the chain reaction of ecological destruction caused by land development. With this in mind, the reports of multiple species eradication loom that much larger.

The earlier examples of architecture, in both autochthonous and colonial cultures, exhibit tremendous adaptation to both site and climate. But as buildings evolved from dwellings necessary for survival to conveyors of status and wealth, architectural planning and forms increasingly ignored the existing environment. A study of contemporary architecture, particularly housing developments, shows the mass production of styles transplanted anywhere. These styles originally became categorized because they evolved from an architectural response to climatic conditions. The stick style's steep roofs and projected eaves respond to climatic conditions while its diagonal "stick work" suggestively diagrams the structural frame. But when these architectural elements appear on the surface of an airtight, concrete box in a development in Dallas, they cease to have any real function. In order to convey sociocultural meaning, the architect/developer and homeowner lose the opportunity to have a building that responds to and respects the natural environment.

The quintessential American architecture—the suburban house complemented by a lawn, paved driveway, and two-car garage—evolved from a long history of antiurban development celebrating a frontier sense of independence and isolation. Unfortunately, this evolution of American housing, combined with the mass production and purchasing of the car, led to the present day condition of major suburban and extraurban growth. Necessitating car use for practically every activity outside the home, the suburban house's auto-reliance causes massive fossil fuel consumption, road building, and parking paving. The extensive development of the American suburb has spawned other enclaved architectural forms: the mall, the retail park, the industrial park, the business park, and the leisure complex. All create a greater dependence on the car and disturb more land. The ecological repercussions are enormous. Considering net hours spent in today's home—families are smaller, more households have both partners working, more people live alone—the increase in square area of living space per person exemplifies society's tendency toward excessive expenditures of money, energy, and other resources. These wastes extend to the land. Each house typically occupies a cleared lot of land, destroying an enormous portion of existing ecological environments. Because an individual normally does not need or use that much land, current efforts encourage a reduction of that private land while increasing community land in the form of public green spaces like parks and undeveloped zones.

The desire for more (land, space, money, things, and so on) seems human, but in fact identifies the most important environmental concern. Reduction represents the most significant means of addressing environmental problems. Whether it be car use, private green space, or total built square footage, *less is environmentally more*. Beginning with less built space starts a whole chain of environmental reductions in energy and materials consumption.

36.3   THE CURRENT DEBATE

Reviewing actions of current political, governmental, and legislative bodies reflects the desire and urgency for change. Green parties, groups, and leaders with environmental agendas aid in public awareness and implementing change. For example, both the former Vice President of the United States Al Gore and the recent Presidents of the American Institute of Architects have raised many concerns to the national level. Within the government, the Environmental Protection Agency (EPA) has researched and implemented change in a broad range of issues from hazardous materials found in the built environment, like asbestos, lead, radon, mercury (found in paints) to energy sources and consumption. Particularly significant and innovative are the new city ordinances, like that of Austin, which encouraged energy conservation through financial incentives. The Green Builder Program, sponsored by the Environmental and Conservation Services Department of Austin, Texas, uses a rating system encouraging environmentally sensitive building practices and products in new homes. Large organizations, like the North Carolina Recycling Association and the National Audubon Society, have publicized environmental concerns and new practices through the design of their buildings. Both aim to conserve natural resources and to be as energy efficient and nontoxic as possible.

As architects have struggled to come to terms with the environmental implications of their buildings, the term "sustainability" has become the catchword. While "sustainability" will not answer all environmental concerns, it provides a program to address current practices. With the present rates of fossil fuel consumption and ozone depletion, the earth's systems may not be able to support life. This risk of extinction necessitates examination and change. As Solow [1] states. "... it is an obligation to conduct ourselves so that we leave to the future the option or the capacity to be as well off as we are ... Sustainability is an injunction not to satisfy ourselves by impoverishing our successors." How can the species sustain itself, i.e., secure a viable environment for future generations? Through analyzing the environmental impact of architectural siting, design, construction, and use, a greater understanding can suggest ways to alter practices in order to do the least possible damage to the environment. The following discussion suggests environmentally conscious practices specifically related to architecture.

Recycling another building normally offers the most significant environmental savings. Particularly in places where there are unused and vacant buildings, to build more of the same represents one of the greatest environmental wastes. The initial energy spent in construction—through preparing the site, manufacturing the building materials, transporting them to the location, and then assembling them in construction—normally exceeds years of operational costs. Therefore, barring the least efficient structures, reuse—even with renovation—is the most sustainable choice. But if this is not possible, there are many ways in which the traditional building process can be readdressed with sustainability in mind.

36.4   SITING

As described above, the sprawl of development has threatened or destroyed many ecosystems. Therefore, one of the first site concerns is to avoid clearing previously untouched land. Once the land or place has been chosen, the existing landscape, the topography, wind movements, and context should be analyzed. First, one should use the given resources. Retaining existing trees and other plant life does the least ecological damage, while additionally saving later expenditures for artificial landscaping. Next, topography and wind movements can be used to naturally assist in creating a more comfortable environment. Careful siting, in relation to the given landscape, reduces the building's heating and cooling loads. For example, tree groupings can provide wind barriers in the winter, while others can direct winds into the building during the summer. Along the same lines, deciduous trees offer a building summer shading, while still allowing for passive solar heating in the winter. Additional concern for solar orientation can provide the building with natural lighting. With a balance in relation to heating/cooling gains and losses, fenestration uses daylight to produce a more comfortable, healthy, and energy-efficient space. The building's context must also be examined as a potential source of environmental hazards and opportunities. The surrounding buildings and structures can significantly influence siting. Like the natural elements mentioned above, built forms create shade and redirect wind. They also effect site hydrology. Buildings and nonporous surfaces (like asphalt) change how water moves through and drains from a site. In general, a site should be well drained with adequate flooding and erosion control for proper building maintenance and a healthy living environment. The best siting will not disturb the normal patterns of water flow and drainage. But if this is unavoidable, the effect of redirected water should be analyzed to avoid upsetting existing ecologies and conditions.

Another important site consideration is potential pollution sources. For the most part, industry and transportation create the greatest amounts of air, noise, and water pollution: roadways, cars, airports, oil refineries, power plants, and so on. Siting analysis of these potential sources should either suggest the use of another site or a way to avoid exposure to the hazards.

36.5 DESIGN

Environmentally conscious design offers perhaps the greatest opportunity for ecological improvement. For the most part, the current "green" trends concentrate on materials, products, and energy systems. This emphasis allows architects to ignore the environmental implications of their buildings. Responsible behavior requires more than a substitution of traditional building materials with recycled or nontoxic products. An ethical response to the environmental concerns necessitates change at the core, i.e., in the architectural theory of design. Environmental concerns must be completely incorporated into architectural thinking. Then, as an integral aspect of the architectural process, sustainability can shape design decisions and form buildings.

Several key concerns shape an environmentally conscious design strategy: minimizing the building's effect on the existing ecosystem, minimizing the use of new resources, increasing the energy efficiency of the building in its form and operation, and creating a healthy environment for the users.

Minimizing the building's effect on the existing environment has been discussed above, specifically in the site analysis section. Additionally, those aspects of the designed landscape can complement and enhance the viability of the existing ecosystems. The use of the traditional turf lawn represents a seriously destructive design practice. It removes the existing, natural environment at the risk of plant and animal biodiversity. Also, lawn maintenance requires irrigation and mowing, which increases water and fossil fuel consumption. Mowing and the use of pesticides both contribute to pollution. Instead, to complement the existing landscape, drought resistant native plantings enhance an outdoor environment to the benefit of resource management and ecosystems. Native plants thrive with a minimum of watering, chemicals (pesticides and fertilizers), and cutting. They also aid in maintaining or restoring an ecosystem's biodiversity. In areas that must be cleared for parking and walkways, the substitution of pervious paving materials (gravel, crushed stone, open paving blocks, and pervious paving blocks) minimizes runoff and increases infiltration and groundwater recharge.

Minimizing the use of resources, particularly new resources, can be achieved in several ways. Again, to recycle that modernist line, *less is more* and, *smaller is better.* Beginning with the preliminary design, the interior space should be kept to a minimum. This reduces land use, building materials, and operational energy expenditures.

Increasing energy efficiency through reduced operational expenditures can be achieved in several ways. Passive systems, such as solar heating, daylighting, and natural cooling (berms, shade, and ventilation), produce a more energy-efficient building with minimal expenditures. As suggested above in the site considerations, a building should be designed to work with the climate and natural energy sources. A building that responds to and takes advantage of what is naturally given results in a more sustainable design. (Climate, solar energy, topology, and on-site materials all qualify as givens.) A multitude of opportunities exist. To begin, what will create a comfortable environment? Orientation, built forms (like shading devices), and window and door locations can reduce heating and cooling loads while simultaneously enhancing living conditions. Also, a more systematized address of heating and cooling loads reduces the operating energy expenses of the building. High levels of insulation, high performance windows, and a tight construction (but not at the expense of indoor air quality) create a more energy-efficient building.

Considering that people generally spend about 90% of their time indoors, the quality of indoor air crucially impacts well-being and comfort. (The reader is referred to Chapter 14 for additional details on indoor air quality.) Indoor air pollution comes from many different sources, both indoor and out. One of the more serious threats to indoor air quality and health is radon. Radon rises from subsurface uranium deposits through and into buildings. Posing a tremendous threat, radon is the nation's second cause of lung cancer [2]. Other outdoor pollutants—like pesticides and car exhaust—threaten many buildings' indoor air quality. All three pollutants—radon, pesticides, and car exhaust—can be significantly reduced by good planning and design. First, an adequate ventilation system prevents accumulation within the building. While the building should open up for natural ventilation, an airtight construction will avoid many problems. For example, radon usually enters a building through cracks in the foundation. In the case of pesticides, the building's envelope works doubly. Careful detailing of the building, particularly in its corners and where it meets the ground, prevents many pests from entering. As a result, toxic interior pesticides and fumigants become unnecessary. Additionally, an airtight construction through detailing prevents many pollutants, particularly exterior pesticides and car exhaust, from entering the interior. Another preventive measure, the removal or avoidance of the pollution source, improves indoor air quality. Detaching a garage or parking structure from inhabited spaces eliminates direct exhaust infiltration into the building. In the cases where the pollution sources cannot be removed, ventilation intakes should be situated to avoid contaminants: other building's exhausts, car pollution, and pesticides.

Indoor air pollutants, like outdoor pollutants, pose more serious problems when buildings have inadequate, poorly maintained, or improperly located ventilation systems. An adequate ventilation system lessens the harmful effects of pollutants like lead, formaldehyde, carcinogenic wood finishes, smoke, and biological contaminants (bacteria, molds, mildew, and viruses). Especially in the case of lead dust and biological contaminants, keeping interiors clean and dust-free improves indoor air quality. In addition to increased ventilation and maintenance, source removal eliminates many problems. Smoking, a major indoor air pollution source, should be prohibited in interiors. Exposure to other pollutants, like lead, mercury, and volatile organic compounds (VOCs), can be more easily avoided through the greater availability of nontoxic building materials and finishes.

In smaller scale residential projects where there is little threat of on-site or near-site pollution sources, natural ventilation may suffice. But with larger scale projects, or those that are exposed to other sources of pollution (traffic, the exhausts of other buildings, the outgassing of building materials) conditions necessitate mechanical ventilation systems. Particularly in buildings like offices, with a large number of users, successful mechanical ventilation becomes crucial to maintaining indoor air quality. Without proper ventilation or systems maintenance, problems like outgassing or sick building syndrome can significantly affect the health and productivity of the building's users.

The building's design should incorporate recycling into the program so that it is easy and available. For example, a kitchen or an office can be designed to include recycling containers or cabinets for glass, aluminum, plastic, and paper. Composting systems for waste and sewage can be specified and located. Also, saving water can serve as a recycling opportunity. The recycled water from clothes washers, baths, showers, and nonkitchen sinks can be redirected for irrigation use. Another way to save water is through harvesting rainwater with a water catchement system.

Finally, the greatest recycling opportunity exists in the building itself. It should be designed with reuse in mind. A building should be adaptable with no or minimal renovation. As mentioned above, this may help avoid the enormous energy and material expenditures required by a new building's construction.

36.6 MATERIALS

Educated material selections greatly enhance the resource and energy savings created by an ecologically aware design. The use of each building material impacts both the global and local environments through its extraction, its manufacture, and its use. For example, the lumbering and strip mining industries have devastated ecological systems. Therefore, the selection of a material should be made only after an impact analysis of its removal or extraction. This type of thinking has led to some changes in the lumber industry. For example, to minimize the use of old-growth timber, sustainably produced lumber or recycled plastic lumber products have been introduced. While this represents an improvement, the greatest ecological savings occurs through using less.

In addition to the raw material itself, the material's processing should be considered. Thinking in terms of the total environmental cost has led to the analysis of materials' embodied energy. Manufacturing a building material often requires large expenditures of water, fossil fuels for energy and transportation, and human labor. Many environmental experts suggest choosing low embodied energy materials, i.e., materials that need less energy to make them usable. As a result, the building product uses less resources and generates less pollution in its manufacture. Normally, the material is closer to its natural state. For example, natural stone has a low embodied energy, while plastics, steel, and aluminum have high embodied energies. Additionally, when available, using materials found on or near the site normally reduces energy expenditures. (This is not to suggest cutting the site's trees for lumber.) For example, using local stone in the place of brick eliminates not only the manufacturing energies and pollution but also larger transportation costs. Regarding materials with low embodied energy and from local sources, it is crucial to consider the net energy calculated with use. One should always try to envision the total picture of chained actions and reactions. For example, a certain type of insulation may be completely synthetic; it requires a large amount of energy to manufacture and also must be transported from elsewhere. But the energy savings resulting from its installation may exceed the preliminary expenditures.

Many building materials outgas, i.e., release harmful, airborne materials that pose a risk to the local environment's air quality. The VOCs most often found in floor finishes, paints, stains, adhesives, synthetic wallpapers, plywood, and chipboards should be avoided to maintain a healthy environment. In substitution, many new VOC-free products are now available. In addition to VOCs, building materials emitting CFCs and HFCs, like insulation, should not be used.

As always, recycling represents an important concern. When possible, salvaged building materials should be used. On the other end of the building process, using building materials that can eventually be recycled will eliminate further resource expenditures. Along similar lines, products and materials need to last; durability increases net energy savings.

36.7   BUILDING SYSTEMS AND EQUIPMENT

Selecting energy-efficient systems and equipment greatly reduces the environmental impact of a building's operation. These systems reduce not only operational costs, but create a whole chain of environmental savings. The lower operational costs normally reflect reduced operational energies and fuel expenditures. For example, the use of low energy bulbs lowers not only the building's electricity requirement, but also the power plant's load. As a result, savings occur on both the local and larger levels. Greater use of this type of equipment would reduce the number of power plants, saving more resources and reducing pollution. Other systems, like high efficiency heating and cooling equipment, have similar advantages. For example, a photovoltaic electric generating system creates an on-site power source with a significant reduction of pollution. Other building equipment, like high-efficiency appliances, significantly reduce electricity expenditures. Water-efficient equipment, such as shower heads and toilets, decrease water use. Studies in California have shown retrofitting buildings and homes with energy-efficient lighting, pumps, fans, refrigerators, etc., can lead to at least a 75% reduction in energy use [3].

36.8   CONSTRUCTION

Construction concerns reiterate many of the points and themes discussed above. The environmental impact of the building's construction should be as minimal as possible. The existing landscape, particularly on-site trees, should be protected. The use of pesticides and other chemicals should be restricted to avoid polluting the groundwater supply. Construction debris should be minimized and recycled. As always, durability and longevity generates the greatest energy savings: *Build to last*. Finally, like the building itself, the construction site should never be a hazard for those who use it.

36.9   FUTURE TRENDS

A program for the future consists of the difficult obligation to implement change. A brief survey like this chapter serves to heighten awareness, but the crisis requires a more significant response. Of course, the most responsible and helpful behavior is to commit to lessening the environmental impact of general practices. From the individual, to corporations, to governmental bodies, every bit counts. Unfortunately, the only proven way to get widespread change is through economic incentives. With penalties issued for unsustainable building practices, awareness and change could possibly extend to all sectors of the building industry. Presently, resistance exists on many levels of the building process—maintaining the status quo is far easier than switching to an unknown. Using economic incentives will significantly empower the environmental cause, especially among groups without obvious reasons for changing their practices (only reason: helping the world). Already, certain programs give lower interest rates for "green" home improvements or utility rebates for high efficiency equipment use. In addition to economic incentives, governmental- and corporate-sponsored projects have the capability to mainstream environmentally conscious practices. Through showcasing sustainable design, work that perhaps would not be funded otherwise, ideas and research turn into practice precedents.

36.10   SUMMARY

1. Architectural planning, building, and use contribute tremendously to the environmental crises; therefore, theory and practice must be analyzed to implement change.
2. Historically and presently, architecture has developed a pattern of wastefulness and indifference in relation to the natural environment.
3. As more attention has been given to architectural concerns in ecological discourse, sustainability has emerged as a program for improved practices.

4. Once a site has been chosen, the existing landscape, given resources, topography, wind movements, and context should be used to form the design.
5. Several key concerns shape an environmentally conscious design strategy: minimizing the building's effect on the existing ecosystem, minimizing the use of new resources, increasing the energy efficiency of the building in its form and operation, and creating a healthy environment for the users.
6. Educated material selections greatly enhance the resource and energy savings created by an ecologically aware design.
7. Energy-efficient systems and equipment reduce operational costs, fuel and other resource expenditures, and pollution; green designed are attempting to produce more energy than they produce.
8. The environmental impact of the building's construction should be as minimal as possible.
9. For the future, economic incentives and other mainstreaming practices will increase awareness and implement change.

REFERENCES

1. Solow, R.M. Sustainability: An economist's perspective, Eighteenth J. Seward Johnson Lecture in Marine Policy, Marine Policy Center, Woods Hole Oceanographic Institution. June 14, 1991.
2. National Council on Radiation Protection and Measurements.
3. Ledger, B. Architecture and the environment: Where do we stand now. *The Canadian Architect*, June 1994, 14.

Part VI

Environmental Risk

Part VI comprises five chapters, in which Chapter 37 serves as an introduction to environmental risk assessment. Chapter 38 is concerned with the general subject of health risk assessment, while Chapter 39 examines hazard risk assessment and the risk evaluation of accidents. Chapter 40 focuses on the important area of public perception of risk. Part VI concludes with Chapter 41 that addresses risk communication issues.

37 Introduction to Environmental Risk Assessment

CONTENTS

37.1 INTRODUCTION

In the 1980s, to satisfy the need to start corrective action programs quickly, many regulatory agencies decided to uniformly apply, at underground storage tank (UST) cleanup sites, regulatory cleanup standards developed for other purposes. It became increasingly apparent that applying such standards without consideration of the extent of actual or potential human and environmental exposure was an inefficient means of providing adequate protection against the risks associated with UST releases. The Environmental Protection Agency (EPA) now believes that risk-based corrective-action processes are tools that can facilitate efforts to clean up sites expeditiously, as necessary, while still assuring protection of human health and the environment [1].

Risk-based decision making and risk-based corrective action (RBCA) are decision-making processes for assessing and responding to a chemical release. The processes take into account effects on human health and the environment, in as much as chemical releases vary greatly in terms of complexity, physical, and chemical characteristics, and in the risk that they may pose. RBCA was initially designed by the American Society for Testing and Materials (ASTM) to assess petroleum releases, but the process may be tailored for use with any chemical release.

The EPA and several state environmental agencies have developed similar decision-making tools. The EPA refers to the process as "risk-based decision making." While the ASTM RBCA standard deals exclusively with human health risk, the EPA advises that, in some cases, ecological goals must also be considered in establishing cleanup goals.

Risk-based decision making and the RBCA process integrate risk and exposure assessment practices, as suggested by the EPA. The processes help to identify which assessment and remediation activities protect both human health and the environment. If a chemical release occurs, or is even suspected, risk-based decision making may be implemented. When utilizing these processes, it is important to establish appropriate safety and health practices and to determine any regulatory limitations prior to their use.

The chapter to follow treats in greater detail "how" to evaluate risks to health and the environment. For the purposes of this chapter, a few definitions of common terms will suffice. Risk is the probability that persons or the environment will suffer adverse consequences as a result of an exposure to a substance. The amount of risk is determined by a combination of the concentration of the chemical the person or the environment is exposed to, the rate of intake or dose of the substance, and the toxicity of the substance. Risk assessment is the procedure used to attempt to quantify or estimate risk. Risk-based decision making distinguishes between the "point of exposure" and the "point of compliance." The point of exposure is the point at which the environment or the individual comes into contact with the chemical release. A person may be exposed by methods such as inhalation of vapors, as well as physical contact with the substance. The point of compliance is a point in between the point of release of the chemical (i.e., its source), and the point of exposure. The point of compliance is selected to provide a safety buffer for effected individuals and/or environments.

37.2   RISK VARIABLES

Placing the health risk in perspective entails translating the myriad technical health risk analyses into concepts of risk both the technical community and the general public can understand. The most effective techniques for presenting risks in perspective is to contrast risks to other, similar risks. There are several variables that affect acceptance of risk. Ten such variables include

1. Voluntary vs. involuntary
2. Delayed vs. immediate
3. Natural vs. man-made
4. Controllable vs. uncontrollable
5. Known vs. unknown
6. Ordinary vs. catastrophic
7. Chronic vs. acute
8. Necessary vs. luxury
9. Occasional vs. continuous
10. Old vs. new

The public generally accepts voluntarily assumed risk more easily than an involuntarily imposed risk. Similarly a naturally occurring risk is more easily accepted than a man-made risk. The more similar risks are with regards to these variables, the more meaningful it is to compare those risks. Walking on a busy street is classified as ordinary, necessary, voluntary, and a known risk. A nuclear meltdown presents a catastrophic, perhaps unnecessary, involuntary, unknown risk. Contrasting the two types of risk is like comparing kangaroos and oranges.

A useful technique is to compare the risks associated with each alternative. Another effective technique is to compare the risks for each technique with federal or state where applicable. If possible one should draw on subsequent health effects at similarly remediated sites and void contrasting the health risks at the site to completely unrelated risks.

Mechanisms for providing information on health risks to the community could include:

1. *Fact sheets*: Presenting detailed information on the site, proposed remediation techniques, health risk analyses, and other information in a readily understandable format to be mailed to libraries, schools, business organizations, and local residents.
2. *Newsletters*: Presenting information similar in content to the fact sheet, but including additional, more general information on hazardous waste management.
3. *Direct contact*: Walking door-to-door to discuss the proposed remediation project and other issues related to the site. This approach gives the risk communicator an opportunity to directly address local residents concerns, thereby giving the residents the feeling that they genuinely do have a say. In some cases, this method is the only way to understand the opinions of the residents regarding remediation.

Of course, it is crucial that each of these community outreach mechanisms be presented in the correct language. An English language fact sheet, no matter how clearly written, does little benefit to a primarily foreign-speaking population.

37.3   WHY USE RISK-BASED DECISION MAKING?

The use of the risk-based decision making process allows for efficient allocation of limited resources, such as time, money, regulatory oversight, and qualified professionals. Advantages of using this process include:

1. Decisions are based on reducing the risk of adverse human or environmental impacts.
2. Site assessment activities are focused on collecting only that information that is necessary to make RBCA decisions.
3. Limited resources are focused on those sites that pose the greatest risk to human health and the environment at any time.
4. Compliance can be evaluated relative to site-specific standards applied at site-specific point(s) of compliance.
5. Higher quality, and in some cases faster, cleanups may be achieved than are currently possible.
6. Documentation is developed that can demonstrate that the remedial action is protective of human health, safety, and the environment.

By using risk-based decision making, decisions are made in a consistent manner. Protection of both human health and the environment is accounted for.

A variety of EPA programs involved in the protection of groundwater and cleanup of environmental contamination utilize the risk-based decision making approach. Under the EPA's regulations dealing with cleanup of UST sites, regulators are expected to establish goals for cleanup of UST releases based on consideration of factors that could influence human and environmental exposure to contamination. Where UST releases affect the groundwater being used as public or private drinking water sources, EPA generally recommends that cleanup goals be based on health-based drinking water standards; even in such cases, however, risk-based decision making can be employed to focus corrective action [1]. (For more on USTs, refer to Chapter 26.)

In the Superfund program, risk-based decision making plays an integral role in determining whether a hazardous waste site belongs on the National Priorities List. Once a site is listed, qualitative and quantitative risk assessments are used as the basis for establishing the need for action and determining remedial alternatives. To simplify and accelerate baseline risk assessments at Superfund sites, EPA has developed generic soil screening guidance that can be used to help distinguish between contamination levels that generally present no health concerns and those that generally require further evaluation. (For more on Superfund, refer to Chapter 27). The Resource Conservation and Recovery Act (RCRA) Corrective Action program also uses risk-based decision making to set priorities for cleanup so that high-risk sites receive attention as quickly as possible to assist in the determination of cleanup standards, and to prescribe management requirements for remediation of wastes.

37.4   THE RISK-BASED CORRECTIVE ACTION APPROACH [2,3]

The RBCA process is implemented in a tiered approach, with each level involving increasingly sophisticated methods of data collection and analysis. As the analysis progresses, the assumptions of earlier tiers are replaced with site-specific data and information. Upon evaluation of each tier, the results and recommendations are reviewed, and it is determined whether more site-specific analysis is required. Generally, as the tier level increases, so do the costs of continuing the analysis. The application of this approach to the remediation follows.

The first step is the site assessment, which is the identification of the sources of the chemical(s) of concern, any obvious environmental impacts, any potentially impacted human and environmental receptors (e.g., workers, residents, lakes, streams, etc.), and potentially significant chemical transport pathways (e.g., groundwater flow, atmospheric dispersion, etc.). The site assessment also includes information collected from historical records and a visual inspection of the site. An example of criteria used for a site classification in outline form follows.

Example of site classification—criteria and prescribed scenarios [2]:

1. Immediate threat to human health, safety, or sensitive environmental receptors.
2. Short-term (0–2 years) threat to human health, safety, or sensitive environmental receptors.

3. Long-term (>2 years) threat to human health, safety, or sensitive environmental receptors.
4. No demonstrable long-term threat to human health or safety or sensitive environmental receptors. Priority 4 scenarios encompass all other conditions not described in priorities 1, 2, and 3 and that are consistent with the priority description given above.

Once the applicable criteria are met, the site is then classified according to the urgency of need for initial response action, based on information collected during the site assessment. Associated with site classifications are initial response actions that are to be implemented simultaneously with the RBCA process. Sites should be reclassified as actions are taken to resolve concerns or as better information becomes available.

A Tier 1 evaluation is then conducted using a "lookup table." The lookup table contains screening level concentrations for the various chemicals of concern. The "lookup table" is defined as a tabulation for potential exposure pathways (e.g., inhalation, digestion, etc.), media (e.g., soil, water, and air), a range of incremental carcinogenic risk levels which are used as target levels for determining remediation requirements, and potential exposure scenarios (e.g., residential, commercial, industrial, and agricultural). If a lookup table is not provided by the regulatory agency or available from a previous evaluation, the person conducting the RBCA analysis must develop the lookup table. If a lookup table is available, the user is responsible for determining that the risk-based screening levels (RBSLs) in the table are based on currently acceptable methodologies and parameters.

The RBSLs are determined using typical, nonsite-specific values for exposure parameters and physical parameters for media. The RBSLs are calculated according to methodology suggested by the EPA [4,5]. The value of creating a lookup table is that users do not have to repeat the exposure calculations for each site encountered. The lookup table is only altered when reasonable maximum exposure parameters, toxicological information, or recommended methodologies are updated. Some states have compiled such tables that, for the most part, contain identical values (as they are based on the same assumptions). The lookup table is used to determine whether site conditions satisfy the criteria for a quick regulatory closure or warrant a more site-specific evaluation.

If further evaluation is required, a Tier 2 evaluation provides the user with an option to determine site-specific target levels (SSTLs) and point(s) of compliance. It is important to note that both Tier 1 RBSL and Tier 2 SSTLs are based on achieving similar levels of protection of human health and the environment. However, in Tier 2 the nonsite-specific assumptions and point(s) of exposure used in Tier 1 are replaced with site-specific data and information. Additional site-assessment data may be needed. For example, the Tier 2 SSTL can be derived from the same equations used to calculate the Tier 1 RBSL, except that site-specific parameters are used in the calculations. The additional site-specific data may support alternate fate and transport analysis. At other sites, the Tier 2 analysis may involve applying Tier 1 RBSLs at more probable point(s) of exposure.

At the end of Tier 2, if it is determined that more detailed evaluation is again warranted, a Tier 3 evaluation is then conducted. A Tier 3 evaluation provides the user with an option to determine SSTLs for both direct and indirect pathways using site-specific parameters and point(s) of exposure and compliance when it is judged that Tier 2 SSTLs should not be used as target levels. Tier 3, in general, can be a substantial incremental effort relative to Tiers 1 and 2, as the evaluation is much more complex and may include additional site assessment, probabilistic evaluations, and sophisticated chemical fate/transport models.

With the RBCA process, the user compares the target levels (RBSLs or SSTLs) to the concentrations of the chemical(s) of concern at the point(s) of compliance at the conclusion of each tier evaluation. If the concentrations of the chemical(s) of concern exceed the target levels at the point(s) of compliance, then either remedial action, interim remedial action, or further tier evaluation should be conducted. When it is judged that no further assessment is necessary or practicable, a remedial alternatives evaluation should be conducted to confirm the most cost-effective option for achieving the final remedial action target levels (RBSLs or SSTLs, as appropriate).

Detailed design specifications may then be developed for installation and operation of the selected measures. The selected measures may include some combination of source removal, treatment, and containment technologies, as well as engineering and institutional controls. Examples of these include the following: soil venting, bioventing, air sparging, "pump-and-treat," and natural attenuation/passive remediation. The remedial action must continue until such time as monitoring indicates that concentrations of the chemical(s) of concern are not above the RBSL or SSTL, as appropriate, at the points of compliance or source area(s), or both. When concentrations of chemical(s) of concern no longer exceed the target levels at the point of compliance, then the user may elect to deem the RBCA process complete. If achieving the desired risk reduction is impracticable due to technology or resource limitations, an interim remedial action, such as removal or treatment of "hot spots," may he conducted to address the most significant concerns, change the site classification, and facilitate reassessment of the tier evaluation.

37.5 COMMUNICATING RISK

Unfamiliar chemicals tend to cause more concern and require more communication (the fear of the unknown) than familiar chemicals. Similarly, individuals are more likely to be concerned with a facility that is perceived to operate mysteriously behind closed gates than with a facility that provides opportunities to tour. Individuals are generally less concerned when risks are easily detectable than they are with risk which are invisible or undetectable (i.e., odorless gas releases).

    I. The following is a summary of important things to communicate about risk and risk data information:
  1. What exposure routes are particularly problematic?
  2. Who is especially at risk (children, the elderly, fetuses, and animals asthmatics)?
  3. What concentration for how long will cause a risk problems?
  4. How sound or reliable the data is (presentation)?
  5. What other data presently is being collected?
  6. Quantity of chemical stored and emitted each year

7. Concentration level measured at/in
   a. The air, water, and ground
   b. At the fenceline
   c. At various locations in the community
   d. On a typical day
   e. For various meteorological condition
8. Chronic vs. acute risks

II. Other factors to communicate in addition to risk data/information include:
   1. Whether the risk is higher or lower than in the past?
   2. Whether the risk is likely to get higher or lower in the future?
   3. Comparison of risk data with a standard (EPA or State) values
   4. What the company is doing to monitor/reduce risk?
   5. Timetable for completing risk management measures and safeguards, if applicable
   6. How the company will provide risk information to the community?
   7. Who can be contacted for more information or to report risk information?

Addition information is provided in subsequent chapters in this Part.

The most common types of numbers and statistics used in risk communications are: concentrations (i.e., parts per million); probabilities (i.e., likelihood of an event); and quantities (i.e., how much water on the soil is contaminated). However, risk comparisons, if used appropriately, can be very useful for putting risk into perspective. The following types of risk comparisons in five categories from first rank risk comparisons (which are the most desirable types of comparisons) to the fifth rank risk comparisons (which is considered rarely acceptable) to listed below:

1. *First rank risk comparisons*
   (First choice–Most desirable)
   • Comparison of the same risk at two different times
   • Comparison with a standard
   • Comparison of different estimates of the same risk
2. *Second rank risk comparisons*
   (Second choice—Less desirable)
   • Comparison of the risk of doing something vs. the risk of not doing it
   • Comparison of alternative solutions to same problem
   • Comparison with the same risk in other places
3. *Third rank risk comparisons*
   (Third choice—Even less desirable)
   • Comparison of the average risk with peak risk at particular time or location
   • Comparison of risk from one source of a particular adverse effect with the risk from all sources of the same adverse effect
4. *Fourth rank risk comparisons*
   (Fourth choice—marginally acceptable)

Additional details on this topic are discussed in Chapter 41.

37.6 FUTURE TRENDS

Understanding risk communications dynamics is essential to successful risk communication efforts. Two-way communication with stakeholders (regulatory agencies, local residents, employees, etc.) prevents costly rework and permit delays and provides information useful for prioritizing risk management efforts. As communities have become more interested and concerned about environmental issues in recent years, the role of the environmental manager has expanded to include communications with key audiences. This interest and concern is certain to expand in the future. In addition to addressing the technical aspects of environmental and health risks; efforts to address process, health and lifestyle concerns has become more critical to the success of environmental projects and risk management.

37.7 SUMMARY

1. Risk-based decision making and RBCA are decision-making processes for assessing and responding to a chemical release which take into account effects on human health and the environment in as much as chemical releases vary greatly in terms of complexity, physical, and chemical characteristics, and in the risk that they may pose.
2. Risk is the probability that individuals or the environment will suffer adverse consequences as a result of an exposure to a substance. Risk assessment is the procedure used to attempt to quantify or estimate risk.
3. The use of the risk-based decision making process allows for efficient allocation of limited resources, such as time, money, regulatory oversight, and qualified professionals.
4. The RBCA process is implemented in a tiered approach, with each level involving increasingly sophisticated methods of data collection and analysis. As the analysis progresses, the assumptions of earlier tiers are replaced with site-specific data and information.

REFERENCES

1. U.S. EPA. *Use of Risk-Based Decision Making*, OSWER Directive 9610.17, U.S. Environmental Protection Agency, Washington, DC, March 1995.
2. ASTM, *Standard Guide for Risk-Based Corrective Action Applied to Petroleum Release Sites*, ASTM E1739-95, American Society for Testing and Materials, Philadelphia, PA.
3. U.S. EPA. *Ecological Assessment of Hazardous Waste Sites: A Field and Laboratory Reference Document*, EPA/600/3-89/013, NTIS No. PB-89205967, U.S. Environmental Protection Agency, Washington, DC, March 1989.
4. U.S. EPA. *Integrated Risk Information System (IRIS)*, U.S. Environmental Protection Agency, Washington, DC, October 1993.
5. U.S. EPA. *Health Effects, Assessment Summary Tables (HEAST)*, OSWER OS-230, U.S. Environmental Protection Agency, Washington, DC, March 1992.

38 Health Risk Assessment

CONTENTS

38.1 INTRODUCTION

As noted in Chapter 37, there are many definitions for the word risk. It is a combination of uncertainty and damage; a ratio of hazards to safeguards; a triplet combination of event, probability, and consequences; or even a measure of economic loss or human injury in terms of both the incident likelihood and the magnitude of the loss or injury [1]. People face all kinds of risks every day, some voluntarily and others involuntarily. Therefore, risk plays a very important role in today's world. Studies on cancer caused a turning point in the world of risk because it opened the eyes of risk scientists and health professionals to the world of risk assessments.

Since 1970 the field of risk assessment has received widespread attention within both the scientific and regulatory committees. It has also attracted the attention of the public. Properly conducted risk assessments have received fairly broad acceptance, in part because they put into perspective the terms toxic, hazard, and risk. Toxicity is an inherent property of all substances. It states that all chemical and physical agents can produce adverse health effects at some dose or under specific exposure conditions. In contrast, exposure to a chemical that has the capacity to produce a particular type of adverse effect, represents a hazard. Risk, however, is the probability or likelihood that an adverse outcome will occur in a person or a group that is exposed to a particular concentration or dose of the hazardous agent. Therefore, risk is generally a function of exposure or dose. Consequently, health risk assessment is defined as the process or procedure used to estimate the likelihood that humans or ecological systems will be adversely affected by a chemical or physical agent under a specific set of conditions [2].

The term risk assessment is not only used to describe the likelihood of an adverse response to a chemical or physical agent, but it has also been used to describe the likelihood of any unwanted event. This subject is treated in more detail in Chapter 39. These include risks such as: explosions or injuries in the workplace; natural

407

catastrophes; injury or death due to various voluntary activities such as skiing, ski diving, flying, and bungee jumping; diseases; death due to natural causes; and, many others [2].

Risk assessment and risk management are two different processes, but they are intertwined. Risk assessment and risk management give a framework not only for setting regulatory priorities but also for making decisions that cut across different environmental areas. Risk management refers to a decision-making process that involves such considerations as risk assessment, technology feasibility, economic information about costs and benefits, statutory requirements, public concerns, and other factors. Therefore, risk assessment supports risk management in that the choices on whether and how much to control future exposure to the suspected hazards may be determined [3]. Regarding both risk assessment and risk management, this chapter will primarily address this subject from a health perspective; Chapter 39 will primarily address this subject from a safety and accident perspective.

38.2   THE HEALTH RISK EVALUATION PROCESS

Health risk assessments provide an orderly, explicit, and consistent way to deal with scientific issues in evaluating whether a health hazard exists and what the magnitude of the hazard may be. This evaluation typically involves large uncertainties because the available scientific data are limited, and the mechanisms for adverse health impacts or environmental damage are only imperfectly understood. When one examines risk, how does one decide how safe is safe, or how clean is clean? To begin with, one has to look at both sides of the risk equation—that is, both the toxicity of a pollutant and the extent of public exposure. Information is required at both the current and potential exposures, considering all possible exposure pathways. In addition to human health risks, one needs to look at potential ecological or other environmental effects. In conducting a comprehensive risk assessment, one should remember that there are always uncertainties, and these assumptions must be included in the analysis [3].

In recent years, several guidelines and handbooks have been produced to help explain the approaches for doing health risk assessments. As discussed by a special National Academy of Sciences committee convened in 1983, most human or environmental health hazards can be evaluated by dissecting the analysis into four parts: hazard identification, dose–response assessment or hazard assessment, exposure assessment, and risk characterization (see Figure 38.1). For some perceived health hazards, the risk assessment might stop with the first step, hazard identification, if no adverse effect is identified or if an agency elects to take regulatory action without further analysis [2]. Regarding hazard identification, a hazard is defined as a toxic agent or a set of conditions that has the potential to cause adverse effects to human health or the environment. Hazard identification involves an evaluation of various forms of information in order to identify the different hazards. Dose–response or toxicity assessment is required in an overall assessment; responses/effects can vary widely since all chemicals and contaminants vary in their capacity

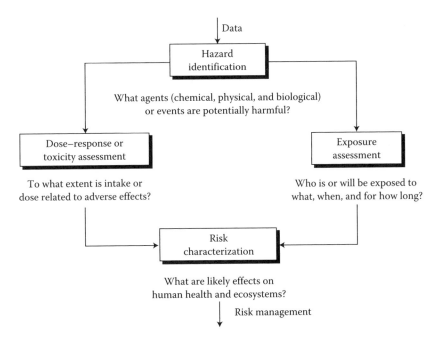

FIGURE 38.1   The health risk evaluation process.

to cause adverse effects. This step frequently requires that assumptions be made to relate experimental data for animals and humans. Exposure assessment is the determination of the magnitude, frequency, duration, and routes of exposure of human populations and ecosystems. Finally, in risk characterization, toxicology and exposure data/information are combined to obtain a qualitative or quantitative expression of risk.

Risk assessment involves the integration of the information and analysis associated with the above four steps to provide a complete characterization of the nature and magnitude of risk and the degree of confidence associated with this characterization. A critical component of the assessment is a full elucidation of the uncertainties associated with each of the major steps. Under this broad concept of risk assessment are encompassed all of the essential problems of toxicology. Risk assessment takes into account all of the available dose–response data. It should treat uncertainty not by the application of arbitrary safety factors, but by stating them in quantitatively and qualitatively explicit terms, so that they are not hidden from decision makers. Risk assessment, defined in this broad way, forces an assessor to confront all the scientific uncertainties and to set forth in explicit terms the means used in specific cases to deal with these uncertainties [4]. An expanded presentation on each of the four health risk assessment steps is provided below.

The reader should note that this topic also receives treatment in Chapter 48—Environmental Implications of Nanotechnology, Section 48.4—Health Risk Assessment.

38.3 HAZARD IDENTIFICATION

Hazard identification is the most easily recognized of the actions of regulatory agencies. It is defined as the process of determining whether human exposure to an agent could cause an increase in the incidence of a health condition (cancer, birth defect, etc.) or whether exposure by a nonhuman receptor, e.g., fish, birds, or other wildlife, might adversely be affected. It involves characterizing the nature and strength of the evidence of causation. Although the question of whether a substance causes cancer or other adverse health effects in humans is theoretically a yes–no question, there are few chemicals or physical agents on which the human data are definitive. Therefore, the question is often restated in terms of effects in laboratory animals or other test systems: "Does the agent induce cancer in test animals?" Positive answers to such questions are typically taken as evidence that an agent may pose a cancer risk for any exposed human. Information for short-term in vitro tests and structural similarity to known chemical hazards may, in certain circumstances, also be considered as adequate information for identifying a health hazard [2].

A hazards identification for a chemical plant or industrial application can include information about

1. Chemical identities.
2. The location of facilities that use, produce, process, or store hazardous materials.
3. The type and design of chemical containers or vessels.
4. The quantity of material that could be involved in airborne release.
5. The nature of the hazard (e.g., airborne toxic vapors or mist, fire, explosion, large quantities stored or processed, handling conditions, etc.) most likely to accompany hazardous materials spills or releases [5].

An important aspect of hazards identification is a description of the pervasiveness of the hazard. For example, most environmental assessments require knowledge of the concentration of the material in the environment, weighted in some way to account for the geographical magnitude of the site affected; that is, a 1-acre or 300-acre site, a 1,000–1,000,000 gal/min stream. All too often environmental incidents regarding chemical emission have been described by statements like "concentrations as high as 150 ppm" of a chemical were measured at a 1000-acre waste site. However, following closer examination, one may find that only 1 of 200 samples collected on a 20-acre portion of a 1000-acre site showed this concentration and that 2 ppm was the geometric mean level of contamination in the 200 samples.

An appropriate sampling program is critical in the conduct of a health risk assessment. This topic could arguably be part of the exposure assessment, but it has been placed within hazard identification because, if the degree of contamination is small, no further work may be necessary. Not only it is important that samples be collected in a random or representative manner, but the number of samples must be sufficient to conduct a statistically valid analysis. The number needed to insure statistical validity will be dictated by the variability between the results. The larger the variance, the greater the number of samples needed to define the problem [2].

The means of identifying health hazards is complex. Different methods are used to collect and evaluate toxic properties (those properties that indicate the potential to cause biological injury, disease, or death under certain exposure conditions). One method is the use of epidemiological studies that deal with the incidence of disease among groups of people. Epidemiological studies attempt to correlate the incidence of cancer from an emission by an evaluation of people with a particular disease and people without the disease. Long-term animal bioassays are the most common method of hazard determination. (A bioassay as referred to here is an evaluation of disease in a laboratory animal.) Increased tumor incidence in laboratory animals is the primary health effect considered in animal bioassays. Exposure testing for a major portion of an animal's lifetime (2–3 years for rats and mice) provides information on disease and susceptibility, primarily for carcinogenicity (the development of cancer).

The understanding of how a substance is handled in the body, transported, changed, and excreted, and of the response of both animals and humans, has advanced remarkably. There are many questions concerning these animal tests as to what information they provide, which kinds of studies are the best, and how the animal data compares with human data. In an attempt to answer these questions, epidemiological studies and animal bioassays are compared to each other to determine if a particular chemical is likely to pose a health hazard to humans. Many assumptions are made in hazard assessments. For example, it is assumed that the chemical administered in a bioassay is in a form similar to that present in the environment. Another assumption is that animal carcinogens are also human carcinogens. An example is that there is a similarity between animal and human metabolisms, and so on. With these and other assumptions, and by analyzing hazard identification procedures, lists of hazardous chemicals have been developed [3].

38.4 DOSE–RESPONSE

Dose–response assessment is the process of characterizing the relation between the dose of an agent administered or received and the incidence of an adverse health effect in exposed populations, and estimating the incidence of the effect as a function of exposure to the agent. This process considers such important factors as intensity of exposure, age pattern of exposure, and possibly other variables that might affect response, such as sex, lifestyle, and other modifying factors. A dose–response assessment usually requires extrapolation from high to low doses and extrapolation from animals to humans, or one laboratory animal species to a wildlife species. A dose–response assessment should describe and justify the methods of extrapolation used to predict incidence, and it should characterize the statistical and biological uncertainties in these methods. When possible, the uncertainties should be described numerically rather than qualitatively.

Toxicologists tend to focus their attention primarily on extrapolations from cancer bioassays. However, there is also a need to evaluate the risks of lower doses to see how they affect the various organs and systems in the body. Many scientific papers focused on the use of a safety factor or uncertainty factor approach since all adverse effects other than cancer and mutation-based developmental effects are believed to have a threshold—a dose below which no adverse effect should occur. Several researchers have discussed various approaches to setting acceptable daily intakes or exposure limits for developmental and reproductive toxicants. It is thought that an acceptable limit of exposure could be determined using cancer models, but today they are considered inappropriate because of thresholds [2].

For a variety of reasons, it is difficult to precisely evaluate toxic responses caused by acute exposures to hazardous materials. First, humans experience a wide range of acute adverse health effects, including irritation, narcosis, asphyxiation, sensitization, blindness, organ system damage, and death. In addition, the severity of many of these effects varies with intensity and duration of exposure. Second, there is a high degree of variation in response among individuals in a typical population. Third, for the overwhelming majority of substances encountered in industry, there are not enough data on toxic responses of humans to permit an accurate or precise assessment of the substance's hazard potential. Fourth, many releases involve multicomponents. There are presently no rules on how these types of releases should be evaluated. Fifth, there are no toxicology testing protocols that exist for studying episodic releases on animals. In general, this has been a neglected area of toxicology research. There are many useful measures available to use as benchmarks for predicting the likelihood that a release event will result in serious injury or death. Several references [6,7] review various toxic effects and discuss the use of various established toxicological criteria.

Dangers are not necessarily defined by the presence of a particular chemical, but rather by the amount of that substance one is exposed to, also known as the dose. A dose is usually expressed in milligrams of chemical received per kilogram of body weight per day. For toxic substances other than carcinogens, a threshold dose must be exceeded before a health effect will occur, and for many substances, there is a dosage below which there is no harm. A health effect will occur or at least be detected at the threshold. For carcinogens, it is assumed that there is no threshold, and, therefore, any substance that produces cancer is assumed to produce cancer at any concentration. It is vital to establish the link to cancer and to determine if that risk is acceptable. Obviously, analyses of cancer risks are much more complex than noncancer risks [3].

Not all contaminants or chemicals are created equal in their capacity to cause adverse effects. Thus, cleanup standards or action levels are based in part on the compounds' toxicological properties. Toxicity data are derived largely from animal experiments in which the animals (primarily mice and rats) are exposed to increasingly higher concentrations or doses. Responses or effects can vary widely from no observable effect to temporary and reversible effects, to permanent injury to organs, to chronic functional impairment to ultimately, death.

38.5   EXPOSURE ASSESSMENT

Exposure assessment is the process of measuring or estimating the intensity, frequency, and duration of human or animal exposure to an agent currently present in the environment or of estimating hypothetical exposures that might arise from the release of new chemicals into the environment. In its most complete form, an exposure assessment should describe the magnitude, duration, schedule, and route of exposure; the size, nature, and classes of the human, animal, aquatic, or wildlife populations exposed; and, the uncertainties in all estimates. The exposure assessment can often be used to identify feasible prospective control options and to predict the effects of available control technologies for controlling or limiting exposure [2].

Much of the attention focused on exposure assessment has come recently. This is because many of the risk assessments performed in the past used too many conservative assumptions which caused an overestimation of the actual exposure. Without exposures there are no risks. To experience adverse effects, one must first come into contact with the toxic agent(s). Exposures to chemicals can be via inhalation of air (breathing), ingestion of water and food (eating and drinking), or absorption through the skin. These are all pathways to the human body.

Generally, the main pathways of exposure considered in this step are atmospheric transport, surface and groundwater transport, ingestion of toxic materials that have passed through the aquatic and terrestrial food chain, and dermal absorption. Once an exposure assessment determines the quantity of a chemical with which human populations may come in contact, the information can be combined with toxicity data (from the hazard identification and dose–response process) to estimate potential health risks [3]. The primary purpose of an exposure assessment is to determine the concentration levels over time and space in each environmental media where human and other environmental receptors may come into contact with chemicals of concern. There are four components of an exposure assessment: potential sources, significant exposure pathways, populations potentially at risk, and exposure estimates [2].

The two primary methods of determining the concentration of a pollutant to which target populations are exposed are direct measurement and computer analysis, also known as computer dispersion modeling. Measurement of the pollutant concentration in the environment is used for determining the risk associated with an exiting discharge source. Receptors are placed at regular intervals from the source, and the concentration of the pollutant is measured over a certain period of time (usually several months or a year). The results are then related to the size of the local population. This kind of monitoring, however, is expensive and time-consuming. Many measurements must be taken because exposure levels can vary under different atmospheric conditions or at different times of the year. Computer dispersion modeling predict environmental concentrations of pollutants (see Chapters 10 and 15 for more information on dispersion modeling). In the prediction of exposure, computer dispersion modeling focuses on discharge of a pollutant and the dispersion of that discharge by the time it reaches the receptor. This method is primarily used for assessing risk from a proposed facility or discharge. Sophisticated techniques are employed to relate reported or measured emissions to atmospheric, climatological, demographic, geographic, and other data in order to predict a population's potential exposure to a given chemical [3].

38.6   RISK CHARACTERIZATION

Risk characterization is the process of estimating the incidence of a health effect under the various conditions of human or animal exposure described in the exposure assessment. It is performed by combining the exposure and dose–response assessments. The summary effects of the uncertainties in the preceding steps should be described in this step. The quantitative estimate of the risk is the principal interest to the regulatory agency or risk manager making the decision. The risk manager must consider the results of the risk characterization when evaluating the economics, societal aspects, and various benefits of the risk assessment. Factors such as societal pressure, technical uncertainties, and severity of the potential hazard influence how the decision makers respond to the risk assessment. There is room for improvement in this step of the risk assessment [2].

A risk estimate indicates the likelihood of occurrence of the different types of health or environmental effects in exposed populations. Risk assessment should include both human health and environmental evaluations (i.e., impacts on ecosystems). Ecological impacts include actual or potential effects on plants and animals (other than domesticated species). The number produced from the risk characterization, representing the probability of adverse health effects being caused, must be evaluated. This is done because certain agencies will only look at risks of specific numbers and act on them.

There are two major types of risk: maximum individual risk and population risk. Maximum individual risk is defined exactly as it implies, that is the maximum risk to an individual person. This person is considered to have a 70-year lifetime of exposure to a process or a chemical. Population risk is again the risk to a population. It is expressed as a certain number of deaths per thousand or per million people. For example, a fatal annual risk of 2×10^{-6} refers to 2 deaths per year for every million individuals. These risks are often based on very conservative assumptions, which may yield too high a risk.

38.7   FUTURE TRENDS

For the most part, future trend accruals will probably be found in expanding the dose–response database. To help promote health risk prevention, companies should start employee training programs. These programs should be designed to alert the technical staff and employees about the health risks they are exposed to on the job surrender type activities will probably take place as the domestic level.

38.8   SUMMARY

1. Health risk assessment is defined as the process or procedure used to estimate the likelihood that humans or ecological systems will be adversely affected by a chemical or physical agent under a specific set of conditions.
2. The health risk evaluation process consists of four steps: hazard identification, dose–response assessment or hazard assessment, exposure assessment, and risk characterization.
3. In hazard identification, a hazard is a toxic agent or a set of conditions that has the potential to cause adverse effects to human health or the environment.
4. In dose–response assessment, effects are evaluated and these effects vary widely because their capacities to cause adverse effects differ.

5. Exposure assessment is the determination of the magnitude, frequency, duration, and routes of exposure to human populations and ecosystems.
6. In risk characterization, the toxicology and exposure data are combined to obtain a quantitative or qualitative expression of risk.
7. A major avenue for reducing risk will involve source reduction of hazardous materials.

REFERENCES

1. AIChE. *Guidelines for Chemical Process Quantitative Risk Analysis*. New York: Center for Chemical Process Safety of the American Institute of Chemical Engineers, 1989.
2. Paustenbach, D. *The Risk Assessment of Environmental and Human Health Hazards: A Textbook of Case Studies*. John Wiley & Sons, Hoboken, NJ, 1989.
3. Burke, G., Singh, B., and Theodore, L. *Handbook of Environmental Management and Technology*, 2nd edn., John Wiley & Sons, Hoboken, NY, 2000.
4. Rodricks, J. and Tardiff, R. *Assessment and Management of Chemical Risks*. Washington, DC: American Chemical Society, 1984.
5. U.S. EPA. *Technical Guidance for Hazards Analysis*. Washington, DC: EPA/FEMA/DOT, December 1987.
6. Clayson, D. B., Krewski, D., and Munro, I. *Toxicological Risk Assessment*. CRC Press, Boca Raton, FL: 1985.
7. Foa, V., Emmett, E. A., Maron, M., and Colombi, A. *Occupational and Environmental Chemical Hazards*. Chichester, U.K.: Ellis Horwood Limited, 1987.

39 Hazard Risk Assessment

CONTENTS

39.1 INTRODUCTION

Risk evaluation of accidents serves a dual purpose. It estimates the probability that an accident will occur and also assesses the severity of the consequences of an accident. Consequences may include damage to the surrounding environment, financial loss, or injury to life. This chapter is primarily concerned with the methods used to identify hazards and the causes and consequences of accidents. Issues dealing with health risks have been explored in Chapter 38. Risk assessment of accidents provides an effective way to help ensure either that a mishap does not occur or reduces the likelihood of an accident. The result of the risk assessment allows concerned parties to take precautions to prevent an accident before it happens.

Regarding definitions, the first thing an individual needs to know is what exactly an accident is. An accident is an unexpected event that has undesirable consequences [1]. The causes of accidents have to be identified in order to help prevent accidents from occurring. Any situation or characteristic of a system, plant, or process that has the potential to cause damage to life, property, or the environment is considered a hazard. A hazard can also be defined as any characteristic that has the potential to cause an accident. The severity of a hazard plays a large part in the potential amount of damage a hazard can cause if it occurs. Risk is the probability that human injury, damage to property, damage to the environment, or financial loss will occur. An acceptable risk is a risk whose probability is unlikely to occur during the lifetime of the plant or process. An acceptable risk can also be defined as an accident that has a high probability of occurring, with negligible consequences. Risks can be ranked qualitatively in categories of high, medium, and low. Risk can also be ranked quantitatively as annual number of fatalities per million affected individuals. This is normally denoted as a number times one millionth that is, 3×10^{-6}; this representation indicates that on the average, for every million individuals three individuals will die every year. Another quantitative approach that has become popular in industry is the fatal accident rate (FAR) concept. This determines or estimates the number of fatalities over the lifetime of 1000 workers. The lifetime of a worker is defined as 10^5 hours, which is based on a 40 hour work week for 50 years. A reasonable FAR for a chemical plant is 3.0 with 4.0 usually taken as a maximum. The FAR for an individual at home is approximately 3.0. A FAR of 3.0 means that there are three deaths for every 1000 workers over a 50 year period.

39.2   RISK EVALUATION PROCESS FOR ACCIDENTS

There are several steps in evaluating the risk of an accident (see Figure 39.1). These are detailed below if the system in question is a chemical plant.

1. A brief description of the equipment and chemicals used in the plant is needed.
2. Any hazard in the system has to be identified. Hazards that may occur in a chemical plant include:

Fire	Explosions
Toxic vapor release	Rupture of a pressurized vessel
Slippage	Runaway reactions
Corrosion	

3. The event or series of events that will initiate an accident has to be identified. An event could be a failure to follow correct safety procedures, improperly repaired equipment, or failure of a safety mechanism.

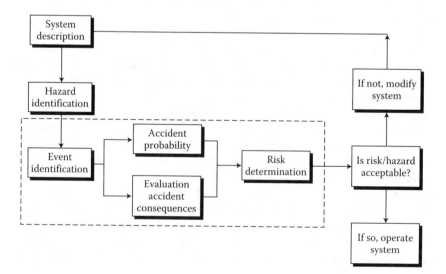

FIGURE 39.1   Hazard risk assessment flowchart.

4. The probability that the accident will occur has to be determined. For example, if a chemical plant has a 10 year life, what is the probability that the temperature in a reactor will exceed the specified temperature range? The probability can be ranked from low to high. A low probability means that it is unlikely for the event to occur in the life of the plant. A medium probability suggests that there is a possibility that the event will occur. A high probability means that the event will probably occur during the life of the plant.
5. The severity of the consequences of the accident must be determined. This will be described later in detail.
6. If the probability of the accident and the severity of its consequences are low, then the risk is usually deemed acceptable and the plant should be allowed to operate. If the probability of occurrence is too high or the damage to the surroundings is too great, then the risk is usually unacceptable and the system needs to be modified to minimize these effects.

The heart of the hazard risk assessment algorithm provided is enclosed in the dashed box (Figure 39.1). This algorithm allows for reevaluation of the process if the risk is deemed unacceptable (the process is repeated starting with either step one or two).

The reader should note that this topic also receives treatment in Chapter 48—Environmental Implications of Nanotechnology, Section 48.5—Hazard Risk Assessment.

39.3 HAZARD IDENTIFICATION

Hazard or event identification provides information on situations or chemicals and their releases that can potentially harm the environment, life, or property. Information that is required to identify hazards includes, chemical identities, quantities, and location of chemicals in question, chemical properties such as boiling points, ignition temperatures, and toxicity to humans. There are several methods used to identify hazards. The methods that will be discussed are the process checklist and the hazard and operability study (HAZOP).

A process checklist evaluates equipment, materials, and safety procedures [1]. A checklist is composed of a series of questions prepared by an engineer who knows the procedure being evaluated. It compares what is in the actual plant to a set of safety and company standards. Some questions that may be on a typical checklist are:

1. Was the equipment designed with a safety factor?
2. Does the spacing of the equipment allow for ease of maintenance?
3. Are the pressure relief valves on the equipment in question?
4. How toxic are the materials that are being used in the process and is there adequate ventilation?
5. Will any of the materials cause corrosion to the pipe(s)/reactor(s)/system?
6. What precautions are necessary for flammable material?
7. Is there an alternate exit in case of fire?

8. If there is a power failure what fail-safe procedure(s) does the process contain?
9. What hazard is created if any piece of equipment malfunctions?

These questions and others are answered and analyzed. Changes are then made to reduce the risk of an accident. Process checklists are updated and audited at regular intervals.

A hazard and operability study is a systematic approach to recognizing and identifying possible hazards that may cause failure of a piece of equipment [2]. This method utilizes a team of diverse professional backgrounds to detect and minimize hazards in a plant. The process in question is divided into smaller processes (subprocesses). Guide words are used to relay the degree of deviation from the subprocesses' intended operation. The guide words can be found in Table 39.1. The causes and consequences of the deviation from the process are determined. If there are any recommendations for revision they are recorded and a report is made.

A summary of the basic steps of a HAZOP study is [2]:

1. Define the objectives.
2. Define the plant limits.
3. Appoint and train a team.
4. Obtain the complete preparative work (i.e., flow diagrams, sequence of events, etc.).
5. Conduct the examination meetings that select subprocesses, agree on intention of subprocesses, state and record intentions, use guide words to find deviations from the intended purpose, determine the causes and consequences of deviation, and recommend revisions.
6. Issue the meeting reports.
7. Follow up on revisions.

There are other methods of hazard identification. A "what–if" analysis presents certain questions about a particular hazard and then tries to find the possible consequences of that hazard. The human error analysis identifies potential human errors

---

TABLE 39.1

Guide Words Used to Relay the Degree of Deviation from Intended Subprocess Operation

Guide Word	Meaning
No	No part of intended function is accomplished
Less	Quantitative decrease in intended activity
More	Quantitative increase in intended activity
Part of	The intention is achieved to a certain percent
As well as	The intention is accomplished along with side effects
Reverse	The opposite of the intention is achieved
Other than	A different activity replaces the intended activity

that will lead to an accident. They can be used in conjunction with the two previously described methods.

39.4　CAUSES OF ACCIDENTS

The primary causes of accidents are mechanical failure, operational failure (human error), unknown or miscellaneous process upset, and design error. Figure 39.2 provides the *relative* number of accidents that have occurred in the petrochemical field [3].

There are three steps that normally lead to an accident:

1. Initiation
2. Propagation
3. Termination

The path that an accident takes through the three steps can be determined by means of a fault tree analysis [1]. A fault tree is a diagram that shows the path that a specific accident takes. The first thing needed to construct a fault tree is the definition of the initial event. The initial event is a hazard or action that will cause the process to deviate from normal operation. The next step is to define the existing conditions needed to be present in order for the accident to occur. The propagation event (e.g., the mechanical failure of equipment related to the accident) is discussed. Any other equipment or components that need to be studied have to be defined. This includes safety equipment that will bring about the termination of the accident. Finally, the normal state of the system in question is determined. The termination of an accident is the event that brings the system back to its normal operation. An example of an accident would be the failure of a thermometer in a reactor. The temperature in the reactor could rise and a runaway reaction might take place. Stopping the flow to the reactor and/or cooling the contents of the reactor could terminate the accident.

Event trees are diagrams that evaluate the consequences of a specific hazard. The safety measures and the procedures designed to deal with the event are presented. The consequences of each specific event that led to the accident are also presented. An event tree is drawn (sequence of events that led up to the accident). The accident

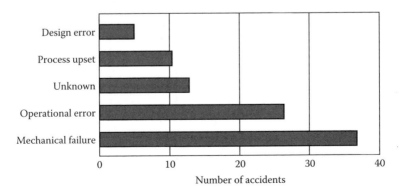

FIGURE 39.2　Causes of accidents in the petrochemical field.

is described. This allows the path of the accident to be traced. It shows what could be done along the way to prevent the accident. It also shows other possible outcomes that could arise had a single event in the sequence been changed.

39.5 CONSEQUENCES OF ACCIDENTS

Consequences of accidents can be classified qualitatively by the degree of severity. A quantitative assessment is beyond the scope of the text; however information is available in the literature. Factors that help to determine the degree of severity are the concentration that the hazard that is released, length of time that a person or the environment is exposed to a hazard, and the toxicity of the hazard. The worst-case consequence or scenario is defined as a conservatively high estimate of the most severe accident identified [1]. On this basis one can rank the consequences of accidents into low, medium, and high degrees of severity [4]. A low degree of severity means that the hazard is nearly negligible, and the injury to person, property, or the environment is observed only after an extended period of time. The degree of severity is considered to be medium when the accident is serious, but not catastrophic, the toxicity of the chemical released is great, or the concentration of a less toxic chemical is large enough to cause injury or death to persons and damage to the environment unless immediate action is taken. There is a high degree of severity when the accident is catastrophic or the concentrations and toxicity of a hazard is large enough to cause injury or death to many persons, and there is long-term damage to the surrounding environment. Figure 39.3 provides a graphical qualitative representation of the severity of consequences [4].

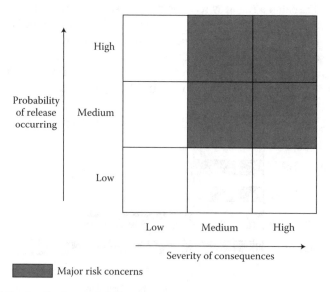

FIGURE 39.3   Qualitative probability–consequence analysis.

39.5.1 CAUSE–CONSEQUENCE ANALYSIS

Cause–consequence risk evaluation combines event tree and fault tree analyses to relate specific accident consequences to causes [1]. The process of cause–consequence evaluation usually proceeds as follows:

1. Select an event to be evaluated.
2. Describe the safety system(s)/procedure(s)/factor(s) that interfere with the path of the accident.
3. Perform an event tree analysis to find the path(s) an accident may follow.
4. Perform a fault tree analysis to determine the safety function that failed.
5. Rank the results on a basis of severity of consequences.

As its name implies cause–consequence analysis allows one to see how the possible causes of an accident and the possible consequences that result from that event interact with each other.

39.6 FUTURE TRENDS

For the most part, future trends will be found in hazard accident prevention, not hazard analysis. To help promote hazard accident prevention, companies should start employee training programs. These programs should be designed to alert the technical staff and employees about the hazards they are exposed to on the job. Training should also cover company safety policies and the proper procedures to follow in case an accident does occur. A major avenue to reducing risk will involve source reduction of hazardous materials. Risk education and communication are two other areas that will need improvement [5].

39.7 SUMMARY

1. Risk assessment of accidents estimates the probability that hazardous materials will be released and also assesses the severity of the consequences of an accident.
2. The risk evaluation process defines the equipment, hazards, and events leading to an accident. It determines the probability that an accident will occur. The severity and acceptability of the risk are also evaluated.
3. Hazard identification provides information on situations or chemicals that can potentially harm the environment, life, or property. The processes include process checklist, event tree, and hazard and operability study.
4. Accidents occur in three steps: initiation, propagation, and termination. The primary causes of accidents are mechanical failure, operational failure (human error), unknown or miscellaneous process upset, and design error.
5. Consequences of accidents are qualitatively classified by degree of severity into low, medium, and high.

6. Cause–consequence analysis allows one to see the possible causes of an accident and the possible accident that results from a certain event.
7. For the most part future trends can be found in hazard prevention.

REFERENCES

1. AIChE. *Guidelines for Hazard Evaluation Procedures.* New York: Batelle Columbus Division for the Center for Chemical Process Safety of the American Institute of Chemical Engineers, 1985.
2. Theodore, L. et al. *Accident and Emergency Management.*: John Wiley & Sons, Hoboken, NJ, 1989.
3. Crowl, J. and Louvar, J. *Chemical Safety Fundamentals with Applications.* Prentice-Hall, Englewood, NJ:, 1990.
4. U.S. EPA, FEMA, USDOT. *Technical Guidance for Hazard Analysis.* Washington, DC: EPA, FEMA, USDOT, 1978.
5. Burke, G., Singh, B., and Theodore, L. *Handbook of Environmental Management and Technology*, 2nd ed. John Wiley & Sons, Hoboken, NJ, 2000.

40 Public Perception of Risk

CONTENTS

40.1 INTRODUCTION

Public concern about risk stems from earthquakes, fires, and hurricanes to asbestos, radon emissions, ozone depletion, toxins in food and water, and so on. Many of the public's worries are out of proportion, with the fear being overestimated or at times underestimated. The risks given the most publicity and attention receive the greatest concern, while the ones that are more familiar and accepted are given less thought.

A large part of what the public knows about risk comes from the media. Whether it is newspapers, magazines, radio, or television, the media provides information about the nature and extent of specific risks. It also helps to shape the perception of the danger involved within a given risk.

Laypeople and experts disagree on risk estimates for many environmental problems. This creates a problem, since the public generally does not trust the experts. This chapter concentrates on how the public views risk and what the future of public risk perception will be.

40.2 EVERYDAY RISKS

The public often worries about the largely publicized risks and thinks little about those that they face regularly. A study was performed that compared the responses of two groups, 15 national risk assessment experts and 40 members of the League of Women Voters on the risks of 30 activities and technologies [1]. This search produced striking discrepancies, as presented in Table 40.1. The League members rated nuclear power as the number 1 risk, while experts numbered it at 20 and the League ranked x-rays at 22 while the experts gave it a rank of 7.

There are various reasons for the differences in risk perception. Government regulators and industry officials look at different aspects in assessing a given risk

TABLE 40.1
Ranking Risks: Reality and Perception

League of Women Voters	Activity or Technology	Experts
1	Nuclear power	20
2	Motor vehicles	1
3	Handguns	4
4	Smoking	2
5	Motorcycles	6
6	Alcoholic beverages	3
7	General (private) aviation	12
8	Police work	17
9	Pesticides	8
10	Surgery	5
11	Fire fighting	18
12	Large construction	13
13	Hunting	23
14	Spray cans	26
15	Mountain climbing	29
16	Bicycles	15
17	Commercial aviation	16
18	Electric power (nonnuclear)	9
19	Swimming	10
20	Contraceptives	11
21	Skiing	30
22	X-rays	7
23	High school and college football	27
24	Railroads	19
25	Food preservatives	14
26	Food coloring	21
27	Power motors	28
28	Prescription antibiotics	24
29	Home appliances	22
30	Vaccinations	25

Source: Goleman, D., *New York Times*, February 1, 1994.

The rankings of perceived risks for 30 activities and technologies, based on averages in a survey of a group of experts and a group of informed laypeople, members of the League of Women Voters. A ranking of 1 denotes the highest level of perceived risk.

than would members of the community. The "experts" will look at the mortality rates to assess risk, while the "laypeople" worry about their children and the potential long-term health risks. Another reason for the difference is that people take reports of bad news more to heart than they would a report that might increase their trust.

Problems exist with risk estimates because the substance or process in question may be calculated to present too high a risk. To understand the significance of risk analyses, a list of everyday risks derived from actual statistics and reasonable estimates is presented in Table 40.2 [2]. A lifetime risk of 70×10^{-6} means that 70 out of 1 million people will die from that specific risk.

Risk managers in government and industry have started turning to risk communication to bridge the gap between the public and the "experts." (Chapter 41 treats

TABLE 40.2
Lifetime Risks to Life Commonly Faced by Individuals

Cause of Risk	Lifetime (70-Year) Risk per Million
Cigarette smoking	252,000
All cancers	196,000
Construction	42,700
Agriculture	42,000
Police killed in the line of duty	15,400
Air pollution (Eastern United States)	14,000
Motor vehicle accidents (traveling)	14,000
Home accidents	7,700
Frequent airplane traveler	3,500
Pedestrian hit by motor vehicle	2,940
Alcohol, light drinker	1,400
Background radiation at sea level	1,400
Peanut butter, four tablespoons per day	560
Electrocution	370
Tornado	42
Drinking water containing chloroform at allowable EPA limit	42
Lightning	35
Living 70 years in zone of maximum impact from modern municipal incinerator	1
Smoking 1.4 cigarettes	1
Drinking 0.5 L of wine	1
Traveling 10 miles by bicycle	1
Traveling 30 miles by car	1
Traveling 1000 miles by jet plane (air crash)	1
Traveling 6000 miles by jet plane (cosmic rays)	1
Drinking water containing trichloroethylene at maximum allowable EPA limit	0.1

Sources: Data from Charles T. Main, Inc. *Health Risk Assessment for Air Emissions of Mends and Organic Compounds from the PERC Municipal Waste to Energy Facility.* Prepared for Penobscot Energy Recovery Company (PERC), Boston, MA, December 1985; and Wilson, R. and Crouch, E.A., *Science*, 236, 267–270, April 1987.

the general subject of risk communication in more detail.) Table 40.2 enables the public to see that certain everyday risks are higher than some dreaded environmental risks. It shows that eating peanut butter possesses a greater risk than toxins in the air or water.

40.3 ENVIRONMENTAL RISKS

In 1987, the EPA released a report titled "Unfinished Business: A Comparative Assessment of Environmental Problems" in order to apply the concepts of risk assessment to a wide array of pressing environmental problems. It is difficult to make direct comparisons of different environmental problems, because most of the data is usually insufficient to quantify risks. Also, risks associated with some problems are incomparable with risks of others. The study was based on a list of 31 environmental problems. Each was analyzed in terms of four different types of risks: cancer risks, noncancer health risks, ecological effects, and welfare effects (visibility impairment, materials damage, etc.)

The ranking of cancer was probably the most straightforward part of the study since the EPA had already established risk assessment procedures and there are considerable data already available from which to work. Two problems were considered at the top of the list: the first was worker exposure to chemicals which does not involve a large number of individuals but does result in high individual risks to those exposed; the other problem was radon exposure, which is causing considerable risk to a large number of people. The results from the cancer report are provided in Table 40.3 [4].

The other working groups had greater difficulty when ranking the 31 environmental problem issues because there are no accepted guidelines for quantitatively assessing relative risks. As noted in the EPA's study, the following general results were produced for each of the four types of risks described [5]:

1. No problems rank high in all four types of risk, or relatively low in all four.
2. Problems that rank relatively high in three of the four types, or at least medium in all four, include criteria air pollutants, stratospheric ozone depletion, pesticide residues on food, and other pesticide risks (runoff and air deposition of pesticides).
3. Problems that rank relatively high in cancer and noncancer health risks but low in ecological and welfare risks include hazardous air pollutants, indoor radon, indoor air pollution other than radon, pesticide application, exposure to consumer products, and worker exposures to chemicals.
4. Problems that rank relatively high in ecological and welfare risk but low in both health risks include global warming, point and nonpoint sources of surface water pollution, physical alteration of aquatic habitats (including estuaries and wetlands), and mining wastes.
5. Areas related to groundwater consistently rank medium or low.

Although there were great uncertainties involved in making these assessments, the divergence between the EPA effort and relative risks is noteworthy. From this study, areas of relatively high risk but low EPA effort/concern include indoor radon,

TABLE 40.3
Consensus Ranking of Environmental Problem Areas on the Basis of Population Cancer Risk

Rank	Problem Area	Selected Comments
1 (tied)	Worker exposure to chemicals	About 250 cancer cases per year estimated based on exposure to four chemicals; but workers face potential exposures to over 20,000 substances. Very high individual risk possible.
1 (tied)	Indoor radon	Estimated 5,000–20,000 lung cancers annually from exposure in homes.
3	Pesticide residues on foods	Estimated 6000 cancers annually, based on exposure to 200 potential oncogens.
4 (tied)	Indoor air pollutants (nonradon)	Estimated 3500–6500 cancers annually, mostly due to tobacco smoke.
4 (tied)	Consumer exposure to chemicals	Risk from four chemicals investigated is about 100–135 cancers annually; an estimated 10,000 chemicals in consumer products. Cleaning fluids, pesticides, particleboard, and asbestos-containing products especially noted.
6	Hazardous/toxic air pollutants	Estimated 2000 cancers annually based on an assessment of 20 substances.
7	Depletion of stratospheric ozone	Ozone depletion projected to result in 10,000 additional annual deaths in the year 2100; not ranked higher because of the uncertainties in future risk.
8	Hazardous waste sites, inactive	Cancer incidence of 1000 annually from six chemicals assessed. Considerable uncertainty since risk based on extrapolation from 35 sites to about 25,000 sites.
9	Drinking water	Estimated 400–1000 annual cancers, mostly from radon and trihalomethanes.
10	Application of pesticides	Approximately 100 cancers annually; small population exposed but high individual risks.
11	Radiation other than radon	Estimated 360 cancers per year. Mostly from building materials. Medical exposure and natural background levels not included.
12	Other pesticide risks	Consumer and professional exterminator users estimated cancers of 150 annually. Poor data.
13	Hazardous waste sites, active	Probably fewer than 100 cancers annually; estimates sensitive to assumptions regarding proximity of future wells to waste sites.
14	Nonhazardous waste sites, industrial	No real analysis done, ranking based on consensus of professional opinion.
15	New toxic chemicals	Difficult to assess; done by consensus.
16	Nonhazardous waste sites, municipal	Estimated 40 cancers annually not including municipal surface impoundments.
17	Contaminated sludge	Preliminary results estimate 40 cancers annually, mostly from incineration and landfilling.

(continued)

TABLE 40.3 (continued)

Consensus Ranking of Environmental Problem Areas on the Basis of Population Cancer Risk

Rank	Problem Area	Selected Comments
18	Mining waste	Estimated 10–20 cancers annually, largely due to arsenic. Remote locations and small population exposure reduce overall risk though individual risk may be high.
19	Releases from storage tanks	Preliminary analysis, based on benzene, indicates low cancer incidence (<1).
20	Nonpoint-source discharges to surface water	No quantitative analysis available; judgment.
21	Other groundwater contamination	Lack of information; individual risks considered less than 10^{-6}, with rough estimate of total population risk at <1.
22	Criteria air pollutants	Excluding carcinogenic particles and VOCs (included under hazardous/toxic air pollutants); ranked low because remaining criteria pollutants have not been shown to be carcinogens.
23	Direct point–source discharges to surface water	No quantitative assessment available. Only ingestion of contaminated seafood was considered.
24	Indirect point–source discharges to surface water	Same as above.
25	Accidental releases, toxics	Short-duration exposure yields low cancer risk; noncancer health effects of much greater concern.
26	Accidental releases, oil spills	See above. Greater concern for welfare and ecological effects.

Source: Based on data from USEPA, 1987.

Not ranked: Biotechnology; global warming; other air pollutants; discharges to estuaries, coastal waters, and oceans; and discharges to wetlands.

indoor air pollution, stratospheric ozone depletion, global warming, nonpoint sources, discharges to estuaries, coastal waters and oceans, other pesticide risks, accidental releases of toxics, consumer products, and worker exposures. The EPA gives high concern but relatively medium or low risks to RCRA sites, Superfund sites, underground storage tanks, and municipal nonhazardous waste sites.

40.4 OUTRAGE FACTORS

The perception of a given risk is amplified by what are known as "outrage" factors. These factors can make people feel that even small risks are unacceptable. More than 20 outrage factors have been identified; a few of the main ones are defined below [6]:

1. *Voluntariness.* A voluntary risk is much more acceptable to people than an imposed risk. People will accept the risk from skiing, but not from food preservatives, although the potential for injury from skiing is 1000 times greater than from preservatives.
2. *Control.* Risks that people can take steps to control are more acceptable than those they feel are beyond their control. When prevention is in the hands of the individual, the risk is perceived much lower than when it is in the hands of the government. You can choose what you eat, but you cannot control what is in your drinking water.
3. *Fairness.* Risks that seem to be unfairly shared are believed to be more hazardous. People who endure greater risk than their neighbors and do not attain anything from it are generally outraged by this. If one is not getting anything from it, why should others benefit?
4. *Process.* The public views the agency: Is it trustworthy or dishonest, concerned or arrogant? If the agency tells the community what's going on before decisions are made, the public feels more at ease. They also favor a company that listens and responds to community concerns.
5. *Morality.* Society has decided that pollution is not only harmful, it is evil. Talking about cost-risk tradeoffs sounds cold-hearted when the risk is morally relevant.
6. *Familiarity.* Risks from exotic technologies create more dread than do those involving familiar ones. "A train wreck that takes many lives has less impact on people's trust of trains than would a smaller, hypothetical accident involving recombinant DNA, which is only perceived to have catastrophic mishaps" [1].
7. *Memorability.* An incident that remains in the public's memory makes the risk easier to imagine and is, therefore, more risky.
8. *Dread.* There are some illnesses that are feared more than others. Today there is greater fear given to AIDS and cancer than there is to asthma.

These outrage factors are not distortions in the public's perception of risk. They are inborn parts of what is interpreted as risk. They are explanations of why the public fears pollutants in the air and water more than they do geological radon. The problem is that many risk experts resist the use of the public's "irrational fear" in their risk management.

40.5 FUTURE TRENDS

A problem exists in the perception of risk because the experts' and laypeople's views differ. The experts usually base their assessment on mortality rates, while the laypeople's fears are based on "outrage" factors. In order to help solve this problem, in the future, risk managers must work to make truly serious hazards more outrageous. One example is the recent campaign for the risk involved in cigarette smoke. Another effort must be made to decrease the public's concern with low to modest hazards, i.e., risk managers must diminish "outrage" in these areas. In addition, people must be treated fairly and honestly.

40.6 SUMMARY

1. Public concern of risk spans a wide range, from fires and hurricanes to radon emissions and toxins in water and air. Most of what the public knows of risk comes from the media.
2. People often overestimate risks that are highly publicized and worry little about the familiar risks that they face everyday.
3. The EPA released a study entitled "Unfinished Business: A Comparative Assessment of Environmental Problems," which compared four different types of risks for problem areas. It helped the EPA decide which problems should be given priority.
4. The public pays a great deal of attention to "outrage" factors which are all deciding factors of risk except the death rate. They are innate parts of what is meant by risk.
5. The experts need to pay more attention to the public's outrage factors. More emphasis must be placed on the higher hazardous risks and an effort is needed to decrease the public's concern on medium to light risks.

REFERENCES

1. Goleman, D. Assessing risk: Why fear may outweigh harm. *New York Times*, February 1, 1994.
2. Adapted from: Charles T. Main, Inc. *Health Risk Assessment for Air Emissions of Mends and Organic Compounds from the PERC Municipal Waste to Energy Facility.* Prepared for Penobscot Energy Recovery Company (PERC), Boston, MA, December 1985.
3. Wilson, R. and Crouch, E. A. Risk assessment and comparisons: An introduction. *Science*, 236, 267–270, April 1987.
4. Based on data from U.S. EPA, 1987.
5. Masters, G. *Introduction to Environmental Engineering and Science.* Prentice Hall, Englewood Cliffs, NJ, 1991.
6. Burke, G., Singh, B., and Theodore, L. *Handbook of Environmental Management and Technology*, 2nd ed. John Wiley & Sons, Hoboken, NJ, 2000.

41 Risk Communication

CONTENTS

41.1 INTRODUCTION

Environmental risk communication is one of the more important problems that this country faces. Since 1987, public concerns about the environment have grown faster than concern about virtually any other national problem [1]. Some people are suffering (and in some cases dying) because they do not know when to worry and when to calm down. They do not know when to demand action to reduce risk or when to relax because risks are trivial or even nonexistent. The key, of course, is to pick the right worries and right actions. Unfortunately, when it comes to health and the environment, society does not do that well. The government and media together have failed to communicate clearly what is a risk and what is not a risk.

There are two major categories of risk: nonfixable and fixable. Nonfixable risks can never substantially be reduced, such as cancer-causing sunlight or cosmic radiation. Fixable risks can be reduced, and include those risks that are both large and small. There are so many of these fixable risks that all of them can never be successfully attacked, so choices must be made. When it comes to risk reduction, the outcome should be to obtain the most reduction possible, taking into account that people fear some risks more than others. This essentially means that the technical community should concentrate on the big fixable targets, and leave the smaller ones to later.

Risk communication comes into play because citizens ultimately determine which risks government agencies attack. On the surface, it appears practical to remedy the most severe risks first, leaving the others until later or perhaps, if the risks are small enough, never remedying the others at all. However, the behavior of individuals in everyday life often does not conform with this view. Consider now two environmental issues: gasoline that contains lead, and ocean incineration [2].

According to the Environmental Protection Agency (EPA), lead in gasoline poses very large risks: risks of learning disabilities, mental retardation, and worse to hundreds of thousands of children. The EPA's decision to reduce lead in gasoline is the most significant protective action the agency has undertaken in a long time. The only difference encountered on this issue was that the public acted with virtual indifference.

On the other hand, citizens threatened to lie down bodily in front of trucks and blockade harbors to stop the EPA's proposal to allow final testing of ocean incineration. The public reacted irrationally here. Every indication showed that the risk involved was small, and that the technology would be replacing more risky alternatives now in use [2].

Why is there such an imbalance on the perception of risk? Ironically, part of the reason lies in the fact that the people responsible for communicating this information did their job too well. They achieved their objective to get the information out to the public. Unfortunately, their objectives did not include the effective communication of risk.

The professionals at the EPA are quite precise in the statements they deliver concerning risks and their apparent hazards. Their job is to present a scientifically defensible product, so they add qualifiers and use scientific terms. The problem with this is that often the public receives a misunderstanding of the actual risk [2].

The challenge of risk communication is to provide the information in ways that can be incorporated in the views of people who have little time or patience for arcane scientific discourse. Success in risk communication is not to be measured by whether the public chooses to set the outcomes that minimize risk as estimated by the experts; it is achieved instead when those outcomes are knowingly chosen by a well-informed public [1].

When citizens understand a risk, and the cost of reducing it, they can determine for themselves if control actions are too lax, too stringent, or just right. The two previous cases that were used as an example demonstrate that the risk message is not getting through to people who need to know when to demand action and when to calm down. The answer is not to communicate more information, but more pertinent and understandable information. All the public needs to know is the following three pieces of information: How big is the risk? What is being done about it? What will it cost? [2].

This chapter will focus upon the communication of more pertinent risk information, how to get the message across in terms that are easy for an average citizen to understand, what concerned citizens can do to have a role in vital environmental solutions, and the accessibility in environmental communication to keep the public informed.

41.2   SEVEN CARDINAL RULES OF RISK COMMUNICATION

There are no easy prescriptions for successful risk communication. However, those who have studied and participated in recent debates about risk generally agree on seven cardinal rules. These rules apply equally well to the public and private sectors. Although many of these rules may seem obvious, they are continually and consistently violated in practice. Thus, a useful way to read these rules is to focus on why they are frequently not followed [3].

1. **Accept and involve the public as a legitimate partner.** A basic tenet of risk communication in democracy is that people and communities have a right to participate in decisions that affect their lives, their property, and the things they value.

   Guidelines: Demonstrate your respect for the public and underscore the sincerity of your effort by involving the community early, before important decisions are made. Involve all parties that have an interest or stake in the issue under consideration. If you are a government employee, remember that you work for the public. If you do not work for the government, the public still holds you accountable.

   Point to Consider: The goal in risk communication in a democracy should produce an informed public that is involved, interested, reasonable, thoughtful, solution-oriented, and collaborative; it should not diffuse public concerns or replace action.

2. **Plan carefully and evaluate your efforts.** Risk communication will be successful only if carefully planned.

   Guidelines: Begin with clear, explicit risk communication objectives, such as providing information to the public, motivating individuals to act, stimulating response to emergencies, and contributing to the resolution of conflict. Evaluate the information you have about the risks and know its strengths and weaknesses. Classify and segment the various groups in your audience. Aim your communications at specific subgroups in your audience. Recruit spokespeople who are good at presentation and interaction. Train your staff, including technical staff, in communication skills; reward outstanding performance. Whenever possible, pretest your messages. Carefully evaluate your efforts and learn from your mistakes.

   Points to Consider: There is no such entity as "the public"; instead, there are many publics, each with its own interests, needs, concerns, priorities, preferences, and organizations. Different risk communication goals, audiences, and media require different risk communication strategies.

3. **Listen to the public's specific concerns.** If you do not listen to the people, you cannot expect them to listen to you. Communication is a two-way activity.

   Guidelines: Do not make assumptions about what people know, think, or want done about risks. Take the time to find out what people are thinking: use techniques such as interviews, focus groups, and surveys. Let all parties that have an interest or stake in the issue be heard. Identify with your audience and try to put yourself in their place. Recognize people's emotions, Let people know that you understand what they said, addressing their concerns as well as yours. Recognize the "hidden agendas," symbolic meanings, and broader economic or political considerations that often underlie and complicate the task of risk communication.

Point to Consider: People in the community are often more concerned about such issues as trust, credibility, competence, control, voluntariness, fairness, caring, and compassion than about mortality statistics and the details of quantitative risk assessment.

4. **Be honest, frank, and open.** In communicating risk information, trust and credibility are your most precious assets.

Guidelines: State your credentials; but do not ask or expect to be trusted by the public. If you do not know an answer or are uncertain, say so. Get back to people with answers. Admit mistakes. Disclose risk information as soon as possible (emphasizing any reservations about reliability). Do not minimize or exaggerate the level of risk. Speculate only with great caution. If in doubt, lean toward sharing more information, not less, or people may think you are hiding something. Discuss data uncertainties, strengths, and weaknesses, including the ones identified by other credible sources. Identify worst-case estimates as such, and cite ranges of risk estimates when appropriate.

Point to Consider: Trust and credibility are difficult to obtain. Once lost they are almost impossible to regain completely.

5. **Coordinate and collaborate with other credible sources.** Allies can be effective in helping you communicate risk information.

Guidelines: Take time to coordinate all interorganizational and intraorganizational communications. Devote effort and resources to the slow, hard work of building bridges with other organizations. Use credible and authoritative intermediates. Consult with others to determine who is best able to answer questions about risk. Try to issue communications jointly with other trustworthy sources (for example, credible university scientists and/or professors, physicians, or trusted local officials).

Point to Consider: Few things make risk communication more difficult than conflicts or public disagreements with other credible sources.

6. **Meet the needs of the media.** The media are a prime transmitter of information on risks; they play a critical role in setting agendas and in determining outcomes.

Guidelines: Be open and accessible to reporters. Respect their deadlines. Provide risk information tailored to the needs of each type of media (for example, graphics and other visual aids for television). Prepare in advance and provide background material on complex risk issues. Do not hesitate to follow up on stories with praise or criticism, as warranted. Try to establish long-term relationships of trust with specific editors and reporters.

Point to Consider: The media are frequently more interested in politics than in risk; more interested in simplicity than in complexity; more interested in danger than in safety.

7. **Speak clearly and with compassion.** Technical language and jargon are useful as professional shorthand, but they are barriers to successful communication with the public.

Guidelines: Use simple, nontechnical language. Be sensitive to local norms, such as speech and dress. Use vivid, concrete images that communicate on a personal level. Use examples and anecdotes that make technical risk data come alive. Avoid distant, abstract, unfeeling language about deaths, injuries, and illnesses. Acknowledge and respond (both in words and with action) to emotions that people express anxiety, fear, anger, outrage, helplessness, etc. Acknowledge and respond to the distinctions that the public views as important in evaluating risks, e.g., voluntariness, controllability, familiarity, dread, origin (natural or man-made), benefits, fairness, and catastrophic potential. Use risk comparisons to help put risks in perspective, but avoid comparisons that ignore distinctions which people consider important. Always try to include a discussion of actions that are under way or can be taken. Tell people what you cannot do. Promise only what you can do, and be sure to do what you promise.

Points to Consider: Regardless of how well you communicate risk information, some people will not be satisfied. Never let your efforts to inform people about risks prevent you from acknowledging and saying that any illness, injury, or death is a tragedy. And finally, if people are sufficiently motivated, they are quite capable of understanding complex risk information, even if they may not agree with you.

The preceding seven cardinal rules of risk communication only seem logical. It is when they are violated that the proper and necessary communication will fail. Because it is the public that determines which risks will be remedied first, it is important to work with them, getting them involved in the decision-making process before it is too late. When one has the cooperation of the public, carefully state the objectives. Work with these objectives to provide necessary information, and motivate the involved individuals to act. Be a listener as well as a talker. Find out what the people want to know and let their voices be heard. Be honest with the issues at hand. State the truth and do not tell people what they may want to hear; they usually only want to know the truth. Work with, not against or in competition with, other credible sources. Get the message across in all possible ways, whether it be pamphlets, radio, or television. Most importantly, speak clearly and in terms that can be understood by everyone. Take into account the concerned people, and work on a personal level. All this is necessary for communicating and being heard. Successful communication will surely follow if the seven rules are enacted.

41.3 COMMUNICATING RISK TO THE PUBLIC: GETTING THE MESSAGE ACROSS

A comprehensive community outreach program covers one or more of the following categories: information and education, the development of a receptive audience, disaster warning and emergency information, and/or mediation and conflict resolution. In developing a community outreach program, three steps must be understood and addressed [4]:

1. Knowing the audience and its concerns. What is their "perception" of the risks involved?
2. Understanding the issue objectives. What is the "message" to get across?
3. Implementing the communication plan. How do we get the message out?

Each of these steps is vital to a successful community outreach plan. The answers to each of these questions will differ depending on the scenario at hand, and the mechanisms for communicating the risks involved will also differ.

The material below will examine four example scenarios in hazardous material/waste management that require risk communication: emergency response, remediation, facility siting, and ongoing plant operations [4].

41.3.1 EMERGENCY RESPONSE

A fire in a chemical storage warehouse is burning out of control. The fire department is on the scene, the nearby highway is closed, traffic is at a standstill, and the radio and television stations are getting up-to-date reports on the scene. Many of the local residents have heard about the fire, and few know what to do, and most are in a state of panic.

Communicating risk to this audience of potentially affected local residents and merchants means providing emergency information and guidance to a frightened public. The immediate concerns of this audience are to learn the threat to their health and safety, and to do something to protect themselves.

The central risk communication issue for an emergency response is generally to disseminate health and safety information. In conjunction with specific emergency information, the risk communicator has an even greater risk reassurance. While a company may be anxious about legal liability, it should still strive to address the community's concerns as completely as possible.

Getting the message across of reassurance, coupled with emergency directives, requires directly addressing the community's concerns. The company must work directly with health and safety personnel to immediately ascertain the risks involved, to inform the community of those risks, and present steps or actions to mitigate them.

The mechanisms for getting the message out during an emergency situation are straightforward. The news media will carry the story, and so will local radio and television stations. In the case of an extreme emergency, direct contact will be made via a mobile public address system, or through door-to-door notification.

41.3.2 REMEDIATION

For years, local industrial companies had been dumping their hazardous wastes into an unlined lagoon. At the time the practice was legal, but now the lagoon is a designated Superfund site. Residents have always fretted over the smell, and now their concerns are more severe. Research indicates possible cancer risks, but no cases have been noted. Local residents are perplexed about these reports, and are worried about property value, their children, and so on. …

Communicating risk to individuals living or working near a hazardous waste site involves a different message, and different methods for getting the message out than that required for an emergency response situation.

The risk communicator in this case must be able to communicate health risk and offer reassurance. The problem here is that the situation does not require immediate action. Health problems can only be assessed after studying the site for some time. This gives the public time to ponder the possible negative effects it will have on them.

The process of site remediation requires site investigation, risk assessment, determination of alternate remediation methods, and subsequent selection of the best alternative. Implementation of various remediation alternatives has various levels and types of risk. The association of risk with each alternative introduces the concept of risk acceptance to the public. The message of the risk communicator must be to place these risks in perspective, while assuring the community that the safest remediation is being sought. Encouraging the community to participate in the remediation planning process may be the best way of assuring the community that the safest solution is indeed being sought.

Mechanisms for getting information on health risks out to the community could include fact sheets, newsletters, and direct contact. How each of these is an effective means for communicating the risks involved is discussed later in this chapter.

41.3.3   FACILITY SITING

A small community is concerned. Their understanding of the risks associated with hazardous wastes is limited, but they should not need to accept any risk. They see no benefits from siting a hazardous waste treatment facility in their community. They only foresee big problems. …

The siting of a hazardous waste treatment, storage, and disposal facility is viewed as a voluntary option to the local community. If they choose to oppose this facility, they can impose cost delays on the project, even to the point of prohibiting the new facility altogether.

Such was the case in a small town on the south shore of Long Island. Since the county's sewage treatment plant was located in East Rockaway, the county also sought to build a sludge dewatering facility in this small, budding young community. Local residents were fearful of bringing in the contaminated sludge to be treated. They opposed the trucks traveling on the main roads, past the schools and local shops. Opposition arose, various town meetings were held, and a delay was imposed on the building of this site.

The local community will act as a friend or a foe to the siting plan, depending on its perspective risks and benefits of the new facility. The important message to get across to this audience is that the benefits of siting this new facility will outweigh the actual risks.

Risks to the community should be presented in the same way as mentioned above, except that the information should include information on the benefits of the project. Some of the possible benefits could be new jobs, and a safer, less costly method for treating sludge. The site will also bring in a profit to the town, and the money will be used to enhance the quality of the existing neighborhoods.

Education is imperative in appraising the local community of the benefits of hosting a waste facility in their area. Education includes information on wastes generated in their area, current management practices, current disposal options, and waste reduction and recycling potential.

41.3.4 ONGOING PLANT OPERATIONS

California's Proposition 65, the OSHA "Right-to-Know" standard, SARA Title 313, and California's AB 2588 the Hot Toxics Spots bill all require companies handling hazardous materials to appraise their workers and/or the local community of the hazards present at their facilities. A consequence of these information dissemination standards may be to alienate the local community from a "risky" facility in their area. A highly publicized tragedy, such as the accidental release of a poisonous gas from a chemical manufacturing plant in Bhopal, India, could further damage the public's image of similar facilities.

The concern for facilities that handle hazardous wastes is to foster a positive image, particularly within the local community. Maintaining this positive image fosters community support for the company's activities, and helps dissuade negative public reaction in the event of an accident.

As in any risk communication effort, the primary task is to establish and maintain credibility. A facility open house is useful for introducing the public to the facility, and for building understanding and confidence in facility operations and safety precautions. Convenient public access to release and emissions data, as well as public information hotlines for disseminating that data and answering other questions are other effective measures for fostering public trust.

41.4 SPECIFIC METHODS OF COMMUNICATING RISK

Regarding understanding and cooperation in risk communication, the content of the message is determined by the intention of the risk communicator. For each of the scenarios discussed above, the intention of the communicator varied. The basic principle is: Whatever it is that is to be conveyed to a worried public, the risk communicator must understand the concerns of the audience and respond directly to these concerns.

Environmental communication takes on many forms; it normally depends on the audience it is intended to serve. At times, it involves either (1) testifying in court, (2) testifying before the Congress of the United States or appearing before committees representing the Congress, (3) delivering speeches at educational facilities, ranging from kindergarten to colleges and universities, (4) addressing seminars, conferences, or workshops, (5) responding to a press release or making appearances on television, and/or (6) preparing technical documents, handbooks, guides, or pamphlets on environmental issues in both the public and private sectors. These are just a few of the activities involving environmental communications. The EPA established the Office of External Programs to assist individuals, particularly within the Agency, to accomplish this complex and critical activity [1].

Environmental communication can also take on other significant means of communication that are not as complex. They include printed material in the form of technical documents, pamphlets, handouts, brochures, magazines, journals, issue papers, newspaper articles, editorials, and the like.

The 1990 National Environmental Education Act contains a provision that specifically calls on the EPA to work with "noncommercial educational broadcasting entities" to educate Americans on environmental problems. Educational or public broadcasting reaches vast numbers of Americans. Many of the public television and radio stations are locally based and independently owned. As such, they are aware of the needs and concerns of their local communities. And, because public broadcasting considers all its programs to be educational, they are reached not only in homes, but into the schools as well. Almost all public television stations provide outreach activities to supplement and support all their programming.

Spurred by growing concerns about global environmental problems, the entertainment industry is in the midst of a massive conscience raising effort on a variety of environmental issues. Although it is not the first social issue adopted by the show business industry, it just might be the catalyst for the most far-reaching public-interest campaign yet launched by the industry [1].

Fact sheets present detailed information on the site, proposed remediation techniques, health risk analyses, and other information in a readily understandable format to be mailed to libraries, schools, business organizations, and local residents. Newsletters present information similar in content to the fact sheet, but include additional, more general information on hazardous waste management. Direct contact entails walking door-to-door to discuss the proposed remediation project and other issues related to site. This approach gives the risk communicator an opportunity to directly address local residents concerns, thereby giving residents the feeling that they do have a say. In some cases, this method is the only way to understand the opinions of the residents regarding remediation [2].

Largely as a result of the 1985 disaster in Bhopal, India, and the releases of other toxic chemicals in Institute, West Virginia, Congress shortly thereafter added the Emergency Planning and Community Right-to-Know Act (EPCRA) to the Superfund Amendments and Reauthorization Act of 1986 (SARA). As discussed in earlier chapters, EPCRA, often referred to as SARA Title III, establishes emergency planning districts to prepare for the appropriate response to releases of hazardous chemicals. It also suggests recommendations for communicating these risks. In addition, EPCRA requires the annual reporting of releases of hazardous chemicals to the environment. This annual inventory has spurred action through legislation and public pressure to force companies to reduce releases to the environment [5]. If one believes that a business which is subject to the EPCRA requirements failed to report to the Toxics Release Inventory, they should immediately contact: Office of Compliance Monitoring (EN342), U.S. Environmental Protection Agency, Washington, DC 20460 [1].

SARA Title III also contains strategies for explaining very small risks in a community context. The objectives of Title III are to improve local chemical emergency response capabilities, primarily through improved emergency planning and notification, and to provide citizens and local governments with access to information about chemicals in their localities. Title III has four major sections that aid in the development of contingency plans. They are as follows [1]:

1. Emergency Planning (Sections 301–303)
2. Emergency Notification (Section 304)
3. Community Right-to-Know Reporting Requirements (Sections 311 and 312)
4. Toxic Chemicals Release Reporting Emissions Inventory (Section 313)

Title III has also developed timeframes for the implementation of the Emergency Planning and Community Right-to-Know Act of 1986.

The relationship between Title III data and community action can best occur at the local level, through the work of the Local Emergency Planning Committee (LEPC). LEPCs are crucial to the success of the Emergency Planning and Community Right-to-Know Act. Appointed by the State Emergency Response Commissions (SERCs), local planning committees must consist of representatives of all the following groups and organizations: elected state and local officials; law enforcement, civil defense, firefighting, first aid, health, and local environmental and transportation agencies; hospitals; broadcast and print media; community groups; and, representatives of facilities subject to the Emergency Planning and Community Right-to-Know Requirements [6].

It was clear from the outset that the public could not put persistent and informed pressure on the EPA without a steady flow of information and guidance from the Agency. Meeting that need has been the purpose of the EPA's public participation programs. Their mission is threefold [1]:

1. To keep the public informed of important developments in the EPA's program areas.
2. To provide technical information and, if necessary, translate that information into plain English.
3. To ensure that the EPA takes community viewpoints into account in implementing these programs.

41.5   THE CITIZEN'S ROLE IN ENVIRONMENTAL ENFORCEMENT

An important part in the communication of risk is to get involved. When information is put forth, take notice and take action. No one will know of one's concerns or problems if one keeps quiet. Communication is a two-way street. It often must be crossed several times in order to be heard. There are two important things to do when one becomes aware of a potential pollution problem: (1) make careful observations of the problem, and (2) report it to the proper authorities.

On sighting a potential problem, fully record any observations, including the date and time, where notice of the problem took place, and how information came about the problem. Try to identify the responsible parties. In the case of dumping, write down the license plate number. If the pollution aspect is visible, take pictures. In the event of no other witnesses, pictures can only tell what they saw; they do not lie.

Once this information has been gathered, call and inform the local or state authorities. When in contact with a representative, carefully give them all the information observed and ask them to look further into the problem. Follow up with phone calls to that person to be sure it has been taken into consideration, and that something is going to be done about it. If that should fail, call them back and ask to speak to the official supervisor or boss. If phone conversation is not possible, write them. If all the above fails, call the EPA regional office that covers the area of concern. If the pollution problem persists and the local, state, and regional EPA offices appear unwilling or unable to help, then contact the EPA Headquarters in Washington, D.C.

Finally, if told that the pollution problem observed is legal, but one firmly believes that it should not be, feel free to suggest changes in the law by writing to the appropriate U.S. Senator or Representative in Washington, D.C., or to the state governor or state legislatures to inform them of the problem. Local libraries should have the names and addresses of these elected officials.

41.6 ACCESSIBILITY IN ENVIRONMENTAL COMMUNICATION

The availability and accessibility of means to ensure environmental communication is crucial in establishing effective communications. Hotlines, toll-free numbers, and information lines provide the consumer a vital link with the EPA's environmental programs, technical capabilities, and services. The EPA is among several environmental agencies currently using these state-of-the-art means of information dissemination and service to the public [1].

Toll-Free Numbers Offered by EPA Headquarters

RCRA/Superfund Hotline National Toll-Free 1-800-424-9346; Washington, D.C. Metro, 1-202-382-3000. The EPA's largest and busiest toll-free number, the RCRA/Superfund Hotline answers nearly 10,000 questions and document requests each year. The RCRA/Superfund Hotline can be reached Monday through Friday from 8:30 AM to 4:30 PM Eastern Standard Time (EST).

National Response Center Hotline, National Toll-Free 1-800-424-8802; Washington, D.C. Metro, 1-202-426-2675. Operated by the U.S. Coast Guard, this hotline responds to all kinds of accidental releases of oil and hazardous substances. Call this number 24 h a day, 7 days a week, every day of the year to report chemical spills.

Chemical Emergency Preparedness Program (CEPP) Hotline, National Toll-Free 1-800-535-0202; Washington, D.C. Metro, and Alaska, 1-202-479-2449. Responds to questions concerning community preparedness for chemical accidents. The recent Superfund Amendments and Reauthorization Act (SARA) has increased the CEPP Hotline's responsibilities. Calls are answered Monday through Friday from 8:30 AM to 4:30 PM EST.

Asbestos Hotline, National Toll-Free 1-800-334-8571, extension 6741. The Asbestos hotline is now available to meet the asbestos information needs of private individuals, government agencies, and regulated industry. This hotline handles about 10,000 calls each year, and it operates Monday through Friday from 8:15 AM to 5:00 PM EST.

Commercial Numbers Offered by EPA Headquarters

Public Information Center (PIC), 1-202-829-3535. Answers inquiries from the public about the EPA's programs and activities, and it offers a variety of general, nontechnical information materials. The public is encouraged to call its commercial telephone lines.

Center for Environmental Research Information, 1-513-569-7391. Central point of distribution for EPA results and reports.

National Small Flows Information Clearinghouse, 1-800-624-8301. Provides information on wastewater treatment technologies for small communities.

Pollution Prevention Information Clearinghouse, 1-703-821-4800. Provides information on reducing waste through source reduction and recycling.

Radon Information. For information about radon, call the state radon office. The Radon Office at EPA Headquarters also responds to requests for information: 1-202-260-9605.

Safe Drinking Water Hotline. Provides information on the EPA's drinking water regulations. This hotline operates Monday through Friday from 8:30 AM to 4:30 PM EST: 1-800-426-4791. In the Washington, D.C. area the number is 1-202-260-5534.

41.7   FUTURE TRENDS

The growing concern that risk communication was becoming a major problem led to the chartering of a National Research Council committee (May 1987 through June 1988) to examine the possibilities for improving social and personal choices on techno-logical issues by improving risk communication. The National Research Council offers advice from governments, private and nonprofit sector organizations, and concerned citizens about the process of risk communication, about the content of risk messages and ways to improve risk communication [7]. The committee's recommendations, if followed, will significantly improve the risk communication process.

Future goals are not to make those who disseminate formal risk messages more effective by improving their credibility, understanding, and so on; it is to "improve" their techniques. "Improvement" can only occur if recipients are also enabled to solve their problems at the same time. Generally, this means obtaining relevant information for better-informed decisions [7]. Implementation of many recommen-dations requires organizational resources of several kinds. One of these resources in particular is time, especially during the most difficult risk communication efforts, as when emergency conditions leave no possibility for consulting with the people concerned, or to assemble the vital information that would be necessary for them.

The committee came forward with three general conclusions that may bring to light why the task of communicating risk does not seem to be working [7]:

Conclusion 1. Even great improvement in risk communication will not resolve the problems or end the controversy. Sometimes, they will tend to create them through poor communication. There is no ready shortcut to improving the nation's risk communication efforts. The needed improvement in performance can only come incrementally and only from constant attention to many details. For example, more interaction with the audience and the intermediaries involved is necessary to fully understand the issue at hand.

Conclusion 2. Better risk communication should not only be about improving procedures, but about improving the content of the risk message. It would be a mistake to believe that better communication is only a matter of a better message. To enhance the credibility, to ensure accuracy, to understand the concerned citizens and their worries, and to gain the insight necessary into how messages are actually perceived, the communicator must ultimately seek procedural solutions.

Conclusion 3. Communication should be more systematically oriented to specific audiences. The concept of openness is the best policy. It is true that the most effective risk messages are those that consciously address the specific audience's concerns. Similarly, the best procedures for formulating risk messages have been those that involved open interaction with the citizen's and their needs.

In the future, the communication of risk can become more effective. It should be understood that risk communication is a two-way exchange of information and opinion among individuals, groups, and organizations. The written, verbal, or visual message containing information about the risk should include advice on waste reduction or elimination, so that in the future communication, efforts will no longer be necessary.

41.8 SUMMARY

1. Risk communication is one of the most important problems in environmental protection this country faces. The government and media together have failed to communicate clearly what is a risk and what is not a risk. The challenge of risk communication is to provide this information in ways that it can be incorporated in the views of people. Success in risk communication is not to be measured by whether the public chooses to set the outcomes that minimize risk as estimated by the experts; it is achieved instead when those outcomes are knowingly chosen by a well-informed public.

2. There are no easy prescriptions for successful risk communication. However, those who have studied and participated in recent debates about risk generally agree on seven cardinal rules. These rules apply equally well to the public and private sectors. Although many of these rules may seem obvious, they are continually and consistently violated in practice. Thus, a way to use these rules is to focus on why they are frequently not followed.

3. A comprehensive community outreach program covers one or more of the following categories: information and education, the development of a receptive audience, disaster warning and emergency information, and/ or mediation and conflict resolution. In developing a community outreach program, three steps must be understood and addressed:

   a. Knowing the audience and its concerns. What is their "perception" of the risks involved?

   b. Understanding the issue objectives. What is the "message" to get across?

   c. Implementing the communication plan. How does one get the message out?

4. Environmental communication can take on significant forms that are not complex; these include printed material in the form of technical documents, pamphlets, handouts, brochures, magazines, journals, issue papers, newspaper articles, editorials, and the like.

5. There are two important things to do when one sites a potential pollution problem: (a) make careful observations of the problem, and (b) report it to the proper authorities.

6. The availability and accessibility of means to ensure environmental dialogue is crucial in establishing effective communications. Hotlines, toll-free numbers, and information lines provide the consumer a vital link with the EPA's environmental programs, technical capabilities, and services.

7. The National Research Council offered knowledge based on advice from governments, private and nonprofit sector organizations, and concerned citizens about the process of risk communication, the content of risk messages, and ways to improve risk communication in the service of public understanding and better-informed individual choices. The committee's recommendations, if followed, will significantly improve the risk communication process.

REFERENCES

1. Burke, G., Singh, B., and Theodore, L. *Handbook of Environmental Management and Technology*, 2nd edn., John Wiley & Sons, Hoboken, NJ, 2000.
2. Russell, M. Communicating risk to a concerned public, *EPA Journal*, November 1989.
3. U.S. EPA. Seven Cardinal Rules of Risk Communication, U.S. EPA OPA/8700, April 1988, including periodic updates.
4. Beuby, R., Faye, D., and Newlowet, E. *Communicating Risk to the Public: Getting the Message Across*, Ensco Environmental Services, Fremont, CA, 1989.
5. Davenport, G. The ABC's of hazardous waste legislation, *Chemical Engineering Progress*, New York, May 1992.
6. U.S. EPA. *Chemicals in Your Community: A Guide to Emergency Planning and Community Right-to-Know Act*, U.S. EPA, September 1988.
7. National Research Council, *Committee on Risk Perception and Communication. Improving Risk Communication*, National Academy Press, Washington, DC, 1989.

Part VII

Other Areas of Interest

Part VII of the book, comprising nine chapers, examines a host of other important environmental issues. The last Part begins by discussing what the authors have come to describe as "The EPA Dilemma." Information on noise pollution is presented in Chapter 44 while Chapter 43 is concerned with the highly sensitive (recently) issue of electromagnetic fields. Details of used oil receives treatment in Chapter 45. The general subject of environmental audits is treated in Chapter 46, while Chapter 47 addresses the important subject of economics.

The final two chapters of Part VII serve as an introduction to the general subject of ethics. Chapter 49 is concerned with environmental ethics. A comprehensive examination of engineering ethics is also included in Chapter 49. Part VII concludes with Chapter 50 which addresses a relatively new area of concern, environmental justice (sometimes referred to as environmental equity or environmental racism); this subject has been notably absent from the mainstream environmental agenda but clearly requires more attention.

42 The EPA Dilemma

Contributing Author: Anna M. Daversa

CONTENTS

42.1 INTRODUCTION

The problems associated with the regulatory framework of the federal environmental management have always been questioned. As with any government-controlled operation, many steps must often be taken before anything meaningful can be accomplished (this appears to apply to many activities, with the exception of war, where the president can exclusively command the armed forces for immediate action).

To implement an environmental regulation, the problem must be first identified (often in an EPA report), then data must be collected and analyzed (usually in another EPA report), and a goal has to be set, ultimately by congressional legislation. Once the law is in effect, it must be enforced by the EPA. The law has often been amended because of unreasonable goals and lax enforcement.

The present problem that exists with the EPA is an intricate one, consisting of primarily four main concerns:

1. Economically efficient measures are seldom, if ever, adopted, causing little progress in achieving environmental goals.
2. Data collection often has limitations, and when insufficient data is used for legislation, an ongoing string of amendments is attached.
3. The legal issues involving environmental problems have rocketed, brought on mainly by the complex legislation.
4. The EPA is presently primarily a legal organization that is serving the best interests of the law profession rather than the environment.

Sections 42.2 and 42.3 briefly describe the history of the EPA, its functions, and some of the legislation EPA is responsible for enforcing. Much of legislation is examined in more detail in Chapter 2. The remainder of the chapter analyzes EPA's accomplishments and performance. Based on this analysis, the last two sections provide suggestions on how the nation, and society in general, can be better served from an environmental point of view.

42.2 HISTORY OF THE EPA

1970 was a cornerstone year for modern environmental policy. The National Environmental Policy Act (NEPA), enacted on January 1, 1970, was considered a "political anomaly" by Lenten K. Caldwell, Washington Senator Henry Jackson's chief advisor to legislation. NEPA was not based on specific legislation; instead it referred in a general manner to environmental and quality of life concerns. The Council for Environmental Quality (CEQ), created by NEPA, was one of the councils mandated to implement legislation. April 22, 1970 brought Earth Day, where thousands of demonstrators gathered all around the nation. NEPA and Earth Day were the beginning of a long, seemingly never ending debate over environmental issues.

As described in Chapter 2, consumer and political interest movements led by Ralph Nader and growing groups of engineers, scientists, and other environmental experts, including some lawyers, influenced many of the new initiatives on the environmental legislation agenda. Events of the late 1960s, such as the oil burning on the Cuyahoga River in the center of Cleveland and the washing up of dead birds on the oil-slicked shores of Santa Barbara, reflected a sense of crisis and dissatisfaction within the society.

In his 1970 State of the Union message and later speeches, President Nixon declared that air pollution and clean water legislation would be the cornerstone of his environmental stance. Nixon's most likely challenger on these issues was Senator Edmund Muskie, the chair of the Senate Committee on Air and Water Pollution. Muskie was the target of some of the new activists, like Nader's groups, for his unwillingness to challenge industry's position on quality debates and his tendency to view the environmental problem as a problem of conservation rather than pollution [3]. However, environmentalists, government, industry, and society as a whole have come to view Muskie's position in a more favorable light during the recent years.

The Nixon Administration became preoccupied with not only trying to pass more extensive environmental legislation, but also implementing the laws. Nixon's White House Commission on Executive Reorganization proposed in the Reorganizational Plan #3 of 1970 that a single, independent agency be established, separate from the CEQ. The plan was sent to Congress by Nixon on July 9, 1970, and this new U.S. Environmental Protection Agency (EPA) began operation on December 2, 1970.

The EPA was formed by bringing together 15 components from 5 executive departments and independent agencies. Air pollution control, solid waste management, radiation control, and the drinking water program were transferred from the Department of Health, Education, and Welfare (now the Department of Health and Human Services). The federal water pollution control program was taken from the

Department of the Interior, as was part of a pesticide research program. EPA acquired authority to register pesticides and to regulate their use from the Department of Agriculture, and inherited the responsibility to set tolerance levels for pesticides in food from the Food and Drug Administration. EPA was assigned some responsibility for setting environmental radiation protection standards from the Atomic Energy Commission, and absorbed the duties of the Federal Radiation Council. Unfortunately, these groups were, and today essentially remain, compartmentalized [1]. The EPA was set up where each office dealt with a specific problem, and new offices were often created sequentially as individual environmental problems were identified and responded to by legislation.

The EPA's first administrator, William Ruckelshaus, initially sought to convey the impression that his agency would aggressively enforce the new policies, and adopted a systems approach by forming two primary program offices to handle the variety of issue areas and legislative mandates under its jurisdiction plus several function oriented divisions designed to be more responsive to White House concerns as well as fulfill certain agency wide objectives, such as enforcement and research. The new agency, however, was quickly overwhelmed by its rapidly expanding regulatory responsibilities, the conflicting signals from the Nixon, and later Ford Administrations on how aggressively it should pursue such regulations, and effective industry maneuvering, which used scientific uncertainty in the regulation process to delay or counter the establishment and enforcement of standards [2].

42.3   KEY EPA LEGISLATION

In contemporary environmental policy, the first important legislation, aside from NEPA, was the Clean Air Act (CAA) of 1970. Technology-based standards, along with national standards, became the key. A 1971 report by the CEQ was very optimistic. While the Council mentioned economic incentives, the relevant section regarding command and control indicated: "The Federal quality programs changed dramatically when the Clean Air Amendments became law. They embody recommendations contained in the President's 1970 message on the environment and proposed significant control for new pollution sources and for all facilities emitting hazardous substances. It also establishes a framework for the States to set emission standards for existing sources in order to achieve national air quality standards" [3].

Congress decided that the driving force for the enforcement operations by the EPA was to be control technology, the machinery for cleaning emissions at each source. Once identified, a specific technology, such as "reasonable available control technology" (RACT), was the basis of implementation of the legislation. Technology was the object of contracts between the EPA and plant owners. When seeking to prove guilt of environmental trespass, the proof came down to the existence and quality of specific machinery and/or processes. A deadline, June 30, 1975, was set for all air quality regions to meet the national air quality standards. Supporters of this federal movement had visions of numerous monitoring stations all across the United States to measure the levels of emissions and assure the delivery of cleaner air. But, as it turned out, when the deadline came, 102 out of 247 regions had not attained national standards, i.e., they had not achieved attainment [4].

Many other deadlines for specific substances or sources were established in the 1970 CAA. The 1974 and the 1977 CAA amendments extended these deadlines, as well as direct the EPA to study various factors of different pollutants. The 1977 amendments also gave the EPA more power in enforcement by civil penalties.

In addition to air pollution legislation in the 1970s, water pollution was also an important federal concern. The 1972 amendments to the Federal Water Pollution Control Act (FWPCA), better known as the Clean Water Act (CWA), set nationally uniform technology-based effluent limitations established by the EPA from major "point sources" of water pollution, with deadlines in 1977 for compliance, according to "best practice control technology." One of the claimed advantages of having geographically uniform regulations is their supposed speed and simplicity; however, this uniformity created other problems. Natural water is by no means uniform; properties such as temperature, toxicity, acidity, alkalinity, natural radioactivity, and the amount of algae and other aquatic life may vary, depending on the location [5].

Setbacks in FWPCA deadlines, as amended in 1972, required another amendment in 1977, where it became better known as the CWA. Here, the deadlines were extended to 1983. The postponement of deadlines, as seen in the CWA and the CAA, demonstrates the overall ineffectiveness, not only in the enforcement of the legislation, but also in creating unrealistic goals.

The Resource Conservation and Recovery Act (RCRA) was created in 1976 to regulate solid and hazardous waste facilities. Provisions for waste recycling had become a major objective. Thousands of new recycling centers were established, not for business purposes, but for environmental conscientiousness-raising. "Bottle Bills," which mandated a deposit or fee for recycled glass containers, were passed in Vermont in 1971 and in Oregon in 1973. They brought conflicts pitting glass industries, retail food industries, and labor groups against mainstream environmental groups and local organizations. "Ultimately, environmental lobbyists were not able to keep the bottle bill provisions in RCRA, but they were able to establish the principle that recovery and management of wastes needed to be developed more systematically at the national level" [6].

In addition to recycling, RCRA contains regulatory safeguards for operators of landfills and waste sites, which imposed "cradle to grave" rules for generators, carriers, and operators of disposal sites for toxic wastes. Until the 1970s, hazardous waste was treated like any other kind of waste, and the number of new and possible toxic chemicals entering the market each year was creating enormous stresses on this regulatory system. Another toxic-related legislation was the Toxic Substance Control Act (TCSA) of 1976, which provided for the federal review of all new chemicals before their production.

In the late 1970s, the public awareness of the toxic waste problem grew with the incident at Love Canal. Even though Hooker Chemical Company knew exactly what was stored in the sealed canal, had taken precautions that would satisfy even today's high standards, gave public warnings of the hazards, and wrote extensive warnings that precluded any use of the land that would threaten human health in the deed of the sale, they were sued by the Department of Justice (DOJ) on behalf of the EPA. The land was sold the Niagara Falls School Board in 1958, where it was developed for a grammar school, and the rest of the land was sold to real estate developers.

When sewer lines were developed, the seal of the canal was ruptured and toxic sludge appeared everywhere. Either the developers were not aware of the public records (in the deed of sale of Hooker Chemical), or proper safeguards were not taken. In any event, the school board and the developers were let off the hook [6].

To deal with Love Canal and the thousands of potentially contaminated sites, Congress passed the Comprehensive Environmental Response, Compensation and Liability Act (CERCLA) on December 11, 1980, and with it established Superfund. Superfund was set at $1.6 billion and was to receive 87.5% of the revenues from taxes on petroleum and 42 listed chemical feedstocks, and the rest from general tax revenues. To implement Superfund, the EPA had to establish a list of at least 400 sites, and they called on the states to give candidate sites. The 400 sites were to include each state's top priority, but a major problem that emerged was that one state might be filled with toxic sites far worse than another state's worst site.

To many, the Ragan Administration brought a decline to the environmental movement. This brought more bureaucratic red tape in the main issue of hazardous waste clean up. The CWA, CAA, and RCRA were all further amended with new deadlines. And the acts were getting lengthier, for example, the 1970 CAA had 50 pages, while the 1990 version has 800 (the effects of this will be discussed later in this chapter).

Just as the 1970s will be remembered for efforts to clean up hazardous waste, the 1990s may be remembered as the decade for pollution prevention. First among pollution-reducing laws is the 1990 Pollution Prevention Act, which *requests* companies to focus on ways to reduce emissions rather than treat wastes. Among other things, the law establishes a source of reduction clearinghouse on pollution prevention information; provide for the development, testing, and disseminating of auditing procedures designed to identify source reduction opportunities; sets up standard methods to measure pollution reduction; expands the toxic release inventory reporting requirements to include questions about source reduction and recycling; and, requires EPA to report to Congress on the progress of the reduction programs [7].

Other laws affecting pollution reduction are the 1990 amendments to the CAA and the TSCA. One such law requires EPA to conduct an engineering research program to develop new technologies for air pollution prevention, and take in consideration process changes or material substitutions when setting emission standards for hazardous chemicals. The idea that EPA could force a facility to change its process or materials to reduce emissions worried some chemical companies [8]. More recent laws and rules include the Clean Air Nonroad Diesel Rule of 2004, the Clean Air Interstate rule of 2005, the Clean Air Mercury Rule of 2005, and the strengthening of PM 2.5 particle pollution standards in 2005 [9].

42.4   IS THE EPA COST EFFECTIVE?

A major criticism of the present regulatory approach to solving environmental problems (and pollution) is its economic inefficiency. The EPA's Annual Performance Plan and Congressional Justification request budget for 2008 is $7.2 billion in discretionary budget authority and 17,324 Full Time Equivalents (FTE). This budget was broken down into five major goals [10]:

1. $912,000 (13% of the budget) on clean air and global climate change
2. $2,714,000 (38% of the budget) on clean and safe water
3. $1,663,000 (23% of the budget) on land preservation and restoration
4. $1,172,000 (16% of the budget) on healthy communities and ecosystems
5. $744,000 (10% of the budget) on compliance and environmental stewardship

Much of the early EPA legislation favored older existing plants over new plants. Existing firms could postpone or avoid enforcement actions, while new firms could not. This had the potential to reduce economic growth and strengthen monopolies in affected industries. And, when the CAA was amended in 1975, postponing the deadlines and restricting growth in nonattainment (national air quality standards were not met) areas, existing plants in dirty regions were helped more. In addition to the higher costs of control equipment, entry to these nonattainment areas was effectively barred.

A case involving Chevron in California led to a change in legislation. Chevron wanted to replace two smaller refineries with one large one, with a capacity of 315,000 barrels per day compared to 90,000 for each of the two older plants. Chevron officials agreed to shut down the two plants for the new one, and a deal with the Bay Area Pollution Control District was made. When the new plant was finished, however, Chevron did not shut down the two older plants. Chevron argued that the total emissions were less than the 1974 level, so the plants should stay open. Out of controversies such as this, transferable pollution rights were emerging. The EPA responded with the "offset policy," where a new facility could be constructed if emission reductions for the same pollutant from existing polluters were realized in an amount that would more than offset the pollution added by the new source. The "bubble concept" also emerged, where sources would have the opportunity to come forward with alternative abatement strategies that would result in the same air quality impact, but at less expense, by placing relatively more control on emission points with a low marginal cost of control and less on emission points with a high cost [6]. Although condoning this concept of allowing polluters to minimize costs while cleaning air, EPA quickly added that these rules would not affect clean areas (in attainment), nor would they be allowed for new sources. Through 1986, there have only been 40 air pollution bubbles approved by the EPA for the entire country, and an additional 89 bubbles were approved by states. Control cost savings generated a total of about $435 million. These savings are not trivial, but the number of bubbles established is small considering 10 years of establishment at that time [6].

In water pollution, economic inefficiencies are just as evident as in air pollution. Consider the example of a large plant with various wastewater streams, each with a different level of toxicity. Common practice is to combine these streams into one and treat everything together. If it is much cheaper to clean the less toxic streams, the costs are increased by combining them all.

Referring to EPA's effluent guidelines for steelmaking, the report noted that the additional cost of removing one unit of the same pollutant was $18,000 in one production process and $2,000 for another. Source-by-source limitations were the basis of the control mechanism, and opportunities for reducing costs existed.

Probably the biggest example of cost ineffectiveness by the federal government is Superfund (see Chapter 27 for more details). As indicated earlier, CERCLA established a $1.6 billion fund made up of taxes on crude oil and commercial chemicals. At the time, it was expected that this amount would be sufficient. When the EPA began the process of site discovery and evaluation, thousands of supposedly potential hazardous waste sites existed, presenting the nation with some of the most challenging pollution problems ever. In 1986, Congress reauthorized another $8.5 billion to the fund in the Superfund Amendments Reauthorization Act (SARA). Superfund was again reauthorized by Congress in 1991.

The EPA is trying to make polluters pay for the cleanup. But, this is how EPA goes about it. EPA goes after potentially responsible parties, or PRPs, to pay for or to conduct the cleanup of a site. When a PRP refuses to pay, it is sued by the EPA, and the EPA may seek "treble damages," where the PRP can pay up to three times the original cost. The major problem with seeking funds from PRPs is that often, when the wastes were dumped, there was no law against it. Also, if more than one party was responsible for the damages, the PRP usually becomes the richest company, and leaves further liability questions up to the PRP (the PRP must then sue the smaller companies). This creates endless jobs for lawyers, both for the EPA and for industries.

In 1992, the EPA proposed a rule that was designed to reduce some of the time and cost burden "incurred by the United States and responsible parties in preparing for, negotiating, and litigating these cases" [11]. Its goal was to streamline the process of cost recovery, which would in turn reduce transaction costs. This sounds like a favorable idea. However, this would be done by decreasing court costs by not allowing defendants to protest successfully. The rule proposes that the PRPs in these cost recovery cases cannot avoid payment on the basis that such costs are unnecessary or unreasonable. Therefore, regardless of how minuscule a contaminant release may be, the responsible party is forced to pay for it if the EPA says so. Along with clean up cost, the responsible party is also forced to pay for other indirect charges, such as travel to the site and the price of phone calls made. Although the EPA would be saving money through this rule, the PRPs would be spending much more [11].

Even past EPA administrator William Reilly has reportedly described Superfund as the worst piece of legislation ever passed by the U.S. Congress [8]. Congress reacted harshly to the Love Canal incident, and the public is facing the consequences. Since 1983, the EPA has listed 1,579 sites on the National Priority List and of these sites only 321 have been cleaned up and taken off the list [12]. Between the years 1980 and 1991 alone, the total authorized expenditures were $15.2 billion [11]. Since 2006, the annual funding has been about $1.2 billion. These numbers do not include the billions of dollars that the EPA is forcing PRPs to clean up their own sites. Billions of dollars are being spent, yet there is little to show for it [12].

One of the major reasons that there are so many economic inefficiencies is the EPA's self-examinations. Since the EPA evaluates its own program, it is inherently self-serving. In 1990, lawmakers included Section 812 to the Clean Air Act Amendments, which directs the EPA to report their costs and benefits to Congress. Both the 1997 and 1999 reports display the inadequacy of their self-examinations. In the first report, the EPA fails to analyze a range of policy of alternatives, and do not evaluate any alternatives in the 1999 report. "EPA's neglect of alternatives testifies to the triumph of its institutional interests over responsible policy analysis. The agency's reports to Congress demonstrate how it seeks to control and constrain the role of benefit–cost analysis in public debates about air pollution control policy [13]." The EPA also fails to disaggregate costs and benefits, making their analysis useless since repealing the CAA as a whole is not being considered. Also, the EPA ignores indirect costs, which can be up to 35% of direct costs and they exclude significant costs, such as the CAA provisions setting ozone standards, which are estimated to cost approximately $53 billion per year. The benefits are also exaggerated, illustrated by the EPA's estimate of 90% of benefits coming from the reduction of risks from particulate matter (PM). This estimate is based on a single study by Pope and colleagues [13]. There is little certainty about the risks of PM and one study is unrepresentative and insufficient evidence to prove 90% benefits. Overall, the EPA is very cost ineffective, with their self-evaluations being a major contributor [13].

42.5   ARE EPA'S DECISIONS JUSTIFIED AND CONSISTENT?

In order to decide what pollutants have adverse effects on human health and environmental well-being, extensive research must be done by the EPA. Also, once legislation is passed, the levels of contamination must be measured to assess the outcome. This data collection must be performed by scientists, yet William Reilly admitted in 1991 that "there has been plenty of emotion and politics, but scientific data have not always been featured prominently in environmental efforts and have sometimes been ignored when available" [14].

An EPA-appointed panel supported this view in a March 1992 report, *Safeguarding the Future: Credible Science, Credible Decisions*. The report doubts the quality of science that is used to justify programs within the EPA. Some specific findings are [14]

1. EPA's "science activities to support regulatory development ... do not always have adequate, credible quality assurance, quality assurance, quality control, or peer review." And although the agency receives "sound advice," it "is not always heeded."
2. The EPA "has not always ensured that contrasting, reputable scientific views are well explored and well documented from the beginning to the end of the regulatory process." Instead, "studies are frequently carried out without the benefit of peer review or quality assurance. They sometimes escalate into regulatory proposals with no further science input, leaving EPA initiatives on shaky scientific ground."
3. The agency "does not scientifically evaluate the impact of its regulations," and "scientists at all levels throughout the EPA believe that the agency does not use their science effectively."

Consider, for example, the issue of asbestos. A ruling of a Federal Circuit Court of Appeals states that the efforts to ban this substance have not followed scientific evidence, possibly increasing risk to consumers, workers, and schoolchildren.

A 1989 ban of asbestos by the EPA in the TSCA was overturned by the U.S. Court of Appeals by October, 1991. "The EPA lied when it claimed that its 1989 ban on all asbestos use was prompted by compelling evidence of risk reported by the medical community. No such evidence was ever presented by the medical community" [15]. The Agency had insufficient evidence to justify a ban, and had failed to follow the statutory requirement under the TSCA to adopt the least burdensome regulation. Intermediate regulation, such as warnings and restrictions were rejected by EPA. Also, the court noted that the EPA failed to consider the potential harm from substitutes, even when they are known carcinogens. Finally, the court questioned the EPA's pursuit of "zero risk" with regard to asbestos. The ruling noted, for example, that the proposed ban of three asbestos products would theoretically save seven lives over a span of 13 years, at a cost of up to $300 million. The number of deaths supposedly prevented this way would be roughly half the fatality toll in a similar period with toothpicks, according to the decision [16]. If one considers the recent surge in automobile and truck accidents attributed to break failure, because asbestos is no longer used in many types of break linings [15], the number of lives in danger is much greater because of not using asbestos in breaks than lung cancer from airborne particulates. Studies by the EPA itself have shown that removing asbestos created airborne particles that are more harmful than if asbestos were left alone. The EPA admitted that ripping out the asbestos was a mistake. The hysteria created by the EPA, as well as some citizens and environmentalists, has caused this apparent mistake about asbestos to become truth in the eyes of the American public.

The EPA has also made mistakes with radon. There is no question that radon causes health hazards, particularly lung cancer in miners. However, EPA assumptions, indicating that the number of deaths caused by radon are between 7,000 and 30,000 a year (a rather wide range), are again based on uncertain linear models. In the study of the miners, a majority of them smoked. This was not taken into consideration. There is no evidence that there is any more radon today than there was thousands of years ago. One may wonder what prompted the EPA's urgent policy. Not one scientific study has proven a statistically significant relationship between indoor radon and lung cancer [15]. Nevertheless, the EPA declared that "virtually all scientists agree that radon causes thousands of deaths every year." One may wonder if the government is really protecting anyone but themselves and their bureaucratic jobs. The financial costs to EPA and industry are negligible, but homeowners who follow the "national standards" are paying out of their own pockets.

EPA data on PM, which is a fine material of about 2.5 μm emitted into the atmosphere (PM 2.5), has also been proven wrong. According to a 1997 Citizens for a Sound Economy Foundation study conducted by Kay Jones, who was a top environmental advisor to President Jimmy Carter, EPA has immensely overstated the health risk associated with PM 2.5. The EPA used mislabeled data points to set the level of the standard and its benefit calculations were based on levels below the PM 2.5 standard, which have already been determined as safe. These errors lead to an overestimation of the health risks due to PM by a factor of 15 [17].

With dioxins, study of chemical industry workers with 60 times the normal level of dioxin showed that they had no increase in disease. Still, the EPA has not changed its stance on dioxins. The safe limit remains a 6 trillionths of a gram per kilogram of body weight per day (tg/kg/d). The average industrial work ingests between 1,000 and 10,000 tg/kg/d, or up to 1,700 times the safe limit [14].

Environmental regulations dealing with urban smog are also suspected of lacking good data. A congressionally mandated and EPA-sponsored report by the National Academy of the Sciences stated that it is difficult to know how to reduce smog in certain areas or even know how bad the problems really are. For example, attempts to measure ozone levels are slowed by a method which does not account for the role of weather in ozone formation. This has led to EPA qualitative classifications of "serious" or "severe" ozone problems in areas where it is unjustified. A senior scientist and research manager who served in the EPA said that the EPA intends to enforce all of the serious and severe classification strategies, whether they are needed or not. To do this, the agency first delays release of data that show fewer cities fail to meet the smog standard, and then explains that the law cannot be changed anyway—a law, however, which was based in part on the information EPA supplied to Congress [14]. To further complicate the matter, emissions tests for cars have been questioned. For example, the General Accounting Office, examining the effectiveness of the new vehicle inspections, found that 28% of the vehicles tested failed an initial test, but passed a second test with no repairs. This raises the question of whether the EPA-mandated test is reliable, and if inaccurate identification could lead to unnecessary repairs.

In December of 2007, California proposed its own state standard of reducing tailpipe carbon dioxide emissions by 30% by 2016. Although the EPA has granted California permissions to set its own standards in the past, it has denied California the EPA waiver required to set its own emissions standards. EPA Administrator S. L. Johnson claims that the greenhouse gas problem is not unique to California; therefore, there is no need for them to set their own standard. On January 2, the EPA was sued by 15 other states supporting California. Connecticut Senator Joseph Lieberman told Johnson, "The federal government is not doing nearly enough to reduce America's greenhouse emissions. It should, at the very least, stay out of the road that many state governments are taking" [18]. It is clear that the main concern of the EPA is no longer pollution prevention. Two senators said that they believed the EPA's decision was based protecting businesses rather than the environment [18].

The problems of the environment need to be examined scientifically. If an environmental concern arises, passing regulations before a good scientific basis and peer review are achieved can result in enormous expenditures in legalities, something that this country is presently burdened with. When environmental legislation is passed, it is often so ambiguous that an array of lawyers is needed to translate them. The main reason for this problem is that amendments are made based on premature or simply ill-defined findings. As mentioned previously, scientific data is not always featured predominantly when politics and emotion flare.

In organizing the EPA, members of various environmental movements (see Chapters 2 and 3) received special positions in the agency. Some of these environmental groups receive federal funding, even though they are mostly narrow special interest groups. These groups, even if they felt that certain proposed legislation would be ineffective, do not speak up for fear of losing the funds.

Scientists and engineers in the EPA know of the burdens imposed by legalities. These individuals are now spending more time monitoring contractors since new rules were imposed to combat alleged abuse involving outside researching and consulting. EPA scientists are now required to log every interaction with contractors and carefully follow every regulation [19]. Now, not only are the scientists responsible for research and data collection, they effectively must do secretarial paperwork.

Complicated legislation passed based on insufficient data is by no means a solution to the environmental problem. Costly control measures are taken, and in some cases, the public's risk is increased. Constant amendments are needed, often doing little to alleviate pollution. Regulations can only help if they are based on sound scientific data. When the legislation is unclear, lawyers are often brought in to "clarify" it. Instead, they usually complicate the problems further.

42.6   CAN THE EPA BE ELIMINATED?

The predictable, bureaucratic tendency which feeds on the professional ambitions of dedicated staff and inevitably generates calls for larger budgets, is reinforced by the high costs of litigation and the long delays associated with the process. This centralizing effect feeds the political machinery to Congress. EPA is the whipping boy, never meeting the impossible deadlines and not doing enough to satisfy the politicians. Industry is the villain, and the flaming emotions of innocent people are fanned by the rhetoric that ensues. Heating hearings, more proposed laws, larger budgets, more lawyers, and limited progress is the result. Political demand continues to outstrip political supply [6].

When the EPA was formed in 1970, it was—in a very real sense—a technical organization. The Agency was manned primarily with engineers and scientists. Most of these individuals were dedicated to a common cause: correcting the environmental problems facing the nation and improving the environment. The problems these individuals tackled were technical, and there were little or no legal complications or constraints. The EPA was indeed a technical organization, run and operated by technical people, attempting to solve technical problems. Much was accomplished during these early years ... but something happened on the way to the forum [20].

Nearly 40 years later, the EPA is no longer a technical organization—it is now a legal organization. The EPA is no longer run by engineers and scientists. It is run and operated by lawyers. And, the EPA is no longer attempting to solve technical problems; it is now stalled in a legal malaise [14].

How in the world did this occur? It happened because it served the best interest of the career bureaucrats, in and out of Congress, most of whom are lawyers, and it happened because the technical community did nothing to stop it. The result is that this nation is now paying the price for an environmental organization with nearly 20,000 employees and an annual budget approaching $10 billion that is not serving the best interests of either the nation or the environment [20].

Interestingly, all of the administrators to the EPA have been lawyers. Though lawyers are required in every industry for helping to settle disputes over legalities, protecting the environment is generally beyond their scope. In the EPA today, for every three engineers there is one lawyer; it is indeed (as described above) a legal organization, serving the legal profession and not the environment. Actual proposals for regulations and control, based on good scientific data, should be designed by scientists and engineers, or those who have come to be defined as problem solvers. They can analytically break down a problem, initially assess the damages, then fix them [20].

Creating problems and not solving them has become the mode of operation for the EPA. One need only look at Superfund (see earlier discussion and Chapter 27) for an example of what the professional bureaucrats have accomplished. When one talks about wasting tax dollars, Superfund is at the top of the list, with nearly $10 billion down the drain.

And to think that the Clinton Administration considered raising the EPA to a cabinet level. Proposals to elevate EPA to cabinet level and change it the Department of Environmental Protection, though passed by Senate, was criticized by Congress. The Senate claimed that this would increase the United States' power in international environmental concerns. If the EPA cannot be effective in the country as an agency, how can it be expected to function as a cabinet, and at an international level? It is hope that Congress' objective for future environmental legislation will focus on easing the financial burden of EPA regulations on industries, private property owners, and state and local governments [21].

More recently, with the emergence of nanotechnology, industry looked to EPA (as well as OSHA) for information and guidance on potential regulations. EPA finally issued a Nanotechnology White Paper whose purpose is to "inform EPA management of the science issues and needs associated with nanotechnology, to support related EPA program office needs and to communicate these nanotechnology issues to stakeholder and the public." Despite industry's demand for information on potential regulatory action, EPA failed to honor its earlier promise to deliver nanotechnology environmental regulatory guidance [22,23].

42.7   FUTURE TRENDS

Something has gone afoul. In this society, engineers are the problem solvers, but rarely the decision makers. Although the world known today has been called a product of engineering, engineers play a minor role in important decision making.

By far the most important policy affecting the environmental future of the country, and the planet, is pollution prevention (see Chapters 30 through 34). Past EPA administrator, Carol Browner, "has repeatedly claimed that pollution prevention is the organization's top priority. *Nothing can be further from the truth.* Despite near unlimited resources, the EPA has contributed little to furthering the pollution prevention effort. The EPA offices in Washington, Research Triangle Park, and Region II have exhibited a level of bureaucratic indifference that has surpassed even the traditional attitudes of many EPA employees. Pollution prevention efforts have been successful in industry because they have either produced profits, or reduced costs, or both. The driving force for these successes has primarily been economics, and not the EPA" [23].

Another important way to help solve the environmental problem is training (see Chapter 43). If the public understands the problems scientifically, solving them is much easier. For the future, employees and consumers must be made aware of the causes of pollution and must know how to prevent it. Pollution prevention is the future, and it has been used effectively in industry.

Finally, the role played by the consumer, which ultimately can control industry, is very important. If the average consumer purchased environmentally friendly products such as recycled containers and concentrated products, not only would there be less waste, but industries would have to respond by exclusively offering these products.

The environmental problem is one that developed over many years of civilization by many different sources. To think that the EPA, with its present mode of operation, can solve this problem is ludicrous. However, there is a solution. Dissolve the EPA now! No reorganization will work, since the lawyers and career bureaucrats have a stronghold in the Agency with their ties to Congress and the White House. What is needed is to make the present EPA disappear and start anew. The nation needs an environmental administration that will solve, not create problems [20]. The nation needs technically competent people who can lead an organization in making cost-effective decisions based on the public well-being, not on politicians whose goal is to get reelected or lawyers who cost the nation billions of dollars annually proposing and enforcing ill-defined legislation.

42.8 SUMMARY

1. The problems associated with the regulatory framework of the federal environmental management have always been questioned.
2. The EPA was formed by bringing together many environmental groups primarily to enforce new policies and to research future policies.
3. Though some legislation was technology based, many of the acts and amendments were based on particular incidents (Love Canal), and almost all of it is ambiguous in text.
4. Economics of environmental protection often favored monopolies of older plants.
5. Data collection has proved erroneous in many cases, endangering health, squandering money, and leading to unclear legislation.
6. Nearly 40 years after its formation, the EPA is no longer a technical organization—it is now a legal organization, run and operated not by scientists and engineers, but by career bureaucrats and lawyers. The EPA is no longer attempting to solve technical problems; instead, it is stalled in legal deadlocks.
7. Dissolve the EPA now! No reorganization will work, since the lawyers and career bureaucrats have a stronghold in the agency with their ties to Congress and the White House. The nation needs a new organization that will solve, not create problems.

REFERENCES

1. Burke, G., Singh, B., and Theodore, L. *Handbook of Environmental Technology*, 2nd edition, John Wiley & Sons, Hoboken, NJ, 2000.
2. Gottlieb, R. *Forcing the Spring*, Island Press, Washington, DC, 1993.

3. Council on Environmental Quality. *Environmental Quality*. The second annual report of the CEQ, Government Printing Office, Washington, DC, August 1971.
4. Landau, J. Who owns the air? The emission offset concept and its implications. *Environmental Law*, 9(3), 578, Spring 1979.
5. Stewart, R. and Krier, J. *Environmental Law Policy: Readings, Materials, and Notes*, pp. 514–515, 56.
6. Yandle, B. *The Political Limitations of the Environmental Regulation*, Quorum Books, New York, 1989.
7. Hanson, D. Pollution Prevention Becoming Watchword for Government, Industry. C & EN, pp. 21–22, January 6, 1992.
8. *Journal of the Air and Waste Management Association*, Pittsburgh, PA, 1994.
9. U.S. EPA. Timeline, 2000s, U.S. EPA, www.epa.gov/history/timeline/00.htm (January 10, 2008).
10. U.S. EPA. 2008 EPA Budget in Brief, U.S. EPA, Office of the Chief Financial Officer, www.epa.gov/ocfo
11. Stroup, R.L. and Townsend, B. EPA's New Superfund rule: Making the problem worse. *Regulation*: The Cato Review of Business & Government, Vol.16, No. 3, Washington, DC, 1993.
12. Hogue, C. Superfund slowdown: Lagging pace of cleanups blamed on technical challenges and lack of money. *Chemical and Engineering News: Government & Policy*, American Chemical Society, Washington, DC, 2008.
13. Belzer, B.B. and Lutter, R. EPA pats itself on the back. *Regulation*: The Cato Review of Business & Government, Vol. 23, No. 3, Washington, DC, 2000.
14. Samuel, P. and Spencer, P. Facts catch up with 'Political Science.' *Consumers' Research*, 10–15, May 1993.
15. Moriarty, M. Asbestos: The big lie. *21st Century Science and Technology*, Winter 1993–1994. Radon, *Garbage*, 24–28, Spring 1994.
16. Brimelow, P. and Spencer, L. You can't get there from here, *Forbes*, July 6, 1992.
17. Reiser, M. EPA Overestimated Health Risk by Factor of 15. FreedomWorks: Making Good Policy Good Politics, Washington, DC, May 11, 2007.
18. Fahrenthold, D.A. O'Malley attacks EPA greenhouse gas decision. *Washington Post*, January 25, 2008.
19. Stone, R. New rules squeeze EPA scientists. *Science*, 647, October 29, 1993.
20. Theodore, L. Dissolve the USEPA…NOW, EM, Pittsburgh, 1995.
21. Cushman, J. EPA critics get boost in Congress. *New York Times*, February 7, 1994.
22. Theodore, L. Personal notes, 2007.
23. Theodore, L. *Nanotechnology: Basic Calculations for Engineers and Scientists*, John Wiley & Sons, Hoboken, NJ, 2007.

43 Electromagnetic Fields

CONTENTS

43.1 INTRODUCTION [1,2]

Most individuals are surrounded by low-level electric and magnetic fields from electric power lines, and appliances and electronic devices. During the 1980s, the public became concerned about such fields because of media reports of cancer clusters in residences and schools near electric substations and transmission lines. In addition, a series of epidemiological studies showed a weak association between exposure to power-frequency electromagnetic fields (EMFs) and childhood leukemia or other forms of cancer.

The high standard of living in the United States is due in large measure to the use of electricity. Technological society developed electric power generation, distribution, and utilization with little expectation that exposure to the resultant electric and magnetic fields might possibly be harmful beyond the obvious hazards of electric shocks and burns, for which protective measures were instituted. Today, the widespread use of electric energy is clearly evident by the number of electric power lines and electrically energized devices. Because of the extensive use of electric power, most individuals in the United States are today exposed to a wide range of EMF. It is estimated that at least 100,000 people have been exposed throughout their lives to technology-generated electric and magnetic fields.

Electrical devices act on charged and magnetic objects with electric and magnetic fields in a manner similar to how the moon influences the ocean tides through its gravitational field. Before the advent of man-made electricity, humans were exposed only to the steady magnetic field of the Earth and to the sudden occasional increases caused by lightning bolts. But since the advent of commercial electricity in the last century, individuals have been increasingly surrounded by man-made fields generated by power grids and the appliances run by it, as well as by higher frequency fields from radio and television transmissions.

The most commonly used type of electricity in the home and work place is alternating current (AC). This type of current does not flow steadily in one direction but moves back and forth. In the United States, it reverses direction 60 times per second. The unit to denote the frequency of alternation is called Hertz (Hz) in honor of Heinrich Hertz who discovered radio waves. The presence of electric currents gives rise to both electric and magnetic fields because of the presence of electric charges. Those electric charges with opposite signs attract each other; on the other hand, charges with the same sign repel each other. These charges, if stationary, create what has come to be defined as "electric fields." "Magnetic fields" are created when the charges are nonstationary, that is, are moving. The combination of these two forces is defined as electromagnetic fields, or simply EMFs. Since these types of fields alternate with alternating electric current, a 60 Hz electric power system will generate 60 Hz electric and magnetic fields.

43.2   EXPOSURE COMMENTS

Electric and magnetic fields at a power frequency of 60 Hz are generated by three main sources: production, delivery, and use of electric power. However, sources of public exposure include

1. Power generation
2. Transmission
3. Electric circuits in homes and public buildings
4. Electric grounding systems
5. Electric appliances

Electric and magnetic fields have been measured in selected residences and outside environments to help resolve uncertainties in the interpretation of epidemiological results. Although almost all state and local governments, utilities, private firms, and individuals are currently measuring EMF, these measurements are often conducted without adequate supervision and expertise, and lack the standard quality assurance/quality control (QA/QC) requirements for research and study projects. Thus, these measurements have not been appropriate for determining public exposure in an absolute sense. In addition, no standardized procedure for EMF measurements exists. Notwithstanding this, instrument development has been responsive to the perceived needs for field measurements. In particular, a number of survey instruments that measure electric and magnetic fields that vary with time are now available, and miniaturized pocket-sized recording instruments have also been recently developed.

Mathematical models to estimate EMF exposure have been developed because measurement of fields at all locations and under all conditions of interest is not practical. Two types of models available are theoretical and statistical. The application of theoretical models usually involves a detailed computer program. Statistical models are used to develop statistical estimates of average exposure; thus, statistical modeling does not predict individual exposures, but provides estimates for groups of the population.

EMF coupling to biological objects is another area of concern. Interestingly, an electric field immediately adjacent to a body is strongly perturbed and the intensity of the field may differ greatly from that of the unperturbed field. On the other hand, the magnetic field that penetrates the body is essentially unchanged. Both external electric and magnetic fields that vary with time induce electric fields internally and the electric current generated inside the body is proportional to the induced internal electric field.

43.3 HEALTH EFFECTS [1–5]

Since 1980, research into the possible health effects of low-level EMFs has expanded greatly. Literally thousands of research papers by scientists in both the United States and Europe have been published on the subject. However, the study of potential health effects from these fields is fraught with complexities, contradictions, and what to some observers seem like impossibilities. In many studies, including human epidemiology and laboratory tests on cells and animals, the results obtained with relatively weak EMFs seem contradictory [6]. Nonetheless, there are many epidemiological studies that have reported an association between EMF exposure and health effects. The most frequently reported health effect is cancer. In particular, EMF exposure has been reported to be associated with elevated risks of leukemia, lymphoma, and nervous system cancer in children. Some occupational studies of adults describe an association between EMF exposure and cancer. However, as indicated above, uncertainties remain in the understanding of the potential health effects of EMF.

One should note that EMF analyses are particularly difficult for epidemiologists. The problems that have been attributed by some people to EMFs include several different kinds of cancer, birth defects, behavioral changes, slowed reflexes, and spontaneous abortions. Therefore, the process of deciding which health problems in a community belong to the "cluster" becomes exceedingly difficult. So, the answer to the question, "Can that source be the cause of my problems?" is "maybe or maybe not." The source might be the problem but trying to show that it is can be very difficult, if not impossible.

Despite the complexities and disagreements, scientific opinion has coalesced around a middle ground in recent years. In that middle ground is a great deal of evidence that EMFs do have some biological effects. They are not yet sure whether such fields can produce adverse health effects, but they believe there is some nontrivial chance that low-level fields could pose a problem, and they place a high priority on research aimed at answering that question. However, beyond the middle ground are a few scientists at one extreme who say enough evidence already exists to show that low-level fields, such as those from power lines, do have adverse health effects, in particular cancer, and therefore society should take strong measures to reduce exposures. At the other end of the spectrum is a small group of experts who say that biological effects from such fields would violate the laws of physics, and therefore low-level EMFs cannot cause cancer or any other disease. They also claim that laboratory studies on cells or animals that seem to show biological effects with very low-level fields are flawed in some way, e.g., that there is some other explanation for the results [6].

For years, scientists assumed that the only harm caused by EMFs was thermal, i.e., their ability to heat up an object. Even so, it was a phenomenon shown to exist only at the higher frequencies of several thousand megahertz, the range in which microwave ovens operate. In the late 1970s, scientists began to question if an association between cancer deaths in Denver children and exposure to extremely low-frequency 60 Hz fields existed. This subject has been debated in the literature since then [7].

Regarding breast cancer, women in electrical jobs are 38% more likely to die of breast cancer than other working women, according to a new study. It found an even higher death rate among female telephone installers, repairers, and line workers. "It's the strongest epidemiological evidence so far that breast cancer may be related to EMFs in some way, but it's still not very strong evidence," said University of North Carolina researcher Dana Loomis, chief author of the study published in the *Journal of the National Cancer Institute* [8]. The new study found that the breast cancer death rate was more than twice as high among female telephone installers, repairers, and line workers, compared with women who worked in nonelectrical occupations. The results were statistically adjusted to factor out income, age, race, and marital status.

The above study also indicated that the risk was 70% higher for female electrical engineers, 28% higher for electrical technicians, and 75% higher for other electrical occupations such as electricians and power line workers. All of those jobs involve sustained exposure to EMFs, but so do some nonelectrical jobs, such as computer programmers, computer equipment operators, keyboard data enterers, telephone operators, and air traffic controllers. And for each of those five jobs, the study found that female breast cancer mortality was no higher than for the rest of the work force [8].

Environmental agents that cause reproductive and developmental effects are important because they may directly influence health, lifespan, propagation, and functional and productive capacity of children. Some epidemiological studies have reported reproductive and developmental effects from exposure to EMF generated by devices in the workplace and home. Investigations of women and the outcome of their pregnancies have included operators of visual display terminals (VDTs) and users of specific home appliances (electric blankets, heated water beds, and ceiling electric heat). The reports of increased miscarriages and increased malformations suggest that maternal EMF exposure may be associated with adverse effects.

Other studies have reported an increased incidence of nervous system cancer in children whose fathers had occupations with potential EMF exposure. Regarding nervous system effects, neurotransmitters and neurohormones are substances involved in communication both within the nervous system and in the transmission of signals from the nervous system to other body organs. Neuroregulator chemicals are released in pulses with a distinct daily or circadian pattern. The few studies in which human subjects have been exposed to EMF in controlled laboratory settings describe the following effects: changes in brain activity of possibly slowed information processing, slowed reaction time, and altered cardiovascular function, including slowed heart rate and pulse that may indicate direct action on the heart.

The immune system defends against cancer and other diseases. Environmental agents that compromise the effectiveness of the immune system could potentially increase the incidence of cancer and other diseases. Studies on the effect of 60 Hz electric fields on the immune system of laboratory animals found no effect of chronic exposure of rates and mice. Thus, it may be concluded at this time that power frequencies have small or no effects on the immune systems of exposed animals.

While the hazards from these fields may or may not be significant, the fear of them is. In state after state, nervous citizens have delayed or even killed electric utilities' plans to build or expand high-voltage transmission lines. Real estate brokers report that houses next to power lines sell more slowly than others, and for lower prices. Parents with children in schools near power lines are demanding that either the schools or the lines be moved. Meanwhile, lawsuits by cancer victims against power companies are making their way through the courts in many states [7].

It is important to keep the overall EMF health risk in perspective. For example, there are about 2600 new cases of childhood leukemia every year in the United States. The chance of a given child's developing leukemia in any year is about 1 in 20,000, with the bulk of cases occurring by the age of five. Some epidemiologic studies have suggested that unusually strong magnetic fields may double a child's risk, raising it to 1 in 10,000. But even in those studies, the vast majority of leukemia cases occurred in houses calculated to have low magnetic fields. In the end, parents must make a personal decision about how much to worry—just as they routinely choose to worry about or ignore, other risks in their lives and their children's lives [7].

43.4   MANAGEMENT/CONTROL PROCEDURES [1]

As described earlier, the major source of environmental exposure to EMF is the electric power system, which includes transmission lines, the distribution system (substations, lines, and transformers), and electric circuits in residential and other buildings that provide power to appliances and machinery. Although considerable effort has been focused on the control of EMF from electric utility systems, little work has been done on controlling fields generated by electrically powered appliances, tools, and other devices.

In most circumstances, the strength of low-frequency EMF decreases with distance from the source. One simple mitigation approach is therefore to increase separation distance. There are other methods known to be effective regardless of frequency. They are

1. Shielding
2. Design
3. Location
4. Component choice
5. Filtering

While electric fields can be easily shielded, magnetic fields are much more difficult to shield. Electric fields are shielded to some degree by almost anything such as trees, bushes, walls, and so forth. Magnetic fields can be reduced by enclosing the source in certain types of metal such as a material called Mu metal, which is a special alloy. The fields are still present, but the metal has the capability to contain them. This approach to reducing field levels may not be practical for many sources, including power lines. Magnetic field intensity can also be reduced by placing wires close together so that the field from one wire cancels the field from the other. This is now being done in new designs for electric blankets. To some degree the same thing can be done for power lines, but for safety and reliability reasons power lines have minimum required spacing.

Because of the way appliances are made, they have the potential to have very high localized fields, but then the fields decrease rapidly with distance. For example, typical magnetic field strengths not near an appliance are 0.1–4 mG, but the field from an electric can opener can be 20,000 mG at 3 cm (approximately 1 in.) from the appliance. At 30 cm (approximately 1 ft), appliance fields are usually around 100 times lower. For the can opener mentioned above, the level would probably be around 20 mG. The reader should note that the Gauss is a unit for the strength of a magnetic field, also known as magnetic flux density. Magnetic flux density is measured in terms of lines of force per unit area. Remember the patterns that were generated by iron filings on a piece of paper which was placed over a magnet? These patterns are field lines. One normally speaks of magnetic fields in terms of (one) thousandths (1/1000) of a gauss or milligauss (mG).

When standing under a power line, one is usually at least 20 ft or more away from the line, depending on its height above ground. Under a typical 230 kV transmission line, the magnetic field is probably less than 120 mG. In contrast, if one moves about 100 ft away from the line, the magnetic field is probably about 15 mG, and at 300 ft away from the line, the magnetic field is probably less than 2 mG. From these examples, one can see that distance from the source of the magnetic or electric field can substantially reduce exposure.

Magnetic/control procedures for specific applications are discussed below. Control technology for transmission and distribution lines has been developed and could be applied if warranted. These techniques focus on compaction and shielding of transmission conductors. Compaction is based on the principle that for three-phase, balanced conductor systems, the net field (electric or magnetic) of the three phases is zero. A disadvantage of compaction is that it results in an increase in electrical arcing, which affects system reliability. For situations in which compaction was an ineffective control technology, shielding techniques have been developed that reduce the electric field at the edge of the right-of-way by approximately 10-fold. Compaction techniques that have been developed include gas-insulated transmission lines, superconducting cables, and direct-current cable technologies. In cable or gas-insulated transmission technologies, conductors are inside a metallic sheath in which the electric field exists only between the conductors and the sheath; electric fields external to cable sheaths are essentially zero.

EMF inside the home and schools can be emitted from appliances, wiring systems, including the grounding, underground and overhead distribution lines, and transmission lines. A few appliances, especially electric blankets and heated water beds, have been identified as important sources of magnetic field exposure because of their close proximity to the body for long periods of time. Hair dryers and electric shavers, because they too are used close to the body, expose people to some of the strongest fields but total exposure from these is limited because they are used for only minutes per day. Manufacturers have responded by developing low magnetic field appliances. Some specific steps one can take to reduce EMF exposure at home, the office, or at school are listed below.

1. Sitting at arms length from a terminal or pulling the keyboard back still further; magnetic fields fall off rapidly with distance.
2. Switching VDTs off (not the computer necessarily) when not in use.
3. Spacing and locating terminals in the workplace so that workstations are isolated from the fields from neighboring VDTs. Fields will penetrate partition walls, but do fall quickly with distance.
4. Using electric blankets (or water bed heaters) to warm beds, but unplugging them before sleeping. Magnetic fields disappear when the electric current is switched off. However, electric fields may exist as long as a blanket is plugged in.
5. Not standing close to sources of EMFs such as microwave ovens while in use. Standards are in place to limit microwave emissions. However, the electric power consumption by a microwave oven results in magnetic fields close to the unit that are high. The same is true of other appliances as well.

Existing mass transit systems and emerging technologies such as magnetically levitated trains, electric automobiles, and superconducting magnetic energy storage devices require special consideration. These systems can produce magnetic fields over large areas at different frequencies. Passengers on magnetically elevated trains will be exposed to static fields and to frequencies up to about 1000 Hz. Existing engineering control technologies may not be sufficient to significantly reduce exposure.

As described earlier, another device that merits special concern is the VDT. In addition to being energized by 60 Hz power, VDTs can produce EMF at frequencies of up to 250,000 Hz. VDT manufacturers, however, have begun to reduce fields by shielding techniques. Metal enclosures are used to shield electric fields, while active magnetic shielding techniques are used to reduce magnetic fields.

The reader should note that there is no simple way to completely block EMFs since the fields are generated by electrical systems and devices in the home, including the wiring and appliances. Electric fields from outside the home (power lines, etc.) are shielded to some extent by natural and building materials, but magnetic fields are not. As noted above, the further a building is from an EMF source, the weaker the fields at the building would be. Keeping fields out of the home would mean keeping any electricity from coming into or being used in the home. The fields from sources inside the home (e.g., appliances, wiring, etc.) will often result in higher fields than from sources outside the home.

At this point, enough evidence suggests a possible health hazard to justify taking simple steps to reduce exposure to EMFs. The larger dilemma is whether the risks justify making major changes in huge, complex electric power systems that could disrupt the reliable, relatively inexpensive electric service Americans have come to take for granted [7].

Some questions are too large to be answered by individuals or families at this time. How much should a community spend to route transmission lines away from a school? Should a high-voltage line be put on taller towers to minimize fields at ground level even though the expense will result in higher electricity rates [7]?

It seems sensible to focus on simple ways of reducing exposure to EMFs rather than to make radical changes. M. Granger Morgan, a public-policy expert at Carnegie Mellon University, has proposed a strategy he calls "prudent avoidance," involving simple low-cost or no-cost measures. Although prudent avoidance can be as easy as leaning back from a computer screen, other methods of avoiding EMFs are more difficult such as moving out of a house near a power line. Whether such a move is prudent or paranoid depends largely on one's own feelings about the nature of the risk [7].

43.5   FUTURE TRENDS

If scientists eventually reach a consensus that low-level EMFs do cause cancer or some other adverse health effect, then regulations defining some safe exposure level will have to be written at some later date. But so far the data are not complete enough for regulators.

Future research is also questionable at this time because much of the research into the effects of EMFs, especially that on mechanisms that could cause health effects, is cross-disciplinary, highly complicated, and has raised more questions than it has answered. It may take more than a decade to elucidate the mechanisms. It appears that the technical profession presently does not know if EMF exposure is harmful (aside from the concern for electric shocks and burns for extreme exposure). It does not know if certain levels of EMFs are safer or less safe than other levels. With most chemicals, one assumes exposure at higher levels is worse than less exposure at lower levels. This may or may not be true for EMFs. More research is required to identify dose–response relationships. There is some evidence from laboratory studies that suggest that there may be "windows" for effects. This means that biological effects are observed at some frequencies and intensities but not at others. Also, it is not known if continuous exposure to a given field intensity causes a biological effect, or if repeatedly entering and exiting of the field causes effects. There is no number to which one can point and say "that is a safe or hazardous level of EMF." Many years may pass before scientists have clear answers on cancer or on any other possible health problems that could be caused by EMFs. But over the long run, avoiding research probably will not be acceptable. It appears that the public will continue to demand research funding and answers to these questions.

Finally, the tendency to sensationalize electric and magnetic fields reporting was discussed by McElfresh [9]. Information on the latest 53 publications (mainly newspapers) that reviewed five major EMF studies was presented. Hopefully, the future will provide more objective reporting by the media on EMF issues.

43.6 SUMMARY

1. International and national organizations, industrial associations, federal and state agencies, Congress, and the public have expressed concern about the potential health effects of exposure to EMF.
2. Exposure assessment research is a high priority research area because it is essential to the successful interpretation of the biological response and is critically important for risk assessment studies.
3. Research on human reproductive effects should emphasize the need to attempt replication of isolated reports of increase miscarriages and increased malformations, and reports of increased incidence of nervous system cancer in children whose fathers had occupations with potential EMF exposure.
4. The potential need for future controls to reduce risks from exposure to EMF is the rationale for control technology research. This research presently is a low priority because no firm cause-and-effect relation between human health risk and EMF exposure has been established.
5. If scientists eventually reach a consensus that low-level EMFs do cause cancer or some other adverse health effect, then regulations defining some safe exposure level will have to be written at some later date.

REFERENCES

1. Recupero, S. The danger associated with electric and magnetic fields; drawn, in part, from a term paper submitted to L. Theodore, Manhattan College, 1994.
2. U.S. EPA. EMF: An EPA Perspective, U.S. EPA, Washington, DC, December 1992.
3. Department of Engineering and Public Policy, *EMF from 60 Hertz Electric Power*, Pittsburgh, PA, 1989.
4. Leonard, A., Neutra, R., Yost, M., and Lee, G. *EEMF Measurements and Possible Effects*, California Department of Health Services, 1990.
5. Wilson, R. Currents of concern. *NY Newsday*, May 1, 1990.
6. Hileman, B. Health effects of electromagnetic fields remain unresolved, *Chemical and Engineering News* (C&EN), November 8, 1993.
7. Electromagnetic fields, *Consumer Reports*, May 1, 1994.
8. Fagin, D. *NY Newsday*, June 15, 1994.
9. McElfresh, R. Responsible reporting of environmental issues by the media, Panel discussion, *AWMA Annual Meeting*, Cincinnati, OH, June 1994.

44 Noise Pollution

CONTENTS

44.1 INTRODUCTION

By definition, noise is a sound that is annoying and has a long-term physiological effect on an individual. Noise is a subtle pollutant. Although it can be hazardous to a person's health and well-being, noise usually leaves no visible evidence. Noise pollution has grown to be a major environmental problem today. An estimated 14.7 million Americans are exposed to noises that pose a threat to hearing on their jobs. Another 13.5 million Americans are exposed to dangerous noise levels, such as from trucks, airplanes, motorcycles, and stereos without knowing it. Moreover, noise can cause temporary stress reactions like increasing the heart rate and blood pressure, and produce negative effects on the digestive and respiratory system.

Sound is a disturbance that propagates through a medium having the properties of inertia (mass) and elasticity. The medium by which audible sound is transmitted is air. The higher the wave, the greater its power; the greater the number of waves a sound has, the larger is its frequency or pitch. The frequency can be described as the rate of vibration that is measured in Hertz (Hz, cycles per second). The human ear does not hear all of the frequencies. The normal hearing range for humans is from 20 to 20,000 Hz. In addition, the human ear cannot define all sounds equally. Very low and very high notes sound fainter to the ear than do 1000 Hz sounds of equal strength; that is how the ear functions. The human voice in conversation covers a median range of 300–4000 Hz; and, the musical scale ranges from 30 to 4000 Hz. Hearing also varies widely between individuals.

The unit of the strength of sound is measured in decibels (dB). (The decibel is a dimensionless unit used to describe sound intensity; it is the logarithm of the ratio of the intensity of sound to the intensity of an arbitrary chosen standard sound.) Although the degree of loudness depends on personal judgments, precise measurement of sound is made possible by the use of the decibel scale (see Table 44.1). The decibel

TABLE 44.1

Sound Levels and Human Response

Common Sounds	Noise Level (dB)	Effect
Carrier deck jet operation	140	Painfully loud
Air raid siren	130	
Jet takeoff (200 ft)		Thunderclap
Discotheque	120	Maximum vocal effort
Auto horn (3 ft)		
Pile drivers	110	
Garbage truck	100	
Heavy truck (50 ft)	90	Very annoying
City traffic		Hearing damage (8 h)
Alarm clock (2 ft)	80	Annoying
Hair dryer		
Noisy restaurant	70	Phone use difficult
Freeway traffic		
Man's voice (3 ft)		
Air-conditioning unit (20 ft)	60	Intrusive
Light auto traffic (100 ft)	50	Quiet
Living room		
Bedroom	40	
Quiet office		
Library	30	Very quiet
Soft whisper (15 ft)		
Broadcasting studio	20	
	10	Just audible
	0	Hearing begins

scale ranges from 0 (minimum) to 194 (maximum). Because the decibel scale is in logarithm form, even a small reduction in values at high levels can make a significant difference in noise intensity.

This decibel scale measures sound pressure or energy according to international standards. By comparing some common sounds, the scale shows how they rank in potential harm. Recent scientific evidence showed that relatively continuous exposures to sound exceeding 70 dB can be harmful to hearing. Noise begins to harm hearing at 70 dB; and, each 10 dB increase seems twice as loud [1].

44.2 NOISE LEGISLATION [2]

Because noise pollution has become such a threat to the health of so many lives, many regulations have been established to monitor and control the level of unwanted harmful sounds. The Occupational Safety and Health Act (OSHA) was signed on December 29, 1970 and went into effect April 28, 1971. The purpose of this Act is "to assure so far as possible every working man and woman in the nation safe and healthful working conditions and to preserve our human resources." The OSHA does not apply to working conditions that are protected by other federal occupational safety and health laws such as the Federal Coal Mine Health and Safety Act, the Atomic Energy Act, the Metal and Nonmetallic Mines Safety and Health Standards, and the Open Pit and Quarries Safety and Health Standards. This Act puts all state and federal occupational safety and health enforcement programs under federal control with the goal of establishing more uniform standards, regulations, and codes with stricter enforcement. Several of the major aspects of the Act will maintain federal supervision of state programs to obtain more uniform state inspection under federal standards. The OSHA will also make it mandatory for employers to keep accurate records of employee exposures to harmful agents that are required by safety and health standards. The law provides procedures in investigating violations by delivering citations and monetary penalties upon the request of an employee. The OSHA establishes a National Institute of Occupational Safety and Health (NIOSH) whose members have the same powers of inspection as members of the OSHA. The Act also delegates to the Secretary of Labor the power to issue safety and health regulations and standards enforceable by law. This last provision is implemented by the Occupational Safety and Health Administration.

The OSHA enforces two basic duties which must be carried out by employers. First, it provides each employee with a working environment free of recognized hazards that cause or have the potential to cause physical harm or death. Second, it fully complies with the Occupational Safety and Health Standards under the Act. To carry out the first duty, employers must have proper instrumentation for the evaluation of test data provided by an expert in the area of industrial hygiene. This instrumentation must be obtained because the presence of health hazards cannot be evaluated by visual inspection. This duty can be used by the employees to allege a hazardous working situation without any requirement of expert judgment. It also provides the employer with substantial evidence to disprove invalid complaints. This law also gives employers the right to take full disciplinary action against those employees who violate safe practices in working methods.

Section 50–204.10 of the Act establishes acceptable noise levels and exposures for safe working conditions, and gives various means of actions which must be taken if these levels are exceeded. A 90 dB level of exposure to sound energy absorbed is taken as the limit of exposure that will not cause any type of hearing loss in more than 20% of those exposed. Workers in any industry must not be exposed to sound levels greater than 115 dB for any amount of time. Noise levels must be measured on the A scale of a standard sound level meter at slow response. The sound level meter is a measuring device that indicates sound intensity. "Slow response" is a particular setting on the meter, and when the meter is at this setting it will average out high-level noise of short-lived duration. For impact noise, a higher level of 140 dB is acceptable because the noise impulse due to impacts is over before the human ear has time to fully react to it.

Regulations of variable noise levels are covered under paragraph (c) of Section 50–204.10 in the OSHA. This paragraph states, "if the variations in noise levels involve maxima at intervals of 1 s or less, it is considered to be continuous." Therefore, when the level on the meter goes from a relatively steady reading to a higher reading, at intervals of 1 s or less, the higher reading is taken as the continuous sound level. Sounds of short duration occurring at intervals greater than 1 s should be measured in intensity and duration over the total work day. Such sounds may be analyzed using a sound level meter and should not be treated as impact sound.

The federal Walsh-Healey Public Contracts Act took effect on May 20, 1969. To comply with the Walsh-Healey regulations on industrial noise exposure, industry must measure the noise level of its working environment. It provides valuable data with which an inspector can evaluate working conditions. The data may be obtained by sound survey meters with classified A-, B-, and C-weighted filters; and, all measurements weigh all the frequencies equally in that range.

On April 30, 1976, the Administrator of the U.S. Environmental Protection Agency (EPA) established the Noise Enforcement Division under the Deputy Assistant Administrator for Mobile Source and Noise Enforcement, Office of Enforcement. The division originally had a staff of 21 individuals whose responsibilities were divided into the following four general enforcement areas:

1. General products noise regulations
2. Surface transportation noise regulations
3. Noise enforcement testing
4. Regional (EPA), state, and local assistance

On December 31, 1975, the EPA set forth noise standards and regulations for the control of noise from portable air compressors. These regulations became effective on January 1, 1978. Additional regulations are currently being developed to control noise from truck-mounted solid waste compactors, truck transport, refrigeration units, wheel and track loaders, and dozers, which have been identified pursuant to Section 5 of the Noise Control Act (NCA). The surface transportation group will have similar responsibilities with respect to transportation-related products. The first such products to be regulated under Section 6 for the control of noise are new medium- and heavy-duty trucks (in excess of 10,000 lb GVWR). Regulations for trucks were set forth on April 13, 1976 and became effective on January 1, 1978. Motorcycles and buses are additional major noise sources that have been identified and for which regulations are presently being developed.

Noise enforcement testing is conducted by the EPA Noise Enforcement Facility located in Sandusky, Ohio. The facility is used to conduct enforcement testing; to monitor and correlate manufacturers' compliance testing; and, to train regional, state, and local personnel for noise enforcement. This program defines and develops the EPA enforcement responsibilities under the NCA. It also provides assistance to state and local agencies regarding enforcement of the federal noise control standards and regulations, and enforcement aspects of additional state and local noise control regulations.

To assist state and local governments in drafting noise control ordinances, EPA has published a Model Community Noise Ordinance, which is available in EPA regional offices and in the EPA headquarters in Washington, D.C.

Additional recent (2007, 2008) information is available at:

1. http://www.marinebuzz.com/2008/02/24/in-the-oceans-underwater-noise-pollution-is-as-harmful-as-oil-pollution/
2. http://www.guardian.co.uk/science/2007/aug/23/sciencenews.uknews.

44.3   EFFECTS OF NOISE

It is estimated that between 8.7 and 11.1 million Americans suffer a permanent hearing disability [3]. This section will examine the overall effects of noise on an industrial worker, not only in terms of hearing loss, but also in work quality.

The ear has its own defense mechanism against noise—the acoustic reflex. However, this reflex has vital weak points in its defenses. First of all, the muscles within the middle ear can become fatigued and slow if overused. A person who works in an environment with high noise levels gradually loses the strength in these muscles and thus more noise will reach the inner ear. Secondly, these muscles can be affected by chemicals within the working environment. Finally, the acoustic reflex is an ear-to-brain-to-ear circuit that takes at least nine-thousandths (0.009) of a second to perform. Individuals with poor acoustic reflex are usually subjected to temporary hearing loss when they come in contact with a loud noise. Most of the hearing loss caused by noise occurs during the first hour of exposure. Recovery of hearing can be complete several hours after the noise stops. The period of recovery depends upon individual variation and the level of noise that caused the deafness.

Noises that pose the greatest threat to the human body are those that are the highest pitched, loudest, poorest in tone, and longest lasting. Another dangerous type of sound is the sound of an explosion. Deafness due to noise usually occurs in conjunction with a fairly common hearing disorder known as recruitment of loudness. The person who has this disorder will have a smaller range of zone of hearing. However, the recruitment ear will retain its sensitivity for loud sound levels. Another problem that a person with recruited ears faces is the discomfort of using hearing aids. The hearing aid is a microphone that transmits sounds from the surrounding environment to an amplifier connected to a small loudspeaker built into an earplug and aimed at the eardrum. The major problem is that the sounds entering the hearing aid have to be amplified enough to be heard loudly, and at that level the sound may produce discomfort.

Researchers have analyzed noise and its effects on the human ear and have come up with several properties of noise that contribute to the loss of hearing. They include the "overall sound level the noise spectrum," "the shape of the noise spectrum," and "total exposure duration." A final characteristic of noise that should be mentioned is the temporal distribution of noise. However, energy in noise is distributed across time and its final effect on the threshold shift is a function of total energy. It has been determined that partial noise exposures are related closely to the continuous A-weighted noise level (a means of correlating speech-interference level and NC [Noise Criteria] or PNC [Preferred Noise Criteria] level, and the unit of this scale is dBA) by equal energy amounts. The relation between energy and the amount of exposure is twice the energy is acceptable for every halving of exposure time, without any increase in danger.

Noise affects the mind and changes emotions and behavior in many ways. Most of the time, individuals are unaware that noise is directly affecting their minds. It interferes with communication, disturbs sleep, and arouses a sense of fear. Psychologically, noise stimulates individuals to a nervous peak. Too much arousal makes a person overly anxious and as a result, tends to cause the person to make more mistakes. The effects of noise increase the frequency of momentary lapses in efficiency.

Noise has its effects on manual workers. From a case history from Dr. Jansen, the employees who worked in the quieter surroundings were easier to interview than the employees who worked in the noisier surroundings [2]. Noise also affects a worker's behavior at home. This study revealed that the workers exposed to higher noise levels had more than twice as many family problems. Since noise affects a worker's attitude and personality, it also affects his or her output. It can interfere with communication greatly. Noise also can cause a decrease in the quality of work output when the background noise exceeds 90 dB. The effects of noise on work output depend largely on the type of work. High noise levels tend to cause a higher rate of mistakes and accidents rather than a direct slowdown of production. Results show that a worker's attention to the job at hand will tend to drift as noise levels increase.

Dr. G. Lehmann, Director of the Max Planck Institute, had determined that noise has an explicit effect on the blood vessels, especially the smaller ones known as precapillaries [2]. Overall, noise makes these blood vessels narrow. It was also found that noise causes significant reductions in the blood supply to various parts of the body. Tests were also conducted employing a ballistocardiogram, which is used to measure the heart with each beat. When the test was conducted on a patient in noisy surroundings, the findings led to one conclusion: noise at all levels causes the peripheral blood vessels in the toes, fingers, skin, and abdominal organs to constrict, thereby decreasing the amount of blood normally supplied to these areas. The vasoconstriction is triggered by various body chemicals, predominantly adrenaline, which is produced when the body is under stress. Finally noise affects the nervous system. Noise wears down the nervous system, breaks down the human's natural resistance to disease and natural recovery, thus lowering the quality of general health.

Most are aware of the harmful effects of noise pollution on land. But, what about underwater noise pollution in the oceans? Interestingly, a deep diving dolphin was found dead on the beach of the U.S. Navy's San Nicolas Island in 2008. Similar deaths of other marine species are often reported from different parts of the world. It is reported that the U.S. Navy deploys Low Frequency Active Sonar (LFA) to detect quiet submarines throughout 80% of the world's oceans and there is concern that this may serve as death traps to marine mammals.

44.4 SOURCES OF NOISE [4]

As more and more noise-generating products become available to consumers, the sources of noise pollution are extremely diverse and are constantly increasing. Commonly encountered motor vehicle noise comes from cars, trucks, buses, motorcycles, and emergency vehicles with sirens. Noise levels near major airports have become so intolerable that residents sometimes are forced to relocate, and property values sometimes depreciate because of noise pollution. Airport noise is the most common source of noise pollution that will produce an immediate effect ranging from temporary deafness to a prolonged irritation.

The noise levels a source produces can be separated into four categories. Machines, such as refrigerators and clothes dryers, are in the first group, usually produce sound levels lower than 60 dB. The second group includes clothes washers and food mixers that produce noise from 65 to 75 dB. The third group includes vacuum cleaners and noisy dishwashers, which produce a noise range from 85 to 95 dB. This group also includes yard-care and shop tools. The fourth group includes pneumatic chippers and jet engines, which produce noise levels above 100 dB. Any amount of exposure to such equipment will probably interfere with activities, disrupt a neighbor's sleep, cause annoyance and stress, and may contribute to hearing loss.

44.5   NOISE ABATEMENT

Noise abatement measures are under the jurisdiction of local government, except for occupational noise abatement efforts. It is impossible for an active person to avoid exposure to potentially harmful sound levels in today's mechanized world. Therefore, hearing specialists now recommend that individuals get into the habit of wearing protectors to reduce the annoying effects of noise.

Muffs worn over the ears and inserts worn in the ears are two basic types of hearing protectors. Since ear canals are rarely the same size, inserts should be separately fitted for each ear. Protective muffs should be adjustable to provide a good seal around the ear, proper tension of the cups against the head, and comfort. Both types of protectors are well worth the small inconvenience they cause for the wearer and they are available at most sport stores and drugstores. Hearing protectors are recommended at work and during recreational and home activities such as target shooting and hunting, power tool use, lawn mowing, and snowmobile riding.

One should be aware of major noise sources near any residence, e.g., airport flight paths, heavy truck routes, and high-speed freeways, when choosing a new house or apartment. When buying a house, check the area zoning master plan for projected changes. In some places, one cannot obtain Federal Housing Administration (FHA) loans for housing in noisy locations. Use the Department of Housing and Urban Development (HUD) "walkaway test." By means of this method, potential buyers can assess background noise around a house. Simply have one person stand with some reading material at chest level and begin reading in a normal voice while the other slowly backs away. If the listener cannot understand the words within 7 ft, the noise level is clearly unacceptable. At 7–25 ft, it is normally unacceptable; at 26–70 ft, normally acceptable; and over 70 ft, clearly acceptable.

Furthermore, look for wall-to-wall carpeting. Find out about the wall construction. Staggered-stud interior walls provide better noise control. Studs are vertical wooden supports located behind walls. Staggering them breaks up the pattern of sound transmission. Check the electrical outlet boxes because noise will pass through the wall if the boxes are back-to-back. Also, check the door construction; solid or core-filled doors with gaskets or weather stripping provide better noise control. Make sure sleeping areas are displaced from rooms with noise-producing equipment. Finally, insulating the heating and air-conditioning ducts help control noise.

There are some helpful hints to make a quieter home, including the use of carpeting to absorb noise. Hang heavy drapes over windows closest to outside noise sources. Put rubber or plastic treads on uncarpeted stairs. Use upholstered rather than hard-surfaced furniture to deaden noise. Use insulation and vibration mounts when installing dishwashers. When listening to a stereo, keep the volume down. Place window air conditioners where their hum can help mask objectionable noises. Use caution in buying children's toys that make intensive or explosive sounds. Also, compare the noise outputs of different makes of an appliance before making a selection.

Housing developments often are located near high-speed highways. Poor housing placement is on the increase in many communities across the country. To cope with the problem of lightweight construction and poor planning, HUD has developed "Noise Assessment Guidelines" to aid in community planning, construction, modernization, and rehabilitation of existing buildings. In addition, the Veterans Administration (VA) requires disclosure of information to prospective buyers about the exposure of existing VA-financed houses to noise from nearby airports.

The EPA is preparing a model building code for various building types. The code will spell out extensive acoustical requirements and will make it possible for cities and towns to regulate construction in a comprehensive manner to produce a quieter local environment.

The Noise Control Act of 1972 provides the EPA with the authority to require labels on all products that generate noise capable of adversely affecting public health or welfare and on those products sold wholly or in part for their effectiveness in reducing noise. The EPA also initiated a study to rate home appliances and other consumer products by the noise generated and the impact of the noise on users and other persons normally exposed to it. Results will be used to determine whether noise labeling or noise emission standards are necessary [4].

44.6   FUTURE TRENDS

Since more and more noise generators have been developed in recent years, the chances of noise affecting individuals in this century will certainly increase. Many scientists and engineers are working on different plans or projects to reduce noise in the future. Two of the examples that have been developed are the electric trains and electric automobiles. In Japan and many eastern countries, electric trains are one of the most popular modes of transportation because they do not cause air pollution and produce minimum noise. The electric automobiles also reduce the noise, because they do not need to burn gasoline to run their engines; this development is a good starting point for reducing noise. In addition, the development of new equipment and tools in the future is certain to reduce noise. Individuals should also minimize noises surrounding or caused by them; for example, try not to use noise generators, such as vacuum cleaners, dishwashers, and high-watt stereos. If every individual does his or her level best to help reduce noise, the noise pollution will be lowered in the future and humans will hopefully live in a better environment.

44.7 SUMMARY

1. Urban noise pollution has rapidly grown to be a major environmental problem. Sound is a disturbance that propagates through a medium having the properties of inertia and elasticity.
2. The OSHA was signed on December 29, 1970 and went into effect April 28, 1971. The purpose of this Act is "to assure so far as possible every working man and woman in the nation safe and healthful working conditions and to preserve our human resources."
3. It is estimated that between 8.7 and 11.1 million Americans suffer a permanent hearing disability.
4. The sources of noise pollution are extremely diverse and are constantly increasing as more and more noise-generating products become available to consumers.
5. It is almost impossible for an active person to avoid exposure to potentially harmful sound level in today's mechanized world.
6. The environment will be better if individuals help to reduce noise in their everyday life.

REFERENCES

1. Thumann, A. and Miller, C. *Fundamental of Noise Control Engineering*, The Fairmont Press, Englewood Cliffs, NJ, 1990.
2. Cheremisinoff, P. and Cheremisinoff, P. *Industrial Noise Control Handbook*, Ann Arbor Science Publishers, Ann Arbor, MI, 1977.
3. U.S. EPA, Administrator of the Environmental Protection Agency. *Report to the President and Congress on Noise*, 92nd Congress Document, No. 92-63, February 1982.
4. Burke, G., Singh, B., and Theodore, L. *Handbook of Environmental Management and Technology*, 2nd edition, John Wiley & Sons, Hoboken, NJ, 2000.

45 Used Oil

CONTENTS

45.1 INTRODUCTION

Used oil is a valuable resource, but it can be an environmental problem and financial liability if improperly disposed of. Used oils pose hazards to human health and the environment, and therefore need to be managed safely. The mismanagement of used oil can contaminate air, water, and soil. Contamination primarily occurs from improper storage in containers and tanks, disposal in unlined impoundments or landfills, burning of used oil mixed with hazardous waste, improper storage practices of used oil handling sites and associated facilities, and road oiling for dust suppression.

Twenty years ago, 1.3 million gallons of used oil were generated. Fifty-seven percent of the 1.3 million gallons generated entered the used oil management system and was recycled. Of the remaining used oil, the do-it-yourselfer (DIY) generator population (i.e., generated by homeowners) disposed of approximately 183 million gallons of mostly automotive crankcase oil, while nonindustrial and industrial generators dumped/disposed of 219 million gallons. The Environmental Protection Agency (EPA) believes that the majority of the remaining 43% of used oil that was generated could and should be recycled in an effort to protect the nation's groundwater, to meet the nation's petroleum needs, and to converse natural resources [1]. In addition to preserving a natural resource, it would lessen dependence on foreign oil.

45.2 USED OIL INDUSTRY

Much of the used oil that is generated nationally is as a result of routine replacement of deteriorated lubricating oils or what can be classified as *viscosity breakdown.*

483

Frequently, lubricants are replaced because they no longer meet their performance standards. Lubricants and oil are made of two basic components: (a) a base stock or material and (b) additives comprising up to 20% of the volume. These additives greatly influence the specific performance of the finished product which eventually becomes a trademark of various lubricants. Typical additives include color stabilizer, viscosity improvers, corrosion inhibitors, rust inhibitators, and detergents. These additives contain specific metal and chemical compounds which results in specific performance standards. Typical lubricating oil additives and their functions are represented in Table 45.1.

Lubricants and industrial oils are typically manufactured from chemical feedstocks, mainly petroleum or petroleum products which are derived from crude oil. They contain a wide assortment of hydrocarbons in addition to chemical compounds such as sulfonates, sulfur, chlorine, and nitrogen compounds; they also contain metals such as barium, zinc, and chromium as a result of additives.

Used oils are generated from literally thousands of different resources. These can be broken down into two main categories: (a) automotive oils, which include engine crankcase oils, transmission fluids, diesel engine oils, and automotive hydraulic fluids, and (b) industrial oils, which covers oils and lubricants generated from industrial sources such as metal working processes, hydraulic equipment and machinery, refrigeration equipment, quenching oils, and turbine lubricating oils. In all automotive and industrial applications, the performance of the oil deteriorates over a period of time as additives break down and as contaminants build up in the oil. The oil must be replaced with new oil which results in the steady generation of what is commonly referred to as "used oil." Used oil can therefore be defined as lubricating oil which, through use, has been contaminated by physical or chemical impurities. Used oil is considered to be a waste product because it has served its

TABLE 45.1
Typical Lubricating Oils Additives and Their Function

Name of Additive	Chemical Composition	Function
Corrosion inhibitor	Metal sulfonates and sulfurized terpenes, and barium dithiophosphates	To react with metal surfaces to form a zinc corrosion-resistant film
Rust inhibitor	Sulfonates, alkylamines, or amine phosphates	To react chemically with steel surfaces to form an impervious film
Antiseptic	Alcohols, phenols, and chlorine compounds	To inhibit microorganisms
Antioxidant	Sulfides, phosphates, phenols	To inhibit oxidation of oil
Detergent	Sulfonates, phosphites, alkyl substituted salicylates combined with barium, magnesium, zinc, and calcium	To neutralize acids in crankcase oils to form compounds suspended in oil
Color stabilizer	Amine compounds	To stabilize oil color

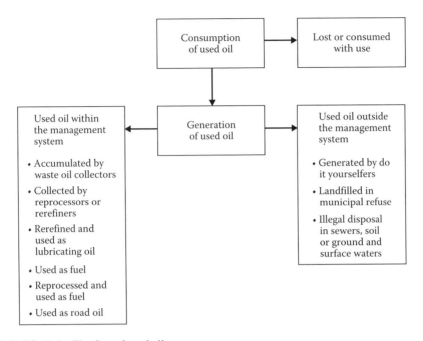

FIGURE 45.1 The fate of used oil.

original intended purpose and must be discarded. However, it is a unique type of waste in that it can be recycled or reused as another product instead of merely being discarded or destroyed [2].

Much of the used oil that is generated nationally as a result of routine replacement of deteriorated lubricating oil enters the "used oil management system." The used oil management system consists of companies that are involved in the generation, collection, transport, processing, and reuse of used oils. These companies interact to provide a mechanism for used oil to flow from its point of generation to its reuse or disposal. Figure 45.1 illustrates the fate of used oil. Many companies are involved in just one used oil function while others participate in more than one activity. An example of this is a reprocessor or re-refiner which collects, transports, and recycles used oil.

45.3 USED OIL RECYCLING AND REUSE

45.3.1 REFINING

Used oil can be re-refined into a base lubricating oil by employing a variety of techniques and processes. With the additives restored, the re-refined oil can be marketed as lubricating oil for industrial and commercial application. Rerefining used oil results in well-defined, marketable products regardless of the type of technology employed. It can produce a refined lubricating base, a distilled light fuel fraction, and distillation bottoms for use as asphalt extenders. The methods of rerefining differ as to the waste material generated, the percent recovered, and by-product marketability.

Common rerefining technologies

1. Acid/clay method. This process involves three steps.
   Step 1. Filtering the used oil to remove water and solids.
   Step 2. Acid treatment—using sulfuric acid treatment to remove toxic impurities.
   Step 3. Clarification—the material is clarified to remove odor and color impurities by filtering through clay.

   Product: Acceptable base with which additives and virgin oil can be blended. Approximately 50% recovery efficiency.

   Disadvantages:

   1. Process is costly.
   2. Batch operation which results in high operating costs.
   3. Environmental problems. Technique yields a considerable amount of acid sludge and clay-like residue from sulfuric acid treatment and clay refiltration. Associated waste disposal costs.

2. Distillation—Clay treatment process. The distillation treatment process involves a five-step process:
   Step 1. Screening to remove solids.
   Step 2. Evaporation to remove water.
   Step 3. Flash vacuum distillation to recover low-boiling components as a distillate fuel.
   Step 4. High-temperature, high-vacuum distillation in a thin film evaporator to separate lubricating oil fraction from the residue and depleted additive, with controlled partial condensation of the distillate to separate lubricating oil into light and heavy fractions.
   Step 5. Final purification of lubricating oil uses clay treatment and filtrations.

   Product:

   1. Lubricating oil base.
   2. High ash content fuel by-product which can be burned.

   Advantages:

   1. Higher recovery 60%–75%.
   2. Continuous operation.
   3. Manageable residue by-product.

Distillation-hydrotreating is a type of rerefining that is quite similar to the distillation-clay process. However, it substitutes hydrotreating for clay treating as a final step. Hydrotreating involves mixing heating oil with recirculating hydrogen to remove impurities. The end product is a lubricating oil base and several by-products which have secondary market value. Recovery can be expected to approach 99%.

Phillips Petroleum developed Phillips Rerefining Oil Process (PROP), a re-refined oil process which combines hydrotreating with chemical demetallization to produce 90% yields of based oil. PROP is a two-stage process. In the first stage, the metal is recovered from the used oil through the use of chemicals. The oil is then filtered to remove the remaining metals. The demetallized and dehydrated oil is hydrotreated in the second major processing stage. Hydrotreating removes unwanted sulfur, nitrogen, oxygen, and chlorine compounds and improves color.

45.3.2 REPROCESSING

Used oil can also be reprocessed to yield fuel oil that can be burned in industrial incinerators or boilers. Reprocessing involves filtering sediment and water from used oil. The technology varies according to the degree of sophistication employed and from facility to facility. Most processing firms produce only fuel.

Reprocessing is a less costly recycling alternative. It does not involve as substantial an investment in its operation as rerefining does. However, the product is of a lesser quality and contains toxic metals and other chemicals that refining techniques can remove.

Minor reprocessors employ fairly simple technology to recycle used oil as a fuel. It includes in-line filtering and gravity settling to remove solids and water. In addition, it may include the addition of a heat source to decrease viscosity and improve gravity settling.

Minor reprocessors market their product by making it available directly to fuel users, fuel oil dealers, road oilers (where allowed), or for purchase by major waste oil reprocessors. In addition, some choose to burn a portion of the oil produced on site to generate heat to induce gravity settling, for space heating, or for some other fuel-consuming process operated at the site.

Major reprocessors utilize comparatively sophisticated processing technology. They go beyond merely filtering and settling used oil and employ treatment devices to further increase oil quality. Some of the devices to improve oil quality include (a) distillation towers to separate and capture light fuel fractions as well as remove water, (b) centrifuges to separate fine solids, and (c) agitators to mix emulsion-breaking chemicals into the oil.

A major portion of used oil which is reprocessed for fuel is blended with virgin fuel oil. Most of this blending is done by virgin fuel oil dealers. However, a small fraction of the blending is done by major reprocessors. The criteria used for blending vary greatly. Some blenders mix used and virgin oil. Others blend to a desired viscosity, moisture content, or any number of other factors, including heat content and percent solids. The criteria are thus a function of the product specifications or characteristics which the blender or his customer have established.

45.4   REGULATIONS GOVERNING USED OIL

On December 18, 1978, the EPA initially proposed guidelines and regulations for the management of hazardous waste as well as specific rules for the identification and listing of hazardous waste under Section 3001 of the Resource Conservation and Recovery Act (RCRA) (43FR 58946). At that time, the EPA proposed to list waste lubricating oil and waste hydraulic and cutting oil as hazardous wastes on the basis of their toxicity. In addition, the EPA proposed recycling regulations to (a) regulate the incineration or burning of used lubricating, hydraulic, transformer, transmission, or cutting oil that was hazardous and (b) the use of waste oils in a manner that constituted disposal. Extensive details are available in the literature [4].

On September 23, 1991, the EPA published a notice in the Federal Register providing information on proposed used oil management standards for recycled oil under Section 3014 of RCRA. In addition, the EPA specifically requested public comments on proposed used oils and residuals to be listed as hazardous, on a number of specific aspects of the newly available data, on specific aspects of the Agency's approach for used oil management standards, and on several aspects of the hazardous waste identification program as related to used oil.

The EPA's overall approach to used oil—as originally developed—consists of three major components. First, the EPA identifies approaches for making a determination whether to list or identify used oil as hazardous waste, as required by Section 3014(b). Second, the EPA proposes a number of alternatives relating to management standards to ensure proper management of used oils that are recycled. The EPA management standards will be issued in two phases. Phase I will consist of basic requirements for used oil generators, transporters, road oilers, and recyclers, including burners and disposal facilities to protect human health and the environment from potential hazards caused by mismanagement of used oil. Once Phase I standards are in place, the EPA may decide to evaluate the effectiveness of these standards in reducing the impact on human health and the environment. Upon such evaluation, the EPA will consider whether or not more stringent regulations are necessary to protect human health and the environment, and propose these regulations as Phase II standards. The third part of the EPA's general approach to used oil is the consideration of nonregulatory incentives and other nontraditional approaches to encourage recycling and mitigate any negative impacts the management standards may have on the recycling of used oil, as provided by Section 3014(a).

The notice presents supplemented information gathered by the EPA and provided to the EPA by individuals commenting on previous notices on the listing of used oil and used oil management standards. Numerous commenters on the 1985 listing of all used oils unfairly subjects them to stringent Subtitle C regulations because their oils are not hazardous. Based on those comments, the EPA has collected a variety of additional information regarding various types of used oil, their management, and their potential health and environmental effects when mismanaged. This notice presents that new information to the public and requests comment on that information, particularly if and how this information suggests new concerns that the EPA may consider in deciding whether to finalize all or part of its 1985 proposal to list used oil as a hazardous waste.

The EPA intends to amend 40 CFR Section 261.32 by adding four waste streams from the reprocessing and rerefining of used oil to the list of hazardous wastes from specific sources. The EPA noted its intention to include these residuals in the definition of used oil in its November 29, 1985 proposal to list used oil as hazardous. The wastes from the reprocessing and rerefining of used oil include process residuals from the gravitational or mechanical separation of solids, water, and oil; spent polishing media used to finish used oil; distillation bottoms; and, treatment residues from primary wastewater treatment.

The notice also includes a description of some of the management standards (in addition to or in place of those proposed in 1985) that the EPA is considering promulgating with the final used oil listing determination. The EPA, under various RCRA authorities, is considering management standards for used oils, whether or not the oil is classified as hazardous waste. When promulgated, the standards may prohibit road oiling; restrict used oil storage in surface impoundments; limit disposal of nonhazardous used oil; require inspection, reporting, and cleanup of visible released of used oil around used oil storage containers and aboveground tanks and during used oil pickup, delivery, and transfer; impose spill cleanup requirements and allow for limited CERCLA liability exemptions; institute a tracking mechanism to ensure that all used oils reach legitimate recyclers; and, require reporting of used oil recycling activities. The used oil burner standards included in 40 CFR Part 266 Subpart E will continue to regulate the burning of used oil for energy recovery. All of the requirements (including those in Part 266, Subpart E) are placed in a new Part (e.g., 40 CFR Part 279). Used oils that are hazardous (either listed or characteristic) and that cannot be recycled are not included in these provisions, but are instead subject to 40 CFR Section 261–270.

On September 10, 1992, the EPA promulgated both a final listing decision for recycled used oil and management standards for used oil pursuant to RCRA Section 3014 (57 FR 42566). Part 279, *Standards for the Management of Used Oil*, was added to codify the management standards. In this rule, the EPA stated that it assumes all used oil will be recycled until the used oil is disposed of or sent for disposal (57 FR 41578). Used oil that is disposed of will need to be characterized like any other solid waste and will need to be managed as hazardous if it exhibits a characteristic of hazardous waste or if it is mixed with a listed hazardous waste.

Standards for the management of used oil (40 CFR Part 279) are a comprehensive set of requirements centered around the various entities involved in the management of used oil. The different subparts incorporates specific requirements for those entities. Of particular importance is the requirements under the following subparts:

1. Subpart C. Standards for used oil generators.
2. Subpart D. Standards for used oil collection centers and aggregation points.
3. Subpart E. Standards for used oil transporters and transfer facilities.
4. Subpart F. Standards for used oil processors and re-refiners.
5. Subpart G. Standards for used oil burners who burn-off specification used oil for energy recovery.
6. Subpart H. Standards for used oil fuel marketers.
7. Subpart I. Standards for use as a dust suppressant and disposal of used oil.

The reader should note that much of the above mentioned material was still applicable at the time of the preparation of this chapter in 2007.

45.5 FACTS ABOUT USED OIL

There are some well-known facts about used oil. Here is a baker's dozen.

1. A gallon of used oil from a single oil change can ruin a million gallons of fresh water—a year's supply for 50 people [5].
2. It takes only 1 gal of used oil to yield the same 2.5 quarts of lubricating oil provided by 42 gal of crude oil [5].
3. Americans who change their own oil throw away 120 million gallons of recoverable motor oil every year [5].
4. If the oil in (3) were recycled, it would save the United States 1.3 million barrels of crude oil per day. This will reduce dependence on foreign oil [5].
5. The damage used oil causes comes from mismanagement.
6. Rerefining used oil takes only about one-third the energy of refining crude oil to lubricant quality.
7. If all used oil improperly disposed of by DIYs were recycled, it could produce enough energy to power 360,000 homes each year or could provide 96 million quarts of high-quality motor oil.
8. One gallon of used oil used as fuel contains about 140,000 Btu of energy.
9. Concentrations of 50–100 parts per million (ppm) of used oil can foul sewage treatment processes.
10. Films of oil on the surface of water prevent the replenishment of dissolved oxygen, impair photosynthetic processes, and block sunlight.
11. Oil dumped onto land reduces soil productivity.
12. Toxic effects of used oil on freshwater and marine organisms vary, but significant long-term effects have been found at concentrations of 310 ppm in several freshwater fish species and as low as 1 ppm in marine life forms.
13. Publicity about used oil recycling can triple DIY participation [3].

45.6 FUTURE TRENDS

Local recycling programs are cooperative efforts between local governments (towns, cities, and counties) and one or more private or semiprivate sponsors, such as (a) environmental or civic groups, or (b) service organizations. Local governments often assist in collecting used oil through collection centers or curbside pickup. Sponsors often help governments design and organize their programs, run the publicity campaigns and outreach, and enlist the help of resourceful and committed volunteers. The future is certain to see more activity in this area.

45.7 SUMMARY

1. Used oil is a valuable resource, but it can be an environmental problem and financial liability if improperly disposed. Used oils pose hazards to human health and the environment, and therefore need to be managed safely. The mismanagement of used oil can contaminate air, water, and soil.
2. Much of the used oil that is generated nationally as a result of routine replacement of deteriorated lubricating oil enters the "used oil management system." The used oil management system consists of companies that are involved in the generation, collection, transport, processing, and reuse of used oils.
3. Used oil can be re-refined into a base lubricating oil by employing a variety of techniques and processes. With the additives restored, the re-refined oil can be marketed as lubricating oil for industrial and commercial application.
4. The regulations governing waste oils were first introduced under the RCRA. At that time, EPA proposed to list waste lubricating oil and waste hydraulic and cutting oil as hazardous wastes on the basis of their toxicity. In addition, the Agency proposed recycling regulations to regulate: (1) the incineration or burning of used lubricating, hydraulic, transformer, transmission, or cutting oil that was hazardous and (2) the use of waste oils in a manner that constituted disposal.
5. Some interesting facts about used oil, e.g., is that one gallon of used oil can ruin a million gallons of fresh water.
6. Recycling used oil can be a rewarding experience. It is an ideal way for interested groups to get constructively involved in environmental action because it deals with an important environmental problem that is best addressed at the local level.

REFERENCES

1. Hazardous Waste Management System; Identification and Listing of Hazardous Waste; Used Oil; Supplemental Notice of Proposed Rulemaking, 40 CFR Parts 261 and 266, Volume 56, No. 184, September 23, 1991.
2. *Used—But Useful: A Review of the Used Oil Management Program in New York State*, Legislative Commission on Toxic Substances and Hazardous Wastes, October 1986.
3. U.S. EPA. How to set up a local program to recycle used oil, EPA/530-SW-89-039A.
4. Burke, G., Singh, B., and Theodore, L. *Handbook of Environmental Management and Technology*, 2nd edition, John Wiley & Sons, Hoboken, NJ, 2000.
5. U.S. EPA. Recycling used oil, 10 steps to change your oil, EPA/530-SW-89-039C.

46 Environmental Audits*

CONTENTS

46.1 INTRODUCTION

Environmental auditing is fast becoming an integral component of a facility's management plan that not only promoting compliance with regulatory requirements but also limiting environmental liabilities in the form of costly penalties and third-party lawsuits. Corporations have come to realize the significant benefits resulting from conducting environmental audits. These benefits range from drastic reduction of fines from federal and state environmental protection agencies through implementation of their audit policies, to participation in the flow of lucrative "green" dollars through businesses that promote and reward other environmentally conscious entities. Consumers often seek out and patronize these businesses for their environmental policies.

Effective environmental auditing can lead to higher levels of overall compliance and reduced risk to human health and the environment. The Environmental Protection Agency (EPA) endorses the practice of environmental auditing and supports its accelerated use by regulated entities to help meet the goals of federal, state, and local environmental requirements. Auditing serves as a quality assurance check to help improve the effectiveness of basic environmental management by verifying that management practices are in place, functioning, and adequate.

Although there are numerous benefits one can derive from an environmental audit, penalties assessed from noncompliance with environmental laws and pollution liability in the form of remediation cost seem to be the two most convincing reasons for conducting environmental audits. In addition, federal and state agencies responsible for enforcing environmental laws offer strong incentives to facilities which voluntarily conduct environmental audits, self-disclose, and promptly correct violations. These include not seeking gravity-based civil penalties or

---

* See Burke et al. [1].

reducing them by 75%, declining to recommend criminal prosecution for regulated entities that self-police, and refraining from routine request for audit reports from those entities.

46.2   DEFINITION OF ENVIRONMENTAL AUDITING

"Environmental auditing" is a systematic, documented, periodic, and objective review by regulated entities of facility operations and practices related to meeting environmental requirements. Audits can be designed to accomplish any or all of the following: verify compliance with environmental requirements to evaluate the effectiveness of environmental management systems already in place; or assess risks from regulated and unregulated materials and practices.

Environmental audits evaluate, and are not a substitute for, direct compliance activities such as obtaining permits, installing controls, monitoring compliance, reporting violations, and keeping records. Environmental auditing may verify but does not include activities required by law, regulation, or permit (e.g., continuous emissions monitoring, composite correction plans at wastewater treatment plants, etc.). Audits do not in any way replace regulatory agency inspections. However, environmental audits can improve compliance by complementing conventional federal, state, and local oversight.

The EPA clearly supports auditing to help ensure the adequacy of internal systems to achieve, maintain, and monitor compliance. By voluntarily implementing environmental management and auditing programs, regulated entities can identify, resolve, and avoid environmental problems.

The EPA does not intend to dictate or interfere with the environmental management practices of private or public organizations. Nor does EPA intend to mandate auditing (though in certain instances EPA may seek to include provisions for environmental auditing as part of settlement agreements, as noted below). Because environmental auditing systems have been widely adopted on a voluntary basis in the past, and because audit quality depends to a large degree upon genuine management commitment to the program and its objectives, auditing should remain a voluntary activity.

An organization's auditing program will evolve according to its unique structures and circumstances. Effective environmental auditing programs appear to have certain discernible elements in common with other kinds of audit programs. These elements are important to ensure project effectiveness [2].

46.3   WHY CONDUCT AN ENVIRONMENTAL AUDIT?

Environmental auditing has been developed for sound business reasons, particularly as a means of helping regulated entities manage pollution control affirmatively over time instead of reacting to crises. Auditing can result in improved facility environmental performance help communication and effect solutions to common environmental problems, focus facilities managers attention on current and opening regulatory requirements, and generate protocols and checklists which help facilities better manage themselves. Auditing also can result in better-integrated management of environmental hazards, since auditors frequently identify environmental liabilities which go beyond regulatory compliance.

One of the most compelling reasons to voluntarily conduct an environmental audit should be to avoid criminal prosecution. Because senior managers of regulated entities are ultimately responsible for taking all necessary steps to ensure compliance with environmental requirements, the EPA has never recommended criminal prosecution of a regulated entity based on voluntary disclosure of violations discovered through audits and disclosed to the government before an investigation was already under way. Thus, EPA will not recommend criminal prosecution for a regulated entity that uncovers violations through environmental audits or due diligence, promptly discloses and expeditiously corrects those violations, and meets all other conditions of Section D of the policy.

There are fundamentally two types of environmental audits:

1. *Compliance audit:* An independent assessment of the current status of a party's compliance with applicable statutory and regulatory requirements. This approach always entails a requirement that effective measures be taken to remedy uncovered compliance problems, and is most effective when coupled with a requirement that the root causes of noncompliance also be remedied.

2. *Management audit:* An independent evaluation of a party's environmental compliance policies, practices, and controls. Such evaluation may encompass the need for (a) a formal corporate environmental compliance policy, and procedures for implementation of that policy; (b) educational and training programs for employees; (c) equipment purchase, operation, and maintenance programs; (d) environmental compliance officer programs (or other organizational structures relevant to compliance); (e) budgeting and planning systems for environmental compliance; (f) monitoring, record-keeping, and reporting systems; (g) in-plant and community emergency plans; (h) internal communications and control systems; and, (i) hazard identification and risk assessment.

46.4   ELEMENTS OF AN EFFECTIVE AUDITING PROGRAM

An effective environmental auditing system will likely include the following general elements:

1. *Explicit top-management support for environmental auditing and commitment to follow up on audit findings.* Management support may be demonstrated by a written policy articulating upper management support for the auditing program and for compliance with all pertinent requirements, including corporate policies and permit requirements as well as federal, state, and local statutes and regulations. Management support auditing program also should be demonstrated by an explicit written commitment to follow up on audit findings in order to correct identified problems and prevent their recurrence.

2. *An environmental auditing function independent of audited activities.* The status or organizational focus of environmental auditors should be sufficient to ensure objective and unobstructed inquiry, observation, and testing. Auditor objectivity should not be impaired by personal relationships, financial or other conflicts of interest, interference with free inquiry or judgment or fear of potential retribution.

3. *Adequate team staffing and auditor training.* Environmental auditors should possess or have ready access to knowledge, skills, and disciplines needed to accomplish audit objectives. Each individual auditor should comply with the company's professional standards of conduct. Auditors, whether full-time or part-time, should maintain their technical and analytical competence through continuing education and training and certification.

4. *Explicit audit program objectives, scope, resources, and frequency.* At a minimum, audit objectives should include assessing compliance with applicable environmental laws and evaluating the adequacy of internal compliance policies, procedures, and personal training programs to ensure continued compliance.

Audits should be based on a process which provides auditors all corporate policies, permits, and federal, state, and local regulations pertinent to the facility; and checklists or protocols addressing specific features that should be evaluated by auditors.

Explicit written audit procedures generally should be used for planning audits, establishing audit scope, examining and evaluating audit findings, communicating audit results, and following up on findings.

5. *A process that collects, analyzes, interprets, and documents information sufficient to achieve audit objectives.* The following information should be collected before and during an on-site visit regarding environmental compliance: (1) environmental management effectiveness (2) and, other matters related to audit objectives and scope. This information should be sufficient, reliable, relevant, and useful to provide a sound basis for audit finds and recommendations. The processes should also include:

   • Sufficient information is factual, adequate, and convincing so that a prudent, informed person would be likely to reach the same conclusions as the auditor.
   • Reliable information is the best attainable through use of appropriate audit techniques.
   • Relevant information supports audit findings and recommendations and is consistent with the objectives for the audit.
   • Useful information helps the organization meet its goals.

The audit process should include a periodic review of the reliability and integrity of this information and the means used to identify, measure, classify, and report it. Audit procedures, including the testing and sampling techniques employed, should be selected in advance to the extent practical and expanded or altered if circumstances warrant. The process of collecting, analyzing, interpreting, and documenting information should provide reasonable assurance that audit objectivity is maintained and audit goals are met.

6. *A process that includes specific procedures to promptly prepare candid, clear, and appropriate written reports on audit findings corrective actions, and schedules for implementation.* Procedures should be in place to ensure such information is communicated to managers, including facility and corporate management, who can evaluate the information and ensure correction of identified problems. Procedures also should be in place for determining what internal findings are reportable to state or federal agencies.

7. *A process that includes quality assurance procedures to assure the accuracy and thoroughness of environmental audits.* Quality assurance may be accomplished through supervision, independent internal reviews, external reviews, or a combination of these approaches [2].

46.5 EPA'S AUDIT POLICY: INCENTIVES FOR SELF-POLICING

The EPA recognized that environmental auditing and sound environmental management generally can provide potentially powerful tools toward greater protection of public health and the environment. The EPA published the *Audit Policy: Incentives for Self-Policing: Discovery, Disclosure, Correction and Prevention of Violations* on December 22, 1995 (60 FR 66706), as excerpted below [3]:

The EPA today issues its final policy to enhance protection of human health and the environment by encouraging regulated entities to voluntarily discover, and disclose and correct violations of environmental requirements. Incentives include eliminating or substantially reducing the gravity component of civil penalties and not recommending cases for criminal prosecution where specified conditions are met, to those who voluntarily self-disclose and promptly correct violations. The policy also restates EPA's long-standing practice of not requesting voluntary audit reports to trigger enforcement investigations. This policy was developed in close consultation with the U.S. Department of Justice, states, public interest groups, and the regulated community and will be applied uniformly by the Agency's enforcement programs.

Section C of EPA's policy identifies the major incentives that EPA will provide to encourage self-policing, self-disclosure, and prompt self-correction. These include not seeking gravity-based civil penalties or reducing them by 75%, declining to recommend criminal prosecution for regulated entities that self-police, and refraining from routine requests for audits. (As noted in Section C of the policy, EPA has refrained from making routine requests for audit reports.

Under Section D(9), the regulated entity must cooperate as required by EPA and provide information necessary to determine the applicability of the policy. This condition is largely unchanged from the interim policy. In the final policy, however, the Agency has added that "cooperation" includes assistance in determining the facts of any related violations suggested by the disclosure, as well as of the disclosed violation itself. This was added to allow the agency to obtain information about any violations indicated by the disclosure, even where the violation is not initially identified by the regulated entity.

EPA will retain its full discretion to recover any economic benefit gained as a result of noncompliance to preserve a "level playing field" in which violators do not gain a competitive advantage over regulated entities that do comply. EPA may forgive the entire penalty for violations which meet conditions 1 through 9 in Section D and, in the Agency's opinion, do not merit any penalty due to the insignificant amount of any economic benefit.

This policy became effective 30 days after publication on December 22, 1995.

Additional documentation relating to the development of this policy is contained in the environmental auditing public docket. Documents from the docket may be obtained by calling (202) 260-7548, requesting an index to docket #C-94-01, and faxing document requests to (202) 260-4400.

46.6   FUTURE TRENDS

The trend for the future appears to indicate that there will be prompt disclosure and correction of violations, including timely and accurate compliance with reporting requirements. In additions, one can expect an increase in corporate compliance programs that are successful in preventing violations, improving environmental performance, and promoting public disclosure an increase in the consistency among state and local can also be expected programs that provide incentives for voluntary compliance can also be expected.

46.7   SUMMARY

1. Environmental auditing is fast becoming an integral component of a facility's management plan, not only promoting compliance with regulatory requirements but also limiting environmental liabilities in the form of costly penalties and third-party lawsuits. Corporations have come to realize the significant benefits resulting from conducting environmental audits.
2. Environmental auditing is a systematic, documented, periodic, and objective review by regulated entities of facility operations and practices related to meeting environmental requirements.
3. Environmental auditing has developed for sound business reasons, particularly as a means of helping regulated entities manage pollution control affirmatively over time instead of reacting to crises.
4. An effective environmental auditing system will likely include explicit top-management support for environmental auditing and commitment to follow-up on audit findings; an environmental auditing function independent of audited activities; explicit audit program objectives, scope, resources, and frequency; a process that collects, analyzes, interprets, and documents information sufficient to achieve audit objectives; and, a process that includes quality assurance procedures to assure the accuracy and thoroughness of environmental audits.

5. The EPA recognized that environmental auditing and sound environmental management generally can provide potentially powerful tools for greater protection of public health and the environment. The EPA published the *Audit Policy: Incentives for Self-Policing: Discovery, Disclosure, Correction, and Prevention of Violations.*

REFERENCES

1. Adapted from: Burke, G., Singh, B., and Theodore, L. *Handbook of Environmental Management and Technology*, 2nd edition, John Wiley & Sons, Hoboken, NJ, 2000.
2. U.S. EPA. *Restatement of Policies Related to Environmental Auditing*, FRL-5021-5, July 28, 1994.
3. U.S. EPA. *Audit Policy: Incentives for Self-Policing*, 60 FR 66706, December 22, 1995.

47 Economics

CONTENTS

47.1 INTRODUCTION

An understanding of the economics involved in environmental management is important in making decisions at both the engineering and management levels. Every engineer or scientist should be able to execute an economic evaluation of a proposed environmental project. If the project is not profitable, it should obviously not be pursued; and, the earlier such a project can be identified, the fewer are the resources that will be wasted.

Economics also plays a role in setting many state and federal air pollution control regulations. The extent of this role varies with the type of regulation. For some types of regulations, cost is explicitly used in determining their stringency. This use may involve a balancing of costs and environmental impacts, costs and dollar valuation of benefits, or environmental impacts and economic consequences of control costs. For other types of regulations, cost analysis is used to choose among alternative regulations with the same level of stringency. For these regulations, the environmental goal is determined by some set of criteria that does not include costs. However, cost-effective analysis is employed to determine the minimum economic way of achieving the goal. For some regulations, cost influences enforcement procedures or requirements for demonstration of progress toward compliance with an environmental quality standard. For example, the size of any monetary penalty assessed for noncompliance as part of an enforcement action needs to be set with awareness of the magnitude of the control costs being postponed or bypassed by the noncomplying facility. For regulations without a fixed compliance schedule, demonstration of reasonable progress toward the goal is sometimes tied to the cost of attaining the goal on different schedules [1].

Before the cost of an environmental project can be evaluated, the factors contributing to the cost must be recognized. There are two major contributing factors, namely, capital costs and operating costs; these are discussed in the next two sections. Once the total cost of a project has been estimated, the engineer must determine whether the process (change) will be profitable. This often involves converting all cost contributions to an annualized basis. If more than one project proposal is under study, this method provides a basis for comparing alternate proposals and for choosing the best proposal. Project optimization is covered later in the chapter, where a brief description of a perturbation analysis is presented.

Detailed cost estimates are beyond the scope of this chapter. Such procedures are capable of producing accuracies in the neighborhood of $\pm 10\%$; however, such estimates generally require many months of engineering work. This chapter is designed to give the reader a basis for preliminary cost analysis only.

47.2 CAPITAL COSTS

Equipment cost is a function of many variables, one of the most significant of which is capacity. Other important variables include equipment type and location, operating temperature and pressure, and degree of equipment sophistication. Preliminary estimates are often made from simple cost–capacity relationships that are valid when other variables are confined to narrow ranges of values; these relationships can be represented by the approximate linear (on log–log coordinates) cost equations of the form [2]:

$$C = \alpha(Q)^\beta \tag{47.1}$$

where
   C represents cost
   Q represents some measure of equipment capacity
   α and β represents empirical "constants" that depend mainly on the equipment type

It should be emphasized that this procedure is suitable for rough estimation only; actual estimates (or quotes) from vendors are more preferable. Only major pieces of equipment are usually included in this analysis; smaller peripheral equipment such as pumps and compressors are not included.

If more accurate values are needed and if old price data are available, the use of an indexing method is better, although a bit more time-consuming. The method consists of adjusting the earlier cost data to present values using factors that correct for inflation. A number of such indices are available; one of the most commonly used in the past is the chemical engineering fabricated equipment cost index (FECI) [2,3], some past outdated values of which are listed in Table 47.1. Other indices for construction labor, buildings, engineering, and so on, are also available in the literature [2,3]. Generally, it is not wise to use past cost data older than 5–10 years, even with the use of the cost indices. Within that time span, the technologies used in the processes have often changed drastically. The use of the indices could cause the estimates to be much greater than the actual costs. Such an error might lead to the choice of alternative proposals other than the least costly.

TABLE 47.1
Fabricated Equipment Cost Index

Year	Index
1999	434.1
1998	435.6
1997	430.4
1996	425.5
1995	425.4
1994	401.6
1993	391.2
1957–1959	100

The usual technique for determining the capital costs (i.e., total capital costs (TCCs), which include equipment design, purchase, and installation) for a project and/or process can be based on the factored method of establishing direct and indirect installation cost as a function of the known equipment costs. This is basically a modi-fied Lang method, whereby cost factors are applied to known equipment costs [4,5].

The first step is to obtain from vendors (or, if less accuracy is acceptable, from one of the estimation techniques previously discussed) the purchase prices of pri-mary and auxiliary equipment. The total base price, designated by X—which should include instrumentation, control, taxes, freight costs, and so on—serves as the basis for estimating the direct and indirect installation costs. The installation costs are obtained by multiplying X by the cost factors, which are available in literature [1,5–8]. For more refined estimates, the cost factors can be adjusted to more closely model the proposed system by using adjustment factors that take into account the complex-ity and sensitivity of the system [4,5].

The second step is to estimate the direct installation costs by summing up all the cost factors involved in the direct installation costs, which include piping, insulation, foundation and supports, and so on. The sum of these factors is designated as the direct installation cost factor (DCF). The direct installation costs are the product of the DCF and X.

The third step consists of estimating the indirect installation costs. The procedure here is the same as that for the direct installation cost—that is, all the cost factors for the indirect installation costs (engineering and supervision, startup, construction fees, etc.) are added. The sum is designated as the indirect installation cost factor (ICF). The indirect installation costs are then the product of ICF and X.

Once the direct and indirect installation costs have been calculated, the TCC [3] may be evaluated as

$$TCC = X + (DCF)(X) + (ICF)(X) \qquad (47.2)$$

This cost is converted to annualized capital costs (ACCs) with the use of the capital recovery factor (CRF), which is described later. The ACC is the product of the CRF and the TCC and represents the total installed equipment cost distributed over the lifetime of the facility.

Some guidelines in purchasing equipment are listed below:

1. Do not buy or sign any documents unless provided with certified independent test data.
2. Previous clients of the vendor company should be contacted and their facilities visited.
3. Prior approval from the local regulatory officials should be obtained.
4. A guarantee from the vendors involved should be required. Startup assistance is usually needed, and an assurance of prompt technical assistance should be obtained in writing. A complete and coordinated operating manual should be provided.
5. Vendors should provide key replacement parts if necessary.
6. Finally, 10%–15% of the cost should be withheld until the installation is completed.

47.3 OPERATING COSTS

Operating costs can vary from site to site because the costs partly reflect local conditions—for example, staffing practices, labor, and utility costs. Operating costs like capital costs may be separated into two categories: direct and indirect costs. Direct costs are those that cover material and labor and are directly involved in operating the facility. These include labor, materials, maintenance and maintenance supplies, replacement parts, waste (e.g., residues after incineration) disposal fees, utilities and laboratory costs. Indirect costs are those operating costs associated with, but not directly involved in operating the facility; costs such as overhead (e.g., building-land leasing and office supplies), administrative fees, local property taxes, and insurance fees fall into this category.

The major direct operating costs are usually associated with the labor and materials costs for the project which often involve the cost of the chemicals needed for operation of the process [8]. Labor costs differ greatly, but are a strong function of the degree of controls and/or instrumentation. Typically, there are three working shifts per day with one supervisor per shift. On the other hand, the plants may be manned by a single operator for only one-third or one-half of each shift; i.e., usually only operator, supervisor, and site manager are necessary to run the facility. Salary costs vary from state to state and depend significantly on the location of the facility. The cost of utilities generally consists of that of electricity, water, fuel, and steam. The annual costs are estimated with the use of material and energy balances. Cost for waste disposal can be estimated on a per-ton-capital basis. Cost of landfilling ash can run significantly upwards of $100/ton if the material is hazardous, and can be as high as $10/ton if it is nonhazardous. The cost of handling a scrubber effluent can vary depending on the method of disposal. For example, if a conventional sewer disposal is used, the effluent probably has to be cooled and neutralized before disposal; the cost for this depends on the solids concentration. Annual maintenance costs can be estimated as a percentage of the capital equipment cost. The annual cost of replacement parts can be computed by dividing the cost of the individual part by its expected lifetime. The life expectancies can be found in the literature [5]. Laboratory costs depend on the number of samples tested and the extent of these tests; these costs can be estimated as 10%–20% of the operating labor costs.

The indirect operating costs consist of overhead, local property tax, insurance, and administration, less any credits. The overhead comprises payroll, fringe benefits, social security, unemployment insurance, and other compensation that is indirectly paid to plant personnel. This cost can be estimated as 50%–80% of the operating labor, supervision, and maintenance costs [7,8]. Local property taxes and insurance can be estimated as 1%–2% of the TCC, while administration costs can be estimated as 2% of the TCC.

The total operating cost is the sum of the direct operating cost and the indirect operating costs, less any credits that may be recovered (e.g., the value of recovered steam). Unlike capital costs, operating costs are calculated on an annual basis.

47.4   HIDDEN ECONOMIC FACTORS [9]

The main problem with the traditional type of economic analysis, discussed above, is that it is difficult—nay, in some cases impossible—to quantify some of the not-so-obvious economic merits of a business and/or environmental program.

Several considerations have just recently surfaced as factors that need to be taken into account in any meaningful economic analysis of a project effort. What follows is a summary of these considerations:

- Long-term liabilities
- Regulatory compliance
- Regulatory recordkeeping
- Dealings with the Environmental Pollution Agency (EPA)
- Dealings with the state and local regulatory bodies
- Fines and penalties
- Potential tax benefits
- Customer relations
- Stockholder support (corporate image)
- Improved public image
- Insurance costs and claims
- Effect on borrowing power
- Improved mental and physical well being of employees
- Reduced health maintenance costs
- Employee morale
- Worker safety
- Rising costs of waste treatment and/or disposal
- Training costs
- Emergency response planning

Many programs have been quenched in their early states because a comprehensive economic analysis was not performed. Until the effects described above are included, the true merits of a project may be clouded by incorrect and/or incomplete economic data. Can something be done by industry to remedy this problem? One approach is to use a modified version of the standard Delphi Panel. In order to estimate these "other" economic benefits, several knowledgeable individuals within and perhaps outside the organization are asked to independently provide estimates, with explanatory details, on these economic benefits. Each individual in the panel is then allowed to independently review all response. The cycle is then repeated until the groups responses approach convergence.

47.5 PROJECT EVALUATION AND OPTIMIZATION

In comparing alternate processes or different options of a particular process from an economic point of view, it is recommended that the TCC be converted to an annual basis by distributing it over the projected lifetime of the facility. The sum of both the ACCs and the annual operating costs (AOCs) is known as the total annualized cost (TAC) for the facility. The economic merit of the proposed facility, process, or scheme can be examined once the total annual cost is available. Alternate facilities or options (e.g., a baghouse versus an electrostatic precipitator for particulate control, or two different processes for accomplishing the same degree of waste destruction) may also be compared. Note that a small flaw in this procedure is the assumption that the operating costs will remain constant throughout the lifetime of the facility.

Once a particular process scheme has been selected, it is common practice to optimize the process from a capital cost and O&M (operation and maintenance) standpoint. There are many optimization procedures available, most of them are too detailed for meaningful application for this chapter. These sophisticated optimization techniques, some of which are routinely used in the design of conventional chemical and petroleum plants, invariably involve computer calculations. Use of these techniques in environmental management analysis is usually not warranted, however.

One simple optimization procedure that is recommended is a perturbation study. This involves a systematic change (or perturbation) of variables, one by one, in an attempt to locate the optimum design from a cost and operation viewpoint. To be practical, this often means that the engineer must limit the number of variables by assigning constant values to those process variables that are known beforehand to play an insignificant role. Reasonable guesses and simple shortcut mathematical methods can further simplify the procedure. More information can be gathered from this type of study because it usually identifies those variables that significantly impact on the overall performance of the process and also helps identify the major contributors to the TAC.

47.6 FUTURE TRENDS

Economic analysis will continue to incorporate costs and benefits environment. Thus trend is continuing to expand with the inclusion of both sustainability effects (see Chapter 9) and green chemistry/engineering (see Chapter 8).

47.7 SUMMARY

1. An understanding of the economics involved in environmental management is important in making decisions at both the engineering and management levels. Every engineer or scientist should be able to execute an economic evaluation of a proposed environmental project.
2. The usual technique for determining the capital costs (i.e., TCC, which includes equipment design, purchase, and installation) for a project and/or process is based on the factored method of establishing direct and indirect installation costs as a function of the known equipment costs.
3. Operating costs can vary from site to site because these costs partly reflect local conditions—for example staffing practices, labor and utility costs. Operating costs like capital costs, may be separated into two categories: direct and indirect costs.
4. The main problem with the traditional type of economic analysis, discussed above, is that it is difficult—nay, in some cases impossible—to quantify some of the not-so-obvious economic merits of a business and/or environmental program. Several considerations have just recently surfaced as factors that need to be taken into account in any meaningful economic analysis of a project effort.
5. In comparing alternate processes or different options of a particular process from an economic point of view, it is recommended that the TCC be converted to an annual basis by distributing it over the projected lifetime of the facility.
6. One simple optimization procedure that is recommended is the perturbation study. This involves a systematic change (or perturbation) of variables, one by one, in an attempt to locate the optimum design for a cost and operation viewpoint.

REFERENCES

1. Burke, G., Singh, B., and Theodore, L. *Handbook of Environmental Science and Technology*, 2nd edition, John Wiley & Sons, Hoboken, NJ, 2002.
2. Matley, J. CE cost indexes set slower pace, *Chem. Eng.*, New York, April 29, 1985, 75–76.
3. McCormick, R.J. and De Rosier, R.J. *Capital and O&M Cost Relationships for Hazardous Waste Incineration*, Acurex Corp., EPA Report 600/2-87-175, Cincinnati, OH, October, 1984.
4. Neveril, R.B. *Capital and Operating Costs of Selected Air Pollution Control Systems*, EPA Report 450/5-80-002, Gard, Inc., Niles, IL, December 1978.
5. Vatavuk, W.M. and Neveril, R.B. Factors for estimating capital and operating costs, *Chem. Eng.*, New York, November 3, 1980, 157–162.
6. Vogel, G.A. and Martin, E.J. Hazardous waste incineration, Part 3—Estimating capital costs of facility components, *Chem. Eng.*, New York, November 28, 1983, 87–90.
7. Ulrich, G.D. *A Guide to Chemical Engineering Process Design and Economics*, John Wiley & Sons, Hoboken, NJ, 1984.
8. Vogel, G.A. and Martin, E.J. Hazardous Waste Incineration, Part 3—Estimating Operating Costs, *Chem. Eng.*, New York, January 9, 1984, 97–100.
9. Dupont, R., Theodore, L., and Ganeson, K. *Pollution Prevention*, CRC Press, Boca Raton, FL, 2000.

48 Environmental Implications of Nanotechnology

CONTENTS

Toxicology data from animal studies do not translate into human studies.
The mere probability of a health/hazard problem does not definitely equate to a problem.
Experimental data on mice is not worthy of front page dark-print news.

48.1 INTRODUCTION

Nanotechnology is the second coming of the industrial revolution. It promises to make the nation that seizes the nanotechnology initiative the technology capital of the world. One of the main obstacles to achieving this goal will be to control, reduce, and ultimately eliminate any environmental and environmental-related problems associated with this technology; unfortunately the success or failure of this new use may well depend on the ability to effectively and efficiently address these environmental issues.

The environmental health and hazard risk associated with both nanomaterials and the applications of nanotechnology for industrial uses are not known. Some early studies indicate that nanoparticles *may* serve as environmental poisons that accumulate in organs. Although these risks *may* prove to be either minor, or negligible, or both, the engineer and scientist is duty bound to determine if there are in fact any health, safety, and environmental impacts associated with nanotechnology. This chapter addresses these issues. Much of the material is drawn from four sources [1–4].

Specific topics covered in this chapter include

1. Nanotechnology
2. Environmental implications
3. Health risk assessment
4. Hazard risk assessment
5. Environmental regulations
6. Future trends

48.2 NANOTECHNOLOGY [1,5]

The term nanotechnology has come to mean different things to different people. The dictionary defines technology as "…the science or study of the practical or industrial arts, applied sciences, etc., …the system by which a society provides its members with those things needed or desired". Nano receives less treatment: "…a combining form meaning a 1,000,000,000th part of (a specified unit) …the factor 10^{-9}." Therefore, in a very generic sense, it is fair to say that nanotechnology is an applied science that is concerned with very tiny substances.

For the technical community, there are three (there are, of course, more) definitions that have regularly appeared in the literature:

1. Molecular manufacturing at the atomic level (atom by atom in a stable pattern); bottom-up approach
2. Research at the 1–100 nm (nanometer) size range
3. Development and uses of nanoparticles in the 1–100 nm range; top-down approach

Terms (1) and (2) deal with futuristic activities that are beyond the scope of this paper, and of little to no concern to the practicing engineer at this time. Applications and peripheral topics of (3) highlight the presentation to follow.

How does (3) above impact the practicing engineer? There are five major areas.

1. Developing new products
2. Improving existing products
3. Cost considerations
4. Health concerns
5. Hazard concerns

The first three areas have the potential to improve the quality of life. However, (4) and (5) can adversely affect society, and these concerns need to be reduced and/or eliminated. These latter two topics receive attention in this chapter.

When familiar materials such as metals, metal oxides, ceramics and polymers, plus novel forms of carbon are reduced into infinitesimally small particle sizes, the resulting particles have some novel and special material properties, especially when compared to macroscopic particles of the same material. These properties can also vary with particle size distribution (PSD), particle shape, particle density, process application, etc. These unique properties and property variations are certain to lead to near infinite number of opportunities and applications.

48.3 ENVIRONMENTAL IMPLICATIONS [2,6]

Any technology can have various and imposing effects on the environment and society. Nanotechnology is no exception, and the results will be determined by the extent to which the technical community manages this technology. This is an area that has, unfortunately, been seized upon by a variety of environmental groups.

There are two thoughts regarding the environmental implications of nanotechnology: one is positive and the other is potentially negative. The positive features of this new technology are well documented. The other implication of nanotechnology has been dubbed by many in this diminutive field as "potentially negative." The reason for this label is as simple as it is obvious: The technical community is dealing with a significant number of unforeseen effects that could have disturbingly disastrous impacts on society. Fortunately, it appears that the probability of such dire consequences actually occurring is near zero ... but *not* zero. This finite, but differentially small, probability is one of the key topics that is addressed in this chapter.

Air, water, and land (solid waste) concerns with emissions from nanotechnology operations in the future, as well companion health and hazard risks, receive attention below. All of these issues arose earlier with the Industrial Revolution, the development/testing/use of the atomic bomb, the arrival of the Internet, Y2K, and so forth; and all were successfully (relatively speaking) resolved by the engineers and scientists of their period. Furthermore, to the author's knowledge there are no documented nano-human health problems; statements in the literature refer to *potential* health concerns.

It should also be noted that nanoenvironmental concerns are starting to be taken seriously around the globe. There are a variety of studies going on into the health and environmental impacts of many applications of nanotechnology. Many believe (this does not include the author) it is in everyone's interest to ensure that any new compound is fully characterized and the long-term implications be studied before it is commercialized. Class action suits in the United States against both tobacco companies and engineering companies, coupled with a new era of corporate responsibility, have ensured that most companies are well aware of this need. Now that potential risks that may have been overlooked are becoming widely known, these companies are more inclined to be proactive than they have been with risks in the past [7].

There has already been a considerable shift in both public and corporate attitudes to the environment. Major scandals such as Enron and WorldCom have led to not only tighter corporate governance but also to calls for greater corporate responsibility. The end result of this shift will be to make companies increase their focus on the environment, and look to leveraging nanotechnology as a way of not only improving efficiency and lowering costs, but doing this by reducing energy consumption and minimizing waste. A typical example would be in the use of nanoparticle catalysts that are not only more efficient, owing to more of the active catalysts being exposed (because of their size), but also require less precious metal (thus reducing cost); it may also increase selectivity, i.e., produce more of the desired reaction product, rather than by-products.

Returning to the positive features of this new activity, nanotechnology will be one of the key technologies used in the quest to improve the global environment in this century. While there will be some direct effects, much of the technology's influence on the environment will be through indirect applications of nanotechnology. Although any technology can always be put to both positive and negative uses, there are many areas in which positive aspects of nanotechnology look promising. These extend from pollution prevention and reduction through environmental remediation to sustainable development.

Though nanotechnology could have some significant effects on environmental technologies, environmental considerations have not historically been given anywhere near the priority in new developments that commercial considerations are given, and this balance, though swinging gradually more toward environmental considerations, still largely dominates. Many of the direct applications of nanotechnology relate to the removal of some element or compound from the environment, through, for example, the use of nanofiltration, nanoporous sorbents (absorbents and adsorbents), catalysts in cleanup operation, and filtering, separating, and destroying environmental contaminants in processing waste products. Most effects, as with other technologies, are likely to be indirect. Some, including one of the authors of this text, have concluded that the major environmental nanotechnology breakthroughs may occur naturally through pollution prevention principles [8].

48.4 HEALTH RISK ASSESSMENT

For some, the rapid progress of nanotechnology-related developments in recent years brings uncertainty. For instance, early studies on the transport and uptake of nanoscaled materials into living systems "suggest" (other popular terms in these studies include "may cause, could cause, alleged, possibly associated with," etc.) that there *may* be harmful effects on living organisms. This has prompted many to call for further study to identify all of the potential environmental and health risks that might be associated with nanosized materials [9].

Studies raise questions about the potential health and environmental effects of nanoscaled materials, and while the initial toxicological data are preliminary, they underscore the need to learn more about how nanoscaled materials are absorbed, how they might damage living organisms, and what levels of exposure create unacceptable hazards.

At the time of the preparation of this chapter, the risks of nanotechnology were not known, and it appears that they will not be known for some time. However, it should also be noted that health benefits, if any, are also not known. Furthermore, there are no specific nano-health-related regulations or rules at Environmental Protection Agency (EPA), Occupational safety and health administration (OSHA), or other organizations, and it may be years before any definite nanoregulations are promulgated (see Regulations Chapter 2). However, the dark clouds on the horizon in this case are the environmental health impacts associated with these new and unknown operations, and the reality is that there is a serious lack of information on these impacts. Risk assessment studies in the future will be that path to both understanding and minimizing these effects.

As described earlier, perhaps the greatest danger from nanomaterials may be their escape and persistence in the environment, the food chain, and human and animal tissues. Although, the potential pollutants and the tools for dealing with them may be different, the methodology and protocols developed for conventional materials will probably be the same, bearing in mind that some instrumentation that is required may not yet be available [2,10]. Thus, environmental risk assessment remains "environmental risk assessment," using the same techniques described earlier regardless of the size of the alleged causative agent.

How is it possible to make decisions dealing with environmental risks from a new application, for example, nanotechnology, in a complex society with competing interests and viewpoints, limited financial resources, and a lay public that is deeply concerned about the risks of cancer and other illness? Risk assessment constitutes a decision-making approach that can help the different parties involved and thus enable the larger society to work out its environmental problems rationally and with good results. It also provides a framework for setting regulatory priorities and for making decisions that cut across different environmental areas. This kind of framework has become increasingly important in recent years for several reasons, one of which is the considerable progress made in environmental control. Nearly 40 years ago, it was not difficult to figure out where the first priorities should be. The worst pollution problems were all too obvious.

The next section addresses primarily "acute" exposures, while this section examines "chronic" exposures. Since both classes of exposure ultimately lead to the subject of risk assessment, the overlap between the two exposures can create problems in the presentation in textual form. For purposes of this chapter, the chronic and acute subjects are described as health risk analysis (HRA) and hazard risk analysis (HZRA), respectively.

Health risk assessments provide an orderly, explicit way to deal with scientific issues in evaluating whether a health problem exists and what the magnitude of the problem may be. Typically, this evaluation involves large uncertainties because the available scientific data are limited, and the mechanisms for adverse health impacts or environmental damage are only imperfectly understood.

As discussed earlier in Chapter 38, most human or environmental health problems can be evaluated by dissecting the analysis into four parts: hazard identification, dose-response assessment or toxicity assessment, exposure assessment, and risk characterization (see Figure 48.1). For many, the heart of a health risk assessment is toxicology. Toxicology is the science of poisons. It has also been defined as the study of chemical or physical agents that produce adverse responses in biological systems. Together with other scientific disciplines (such as epidemiology—the study of the cause and distribution of disease in human populations—and risk assessment), toxicology can be used to determine the relationship between an agent of interest and a group of people or a community [11].

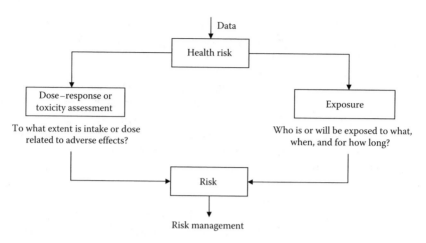

FIGURE 48.1   Health risk evaluation process.

Six primary factors affect human response to toxic substances or poisons. These are detailed below [2,6,12]:

1. *The chemical itself.* Some chemicals produce immediate and dramatic biological effects, whereas other produce no observable effects or produce delayed effects.
2. *The type of contact.* Certain chemicals appear harmless after one type of contact (e.g., skin), but may have serious effects when contacted in another manner (e.g., lungs).
3. *The amount (dose) of a chemical.* The dose of a chemical exposure depends upon how much of the substance is physically contacted.
4. *Individual sensitivity.* Humans vary in their response to chemical substance exposure. Some types of responses that different persons may experience at a certain dose are serious illness, mild symptoms, or no noticeable effect. Different responses may also occur in the same person at different exposures.
5. *Interaction with other chemicals.* Toxic chemicals in combination can produce different biological responses that the responses observed when exposure is to one chemical alone.
6. *Duration of exposure.* Some chemicals produce symptoms only after one exposure (acute), some only after exposure over a long period of time (chronic), and some may produce effects from both kinds of exposure.

In a very real sense, the science of toxicology will be significantly impacted by nanotechnology. Unique properties cannot be described for particles in the nano-size range since properties vary with particle size. This also applies to toxicological properties. In effect, a particle of one size could be carcinogenic while a particle (of the same material) of another size would not be carcinogenic. Alternatively, two different sized nanoparticles of the same substance could have different threshold limit values (TLVs). This problem has yet to be resolved by the toxicologist.

In conclusion, the purpose of the toxicity assessment is to weigh available evidence regarding the potential for particular contaminants to cause adverse effects in exposed individuals and to provide, where possible, an estimate and the increased likelihood and/or severity of adverse effects.

48.5   HAZARD RISK ASSESSMENT [1,2,5]

As indicated in the previous section, many practitioners and researchers have confused health risk with hazard risk, and vice versa. Although both employ a four-step method of analysis, the procedures are quite different, with each providing different results, information, and conclusions.

As with health risk, there is a serious lack of information on the hazards and associated implications of these hazards with nanoapplications. The unknowns in this risk area are both larger in number and greater in potential consequences. It is the authors' judgment that hazard risk has unfortunately received something less than the attention it deserves. However, HZRA details are available and the traditional approaches, e.g., HAZOP, successfully applied in the past are available in the literature (see also Chapter 39). Future work will almost definitely be based on this methodology.

Much has been written about Michael Crichton's powerful science-thriller novel entitled *Prey*. (The book was not only a best seller, but the movie rights were sold for $5 million.) In it, Crichton provides a frightening scenario in which swarms of nanorobots, equipped with special power generators and unique software, prey on living creatures. To compound the problem, the robots continue to reproduce without any known constraints. This scenario is an example of an accident and represents only one of a near infinite number of potential hazards that can arise in any nanotechnology application, particularly for bottom-up systems. Although the probability of the horror scene portrayed by Crichton, as well as other similar events, is extremely low, steps and procedures need to be put into place to reduce, control, and, it is hoped, to eliminate these events from actually happening.

The previous section defined both "chronic" and "acute" problems. As indicated, when the two terms are applied to emissions, the former usually refers to ordinary, round-the-clock, everyday emissions while the latter term deals with short, out-of-the-norm, accidental emissions. Thus, acute problems normally refer to accidents and/or hazards. The Crichton scenario discussed above is an example of an acute problem, and one whose solution would be addressed/ treated by a hazard risk assessment, rather than a health risk approach.

There are several steps in evaluating the risk of an accident (see Figure 48.2). These are detailed below if the system in question is a chemical plant. Note that this material can also be found in Chapter 39.

1. A brief description of the equipment and chemicals used in the plant is needed.
2. Any hazard in the system has to be identified. Hazards that many occur in a chemical plant include fire, toxic vapor release, slippage, corrosion, explosions, rupture of a pressurized vessel, and runaway reactions.
3. The event or series of events that will initiate an accident had to be identified. An event could be a failure to follow correct safety procedures, improperly repaired equipment, or failure of a safety mechanism.

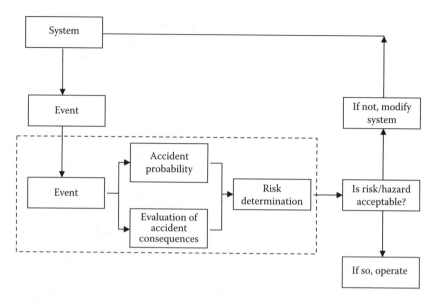

FIGURE 48.2   Hazard risk assessment flowchart.

4. The probability that the accident will occur has to be determined. For example, if a chemical plant has a given life, what is the probability that the temperature in a reactor will exceed the specified temperature range? The probability can be ranked from low to high. A low probability means that it is unlikely for the event to occur in the life of the plant. A medium probability suggests that there is a possibility that the event will occur. A high probability means that the event will probably occur during the life of the plant.

5. The severity of the consequences of the accident must be determined. This will be described later in detail.

6. If the probability of the accident and the severity of its consequences are low, then the risk is usually deemed acceptable and the plant should be allowed to operate. If the probability of occurrence is too high or the damage to the surroundings is too great, then the risk is usually unacceptable and the system needs to be modified to minimize these effects.

The heart of hazard risk assessment algorithm provided is enclosed in the dashed box (see Figure 48.2). The overall algorithm allowed or reevaluation of the process. If the risk is deemed unacceptable (the process is repeated starting with either step 1 or 2).

48.6   ENVIRONMENTAL REGULATIONS [1,2]

Many environmental concerns are addressed by existing health and safety legislation. Most countries require a health and safety assessment for any new chemical before it can be marketed. Further, the European Commission (EC) recently introduced the world's most stringent labeling system. Prior experience with materials such as PCBs and asbestos, and a variety of unintended effects of drugs such as thalidomide, mean that both companies and governments have an incentive to keep a close watch on potential negative health and environmental effects [11].

It is very difficult to predict future nanoregulations. In the past, regulations have been both a moving target and confusing. What can be said (for certain?) is that there will be regulations, and the probability is high that they will be contradictory and confusing. Past and current regulations provide a measure of what can be expected.

As discussed above, completely new legislation and regulatory rulemaking may be necessary for environmental control related to nanotechnology. However, in the meantime, one may speculate on how the existing regulatory framework might be applied to the nanotechnology area as this emerging field develops over the next several years. One experienced Washington DC attorney has done just that, as summarized below in the next five paragraphs [1,2,13–17]. The reader is encouraged to consult the cited references as well as the text of the laws that are mentioned and the applicable regulations derived from them.

Commercial applications of nanotechnology are likely to be regulated under TSCA, which authorizes the EPA to review and establish limits on the manufacture, processing, distribution, use, and/or disposal of new materials that EPA determines to pose "an unreasonable risk of injury to human health or the environment." The term chemical is defined broadly by TSCA. Unless qualifying for an exemption under the law (a statutory exemption requiring no further approval by EPA), low-volume production, low environmental releases along with low volume, or plans for limited test marketing, a prospective manufacturer is subject to the full-blown procedure. This requires submittal of said notice, along with toxicity and other data to EPA at least 90 days before commencing production of the chemical substance.

Approval then involves recordkeeping, reporting, and other requirements under the statute. Requirements will differ, depending on whether EPA determines that a particular application constitutes a "significant new use" or a "new chemical substance." The EPA can impose limits on production, including an outright ban when it is deemed necessary for adequate protection against "an unreasonable risk of injury to health or the environment." The EPA may revisit a chemical's status under TSCA and change the degree or type of regulation when new health/environmental data warrant. If the experience with genetically engineered organisms is any indication, there will be a push for EPA to update regulations in the future to reflect changes, advances, and trends in nanotechnology [19].

Workplace exposure to a chemical substance and the potential for pulmonary toxicity is subject to regulation by the OSHA, including the requirement that potential hazards be disclosed on a Material Safety Data Sheet (MSDS). (An interesting question arises as to whether carbon nanotubes, chemically carbon but with different properties because of their small size and structure, are indeed to be considered the same as or different from carbon black for MSDS purposes.) Both governmental and private agencies can be expected to develop the requisite TLVs for workplace exposure. Also, the EPA may once again utilize TSCA to assert its own jurisdiction, appropriate or not, to minimize exposure in the workplace. Furthermore, the National Institute for Occupational Safety and Health (NIOSH) was anticipated to provide workplace guidance for nanomanufactures and their employees in 2005; this has not occurred as of the writing of this chapter. This was almost definitely wishful thinking given the past performance of similar bureaucratic agencies, e.g., the EPA. Adding to NIOSH's dilemma is the breadth of the nano field and the lack of applicable toxicology and epidemiology data.

Another likely source of regulation would fall under the provisions of the Clean Air Act (CAA) for particulate matter less than 2.5 μm ($PM_{2.5}$). Additionally, an installation manufacturing nanomaterials may ultimately become subject as a "major source" to the CAA's Section 112 governing hazardous air pollutants (HAP).

Wastes from a commercial-scale nanotechnology facility would be classified under the Resource Conservation and Recovery Act (RCRA), provided that it meets the criteria for RCRA waste. RCRA requirements could be triggered by a listed manufacturing process or the act's specified hazardous waste characteristics. The type and extent of regulation would depend on how much hazardous waste is generated and whether the wastes generated are treated, stored, or disposed of on site.

One of the authors has also speculated on the need for future nanoregulations. His suggestions and potential options are provided in Figure 48.3 [18] while noting that the ratio of pollutant nanoparticles (from conventional sources such as power plants) to engineered nanoparticles being released into the environment may be as high a trillion to one [19], i.e., 10^{12}. If this be so, the environmental concerns for nanoparticles can almost certainly be dismissed.

The reader is left to ponder the type, if any, of nanoregulations required at this time and the need to curb/eliminate liability concerns.

48.7 FUTURE TRENDS

The unbridled promise of nanotechnology-based solutions has motivated academic, industrial, and government researchers throughout the world to investigate nanoscaled materials, devices, and systems with hope of commercial-scale production and implementation. Today, the private-sector companies that have become involved run the gamut from established global leaders throughout the chemical process industries, to countless small entrepreneurial startup companies, many of which have been spun off from targeted research and development efforts at universities.

The governments of many industrialized nations are also keenly interested in nanotechnology. This stems in part from their desire to maintain technological superiority in an important evolving field, and from the military recognition that some applications of nanotechnology could have significant implications for national security. As with nuclear energy in the mid-1930s, the authors believe that the nation that conquers nanotechnology will conquer the world [2].

In any event, future developmental efforts and advances will primarily be fueled by economic considerations. The greatest driving force behind any nano project is the promise of economic opportunities and cost savings over the long term. Hence, an understanding of the economics involved is quite important in making decisions at both the engineering and management levels.

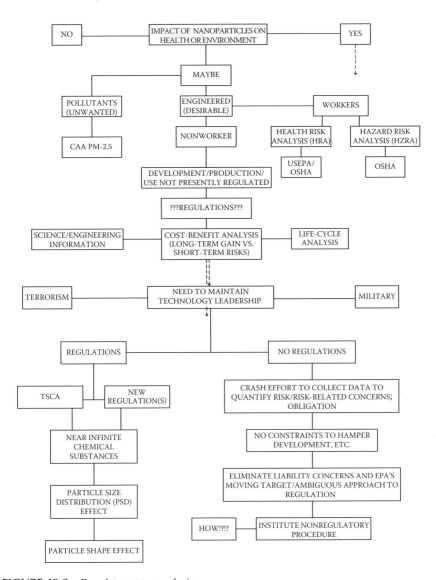

FIGURE 48.3 Regulatory tree analysis.

From an environmental health/safety perspective, this technology promises:

1. Use of less raw materials, some of which are irreplaceable
2. The generation of less waste/pollutants
3. Reduced energy consumption
4. A safer environment with reduced risks

Regarding regulations, both industry and government need to support reasonable policies and regulations. They need to avoid the traditional environmental precautionary approach. The position of many radical environmental organizations and environmental research agencies (particularly at the university level) needs to be ignored until applicable concrete evidence is provided to support their argument(s). In effect, an environmental problem should not be *assumed* to exist.

For nanotechnology's most ardent supporters, the scope of this emerging field seems to be limited only by the imaginations of those who would dream at these unprecedented dimensions. However, considerable technological and financial obstacles still need to be reconciled before nanotechnology's full promise can be realized [4,20].

48.8 SUMMARY

1. Nanotechnology is the second coming of the industrial revolution. It promises to make that nation that seizes the nanotechnology initiative the technology capital of the world.

2. The environmental health and hazard risk associated with both nanomaterials and the applications of nanotechnology for industrial uses are not known. Some early studies indicate that nanoparticles may serve as environmental poisons that accumulate in organs.

3. Any technology can have various and imposing effects on the environment and society. Nanotechnology is no exception, and the results will be determined by the extent to which the technical community manages this technology.

4. Studies raise questions about the potential health and environmental effects of nanoscaled materials, and while the initial toxicological data are preliminary, they underscore the need to learn more about how nanoscaled materials are absorbed, how they might damage living organisms, and what levels of exposure create unacceptable hazards.

5. At the time of the preparation of this chapter, the risks of nanotechnology were not known, and it appears that they will not be known for some time. However, it should also be noted that health benefits, if any, are also not known.

6. As with health risk, there is a serious lack of information on the hazards and associated implications of these hazards with nanoapplications. The unknowns in this risk area are both larger in number and greater in potential consequences.

7. Workplace exposure to a chemical substance and the potential for pulmonary toxicity is subject to regulation by the OSHA, including the requirement that potential hazards be disclosed on a MSDS.

8. The ratio of pollutant nanoparticles (from conventional sources such as power plants) to engineered nanoparticles being released into the environment may be as high a trillion to one [19], i.e., 10^{12}. If this be so, the environmental concerns for nanoparticles can almost certainly be dismissed.

REFERENCES

1. Theodore, L. *Nanotechnology: Basic Calculations for Engineers and Scientists*, John Wiley & Sons, Hoboken, NJ, 2006.
2. Theodore, L. and Kunz, R. *Nanotechnology: Environmental Implications and Solutions*, John Wiley & Sons, Hoboken, NJ, 2005.
3. Theodore, M.K. and Theodore, L. *Major Environmental Issues Facing the 21st Century*, Theodore Tutorials, East Williston, NY, 1996 (originally published by Simon & Schuster).
4. Murphy, M. and Theodore, L. Environmental impacts of nanotechnology: Consumer issues, *A Conference on Nanotechnology and the Consumer, Consumer Reports*, Yonkers, NY, 2006.
5. Theodore, L. Personal notes, 2004.
6. Burke, G., Singh, B., and Theodore, L. *Handbook of Environmental Management and Technology*, 2nd edition, John Wiley & Sons, Hoboken, NJ, 2000.
7. Anonymous. *Nanotechnology Opportunity*, Report, 2nd edition, location unknown, 2003.
8. Dupont, R., Theodore, L., and Ganesan, R. *Pollution Prevention: The Waste Management Approach for the 21st Century*, Lewis/CRC, Boca Raton, FL, 2000.
9. Roco, M. and Tomelini, R. (Eds.). Nanotechnology: Revolutionary opportunities and societal implications, *Summary of Proceedings of the 3rd Joint European Commission-National Science Foundation Workshop on Nanotechnology*, Lecce, Italy, January 21–February 1, 2002.
10. Friedlander, S.K., Workshop Chair. Emerging issues in nanoparticle aerosol science and technology (NAST), NSF Workshop Report, p. 15, University of California, Los Angeles, June 27–28, 2003.
11. National Center for Environmental Research. Nanotechnology and the environment: applications and implication, *STAR Progress Review Workshop*, Office of Research and Development, National Center for Environmental Research, Washington, DC, 2003.
12. U.S. EPA. Report to Congress, Office of Solid Waste, U.S. EPA, Washington, DC, 1530-SW-86–033, 1986.
13. Bergeson, L. Nanotechnology and TSCA, *Chemical Processing*, November 2003.
14. Bergeson, L. Nanotechnology trend draws attention of federal regulators, *Manufacturing Today*, March/April 2004.
15. Bergeson, L. and Auerbach, B. The environmental regulatory implications of nanotechnology, *BNA Daily Enviroment Reporters*, pp. B-1–B-7, April 14, 2004.
16. Bergeson, L. Expect a busy year at EPA, *Chemical Processing*, 17, February 2004.
17. Bergeson, L. Genetically engineered organisms face changing regulations, *Chemical Processing*, March 2004.
18. Theodore, L. Waste management of nanomaterial, USEPA, Washington, DC, 2006.
19. Theodore, L. Personal notes, 2006.
20. Murphy, M. Personal notes, Manhattan College, Bronx, NY, 2006.

49 Environmental Ethics

CONTENTS

49.1 INTRODUCTION

In 1854, President Franklin Pierce petitioned Chief Seattle—the leader of the Coastal Salish Indians of the Pacific Northwest—to sell his tribe's land to the United States. In his response to President Pierce and the white Europeans' pursuit to own and "subdue" the Earth, Chief Seattle penned thoughts as environmentally pensive and poignant as any uttered in more than 140 years since: "Continue to contaminate your bed and you will one day lay in your own waste" [1].

His message fell on the deaf ears of the U.S. government and public. Cries for respect for the Earth such as his remained few and far between for the next century. In the wake of events such as the Industrial Revolution, the First and Second World Wars, and the Cold War, a concern for the environment played little, if any, part in influencing either public policy or private endeavors.

More than 140 years later, however, Chief Seattle's words echoed in every Superfund site, landfill, and oil spill. Public opinion has swung to the green side and a new ethic has evolved: an environmental ethic. As one shall soon see, however, the recent movement toward environmentalism has not created new moral codes. Instead, it has changed the emphasis and expanded the concept of the "common good" that lies at the heart of determining if an action is ethical.

This chapter will first present the variety of moral theories and philosophies that have governed ethics historically. The movement of environmentalism into an influential ethical force is then developed. Once these historical developments have been presented, today's dilemma of coordinating technology with environmental responsibility will be explored. Finally, the future trends evidenced by present and past activities will be discussed.

49.2 MORAL ISSUES

The conflict of interest between Chief Seattle (and Native Americans in general) and President Pierce (and the European/American expansion) provides a perfect example of how ethics and the resulting codes of behavior they engender can differ drastically from culture to culture, religion to religion, and even person to person. This enigma, too, is noted again and again by Seattle [2]:

> I do not know. Our ways are different from your ways ... But perhaps it is because the red man is a savage and does not understand ... The air is precious to the red man, for all things share the same breath ... the white man does not seem to notice the air he breathes ... I am a savage and do not understand any other way. I have seen a thousand rotting buffaloes on the prairie, left by the white man who shot them from a passing train. I am a savage and I do not understand how the smoking iron horse can be more important than the buffalo we kill only to stay alive.

Chief Seattle sarcastically uses the European word "savage" and all its connotations throughout his address. When one finishes reading the work it becomes obvious which viewpoint (President Pierce's or his own) Chief Seattle feels is the savage one. What his culture holds dearest (the wilderness) the whites see as untamed, dangerous, and savage. What the whites hold in highest regard (utilization of the earth and technological advancement) the Native Americans see as irreverent of all other living things. Each culture maintains a distinct and conflicting standard for the welfare of the world. Opposing viewpoints and moralities such as these are prevalent throughout the world and have never ceased to present a challenge to international, national, state, community, and interpersonal peace.

It is generally accepted, however, that any historical ethic can be found to focus on one of four different underlying moral concepts [3]:

1. "Utilitarianism" focuses on good consequences for all
2. "Duties Ethics" focus on one's duties
3. "Rights Ethics" focus on human rights
4. "Virtue Ethics" focus on virtuous behavior

(Note that Duties and Rights Ethics are often considered together as Deontological Ethics.)

Utilitarians hold that the most basic reason why actions are morally right is that they lead to the greatest good for the greatest number. "Good and bad consequences are the only relevant considerations, and, hence all moral principles reduce to one: 'We ought to maximize utility'" [2].

Duties Ethicists concentrate on an action itself rather than the consequences of that action. To these ethicists there are certain principles of duty such as "Do not deceive" and "Protect innocent life" that should be fulfilled even if the most good does not result. The list and hierarchy of duties differs from culture to culture, religion to religion. For Judeo-Christians, the Ten Commandments provide an ordered list of duties imposed by their religion [2].

Often considered to be linked with Duties Ethics, Rights Ethics also assesses the act itself rather than its consequences. Rights Ethicists emphasize the rights of the people affected by an act rather than the duty of the person(s) performing the act. For example, because a person has a right to life, murder is morally wrong. Rights Ethicists propose that duties actually stem from a corresponding right. Since each person has a "right" to life, it is everyone's "duty" not to kill. It is because of this link and their common emphasis on the actions themselves that Rights Ethics and Duty Ethics are often grouped under the common heading: Deontological Ethics [7].

The display of virtuous behavior is the central principle governing Virtue Ethics. An action would be wrong if it expressed or developed vices—for example, bad character traits. Virtue Ethicists, therefore, focus upon becoming a morally good person.

To display the different ways that these moral theories view the same situation one can explore their approach to the following scenario that Martin and Schinzinger [2] present:

> On a midnight shift, a botched solution of sodium cyanide, a reactant in organic synthesis, is temporarily stored in drums for reprocessing. Two weeks later, the day shift foreperson cannot find the drums. Roy, the plant manager, finds out that the batch has been illegally dumped into the sanitary sewer. He severely disciplines the night shift foreperson. Upon making discreet inquiries, he finds out that no apparent harm has resulted from the dumping. Should Roy inform government authorities, as is required by law in this kind of situation?

If a representative of each of the four different theories on ethics just mentioned were presented with this dilemma, their decision-making process would focus on different principles.

The Utilitarian Roy would assess the consequences of his options. If he told the government, his company might suffer immediately under any fines administered and later (perhaps more seriously) due to exposure of the incident by the media. If he chose not to inform authorities, he risks heavier fines (and perhaps even worse press) in the event that someone discovers the cover-up. Consequences are the utilitarian Roy's only consideration in his decision-making process.

The Duties Ethicist Roy would weigh his duties and his decision would probably be more clear-cut than his utilitarian counterpart. He is obliged foremost by his duty to obey the law and must inform the government.

The Rights Ethicist mind frame would lead Roy to the same course of action as the duties ethicist—not necessarily because he has a duty to obey the law but because the people in the community have the right to informed consent. Even though Roy's inquiries informed him that no harm resulted from the spill, he knows that the public around the plant has the right to be informed of how the plant is operating.

Vices and virtues would be weighed by the Virtue Ethicist Roy. The course of his thought process would be determined by his own subjective definition of what things are virtuous, what things would make him a morally good person. Most likely, he would consider both honesty and obeying the law virtuous, and withholding information from the government and public as virtue-less and would, therefore, tell the authorities.

The scenario used here will be revisited later in this chapter through the eyes of environmentalism to illustrate how this movement is changing the focus of old theories about morality.

49.3 MODERN DAY MAINSTREAM ENVIRONMENTALISM

Minds like John Muir and Rachel Carson were unique in their respective generations. Their ideas of respect for all flora and fauna were far from predominant in the American mainstream. Rachel Carson's 1962 benchmark book "Silent Spring" [4] took environmentalism from pure naturalism into the scientific realm. The evidenced claims she made about the harm caused to wildlife by a range of pesticides (most notably DDT) were as controversial as they were groundbreaking. Over the next decade the younger generation embraced a new concern for the environment. The older generation, however, generally dealt with this young movement with opposition rather than cooperation. This was due in large part to the confrontational attitude of many of the youths as well as the perceived threat that the industry-restricting movement itself caused to their economic well-being. As the younger generation grew into positions of power and learned more cooperative tactics, their environmentalist ideas moved from the fringes to the mainstream. On route, the conversion was carried out in the form of both personal growth and government legislation. There seems still to exist, however, two factions of environmentalism: pure environmentalism (environmentalism for its own sake) and environmentalism for humanity's sake. While they share a common concern for the well-being of the natural world, fundamental differences exist.

One of the most common arguments against the destruction of rainforest land is that any one of the plants or insect species destroyed in the process could contain the elusive cure for cancer or AIDS. With this argument, the ultimate concern is for humanity: We should preserve the natural world because it is best for the human race to do so. This could be considered environmentalism for humanity's sake and there are a number of other manifestations of it in today's world. The war against the destruction of ozone in the earth's stratosphere is waged largely in the interest of human welfare. While the greenhouse effect has the potential to harm wildlife also, this effect is secondary to that on humanity—both today and in future generations. This type of environmentalism displays the inherent egocentric attitude of humankind. This faction maintains "an ethic that is secondarily ecological" [5]. Here the natural world should be protected because of humanity's dependence on its homeostasis.

The second, more "extremist" form of environmental morality is "primarily ecological" [5]. As Aldo Leopold proclaimed, "A thing is right when it tends to preserve the integrity, stability, and beauty of the biotic community. It is wrong if it tends otherwise" [6]. Here, humanity has a binding responsibility to protect the homeostasis of the natural world. In this view, humanity is considered a part of the interdependent environment rather than something above it. The Native American's adoration of the Great Spirit—which favored the human species no more than any other—is the religious embodiment of such a viewpoint.

The renewed awareness of the environment and awakened concern for its well-being has influenced the ethical world to the point that it has uprooted the focus of the moral correctness of an action. This effect on ethical theories was predicted by John Passmore in 1974: "What it needs for the most part is not so much a 'new ethic' as a more general adherence to a perfectly familiar ethic. For the major sources of our ecological disasters—apart from ignorance—are greed and short-sightedness" [7].

Aldo Leopold made the following observation on personal ethics in his 1949 *A Sand County Almanac:* "The scope of one's ethics is determined by the inclusiveness of the community with which one identifies oneself" [6]. Leopold parallels the mistreatment of the earth to the mistreatment of slaves that were handled as property. The slave owners were not ethically obliged to the slaves because they considered them outside rather than part of their community. Just as the realm of community grew to include the ex-slaves, it must once again expand to incorporate the whole land community [6]. The incorporation of environmentalism into everyday ethics, therefore, does not require a redefinition of one's ethics, but, rather, a redefinition of one's "community." This can be applied to each of the ethical theories presented above.

For the utilitarian it requires counting the natural world among those affected by bad and good consequences. The focus of utilitarianism is broadened to include effects on future generations and the welfare of living things other than humans. For the deontological ethicists, the recognition of the environment as part of the community gives it inherent rights and, in turn, imposes on humans the duty to respect those rights. For the virtue ethicists, the virtue of respecting all members of the community would bind them to consider the environment when making decisions.

In the scenario presented earlier, Roy's moral thought process would be affected by the inclusion of the environment into his community regardless of the ethical school of thought he associated himself with. Although his discreet inquiries informed him that no apparent harm resulted from the chemical spill, an environmental impact analysis would have to be made for the utilitarian Roy to fully assess good and bad consequences. If future harm were likely, it may be essential to let the government know so that remediation techniques may be employed at the dumping site. The decision of the rights and duties ethicist Roys would be influenced by their obligation to the environment as well as the surrounding human community. During the virtue ethicist Roy's decision-making process, he would consider which option was the most virtuous with respect to the environment. In each of these cases, an ecologically ethical Roy would have to obtain a reasonable estimate, with the help of the government if necessary, of the environmental effects—immediate and long-term—of the dumping.

In each of these new twists upon old theories on ethics, there exists the fundamentals of a "land ethic." The ethical umbrella is expanding to include under its cover all living beings. Fields of conduct such as disposal and treatment of owned property and land are now becoming judged ethically rather than on the grounds of economic feasibility and personal whimsy.

The mainstreaming of environmentalism is by no means worldwide. The countries in which the greatest impact is seen are the same countries where extensive industrialization exists. Industrialization itself has been crucial to the development of the environmental movement. Not only do its environmental problems and pollution generate concern, citizens of industrialized nations enjoy lives with the luxury of free time and options necessary to be able to devote themselves to such a concern. In poorer countries and communities, the struggle of everyday survival far outweighs any aesthetic concern for the environment. Abraham Maslow's concept of a "hierarchy of needs" can be applied in explaining the difficulty of establishing the environmental movement in impoverished communities and third-world countries.

Maslow maintains that there exist the following "hierarchy of needs" for every human being. He finds "five levels of need":

1. Survival (physiological needs): food, shelter, and health
2. Security (safety needs): protection from danger and threat
3. Belonging (social needs): friendship, acceptance, and love
4. Self-esteem (ego needs): self-respect, recognition, and status
5. Self-actualization (fulfillment needs): creativity and realization of individual potentialities

Maslow maintains that these levels form a hierarchy; lower levels must be satisfied before the individual can give attention to higher levels [7]. Until the lower levels of need are at least partially satisfied, a person cannot commit him or herself to the pursuit of higher-level needs. For example, a person who is struggling to find any source of food will not be preoccupied with how environmentally conscious the farmer was in the use of fertilizers or pesticides while cultivating the food.

Consider, for example, a town such as many in the mountains of Appalachia where one industry—coal mining—provides all of the town's employment and generates most of the taxes used by the town in running schools and other municipal operations. When the coal mining company turns to strip mining—a process that essentially rips the mountains to shreds and contaminates groundwater with the heavy metals released—can the miners be expected to jeopardize the welfare of their entire families by protesting because the methods of their employer are environmentally negligent? Their survival needs for food and shelter supersede any idealistic desire they have to preserve the environment. Abuse of this natural hierarchy has been defined as environmental racism (see Chapter 50) and is epitomized by the disproportionately large number of landfills, chemical plants, and toxic dumps in the poorer communities and countries.

49.4 TECHNOLOGY AND ENVIRONMENTALISM

In the ethical theories presented here, established hierarchies of duties, rights, virtues, and desired consequences exist so that situations where no single course of action satisfies all of the maxims can still be resolved. The entry of environmentalism into the realm of ethics raises questions concerning where it falls in this hierarchy. Much debate continues over these questions of how much weight the natural environment should be given in ethical dilemmas, particularly in those where ecological responsibility seems to oppose economic profitability and technological advances. Those wrapped up in this technology/economy/ecology debate can generally be divided into three groups:

1. Environmental extremists
2. Technologists to whom ecology is acceptable provided it does not inhibit technological or economic growth
3. Those who feel technology should be checked with ecological responsibility

Each is briefly discussed below.

After his year-and-a-half of simple living on the shores of Walden Pond, Henry David Thoreau professed "in wildness is the preservation of the world" [7]. He rejected the pursuit of technology and industrialization. While most would agree with his vision of nature as being inspirational, few would choose his way of life. Even so, the movement rejecting technological advances in favor of simple, sustainable, and self-sufficient living is being embraced by more and more people who see technology as nothing but a threat to the purity and balance of nature. Often called environmental extremists by other groups, they even disregard "environmental" technologies that attempt to correct pollution and irresponsibilities, past and present. They see all technology as manipulative and uncontrollable and choose to separate themselves from it. To them, the environment is at the top of the hierarchy.

On the other extreme are the pure technologists. They view the natural world as a thing to be subdued and manipulated in the interest of progress—technological and economic. This is not to say one will not find technologists wandering in a national park admiring the scenery. They do not necessarily deny the beauty of the natural environment, but they see themselves as separate from it. They believe that technology is the key to freedom, liberation, and a higher standard of living. It is viewed, therefore, as inherently good. They see the environmental extremists as unreasonable and hold that even the undeniably negative side effects of certain technologies are best handled by more technological advance. The technologists place environmental responsibility at the bottom of their ethical hierarchy.

Somewhere in the middle of the road travels the third group. While they reap the benefits of technology, they are concerned much more deeply than the technologists with the environmental costs associated with industrialization. It is in this group that most environmental engineers find themselves. They are unlike the environmental extremists since, as engineers, they inherently study and design technological devices and have faith in the ability of such devices to have a positive effect on the condition of the environment. They also differ from the technologists. They scrutinize the effects of technologies much more closely and critically. While they may see a brief, dilute leak of a barely toxic chemical as an unacceptable side effect of the production of a consumer product, the technologists may have to observe destruction—the magnitude of that caused by Chernobyl—before they consider rethinking a technology they view as economically and socially beneficial. In general, this group sees the good in technology but stresses that it cannot be reaped if technological growth goes on unchecked.

49.5   ENGINEERING ETHICS

The ethical behavior of engineers is more important today than at any time in the history of the profession. The engineers' ability to direct and control the technologies they master has never been stronger. In the wrong hands, the scientific advances and technologies of today's engineer could become the worst form of corruption, manipulation, and exploitation. Engineers, however, are bound by a code of ethics that carry certain obligations associated with the profession. Some of these obligations include:

1. Support ones professional society
2. Guard privileged information
3. Accept responsibility for one's actions
4. Employ proper use of authority
5. Maintain one's expertise in a state-of-the-art world
6. Build and maintain public confidence
7. Avoid improper gift exchange(s)
8. Practice conservation of resources and pollution prevention
9. Avoid conflict of interest
10. Apply equal opportunity employment
11. Practice health, safety, and accident prevention
12. Maintain honesty in dealing with employers and clients

There are many codes of ethics that have appeared in the literature. The preamble for one of these codes is provided below:

"Engineers in general, in the pursuit of their profession, affect the quality of life for all people in our society. Therefore, an Engineer, in humility and with the need for divine guidance, shall participate in none but honest enterprises. When needed, skill and knowledge shall be given without reservation for the public good. In the performance of duty and in fidelity to the profession, Engineers shall give utmost [2]."

Regarding environmental ethics, Taback [8] defined ethics as, "the difference between what you have the right to do and the right thing to do." More recently, has added that the environmental engineer/scientists should "recognize situations encountered in professional practice with conflicting interests that test one's ability to take the "right" action. Then, take each situation to a trusted colleague to determine the best course of action consistent with the above percepts and which would have the least adverse impact on all stakeholders."

The Air & Waste Management Association (A&WMA), primarily through the efforts of Taback, developed The Code of Ethics in 1996. Details are provided below in Sections 49.5.1–3.

49.5.1 PREAMBLE

In the pursuit of their profession, environmental professionals must use their skills and knowledge to enhance human health and welfare and environmental quality for all. Environmental professionals must conduct themselves in an honorable and ethical manner so as to merit confidence and respect, as well as to maintain the dignity of the profession. This code is to guide the environmental professional in the balanced discharge of his or her responsibilities to society, employers, clients, coworkers, subordinates, professional colleagues, and themselves.

49.5.2 PLEDGE

As an environmental professional, I shall regard my responsibility to society as paramount and shall endeavor to

1. Direct my professional skills toward conscientiously chosen ends I deem to be of positive value to humanity and the environment; decline to use those skills for purposes I consider to conflict with my moral values.
2. Inform myself and others, as appropriate, of the public health and environmental consequences, direct and indirect, immediate and remote, of projects in which I am involved, consistent with both standards of practice in industry and government, as well as laws and regulations that currently exist.
3. Comply with all applicable statutes, regulations, and standards.
4. Hold paramount the health, safety, and welfare of the public, speaking out against abuses of the public interest that I may encounter in my professional activities, as deemed appropriate per professional standards and existing laws and regulations.
5. Inform the public about technological developments, the alternatives they make feasible, and possible associated problems wherever known.
6. Keep my professional skills up-to-date and endeavor to be aware of current events, as well as environmental and societal issues pertinent to my work.
7. Exercise honesty, objectivity, and diligence in the performance of all my professional duties and responsibilities.
8. Accurately describe my qualifications for proposed projects or assignments.
9. Act as a faithful agent or trustee in business or professional matters, provided such actions conform to other parts of this code.
10. Keep information on the business affairs or technical processes of an employer or client in confidence while employed and later, as required by contract or applicable laws, until such information is properly released and provided such confidentiality conforms to legal requirements and other parts of this code.
11. Avoid conflicts of interest and disclose those known that cannot be avoided.
12. Seek, accept, and offer honest professional criticism, properly credit others for their contributions, and never claim credit for work I have not done.

13. Treat coworkers, colleagues, and associates with respect and respect their privacy.
14. Encourage the professional growth of colleagues, coworkers, and subordinates.
15. Report, publish, and disseminate information freely, subject to legal and reasonable proprietary or privacy restraints, provided such restraints conform to other parts of this code and do not unduly impact public health, safety, and welfare.
16. Promote health and safety in all work situations.
17. Encourage and support adherence to this code, never giving directions that could cause others to compromise their professional responsibilities.

49.5.3 A&WMA BYLAWS

Article XIV—Professional Practice
Section 1—Code of Conduct

It is the duty of every member to adhere to a Code of Conduct as may be adopted by the Board of Directors. Such code shall include the Association's Code of Ethics. The Code of Conduct shall be published periodically by the Association and shall be provided to new members.

Additional details are available at www.awma.org/about/index/html.

49.6 FUTURE TRENDS

Although the environmental movement has grown and matured in recent years, its development is far from stagnant. To the contrary, change in individual behavior, corporate policy, and governmental regulations are occurring at a dizzying pace.

Because of the Federal Sentencing Guidelines, the Defense Industry Initiative, as well as a move from compliance to a value-based approach in the marketplace, corporations have inaugurated company wide ethics programs, hotlines, and senior line positions responsible for ethic training and development (at the time of the writing of this chapter, an Ethics Officers Association of A&WMA was being formed.). The Sentencing Guidelines allow for mitigation of penalties if a company has taken the initiative in developing ethics training programs and codes of conduct.

In the near future, these same Guidelines will apply to infractions of environmental law [9]. As a result, the corporate community will undoubtedly welcome ethics integration in engineering and science programs generally, but more so in those that emphasize environmental issues. Newly hired employees, particularly those in the environmental arena, who have a strong background in ethics education will allay fears concerning integrity and responsibility. Particular attention will be given to the role of public policy in the environmental arena as well as in the formation of an environmental ethic.

Regarding education, the ABET 2000 accreditation guidelines, programs have to show that students are exposed to an ethics education; they also have to do outcome assessments. In spite of indicators that reveal the value of an ethics education, few large universities require an ethics course. Ideally, a student would take an ethics course and would also be exposed in several other courses each year.

Examples of how to integrate ethics into technology problems are available at ethics.iit.edu. [10].

Regulations instituted by federal, state, and local agencies continue to become more and more stringent. The deadlines and fines associated with these regulations encourage corporate and industrial compliance of companies (the letter of the law) but it is in the personal conviction of the corporate individuals that lies the spirit of the law, and the heart of a true ecological ethic.

To bolster this conviction of the heart, there must be the emergence of a new "dominant social paradigm" [7]. This is defined as "the collection of norms, beliefs, values, habits, and survival rules that provide a framework of reference for members of a society. It is a mental image of social reality that guides behavior and expectations" [7]. The general trend in personal ethics is steadily "greener" and is being achieved at a sustainable pace with realistic goals.

A modern day author suggests the following: "The flap of one butterfly's wings can drastically affect the weather [11]." While this statement sounds much like one conceptualized by a romantic ecologist, it is actually part of a mathematical theory explored by the contemporary mathematician Gleick [10]. The "butterfly" theory illustrates that the concept of interdependence, as Chief Seattle professed it, is emerging as more than just a purely environmental one. This embracing of the connectedness of all things joins the new respect for simplified living and the emphasis on global justice, renewable resources, and sustainable development (as opposed to unchecked technological advancement) as the new, emerging social paradigm. The concept of environmentalism is now "widely" held; its future is becoming "deeply" held.

49.7 SUMMARY

1. In 1854, Chief Seattle penned warnings as environmentally pensive and poignant as any uttered in the 140 years since: "Continue to contaminate your bed and you will one day lay in your own waste."
2. It is generally accepted that any historical ethic can be grouped into one of the following:
   a. Utilitarianism
   b. Duties Ethics
   c. Rights Ethics
   d. Virtue Ethics
3. The incorporation of environmentalism into everyday ethics does not require a redefinition of one's ethics, but rather, a redefinition of one's "community" to include nonhuman inhabitants of the land.
4. In the traditional ethical theories, established hierarchies of duties, rights, virtues, and desired consequences exist so that situations where no single course of action satisfies all of the maxims can still be resolved. Debate continues over where the environment falls in this hierarchy.
5. "Engineers in general, in the pursuit of their profession, affect the quality of life for all people in our society. Therefore, an Engineer … shall participate in none but honest enterprises … ."
6. At present, the concept of environmentalism is "widely" held; its future is becoming "deeply" held.

REFERENCES

1. Fahey, J. and Armstrong, R. (Eds.). *A Peace Reader: Essential Readings on War, Justice, Non-violence & World Order*, Paulist Press, Mahwah, NJ, 1987.
2. Martin, M.W. and Schinzinger, R. *Ethics in Engineering*, McGraw Hill, New York, 1989.
3. Wilcox, J. and Theodore, L. *Environmental and Engineering Ethics: A Case Study Approach*, John Wiley & Sons, Hoboken, NJ, 1998.
4. Carson, R. *Silent Spring*, Houghton Mifflin, New York, 1962.
5. Rolston, H., III. *Philosophy Gone Wild*, Prometheus Books, Buffalo, NY, 1986.
6. Leopold, A. *A Sand County Almanac*, Oxford University Press, New York, 1949.
7. Barbour, I. *Ethics in an Age of Technology*, Harper, San Francisco, 1993.
8. Taback, H. A&WMA's Code of Ethics, EM, Pittsburgh, pp. 42–43, September 2007.
9. Presentation by Cartusciello, N. Chief, Environmental Crimes Section, U.S. Department of Justice, May 4, 1994.
10. Daniels, A. *Walking the Line*, Prism, October 2007.
11. Gleick, J. *Chaos, Making a New Science*, Viking, New York, 1987.

50 Environmental Justice

Contributing Author: Francesco Ricci

CONTENTS

50.1 INTRODUCTION

In response to growing environmental concerns, the Environmental Protection Agency (EPA) was created by the federal government. Its agenda has been defined by a series of legislative acts since the late 1960s. (The reader is referred to Chapter 2 for more details.) The environmental policy of the EPA has historically had two main points of focus: defining an acceptable level of pollution and creating the legal rules to reduce pollution to a specified level. To some, it seems that the program has been most concerned with economic costs and efficiency [1]. Consequently, policy seems to lack considerations of equity, both distributional and economic. While EPA's two main points of focus are important considerations, relying on such criteria in the formation of environmental protection policy neglects to account for potential inequalities of capitalism and its effects throughout the policy process.

The history of environmental policymaking illustrates to some the incompatibility of equity and efficiency; it seems unlikely that increases in progressive distribution will come without a loss of efficiency. Economic pressures of environmental regulation have motivated corporations to seek new ways to reduce costs. Industries have attempted to maximize profits by externalizing the environmental costs [2]. It has been suggested that this redistribution of costs is more regressive in its effects than the general sales tax [3]. To date, big corporate polluters often have more to gain financially by continuing pollution practices than in obeying regulations. In some instances, the result of increased environmental costs has paradoxically caused negative impacts on environmental regulations. As long as corporations feel unaffected by such environmental degradation, they have little incentive other than altruism to end debilitating practices.

Generally, poverty and poor living conditions go hand in hand. As a result, lower-class citizens of the United States are more affected by pollution and environmental hazards. Evidence of the effects and concentration of environmental pollution in low-income communities has fueled a grassroots environmental movement since the early 1980s. While this movement can be regarded as a socioeconomic upheaval, intended to improve living conditions for the impoverished, it has taken on serious racial and cultural undertones. The movement calls for grassroots, multiracial, and multicultural activism to redress the distributional inequalities that have resulted from past policy and to prevent the same inequalities from occurring in future policy. The movement advocates that minorities use historically nonexercised political and legal power to push the EPA to address concerns and to oppose policies that impoverish the poor.

The environmental justice movement is committed to political empowerment as a way to challenge inequities and injustices. Empowerment is the inclusive involvement and education of community members by equipping them with skills for self-representation and defense. Organized activism at the grassroots level could circumvent the power structures that underrepresented particular communities in the first place. Community activists want to participate in the decision making that affects their communities. Further, they argue for increased pay for community members who engage in environmentally hazardous labor, for better working conditions in factories, and for the requirement of more on the job safety precautions. Activists contend that industries must contribute to community development if they are to detract from the community in other ways. For example, activists suggest industrial investment in community projects and educational systems [4]. The use of legal power is also encouraged. The most profound way of ensuring that law-breaking corporations are reprimanded is to utilize the American justice system to set legal precedents.

50.2 HISTORY AND SCIENTIFIC RESEARCH

According to a U.S. General Accounting Office study examining population ethnicity and location of off-site hazardous waste landfills in the southeastern region of the United States, African Americans comprise the majority of the population of three out of every four communities with such hazardous waste landfills [5]. While siting supposedly results from technical concerns, there are usually no geological reasons to site environmental waste in low-income areas. However, economic reasons provide a logical explanation for this concentration. In many situations, the cost of land is already cheap due to landfills or hazardous waste located in the vicinity. Therefore, nearby inhabitants are usually low-income since cheap real estate is economically favorable for them. As long as there are correlations between race and poor socioeconomic status, certain secondary correlations will be made between race and unfavorable living conditions. As a result, several studies have investigated environmentally triggered disease in minority groups. Some of them are discussed below.

It has been found that African-American children have a higher percentage of unacceptably high blood lead levels [6]. A common route of exposure occurs in buildings with deteriorating lead-based paint through the ingestion of paint chips and inhalation of paint dust. Lead poisoning is a particularly frightening epidemic because the effects of exposure are not immediately visible. Children, more than adults, are particularly sensitive to the physiological and neurobehavioral effects of lead poisoning at low levels. While lead poisoning is preventable with blood lead testing and abatement, this preventable toxin continues to poison communities where testing is not always available and lead abatement is rarely affordable (the reader is referred to Chapter 29 for more information on lead).

An EPA study conducted in Florida was released in 2002. It tested 571 facilities in 15 counties in order to assess the demographic of people living within one-half, one, and two miles of the aforementioned facilities. The study showed that within a two mile radius of most facilities, there was a disproportionately high number of non-English speaking, low-income minorities and renters. It was not determined how many of the surrounding inhabitants were legally residing in the United States. The report concluded that, "Minority and low-income communities are disproportionately impacted by targeted environmental impacted sites" [7]. It is apparent that the EPA has irrevocably tied a secondary racial correlation to the primary issue: social stature.

A nationwide study of selected pesticides in the milk of mothers found that Hispanic women had higher levels of certain pesticides than White women [8]. This evidence is explained by the fact that most Hispanic women in the study were from the Southwest, where pesticide use is generally higher. It should be noted that agricultural workers are exposed to many toxic substances in the workplace. Such exposure can cause cancer and a wide range of noncancerous health effects [3]. Agricultural workers are predominantly Latinos, and they are therefore disproportionately exposed to pesticides [5]. Once again, those who are of a lower economic class are usually found working and living in nonideal, sometimes perilous conditions.

Dietary exposure to pollutants such as polychlorinated biphenyl (PCBs) and dioxins can occur through fish consumption. One particular group of people who consume large amounts of fish are the decendents of the American Indians [9]. Native American communities tend to consume far more fish for their dietary protein than the average population. Even when concentrations of chemicals in water are below detection limits, damaging levels of pollutants can bioaccumulate in fish tissues and contaminate the fish consumer with toxins. The quantity of fish eaten, the method of fish preparation, and the species of fish eaten contribute to the level of exposure to contaminants. This is not to insinuate that only Native Americans are at risk; to the contrary, anyone who consumes large quantities of toxin-laced fish would certainly elevate his risk.

Environmental justice, by virtue of its illusive characterization, can also be utilized by those of better economic means. One example is a polluted waterway, which does not exclusively effect the impoverished. The Hudson River is one of the most well-known rivers in all of the United States. People from a wide range of economic backgrounds live along its shores. However, those who live in the southern half of the Hudson Valley all share a common burden: a long history of pollution and environmental degradation. Since the advent of PCBs, the General Electric (GE) Company (located in the Hudson Valley) has dumped over an estimated one million pounds of PCBs into the Hudson [10]. Despite GE's claims that these dumps were completely legal, the EPA has worked to remove these harmful "probable carcinogens" [11]. When the ill effects of this dumping reached the public's attention late in the twentieth century, staunch environmentalists and apathetic citizens alike took measures to ensure that the Hudson would be restored to its former vitality. These efforts culminated when the EPA recently led GE to build a Hudson River cleanup facility. This facility, once completed, will treat contaminated sediments which will be dredged from the river beginning in 2009. Groundbreaking for the multimillion dollar complex took place in early 2007 [12]. This demonstrates how environmental justice can take on a variety of activities, and can apply to any peoples (not just minority groups) who have been disenfranchised by pollution.

In response to growing concerns from specifically minority activists, the EPA created an "Environmental Equity Work Group" to review evidence that low-income and minority communities bear a large exposure to environmental risks. The findings of the study indicated "a clear cause for health concerns." The report concluded that "racial minorities may have a greater potential for exposure to some pollutants because they tend to live in urban areas, are more likely to live near a waste site, or exhibit a greater tendency to rely on subsistence fishing." At the same time, EPA claimed that poverty is a more significant factor than race in determining which communities are at high risk. In the case of lead, however, the epidemiological data unequivocally demonstrated that "Black children have disproportionately higher blood levels than White children even when socioeconomic variables are factored in [5]."

The above are some quite interesting findings. It leads one to contemplate what other factors may come into play. One study from New Jersey Medical School correlates calcium intake (a dietary habit) to lowering lead retention in humans. According to the research, most Americans do not drink the recommended amount of milk; this is "particularly true of African-American children and adults" as compared to White Americans [13]. Put plainly, by increasing dietary calcium intake on average, African-American communities would lower lead retention. In this case, as with many scientific inquiries, more than one variable can contribute to the overall result. Without noticing this dietary deficiency, a true portrait of the situation at hand could not be understood, nor properly addressed.

50.3   FEDERAL ACTION TO ADDRESS ENVIRONMENTAL JUSTICE IN MINORITY POPULATIONS

On February 11, 1994, President William Clinton issued Executive Order 12898, "Federal Actions to Address Environmental Justice in Minority Populations," which focused the attention of federal agencies on the environment and human health conditions of minority and low-income communities. The Executive Order directed Federal agencies to develop environmental justice strategies by April 11, 1995, that identify and address disproportionately high exposure and adverse human health or environmental effects on programs, policies, and activities on minority populations and low-income populations. All agency strategies must consider enforcement of statutes in areas of minority populations and low-income populations, greater public participation, improvement of research and identification of different patterns of subsistence use of natural resources. The Executive Order also requires that agencies conduct activities that substantially affect human health or the environment in a nondiscriminatory manner. In addition, better data collection and research is required by the Executive Order, and it declares that whenever practicable and appropriate, future human health research must look at diverse segments of the population and must identify multiple and cumulative exposures. The Executive Order applies equally to Native American programs.

The Executive Order contained six sections:

1. Implementation
2. Federal agency responsibilities for federal programs
3. Research data collection and analysis
4. Subsistence consumption of fish and wildlife
5. Public participation and access to information
6. General provisions

The first section begins with a general mission statement, it is shown below:

Section 1. Implementation

1–101. Agency Responsibilities. To the greatest extent practicable and permitted by law, and consistent with the principles set forth in the report on the National Performance Review, each Federal agency shall make achieving environmental justice part of its mission by identifying and addressing, as appropriate, disproportionately high and adverse human health or environmental effects of its programs, policies, and activities on minority populations and low-income populations in the United States and its territories and possessions, the District of Columbia, the Commonwealth of Puerto Rico, and the Commonwealth of the Marian Islands.

Highlights of the remaining five sections from the Executive Order are available in the literature [14].

50.4   THE CASE FOR ENVIRONMENTAL JUSTICE

Environmental protection policy has attempted to reduce environmental risks overall; however, in the process of protecting the environment, risks have been redistributed and concentrated in particular segments of society. Although federal regulations to protect the environment are not explicitly discriminatory, some argue that environmental protection policies have not been sensitive to distributional inequalities. Others insist that they have not adequately addressed specific minority environmental concerns. As noted above, low-income minority communities are disproportionately exposed to environmental hazards such as toxic waste disposal sites, lead, pesticides, air pollution, and contaminated fish [15].

History has suggested that the tendency to maintain a status quo inhibits attempts at environmental protection; as legislation evolves through the policy process, existing inequities are reinforced. This has caused much speculation as to the efficacy and neutrality of the mainstream environmental protection agenda and has essentially enlisted minority efforts in the movement for environmental justice. Lead provides an example of successful grassroots activism through community empowerment. A related minority response charges that policy has been orchestrated with intentionally racist motives. Domestic accusations of environmental racism are echoed internationally, where third world countries, inhabited predominately by poor "minorities", have become receptacles for the hazardous waste of western countries. Once again, be it accurate or manipulative, a primarily environmental issue has been used as a tool of racial activism.

Environmental justice can be achieved, in part, with a concerted effort on the part of grassroots and mainstream activists. Those directly affected have a responsibility to exercise their rightful political and legal power. At the same time, federal protection policy needs to devote attention to the concerns of the lower-class, to monitor the implementation and enforcement of environmental regulations, and to incorporate considerations of equity into policy.

50.5 THE CASE AGAINST ENVIRONMENTAL JUSTICE

Like many programs of reform and activism, environmental justice was principally started with good intentions. However, ground rules need to be set before any meaningful discussion regarding environmental justice can be presented. One of the problems is that environmental justice has come to mean different things to different people at different times and for different situations. There appears to be no clear-cut decision regarding this term but the EPA defines it as "The fair treatment of people of all races, cultures, and incomes with respect to development, implementation, and enforcement of environmental laws and policies, and their meaningful involvement in the decision-making process of government."

Based on the EPA's definition, there appears to be three major components of environmental justice: environmental racism, environmental equity, and environmental health. Each of these components is briefly discussed below. To simplify the presentation that follows, the adversely affected communities will be assumed to be composed of low-income Hispanics. The environmental action of concern (unless otherwise indicated) will refer to siting of a hazardous waste incinerator.

50.5.1 ENVIRONMENTAL RACISM

As with environmental justice, environmental racism needs to be clearly defined. A reasonable, logical definition is as follows: the act or process of environmentally exploiting individual groups, e.g., American Indians, African Americans, Hispanics, Hasidic Jews, etc., by "others" because of race, color, religion, etc. The "others" normally refers to industry, but can (also) include government agencies, individuals, private clubs and special organizations. Taken in a near absolute sense, this definition effectively states that "others" knowingly and deliberately created environmental problems that would adversely affect the groups in question. This hypothesis is hard to believe. Only one with a Hitler-like mentality would be capable of such action. Individuals who participate in this sort of activity are subject to fines and/or imprisonment (see Chapter 49). Based on this analysis, one can conclude that environmental racism is a nonentity.

50.5.2 ENVIRONMENTAL EQUITY

The situation involved here is not quite as clear since the quest for equity seems noble. However, this nation, and capitalism, is based on opportunity and inequity, not equity. Opportunities provide the mechanism to become displaced from inequity, and ascend the "socioeconomic ladder".

It is indisputable that capitalist societies are inherently unequally distributive in terms of property and capital. The literature abounds with societal inequities. But the question that begs an answer is: Is this wrong? If a chemical complex is moved to a new area and is sited adjacent to a Hispanic community, and property values drop, should the Hispanics be compensated? If the answer is yes, then who gets the money and how much money is involved? There are numerous other similar situations. The point to keep in mind is that the siting (or permit) has been obtained legally; therefore, there should be no compensation. Any other solution would be a veritable "dream-come-true" for the law profession.

Consider the following scenario. Waterfront property (with a water view) was recently purchased for $500,000. The owners of the two adjacent properties sandwiching the waterfront property build large structures shortly thereafter which effectively block the majority of the water view. The aesthetic value of the waterfront property is diminished, resulting in a significant loss. Should the new owner be compensated? Alternatively, if the two adjacent properties require a parking lot that can only (conveniently) be situated on the waterfront property, the value of new property skyrockets. Should the new owner be entitled to the profit? The bottom line is that actions, programs, sitings, etc., that can adversely environmentally impact a group or individual(s) are primarily based on economics. The final decision is almost always the most cost-effective decision. This is as it should be in a democratic, capitalistic system. After all, despite the agenda wishes of some radical environmentalists, it does not make good sense to site potentially questionable environmental facilities on prime waterfront property, or Times Square in New York City, or at the foot of Mt. Rushmore.

50.5.3 ENVIRONMENTAL HEALTH

The debate on environmental health can also be confusing. But, the same basic argument can be applied here that was applied to both environmental racism and equity. The EPA, based on extensive medical, toxicological, and engineering (scientific) data, has set "guidelines" that determine what is safe from both a health and hazard perspective. No facility should be sited if it is deemed (based on regulations) to be unsafe. This does not mean that one is safer near a facility than significantly displaced from the facility. It simply means that the present or proposed facility will have an insignificant impact on the surrounding community. Unfortunately, this is an area where irresponsible organizations and individuals have exploited the health-related aspect of environmental justice since it is difficult to satisfactorily explain the concept of risk (see Chapters 38 and 39) to the layman, i.e., the nontechnical individual.

Some argue that environmental justice creates imbalances other areas of economic life. An article entitled, "Environmental Justice: Deterrent to Economic Justice" enforces this belief [16].

It is my contention that the flame of environmental justice, like some well-intentioned, yet misdirected eco-terrorist, is now burning away an increasing number of tracts of our economic prosperity. Clearly, America stands for 'Life, liberty, and the pursuit of happiness.' No matter what you might think about civil rights in America, you must admit that the American Dream is fundamentally an economic dream. Environmental justice is a cause that hinders the goal of economic justice for all races, genders, and religions. Environmental justice, instead of helping different races, threatens to deprive us of the means by which each race can pursue the American Dream. And even more seriously, the deterrent imposed on economic justice by the environmental justice movement adversely affects public health ... environmental justice deters economic justice by hindering, and sometimes preventing, the influx of significant capital expenditures from private companies into economically depressed areas.

50.6 FUTURE TRENDS

Regardless of one's views on environmental justice, it is imperative to foresee the state of affairs in the near future. To date, environmentalists and liberal activists alike have argued that legislation has not fully addressed the inequities inherent in capitalism, and that economic factors predispose certain segments of the population to increased risk. They also contend that the process of policymaking, implementation, and enforcement of environmental protection regulations have allowed an increase in concentrated risks in lower-income communities. Even though these communities seem to lack political, legal, and economic power, community activism and mobilization have been effective in some instances in combating specific environmental problems.

Although change can only be secured through the diligent efforts of those afflicted, the environmental justice movement must echo with voices other than those of low-income victims globally. A robust, peace-time economy will aid the environmental justice movement in the future, despite lacking logical arguments. The technical community will undoubtedly be addressing this issue, and hopefully decide on its merits in an impartial manner [17].

50.7 SUMMARY

1. The environmental policy of the EPA has historically had two main points of focus: defining an acceptable level of pollution and creating the legal rules to reduce pollution to a specified level.
2. Evidence of the effects and concentration of environmental pollution in low-income communities has fueled a grassroots environmental movement since the early 1980s. The movement calls for grassroots activism to redress the distributional inequalities that have resulted from past policy and to prevent the same inequalities from occurring in future policy.
3. Many theories attempt to explain why minorities are disproportionately affected by environmental pollution. An awareness of the litigation process,

the history of segregation, and the process of siting decisions sheds some light on this sensitive issue.

4. Some have argued that economic pressures of environmental regulations have encouraged a dangerous negotiation process involving "an exchange of money" for health hazards.

5. Others have argued that environmental justice is a cause that hinders the goal of economic justice for all races, genders, and religions. Environmental justice, instead of helping races, threatens to deprive the means by which each race can pursue the American Dream.

6. A robust, peace-time economy will aid the environmental justice movement in the future, despite lacking logical arguments.

7. The technical community will undoubtedly be addressing this issue, and hopefully decide on its merits in an impartial manner.

REFERENCES

1. Lazarus, R. Pursuing environmental justice: The distributional effects of environmental protection, *Northwestern University Law Review*, Spring 1987.

2. Cole, L. Empowerment as the key to environmental protection: The need for environmental poverty law. *Ecology Law Quarterly*, 1992.

3. Dorfman, N. and Snow, A. Who will pay for pollution control? The distribution by income of the burden of the National Environmental Protection Program, *National Tax Journal,* 1972–1980.

4. Marquez, B. Lecture, University of Wisconsin–Madison, May 5, 1994.

5. U.S. EPA. *Environmental Equity: Reducing Risk for All Communities*, U.S. EPA, Washington, DC, 1992.

6. Agency for toxic substances and disease registry. *The Nature and Extent of Lead Poisoning in Children in the United States: A Report to Congress*, Centers for Disease Control, Atlanta, GA, 1988.

7. National Academy of Public Administration. *Models for Change: Efforts by Four States to Address Environmental Justice*, NAPA, Washington, D.C., 2002. http://www.epa.gov/Compliance/resources/publications/ej/napa_epa_model_4_ states.pdf (pp. 54–55).

8. Savage, E. *National Study to Determine Levels of Chlorinated Hydrocarbon Insecticides in Human Milk*, EPA, Fort Collins, CO, 1976.

9. West, P., Fly, J., Larkin, F., and Marans, P. Minority anglers and toxic fish consumption: Evidence of the statewide survey of Michigan, 1989. In Bryant, B. and Mohai, P., Eds., *The Proceedings of the Michigan Conference on Race and the Incidence of Environmental Hazards*, 1989, pp. 108–122.

10. Clearwater. The Hudson River PCB story, http://www.clearwater.org/pcbs/index.html

11. Clearwater. What Are The Human Health Effects Of PCBs? http://www.clearwater.org/news/pcbhealth.html

12. U.S. EPA. Ground-breaking for construction of Hudson River cleanup facility set for the spring (press release), February 8, 2007. http://epa.gov/hudson/2009hudson_press_release.pdf

13. Bogden, J., Oleske, J., Louria, D. Lead poisoning—One approach to a problem that won't go away, *Environmental Health Perspectives*, 105, 1997. http://www.ehponline.org/members/1997/105–12/bogden-full.html

14. Executive Order 12898, President William J. Clinton, The White House, February 11, 1994.

15. Bryant, B. and Mohai, P. *Race and the Incidence of Environmental Hazards*, Westview Press, Boulder, CO, 1992.
16. Heaton, J.S. Environmental justice: Deterrent to economic justice, *Environmental Magazine*, January 1999, pp. 11–12.
17. L. Theodore, Personal notes, 2008.

Index

545